NRM1 Cost Management Handbook

The 'RICS New rules of measurement: Order of cost estimating and cost planning of capital building works' (referred to as NRM1) is the cornerstone of good cost management of capital building works projects – enabling more effective and accurate cost advice to be given to clients and other project team members, while facilitating better cost control.

The *NRM1 Cost Management Handbook* is the essential guide to how to successfully interpret and apply these rules, including explanations of how to:

- quantify building works and prepare order of cost estimates and cost plans;
- use the rules as a toolkit for risk management and procurement;
- analyse actual costs for the purpose of collecting benchmark data and preparing cost analyses;
- capture historical cost data for future order of cost estimates and elemental cost plans;
- employ the rules to aid communication;
- manage the complete 'cost management cycle';
- use the elemental breakdown and cost structures, together with the coding system developed for NRM1, to effectively integrate cost management with Building Information Modelling (BIM).

In the *NRM1 Cost Management Handbook*, Benge explains in clear terms how he intended NRM1 to be used in familiar quantity surveying tasks, as well as a range of activities of crucial importance for professionals in years to come. Worked examples, flow charts, diagrams, templates and checklists ensure readers of all levels will become confident and competent in the use of NRM1. This book is essential reading for anyone working with NRM1 and is the most authoritative guide to practice possible for those preparing to join the industry.

David P. Benge is a Fellow of the Royal Institution of Chartered Surveyors (RICS) with over 30 years experience as a quantity surveyor and educator. As Head of Quality Management at Gleeds, international management and construction consultants, David focuses on service improvement for all aspects of quantity surveying and project management activities – with a worldwide remit. Through Gleeds, he provides quality consulting services and training on the use and application of the NRM suite of measurement rules. David authored both the first and second editions of *NRM1* for the RICS and was co-author and executive technical editor on both *NRM2* and *NRM3*.

NRM1 Cost Management Handbook

The definitive guide to measurement and estimating using NRM1, written by the author of NRM1

David P. Benge

LONDON AND NEW YORK

First published 2014
by Routledge
2 Park Square, Milton Park, Abingdon, Oxon, OX14 4RN

and by Routledge
711 Third Avenue, New York, NY 10017

Routledge is an imprint of the Taylor & Francis Group, an informa business

British Library Cataloguing in Publication Data
A catalogue record for this book is available from the British Library

Library of Congress Cataloging-in-Publication Data
Benge, David P.
NRM1 cost management handbook / David P Benge.
pages cm
Includes bibliographical references and index.
1. Building–Cost control–Handbooks, manuals, etc. I. Royal Institution of Chartered Surveyors.
II. Title.
TH438.15.B46 2014
690.068'1–dc23

2013033979

ISBN13: 978-0-415-72077-9 (pbk)
ISBN13: 978-1-315-84875-4 (cbk)

Typeset in Goudy and Helvetica by
Servis Filmsetting Ltd, Stockport, Cheshire

For Gillian,
my wife, soulmate and best friend,
and
my children
Tim, Rebecca and Chris

Summary of contents

Contents

Part 3 Cost planning

Figures

Tables

Examples

Foreword – Sean Tompkins

Sean Tompkins, Chief Executive Officer of the Royal Institution of Chartered Surveyors

On behalf of the Royal Institution of Chartered Surveyors, I am delighted to introduce the *NRM 1 Cost Management Handbook*. David has masterfully created a user-friendly accompaniment to the NRM 1 standard for measuring capital building works for the purpose of cost planning that speaks in a contemporary language and style to articulate a skill set that has been developed over generations.

This handbook, used in conjunction with the NRM (RICS *New Rules of Measurement*) suite will raise professional standards of quantity surveying and cost management, adding greater value to clients and improving client confidence in the process.

I am grateful that David has acknowledged how NRM 1 can be used as an integrated tool with building information modelling (BIM) enabled projects; this will help place quantity surveyors and cost managers at the heart of construction projects – right where they should be.

This handbook will also support cost managers in setting realistic elemental cost targets so that design teams can 'design to cost' and not 'cost the design', which is particularly apposite given the current UK government's construction strategy that calls for project teams to know what projects 'should cost' rather than 'did cost'. The *NRM 1 Cost Management Handbook* will help instil this new paradigm of cost management and drive this imperative forward.

Overall this book will be an excellent addition to any quantity surveyor's or cost manager's library, and I fully support the publication of the *NRM 1 Cost Management Handbook*.

Foreword – Richard Steer

Richard Steer, Chairman of Gleeds, International Management & Construction Consultants

The *NRM 1 Cost Management Handbook* is a must have for all quantity surveyors and cost managers working in today's fast moving environment. The handbook has been created as an accessible accompaniment to the NRM 1 standard for measuring capital building works for the purpose of cost planning, and it communicates in a manner and style that works at all levels. Whilst appropriate for the current generation of construction management professionals, David has ingeniously drawn on the knowledge and background that would be familiar to quantity surveyors from earlier generations but has given the book a contemporary feel.

The book highlights the importance of measurement in cost management today and furthermore shows the need for colleges and universities to embrace and teach measurement.

On initial review, NRM 1 may appear highly detailed but when cost managers begin to reflect on how the guidance is structured and how each section can be meaningfully applied, I am convinced it will soon become a respected best practice tool, supporting day-to-day cost estimating and cost planning activities.

What David has cleverly shown is that professional cost management does not simply consist of a series of measurement rules. The NRM 1 provides a toolkit for the cost management of construction projects embracing and exemplifying best practice which will help the reader in developing a professional reputation from both an individual and corporate perspective.

It is a strength of the book that David does not just focus on best practice in the modern workplace but he also highlights the common pitfalls and advises how to avoid them. Forewarned is forearmed and it is this practical rather than purely theoretical approach that gives the book such value. I therefore thoroughly recommend this valued work, as the guidance provided by David in the *NRM 1 Cost Management Handbook* has been clearly stated and is accessible to all levels.

Foreword – Richard Steer

Richard Steer, Chairman of Gleeds, International Management & Construction Consultants

Foreword – Stuart Earl

Stuart Earl BA(Hons) BSc(Hons) FRICS, Chairman of RICS Measurement Initiative Steering Group and Director of Gleeds Cost Management Limited

On 24 April 2012, the RICS launched its landmark measurement initiative, the New Rules of Measurement (NRM). The rules are arguably the most significant launch to the construction sector by the RICS in the past 35 years.

Before NRM, the RICS had provided quantity surveyors with rules for the measurement of building works – the 'Standard Method of Measurement (SMM)'. However, these rules were specifically drafted to advise quantity surveyors of how to measure building works in detail for inclusion in bill of quantities that, in turn, were used to obtain tender prices from contractors; they also aided the measurement and valuation of variations issued during the construction phase. A key requirement of the SMM was the need for a full and detailed specification and drawings from the designers – which were and are now seldom provided. This resulted in the overuse of provisional sums and abuse of the use of bill of quantities. The lack of detailed design is primarily caused by the cost of finance and the need for clients to complete and put the building to use as quickly as possible, as well as the impetus to get contractors involved early in the design process to provide input on design, buildability and value for money.

As a consequence, over the past 20 years, the use of design and build contract strategies have come to the forefront; with contracts commonly being awarded based on 'Concept Design' or 'Developed Design' (RIBA Work Stages 2 or 3). Well before the completion of 'Technical Design' (RIBA Work Stage 4), at which stage the use bill of quantities could be considered. For these reasons, cost planning has become an essential cost management tool.

NRM 1 (the RICS *New rules of measurement: Order of cost estimating and cost planning of capital building works*) is an overdue statement of how cost planning is applied in practice and is a significant step forward in improving professional standards. Firstly, it re-establishes measurement as the focus of our professional standing and secondly, it makes it easier to benchmark our cost planning procedures against those which are widely accepted as best practice. It also provides learning establishments with a much clearer statement of the cost planning competencies required by students.

Moreover, NRM 1 presents a clear framework, which facilitates a systematic approach to compiling and managing cost estimates and cost plans. In this *NRM 1 Cost Management Handbook*, David has successfully augmented the rules and provides clear sound advice on how to use the tools that can be found ensconced in them. He must be congratulated for this work.

I am also pleased to see that David has used the in this *NRM 1 Cost Management Handbook* to bring NRM 1 into line with the latest RIBA Plan of Work (i.e. 'RIBA Plan of Work 2013').

For those practitioners who have yet to implement the best practice guidance provided by NRM 1, and those who want to be more informed about it, I am certain that David's *NRM 1 Cost Management Handbook* will be seen as an indispensable aid that greatly assists you in moving forward to embrace best practices in cost estimating and cost planning. Furthermore, understanding and using NRM 1 will be an essential selling point when working in other countries, helping to avoid misunderstandings that often arise due to different countries.

Learning and adopting best practice is essential.

Preface

As part of its commitment to continually raising the standards that its members work to, the RICS launched the RICS *New rules of measurement: Order of cost estimating and cost planning for capital building works* – now commonly called NRM 1. RICS standards are recognised across the globe as the best technical practice in construction.

There has never been a more pressing time to introduce NRM 1. The reforms set out in the UK government's Construction Strategy, as well as the UK's chief construction adviser's efficiency agenda, the increasing focus on building information modelling (BIM) and the economic challenges that face the construction industry at the present time, all demand a step change in working culture – including the working culture of the cost manager.

The relevance of NRM 1 within a BIM-enabled construction industry is particularly pertinent. BIM is intended to address issues of process management and data retention, bringing the collection of co-ordinated data to the forefront. NRM 1 is linked to this, enabling the consistent collection of construction cost data that is synchronised with the design data – as is NRM 3, in respect of building maintenance cost data.

NRM 1 represents the essentials of best practice and forms the 'cornerstone' of good cost management of construction projects. The rules provide a standard of measuring for the purpose of developing order of cost estimates and cost plans, and enable effective, accurate and transparent cost advice to be given. They also facilitate better pre- and post-contract cost control. Consequently, NRM 1 sets out the standards required that cost managers, contractors and any others should follow.

It is important to understand that NRM 1 is a toolkit for cost management, not just a set of rules for the quantification of capital building works. Along with other advice, the rules provide guidance on:

- *How the method of measurement changes as the design of a building project develops* – from high-level measurement of areas and/or functional units to the measurement of more detailed elements, sub-elements and cost-significant components.
- *Total project fees* – considering fees in connection with consultancy services from cost managers, architects, engineers and legal advisers, as well as those in connection with site surveys – both desktop and intrusive investigations – in addition to main contractor's pre-construction fees and main contractor's and subcontractors' design fees.
- *Total building project costs* – how all cost centres, including non-construction items, relating to the building project can be considered and pulled together into a single cost plan for the entire building project.
- *Risk* – based on the properly considered assessment of dealing with risks should they materialise – dispensing with the widely mismanaged concept of contingences.

- *Information requirements* – what information is required by the cost manager from the employer and other project team members at each design stage to enable more certainty in their cost advice.
- *Key decisions* – that the client needs to make at each Work Stage of the RIBA Plan of Work, or OGC Gateway.
- *Codification* – providing a framework for codifying elements, sub-elements and components so that structure of cost plans can be converted from elemental to work package, and vice versa, to facilitate the management of costs through both the bid and construction phases of a building project.
- *Reporting* – providing advice on communicating cost advice to clients.
- *Data collection* – providing a common basis for analysing and collecting real-time cost data that can be used for benchmarking and to estimate the cost of future building projects.

To be effective, a cost manager must be able to understand and use the rules of measurement, as well as being able to apply common sense. In contrast, an effective project manager must have an understanding of how cost managers construct their cost estimates and cost plans to be able to discuss from a position of knowledge.

Use of the rules by cost managers demonstrates a professional and responsible approach to the cost management of building projects. Moreover, it is beneficial to project managers, clients and others involved in financial management who wish to better understand the cost management of building projects.

I hope that this *NRM 1 Cost Management Handbook* will help that understanding.

David P. Benge

About this handbook

My aim in writing this handbook has been to provide a text which is practical enough to be useful to practitioners but which also has enough academic content to meet the requirements of degree courses.

The handbook is designed to take the reader step-by-step though the latest RICS best practice guidance (NRM 1) on preparing and communicating cost estimates and cost plans for capital building projects. In addition to showing how NRM 1 is to be interpreted and used, the handbook provides sound practical advice on the cost management of building projects, as well as on the pitfalls to be avoided.

The unique features of this handbook are:

Style of writing

- The language used in this handbook is lucid, easy to understand and facilitates easy grasp of concepts.
- The chapters have been logically arranged in sequence.
- The handbook is written in a reader-friendly manner both for students and practitioners.
- Explanations are supported by diagrams wherever required.

Content of the handbook – theory

- The handbook explains how to use and apply the principles of measurement and the tools advocated by NRM 1.
- Sufficient worked examples have been included to reinforce understanding.

Content of the handbook – practical

- In addition to explaining how to use and apply the principles of NRM 1, the theory is supported by practical information on the quantification of building works, as well as discussion on the pitfalls to be avoided when preparing cost estimates and cost plans.
- Certain chapters include templates that can be adopted and amended for practical use.

Targeted readership

This handbook has been written keeping in mind both students and practitioners who wish to acquire a practical understanding of the RICS *New rules of measurement: Order of cost estimating and elemental cost planning* (NRM 1) quickly.

Although primarily written for a UK audience, the handbook provides essential guidance for students and practitioners worldwide – the principles and tools within NRM 1 apply in whichever country you are situated.

This handbook is well suited for students studying HNC/D, BSc degrees and MSc degrees in:

- cost management;
- quantity surveying;
- project management;
- commercial management;
- construction management;
- construction procurement management;
- building surveying.

It is essential reading for any graduate embarking on the RICS Assessment of Professional Competence (APC).

Structure of the handbook

This handbook is divided into the following seven parts:

Part 1 (Introduction) comprises two chapters: Chapter 1 describes the basic principles and importance of cost management, outlines the role of the cost manager, explains the purpose and benefits of NRM 1 and its relationship with NRM 2 and NRM 3, explains how NRM 1 is integral to Building Information Management (BIM) and describes the cost management cycle; and Chapter 2 considers the composition of cost estimates and cost plans, defines the relationship between NRM 1 and the RIBA Work Stages and OGC Gateways, describes the formal cost-estimating stages, considers the impact of procurement and contract strategies on the cost-planning process and identifies the responsibilities of the project team in the cost-planning process.

Part 2 (Order of cost estimates) comprises seven chapters. Chapter 3 explains the purpose of order of cost estimates; Chapter 4 describes the planning of an order of cost estimate and sets out the information required by the cost manager to provide more cost certainty at an early stage; Chapters 5 and 6 provide step-by-step guidance on preparing an order of cost estimate; Chapter 7 provides a worked example of an order of cost estimate using a real building project; Chapter 8 explains the purpose of the elemental method of estimating and shows how to prepare an initial elemental cost model using the elemental method of estimating; and Chapter 9 provides step-by-step guidance on how to generate element unit quantities and pull together a elemental cost model.

Part 3 (Cost planning) comprises 18 chapters: Chapter 10 explains the concepts and objectives of cost planning, describes the different types of cost plans and sets out the information required by the cost manager at each RIBA Work Stage and OGC Gateway to provide more cost certainty; Chapter 11 describes the format, structure and codification of cost plans; Chapter 12 gives step-by-step guidance on preparing a cost plan; Chapter 13 shows how to use the tabulated rules within NRM 1, defines and distinguishes between prime cost sums and provisional sums, considers the pitfalls to be avoided when recording dimensions and quantities ascertained when using electronic measuring devices, provides guidance on quantification and the formulation of component descriptions, and advises on how to quantify and describe components for which there is inadequate design information; Chapters 14 to 22 provide step-by-step guidance on how to measure and describe the components with each element and sub-element, giving examples of how to formulate descriptions for components; Chapter 23 illustrates the hierarchical structure of cost data, considers the pitfalls associated with 'in-house' and 'published' cost data, and provides worked examples on how to determine unit rates for components; Chapter 24 shows how to calculate the building works estimate; Chapter 25 defines the items that constitute main contractor's preliminaries, explains how to deal with subcontractor's preliminaries, considers factors that can significantly influence the cost of main contractor's preliminaries, describes how to calculate main contractor's and subcontractors' preliminaries, and provides worked examples on cost checking the adequacy of main contractor's preliminaries; Chapter 26 defines and shows how to calculate both the main contractor's and subcontractor's overheads and profit; and Chapter 27 explains how to calculate the works cost estimate.

Part 4 (Estimating cost targets for non-building works items and risk allowances) comprises six chapters: Chapter 28 describes the categories of project and design team fees in a building project, provides step-by-step guidance on estimating fees, together with worked examples, and explains how to deal with main contractor and subcontractor design fees; Chapter 29 defines and explains the approach to estimating other development and project costs; Chapter 30 discusses the risk management of building projects and provides guidance on the setting and managing of risk allowances; Chapter 31 defines the concept of inflation, illustrates how inflation is dealt with in the context of building projects and gives guidance on estimating the possible effects of inflation; Chapter 32 provides step-by-step guidance on how to establish the cost limit; and Chapter 33 discusses the different types of taxes and incentives applicable to building projects.

Part 5 (Writing cost estimate and cost plan reports) comprises one chapter: Chapter 34 provides guidance on the writing of cost estimate and cost plan reports using a progressive worked example, and discusses how to communicate the cost estimate or cost plan to the employer and other stakeholders.

Part 6 (Designing pricing documents using NRM 1) comprises one chapter: Chapter 35 explains how the NRM 1 framework can be used to formulate a pricing document, together with worked examples.

Part 7 (Analysing bids and collecting data using NRM 1) comprises one chapter: Chapter 36 discusses the impact that changing procurement strategies have had on the ability of the cost manager to collect and analyse cost data for use in future building projects, illustrates the problems with collecting historical cost data, explains how the NRM 1 framework can be used for collecting and analysing historical cost data and considers the way in which benchmarking is used by cost managers.

Learning aids

The book aims to help readers to understand how order of cost estimates and cost plans are prepared in the real world. To assist with this aim I have incorporated a number of learning aids:

- chapter opening summaries providing a list of key points covered by each chapter;
- headings and sub-headings to break up material into clearly defined topics, giving readers quick access to the topics they need;
- diagrams to clearly illustrate the overall logic of the rules;
- tables to present certain information in an easy-to-read format, as well as to support real-time management of cost estimates and cost plans;
- process maps to provide step-by-step guidance on preparing cost estimates and cost plans;
- flow charts to illustrate work flows;
- worked examples using real-time project information;
- templates for use in practice;
- cross-referencing to enable the reader to refer to applicable sub-sections or definitions; and
- definitions of key terms and phrases used in each chapter.

Acknowledgements

I should like to thank the various people who have helped to bring this book to completion. First of all, Stuart Earl (Chair), Joe Martin (Executive Director of BCIS) and the other members of the RICS Measurement Initiative Steering Group for having the confidence in entrusting me with the writing of RICS *New rules of measurement: Order of cost estimating and cost planning for capital building works* (now commonly referred to as NRM 1). Second, I must acknowledge the helpful comments of my colleagues at Gleeds, who reviewed parts of the book as it was written; I hope that they will recognise the effects of their comments in the final result. Third, I must thank Bryan Avery, of Avery Associate Architects, for kindly allowing me to use his drawings to illustrate the use of NRM 1. Last but not least, my long-suffering and patient family – Gillian, Tim, Rebecca and Chris. To all those other people that I have forgotten, my apologies.

About the author

David P Benge BSc, MSc, FRICS, MACostE is a Fellow of the Royal Institution of Chartered Surveyors (RICS) with over 30 years experience as a quantity surveyor and educator. As Head of Quality Management at Gleeds, International Management and Construction Consultants, David focuses on service improvement for all aspects of quantity surveying and project management activities – with a worldwide remit.

David holds a BSc Degree in Quantity Surveying and an MSc Degree in Construction Procurement, obtained from South Bank Polytechnic (now University) and Nottingham Trent University, respectively.

He became a Member of the RICS in 1986 and a Member of the Association of Cost Engineers in 1988. He was made an Honouree Fellow in 2012 at the personal invitation of the RICS Chief Executive and the Chairman of the Quantity Surveying and Construction Professional Group.

His career spans both the private and public sectors, in almost equal periods, from which he draws his significant experience.

His research interests include the quality of service delivery in cost management and construction procurement management.

David authored both the first and second editions of the 'RICS New rules of measurement: Order of cost estimating and cost planning for capital building works (referred to as NRM 1)', which is an RICS best practice publication and the first edition became effective on 1 May 2009. He also co-authored 'NRM 2 (the RICS New rules of measurement: Detailed measurement for building works)' and 'NRM 3 (the RICS New rules of measurement: Order of cost estimating and cost planning for building maintenance works)'.

He lives in St Leonards-on-Sea, East Sussex with his wife Gillian. They have three grown-up children - Tim, Rebecca and Chris - who have absolutely no interest in either quantity surveying or cost management!

Through Gleeds, David provides quality consulting services and training on the use and application of the NRM suite of measurement rules. He can be reached at david.benge@gleeds.co.uk.

Abbreviations

ACMs	asbestos-containing materials
BCIS	Building Cost Information Service
BIM	building information modelling
BMU	building maintenance unit
BQ	bill of quantities
BREEAM	Building Research Establishment Environmental Assessment Method
CAD	computer aided design
CBS	cost breakdown structure
CDP	Contractor's Design Portion
CFA	continuous flight auger
CFP	cased flight auger piles
CHP	combined heat and power
CO_2	carbon dioxide
cost/m^2	cost per square metre
DDA	Disability Discrimination Act
DPM	damp-proof membrane
EU	European Union
EUQ	element unit quantities
EUR	element unit rate
FF&E	fittings, furnishings and equipment
GEA	gross external area
GIA	gross internal area (synonymous with GIFA)
GIFA	gross internal floor area (synonymous with GIA)
GL	ground level
ha	hectare(s)
HSE	Health and Safety Executive
HV	high-voltage
IPT	insurance premium tax
IT	information technology
JCT	Jaint Contracts Tribunal
kg	kilogram(s)
kN	kilonewton
KPI	key performance indicator
kW	kilowatt(s)
LPG	liquefied petroleum gas
LV	low-voltage
m	metre(s)
m^2	square metre(s)
m^3	cubic metre(s)
mm	millimetre(s)
m/sec	metres per second
M&E	mechanical and electrical
MoD	Ministry of Defence

NHBC	National House-Building Council
NIA	net internal area
nr	number
NRM	new rules of measurement
OGC	Office of Government Commerce
QMS	quality management system
PC	prime cost
PCSA	Pre-Construction Services Agreements
RIBA	Royal Institute of British Architects
RICS	Royal Institution of Chartered Surveyors
RVF	regional variation factors
SA	site area
SFCA	Standard Form of Cost Analysis
SMM7	Standard Method of Measurement
SRO	Senior Responsible Owner
SUDS	sustainable urban drainage schemes
t	tonne(s)
TCM	total cost management
TPI	tender price index
UK	United Kingdom
UPS	uninterrupted power supply
VAT	value added tax
VAV	variable air volume
VE	value engineering
VM	value management
VfM	value for money
VRV	variable refrigerant volume
WBS	work breakdown structure
WPC	Working Platform Certificate
£GBP	pound (sterling)

PART 1

Introduction

CHAPTER 1

Cost management

Introduction

This chapter:

- defines cost management in the context of building projects;
- describes the basic principles of cost management;
- explains the importance of cost management;
- identifies the objectives of cost management;
- outlines the role of the cost manager;
- describes the cost management cycle phases along with the activities needed to complete them;
- explains the role that the RICS [Royal Institution of Chartered Surveyors] *New Rules of Measurement: Order of Cost Estimating and Cost Planning for Capital Building Works* (NRM 1) has in effective cost management;
- explains the relationship of the documents that comprise the RICS suite of new measurement rules;
- explains how NRM 1 supports building information modelling (BIM); and
- summarises the key benefits of the rules.

1.1 What is cost management?

Cost management is all about achieving value for money (VfM). It is much more than simply maintaining records of expenditure and issuing cost reports. Management means control, so cost management means all those actions necessary to understand why costs occur and the necessary responses so that decisions controlling costs are taken promptly – in light of all relevant information.

In the context of building projects, cost management involves the overall planning, co-ordination, control and reporting of all cost-related aspects from initiation to operation and maintenance. It is the process of identifying all costs associated with the investment, making informed choices about the options that will deliver best value for money and managing those costs, including costs throughout the life of the building, where whole life costs are being considered.

Cost management of building projects is just one of the specialist services undertaken by the quantity surveyor.

1.2 Basic principles of cost management

Cost management is one of the cornerstones of project management. The establishment and implementation of effective cost management procedures at an early stage in the development of a building project will help ensure success. Established and effective cost control systems and procedures, understood and adopted by all members of the project team, entail less effort than 'crisis management' and will release management effort to other areas of the building project.

The principle areas of cost management can be described as follows:

- **Scope** – defining what is to be included within the building project and limiting expenditure accordingly.
- **Programme** – defining the programme for the building project from inception to completion and ensuring compliance. Estimates and cash flow projections should be consistent with the programme.
- **Design** – ensuring that designs meet the scope and budget (i.e. cost limit); delivering quality that is appropriate to the employer's brief.
- **Risk allowances** – ensuring that all monies are appropriately allocated from risk allowances and are properly authorised. Monitoring the use of risk allowances to forecast the cost limit.
- **Contracts and materials** – ensuring that the contracts provide full and proper control and that all costs are incurred as authorised; ensuring that materials are properly specified so as to meet the scope and design, and that they can be procured effectively.
- **Cash flow** – planning and controlling both commitments and expenditure within budgets so that unexpected cost overruns or underruns do not occur.
- **Cost records and reports** – ensuring that all transactions are properly recorded and authorised and, where appropriate, decisions are justified; and that regular, consistent and accurate reports are available to the employer.

It must be emphasised that cost management procedures need to be varied and flexible. Cost managers should discuss and jointly agree appropriate controls and review mechanisms with the employer and, where appointed, the project manager.

1.3 Why is cost management important?

Cost management is an essential part of effective programme/project management, but when poorly performed can be a barrier to the successful delivery of the building project and can result in failure to achieve value for money. Cost overruns are often caused by the employer through objectives that are unclear and changed during the course of the building project. The other main reasons for cost increases are:

- unrealistic cost estimates (usually too optimistic);
- risk allocation that is ambiguous; and
- inadequate management control.

In building projects, additional problems are frequently caused by:

- design that does not meet planning or statutory requirements;
- design that lacks co-ordination; and
- design that is difficult to build and maintain.

1.4 Objectives of cost management

The objectives of cost management are:

- to deliver the building project at the lowest cost compatible with the specified quality and as closely as practicable to the cost limit (taking account of whole life costs, where appropriate);
- to ensure that, throughout the building project, full and proper accounts are maintained (and kept up to date at all times) of all transactions including commitments, payments and changes; and
- to ensure that all transactions fully accord with the requirements of the employer or, in the case of public sector organisations, with the requirements of public accountability, probity and propriety.

1.5 Responsibility for cost management

The cost management of a building project is the joint responsibility of the whole project team; not the cost manager alone. Cost management requires continuing and active involvement from all project team members. Therefore, all project team members must remain mindful of their joint responsibility for cost management and draw to the attention of the employer, or project manager, anything that might affect cost. In view of this, it is important that the employer clearly sets out each consultant's responsibility in respect of cost management in the consultant's appointment.

1.6 The role of the cost manager

Management of the overall cost of the building project is the responsibility of the cost manager, maintaining effective financial control through the processes of evaluating, estimating, budgeting, monitoring, analysing, forecasting and reporting. The main tasks of the cost manager are to:

- provide initial cost advice on capital investment costs;
- produce cost estimates and cost plans in respect of capital investment costs;
- advise on and estimate whole life costs;
- produce risk allowance estimates;
- manage the base cost estimate and risk allowance during design development and construction;

- undertake cost-in-use studies and option costs;
- produce cost reports, estimates and forecasts;
- maintain an up-to-date estimated outturn cost and cash flow;
- manage expenditure of the risk allowance;
- initiate action to avoid overspend;
- prepare pricing documents for the purpose of tender;
- evaluate tender bids;
- scrutinise actual cost of tenders;
- collect and analyse cost data;
- prepare interim valuations;
- value change instructions;
- ascertain cost implication of contractor's financial claims;
- negotiate and agree final accounts; and
- issue financial reports or statements (i.e. throughout the building project to report the financial status).

1.7 The cost management cycle (the Benge Cycle)

The cost management cycle is shown in Figure 1.1. Proactive cost management takes place throughout the cost management cycle (*the Benge Cycle*). Each cost management cycle phase is described below, along with the activities needed to complete it.

Where required, whole life costs will be considered during each phase of the cost management cycle.

1.8 Relationship of the documents that comprise the RICS suite of new measurement rules

NRM 1 forms part of the RICS suite of publications referred to as the 'new rules of measurement (NRM)'. The RICS Quantity Surveying and Construction Group have developed the NRM suite of documents. The primary aim of the rules is to provide a consistent approach to the measurement and quantification of capital building works and maintenance, supported by a common means of analysing cost data for future use.

Although the NRM suite has principally been based on UK practice, the requirements for a co-ordinated set of rules and their underlying philosophy have worldwide application.

The NRM suite comprises the following three volumes (summarised in Figure 1.2):

NRM 1 – Order of cost estimating and cost planning for capital building works;

NRM 2 – Detailed measurement for building works; and

NRM 3 – Order of cost estimating and cost planning for building maintenance works.

These rules are supported by the BCIS (Building Cost Information Service) 'Elemental Standard Form of Cost Analysis, (NRM) Edition', which sets out

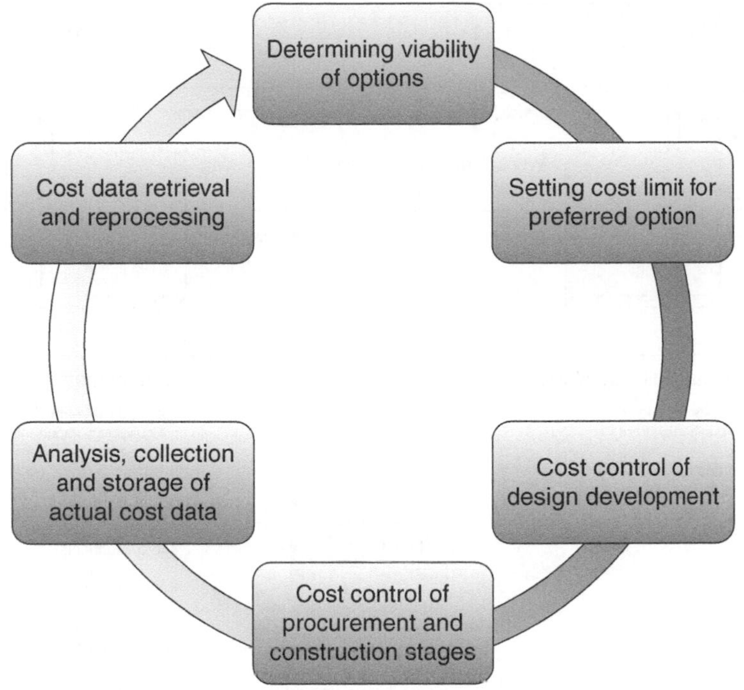

Figure 1.1 The cost management cycle (the Benge Cycle)

Notes:

(1) *Determining viability of options* – preparing order of cost estimates for one or more building or development options to establish the likely cost and determine the viability and affordability of each option

(2) *Setting cost limit for preferred option* – establishing the maximum expenditure that the employer is prepared to make in relation to the completed building or development (i.e. the cost limit)

(3) *Cost control of design development* – determining cost targets for key elements of the building project through cost planning the design, including cost checking and value engineering the developing design to ensure the projected outturn cost remains within the cost limit set by the employer

(4) *Cost control of procurement and construction stages* – preparing pre-tender estimates (derived from the latest cost plan) to predict tender prices; reconciling tender prices against pre-tender estimates and cost plans; establishing contract sums; managing and reporting costs during the construction phases of the building project to ensure the projected outturn cost remains within the cost limit set by the employer; negotiating and agreeing final accounts

(5) *Analysis, collection and storage of actual cost data* – planning for, obtaining, analysing, collecting and storing real-time cost data from building projects, including discrete works packages and trade packages, which can be retrieved and reprocessed for use in future order of cost estimates and cost plans

(6) *Cost data retrieval and reprocessing* – retrieving and reprocessing cost data obtained from previous building projects for use in future order of cost estimates and cost plans

the principles of analysing building costs. Together, these rules deal with the quantification of buildings from 'cradle to grave' – from inception to demolition.

1.8.1 NRM 1 – Order of cost estimating and cost planning for capital building works

NRM 1 provides essential guidance on the quantification of building works for the purpose of preparing cost estimates and cost plans. The rules provide direction on how to quantify other items that form part of the total cost of a construction project,

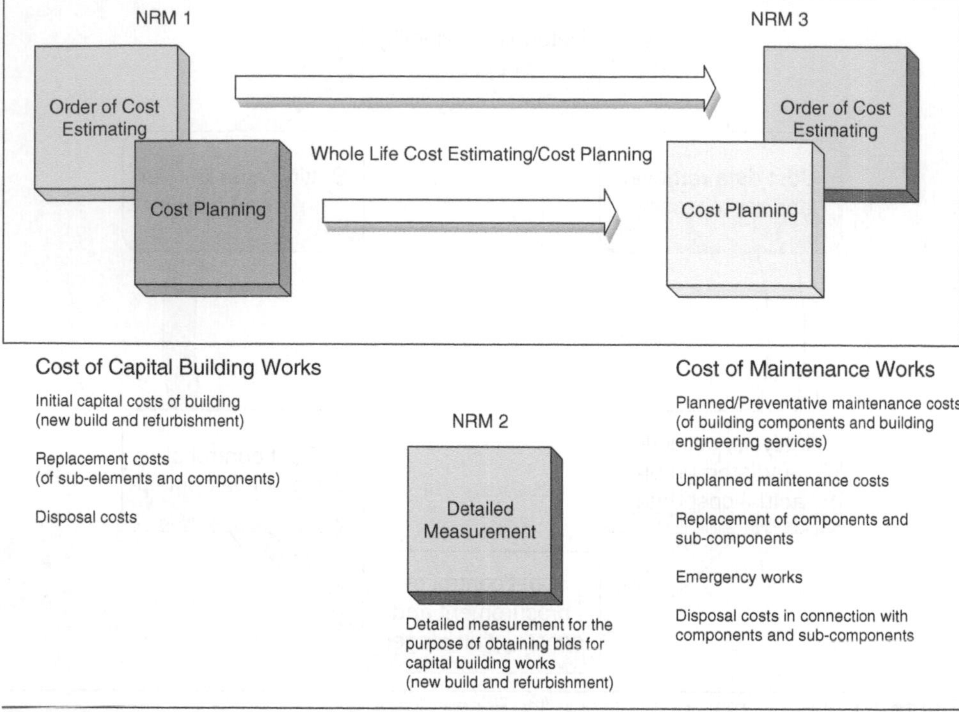

Figure 1.2 Relationship and purpose of the RICS NRM suite of measurement rules

but which are not reflected in the measurable building work items (e.g. preliminaries, overheads and profit, project team and design team fees, risk allowances, inflation, and other development and project costs). Additionally, the rules can be used as a basis for capturing historical cost data in the form required for use in future order of cost estimates and cost plans, thereby completing the 'cost management cycle'. NRM 1 provides not only a uniform basis for measuring building works and other items but also embodies the essentials of best practice.

NRM 1 provides the 'cornerstone' of good cost management of capital building projects – by enabling cost managers to give more effective, accurate and transparent cost advice to employers and other project team members, as well as facilitating better cost control. More importantly, it aids the cost manager to ensure that capital building projects are designed to an agreed cost, rather than allowing designers to set the costs based on their designs.

Although written primarily for the preparation of order of cost estimates and cost plans, the rules can also be used as a guide when preparing approximate estimates, including those for relatively low-value building projects. In addition, they can be used to quantify replacement components (e.g. replacement windows), building maintenance items (e.g. redecoration) and demolition works for the purpose of whole life cycle cost (or value) assessments. However, they do not deal with the maintenance and operation of mechanical and electrical services installation or other works and services carried out throughout the life cycle of a building.

1.8.2 NRM 2 – Detailed measurement for building works

NRM 2 became effective on 1 July 2013, thereby substituting the seventh edition of the Standard Method of Measurement (SMM7), the first edition of which had been published in 1921.

Like its predecessor, NRM 2 provides guidance on the detailed measurement and description of building works for the sole purpose of obtaining prices from contractors. These rules address all aspects of bill of quantities (BQs) production, including setting out the information required from the employer and other construction consultants to enable the preparation of a BQ, as well as dealing with the quantification of non-measurable work items, contractor-designed works (sometimes referred to as 'Contractor's Design Portion (CDP)') and the transference of risks to the contractor. NRM 2 covers the full range of possible building works activities, breaking these down into 41 work sections.

While written primarily for the preparation of BQs, quantified schedules of works and quantified work schedules, NRM 2 provide a basis for designing and developing both standard and 'in-house' schedules of rates.

1.8.3 NRM 3 – Order of cost estimating and cost planning for building maintenance works

NRM 3, although not published at the time of writing this book, provides guidance on the quantification of maintenance works for the purpose of preparing order of cost estimates for building maintenance works during the strategic or business justification stages of a capital building project, and cost plans for building maintenance works during the design and development stages of a capital building project. NRM 3 can also be used in preparing asset-specific cost plans and life cycle cost plans for existing buildings and facilities (i.e. for setting budgets and for procuring building maintenance for existing premises). However, NRM 3 does not address operational costs or other soft costs associated with the operation of buildings or facilities.

NRM 3 follows the same framework, cost code structure and premise as NRM 1. Consequently, NRM 3 also provides direction on how to quantify and measure other items associated with building maintenance that are not reflected in the measurable maintenance works items (e.g. maintenance contractor's management and administration charges, overheads and profit, other maintenance-related costs, consultants' fees and risks in connection with maintenance works.

Unlike capital building works projects, maintenance works are required to be carried out from the day the building is put to use (or handed over to the employer) until the end of its life. Accordingly, while the capital costs of building projects are usually incurred by the building owner or developer over a relatively short term, costs in connection with maintenance are incurred throughout the life of the building – over the long term. In view of this, NRM 3 also provides guidance on the measurement and calculation of the time value of money (i.e. the present value), and guidance on using the measured data to inform life cycle cost plans, as well as providing guidance on value added tax (VAT), and enhanced capital allowances and other financial incentives.

Together, NRM 1 and NRM 3 provide the rules for measurement, as well as the methodology for compiling, whole life cycle cost plans. Moreover, in the same way that NRM 1 does for capital building works, the rules within NRM 3 provide a framework for analysing building maintenance costs so that maintenance cost data is in a form that is readily usable in future cost estimates and cost plans for building maintenance works.

1.8.4 BCIS 'Elemental Standard Form of Cost Analysis (NRM Edition)'

The BCIS 'Elemental Standard Form of Cost Analysis (NRM Edition)' – often referred to as the BCIS SFCA – provides a cost data capture methodology that allows comparisons to be made between the cost of buildings and building elements. The BCIS SFCA was first published in 1961 and was updated using the same elemental and data structure in 1969 and 2008. The BCIS is a business division of the RICS.

In 2012, the BCIS SFCA published a new edition of the SFCA, which now shares the same elemental definitions and cost data structure devised for NRM 1.

1.9 NRM 1 and building information modelling (BIM)

Building information modelling, better known as 'BIM' – is a co-ordinated set of processes, supported by technology that seeks to add value by creating, managing and sharing the properties of a built asset throughout its life cycle. BIM incorporates data – physical, commercial, environmental and operational – on every element of a development's design.

The purported benefits of BIM are:

- *better outcomes through collaboration*. All project team members – different design disciplines, the cost manager, project manager, employer, contractor, specialists and suppliers – use a single, shared 3D model, cultivating collaborative working relationships. This is to ensure that all involved are focused on achieving best value, from project inception to eventual decommissioning.
- *enhanced performance*. BIM makes possible swift and accurate comparison of different design options, enabling development of more efficient, cost-effective and sustainable solutions.
- *optimised solutions*. Through deployment of new generative modelling technologies, solutions can be cost-effectively optimised against agreed parameters.
- *greater predictability*. Capital building works projects can be visualised at an early stage, giving owners, users and operators a clear idea of design intent and allowing them to modify the design to achieve the outcomes they want. It is also advocated that, in advance of construction, BIM will also enable the project team to 'build' the project in a virtual environment, rehearsing complex procedures, optimising temporary works designs and planning procurement of materials, equipment and manpower.
- *faster project delivery*. Significant time savings are achievable by agreeing the design concept early in building project development to eliminate late-stage design changes; using standard design elements when practicable; resolving complex construction details before the project goes on site; avoiding clashes; taking advantage of intelligence and automation within the model

to check design integrity and estimate quantities; producing fabrication and construction drawings from the model; and using data to control construction equipment.

- *reduced safety risk*. Crowd behaviour and fire-modelling capability enable designs to be optimised for public safety. Asset and facilities managers can use the 3D model to enhance operational safety. Contractors can minimise construction risks by reviewing complex details or procedures before going on site.
- *fits first time*. Integrating multidisciplinary design inputs using a single 3D model allows interface issues to be identified and resolved in advance of construction, eliminating the cost and time impacts of redesign. It is also contended that BIM enables new and existing assets to be integrated seamlessly.
- *reduced waste*. Exact quantity take-offs mean that materials are not over-ordered. Precise programme scheduling enables just-in-time delivery of materials and equipment, reducing potential for damage. Use of BIM for automated fabrication of equipment and components enables more efficient materials handling and waste recovery.
- *whole life asset management*. BIM models contain product information that assists with commissioning, operation and maintenance activities – for example sequences for start-up and shut-down, interactive 3D diagrams showing how to take apart and reassemble equipment items, and specifications allowing replacement parts to be ordered.
- *continual improvement*. Project team members can feed back information about the performance of processes and items of equipment, driving improvements on subsequent projects.

How does NRM 1 aid the BIM process? NRM 1 supports the BIM process by:

- providing both a common work breakdown structure (WBS) and a cost breakdown structure (CBS);
- providing a common cost data structure;
- providing a codification framework;
- providing standard element definitions;
- providing transparency of costs; and
- supporting designing to a cost.

1.10 Benefits of the rules

What are the benefits of the rules? First and foremost, NRM 1 exemplifies 'best practice' – it defines what 'good' looks like in terms of cost estimating and cost planning. It provides the basis for improved accuracy of estimating and better cost advice that, in turn, will help establish more effective cost control systems and procedures to instil greater employer and project team confidence in the cost advice received from the cost manager. Additionally, cost estimates and cost plans produced in accordance with NRM 1 should inspire greater trust in banks and other lenders providing funding for construction projects that requests for funds have been properly thought out by those wishing to obtain finance.

Although based on UK practice, the underlying philosophy of NRM 1 has world-wide application.

The rules within NRM 1 are not restrictive, but provide a simple but powerful toolkit for managing the total costs of a building project. They provide the following benefits:

- **Training and education:**
  - promoting common sense;
  - tabulated information acting as an invaluable aide-memoire;
  - helping learning – both measurement and construction knowledge (the rules are aimed at students, learning, as much as practioners – they have been written with all levels of user in mind); and
  - establishing cost-estimating and cost-planning competencies required by students, which universities and other learning establishments need to deliver.
- **Approach:**
  - introducing formal estimating stages – aligned to both the 'RIBA Plan of Work' and the 'OGC [Office of Government Commerce] Gateway Process';
  - presenting a clear framework, which facilitates a systematic approach to compiling cost estimates and cost plans; and
  - providing a comprehensive work breakdown structure (WBS) and a cost breakdown structure (CBS) for cost estimates and cost plans – both for projects comprising a single building and projects encompassing more than one building or structure.
- **Measurement:**
  - recognising that measurement is progressive (i.e. an iterative process which is dependent upon the design information available);
  - providing a uniform basis on which to measure areas;
  - in the measurement of cost-significant items (i.e. components), adopting Pareto's principle (the 80:20 rule); that is, 80% of the cost of the building is in 20% of the items measured; and
  - providing a uniform approach to measurement of components (i.e. building works items – for new works and rehabilitation works, as well as for future replacement).
- **Modern construction:**
  - dealing with modern construction products and methods, including modular units and complete buildings; and
  - considering sustainable construction.
- **Preliminaries:**
  - defining main contractor's preliminaries, which can be used for any procurement or contract strategy (including prime contracting and private finance initiatives).
- **Risk management:**
  - supporting the need for a risk management strategy;
  - promoting use of risk allowances, and the view that risk allowances should not be a standard percentage, but a properly considered assessment of the cost of dealing with risk should it occur, which must be managed; and
  - defining four categories of risk (i.e. design development risk, construction risk, employer's change risks and employer's other risks).

- **Inflation:**
  - defining two categories of inflation (i.e. tender inflation and construction inflation – with unexpected changes in market conditions being treated as a risk); and
  - defining measurement of inflation.
- **Cost management:**
  - promoting budget setting using a cost limit, which includes allowances for risk – its purpose is to provide a realistic cost limit for project cost control within which a construction project can proceed without further resource- and time-intensive authorisation requirements;
  - helping establish more effective cost control systems;
  - providing essential guidance on dealing with non-construction-related costs such as consultants' fees and other development and project costs (the cost of acquiring land and property, fees and charges, planning contributions, decanting and relocation costs, marketing costs and the cost of finance); and
  - providing advice on how to deal with VAT, taxation allowances, taxation relief and grants.
- **Accuracy of estimates:**
  - defining the information required by the cost manager to produce cost estimates and cost plans at each formal estimating stage; and
  - helping improve the accuracy of cost estimates and cost plans.
- **Innovation:**
  - dealing with the total cost of delivering a building project, not simply construction costs (total cost management);
- **Value management and value for money:**
  - performing an essential part of a value-for-money framework – particularly for central government (e.g. Office of Government Commerce and Ministry of Defence (MoD)), local government and other publicly funded organisations that have an obligation to demonstrate that value for money has been achieved in construction projects;
  - helping underpin a business case for a building project;
  - quantification of replacement components and building maintenance items for the purpose of whole life cycle cost/value assessments;
  - aiding value management (VM) and value engineering (VE) processes;
  - affording greater transparency to cost estimates and cost plans (note: transparency can assist with fraud prevention);
  - aiding actual cost scrutiny; and
  - helping improve clear and effective communication between the employer and project team members – thereby making sure that risks associated with the building project are identified, analysed and responded to.
- **Procurement:**
  - providing a method for codifying elemental cost plans so that they can be converted to works packages for procurement, and cost management during the construction stage; and
  - aiding cost management during the construction phase of a project.

- **Data acquisition, analysis and evaluation:**
  - providing a robust basis for capturing historical cost data in the form required for future cost estimates and cost plans, thereby completing the 'cost management cycle'.
- **Usability:**
  - ability to be computerised (i.e. cost-estimating and cost-planning systems, together with integral reporting).
- **Quality control:**
  - ability to be integrated into an organisation's quality management system (QMS).

The main benefits of NRM 1 are recapitulated in Figure 1.3.

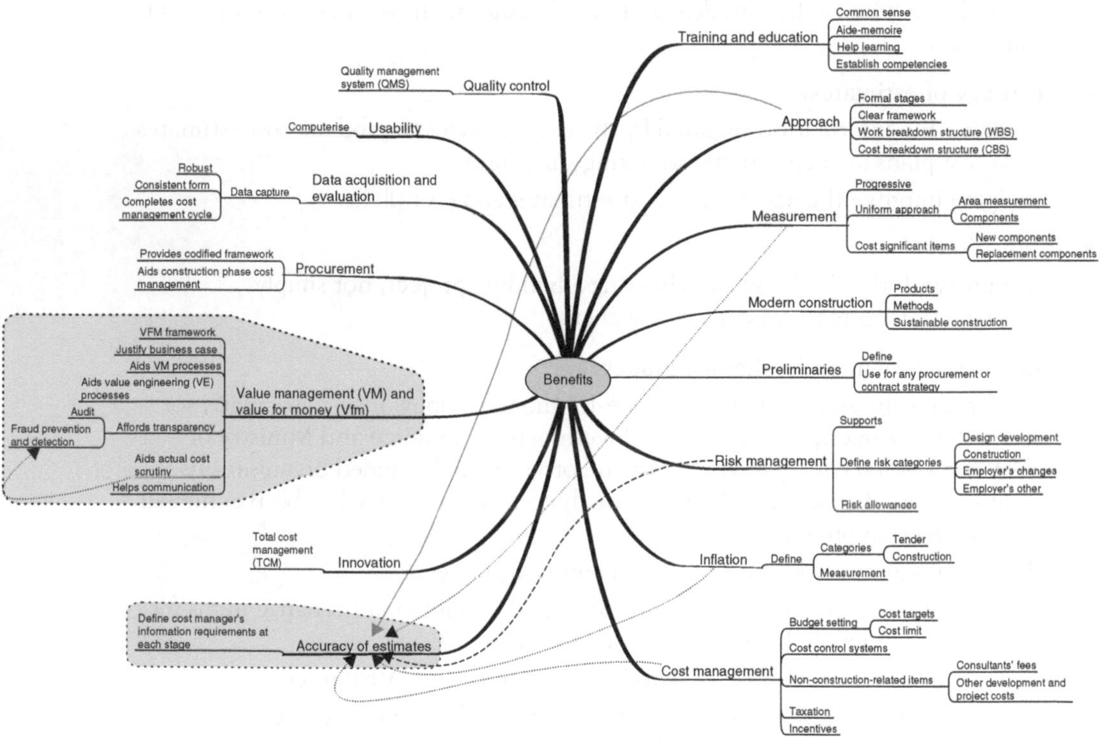

Figure 1.3 Benefits of NRM 1

Box 1.1 Key definitions

- **Value engineering** – an organised approach to the identification and elimination of unnecessary cost. Unnecessary cost is cost which does not provide use, or life, or quality, or appearance, or customer features. The following tasks undertaken by cost managers are involved in VE practice:
  - cost estimating;
  - advising on cost limits and preparing budgets;
  - undertaking order of cost estimates for options (option appraisals);
  - measuring and describing construction work in terms of cost planning;
  - investment appraisal;
  - advising on whole life costs;
  - cost analysis;
  - benchmark analysis;
  - cost/benefit analysis;
  - evaluating alternative designs;
  - advising on cash flow forecasting;
  - forecasting expenditure flows; and
  - preparing and administering maintenance programmes.
- **Value management** – an approach which aims to establish, at the start of a project, the strategic plan by which it should develop. This is partly achieved through the use of a series of workshops at key stages in the development of the project; and it is complemented by value-engineering techniques. These also make use of workshop techniques, but are concerned with obtaining value for money through an organised systematic approach, placing emphasis upon whole life costing.
- **Value for money** – the relationship between economy (price), efficiency and effectiveness. The VfM principle underpins decision making in public sector procurement. Here, the definition of VfM is: 'The optimum combination of whole-life cost and quality (or fitness for purpose) to meet user's requirements. This is rarely synonymous with price' (adapted from 'CUP [Central Unit on Procurement] Guidance Note No. 54').
- **Whole life costs** – the costs of acquiring a building or facility (including consultancy, design and construction costs, and equipment), the costs of operating it and the costs of maintaining it over its whole life through to its disposal – that is, the total ownership costs.

CHAPTER 2

Cost estimating and cost planning

Introduction

This chapter:

- considers the composition of cost estimates and cost plans;
- describes both the RIBA [Royal Institute of British Architects] Plan of Work and OGC Gateway Process;
- illustrates the interrelation between the cost-estimating and cost-planning stages and both the RIBA Work Stages and the OGC Gateways;
- provides an overview of the cost-estimating and cost-planning stages;
- considers the impact of procurement and contract strategies on the cost-planning process; and
- identifies responsibility for cost planning.

2.1 Cost estimates and cost plans

Essentially, cost estimates and cost plans are made up of two distinct elements: the base cost estimate and the risk allowance:

- **Base cost estimate** – an estimate of all those items which are known or for which a degree of certainty exists; and
- **Risk allowance** – an allowance for dealing with the financial impact of the employer's residual risks, should they materialise.

The risk allowance should be steadily reduced over time as the risks or their consequences are minimised through good risk management.

The composition of cost estimates and cost plans will depend on the employer's requirements. For example, an employer might simply require the cost manager to manage construction costs. Equally, the employer might require the cost manager to manage all costs in connection with a building project – total cost management (TCM). As a consequence, NRM 1 takes into consideration the varying degrees of cost management services that might be requested and, as a result, deals with cost estimates under the following five principal headings:

1 works estimate;
2 project and design team fees estimate;
3 other development and project costs estimate;
4 risk allowances estimate; and
5 inflation estimate.

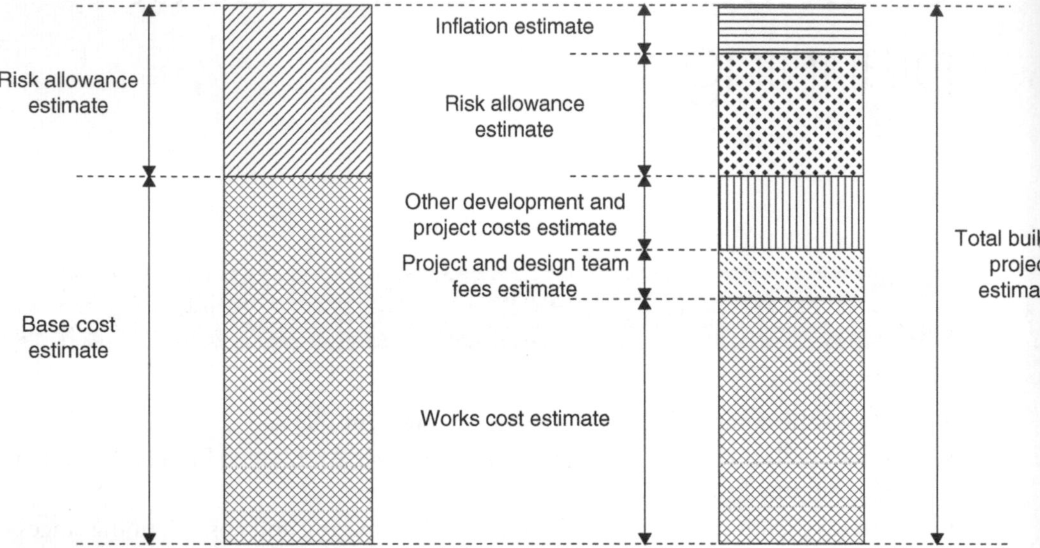

Figure 2.1 Build-up of total building project estimate

The works, project and design team fees, and other development and project costs estimates represent the base cost estimate, while the risk allowances and inflation estimates comprise the risk allowance. The risk allowance, when added to the base cost estimate, will give the cost limit most likely to be required for the building project (refer to Figure 2.1).

2.2 Process models for managing and designing building projects

Throughout the rules, references are made to both the RIBA Plan of Work and the OGC Gateway Process, and to the RIBA Work Stages and OGC Gateways within. These are the two most commonly used process models for managing and designing building projects.

The RIBA Plan of Work is a construction industry-recognised framework that organises the process of managing and designing building projects and administering building contracts into a number of key Work Stages. The RIBA Plan of Work consists of eight sequential stages (identified by the numbers 0 to 7), and the employer is required to 'approve' the design and latest cost estimate, or cost plan, before authorising commencement of the next RIBA Work Stage. Despite this apparent linear nature, the sequence or content of RIBA Work Stages might need to be varied or overlapped to suit the proposed procurement method. Consequently, when two or more Work Stages are combined, it is not always transparent when a building project is moving from one stage to another. As such, it is an ideal tool, provided that it is conceptualised as providing the basic outline of the building project process.

As an alternative to the RIBA Plan of Work, many public sector organisations have adopted the OGC Gateway Process for managing and designing building projects. This process examines programmes and projects at key decision points in their life cycle. It looks ahead to provide assurance that the employer can progress to the next stage. Typically a project will undergo three reviews before commitment to invest and two looking at service implementation and confirmation of the operational benefits.

Prior to the employer authorising the project team to progress to each RIBA Work Stage or OGC Gateway, the cost estimate or cost plan produced for the preceding RIBA Work Stage or OGC Gateway must be reviewed by the employer and the project team to ensure that:

- the latest estimate is compared with the previously approved cost limit (i.e. authorised budget) and does not exceed it without fully reasoned justification;
- the cost target for each element of the building project is reasonable and up to date;
- the latest estimate is made up of the base estimate and the risk allowance; and
- funds are available for the intended expenditure.

Consequently, the rules advocate a cost management strategy that comprises a number of formal cost-estimating and cost-planning stages, which correspond with both the key RIBA Work Stages (i.e. from the RIBA Plan of Work) and the OGC Gateways (i.e. from the OGC Gateway Process). Figure 2.2 shows the relationship between the RICS formal cost-estimating and cost-planning stages and both the RIBA Work Stages and OGC Gateways.

2.3 An overview of the cost-estimating and cost-planning stages

The rules identify a number of formal stages at which cost estimates and cost plans are reported to the employer. These stages are applicable to most building projects and are linked to both the RIBA Work Stages and the OGC Gateways.

- **Order of cost estimates** – prepared at RIBA Work Stages 0 (Strategic Definition) and 1 (Preparation of Brief), or OGC Gateway 2 (Delivery Strategy), to establish if the proposed building project is affordable and, if affordable, to determine a realistic cost limit (budget). A number of order of cost estimates for a number of different building options or development scenarios are normally undertaken and compared.
- **Formal cost plan 1** – prepared for the preferred option from the appraisal, or business case study, at RIBA Work Stage 2 (Concept Design), or OGC Gateway 3A (Design Brief and Concept Approval). This cost plan is prepared at a point where the scope of work is fully defined and key criteria are specified but no detailed design has commenced. Formal cost plan 1 will provide the frame of reference for the subsequent cost plan (formal cost plan 2).
- **Formal cost plan 2** – prepared at RIBA Work Stage 3 (Developed Design), or OGC Gateway 3B (Detailed Design Approval), as a result of the cost checking

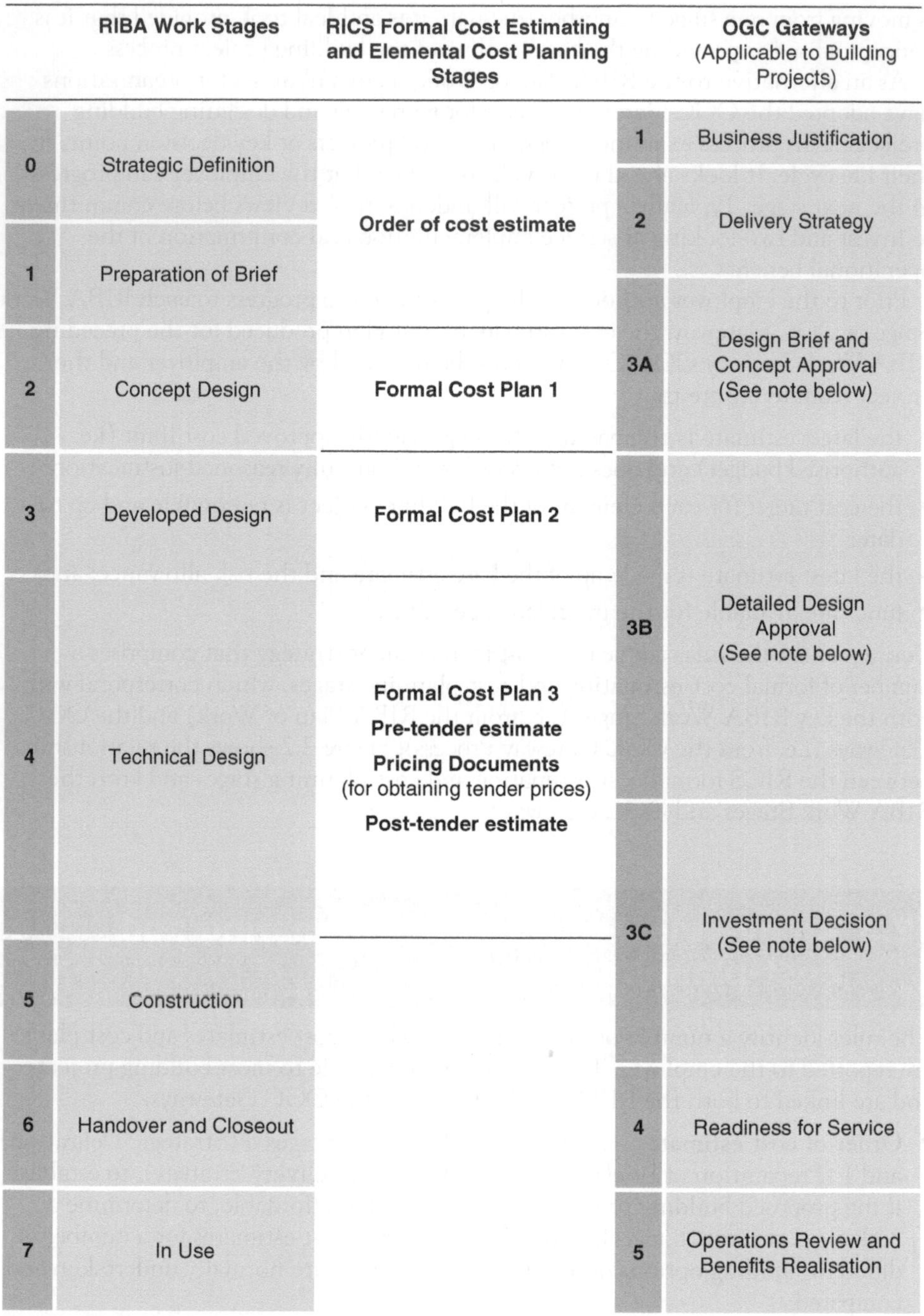

	RIBA Work Stages	RICS Formal Cost Estimating and Elemental Cost Planning Stages		OGC Gateways (Applicable to Building Projects)
0	Strategic Definition	Order of cost estimate	1	Business Justification
1	Preparation of Brief		2	Delivery Strategy
2	Concept Design	Formal Cost Plan 1	3A	Design Brief and Concept Approval (See note below)
3	Developed Design	Formal Cost Plan 2	3B	Detailed Design Approval (See note below)
4	Technical Design	**Formal Cost Plan 3** **Pre-tender estimate** **Pricing Documents** (for obtaining tender prices) **Post-tender estimate**	3C	Investment Decision (See note below)
5	Construction			
6	Handover and Closeout		4	Readiness for Service
7	In Use		5	Operations Review and Benefits Realisation

Figure 2.2 Relationship between the RICS formal cost-estimating and cost-planning stages and both the RIBA Work Stages and OGC Gateways (adapted from the *RIBA Outline Plan of Work 2013*)

Note: A prerequisite of OGC Gateway Review 3: Investment Decision is that the design brief, concept design and detailed design have been approved and signed off by the Senior Responsible Owner (SRO). For the purpose of comparing the OGC Gateway Process with the RIBA Work Stages, these two decision points are referred to as OGC Gateway 3A (Design Brief and Concept Approval) and OGC Gateway 3B (Detailed Design Approval), with OGC Gateway 3C representing the final OGC Gateway Review 3 (Investment Decision) of cost-significant components and cost targets set for cost plan 1 as more design information and further information about the site becomes available. Formal cost plan 2 will provide the frame of reference for formal cost plan 3.

- **Formal cost plan 3** – prepared at RIBA Work Stage 4 (Technical Design) or OGC Gateway 3C (Investment Decision), as a result of cost checking the developing design against the cost targets established in formal cost plan 2.
- **Pre-tender estimate** – prepared immediately before calling the first tenders for construction. This is the final cost check undertaken by the cost manager before tender bids for the building project, or any part of the building project, are obtained. It is most likely that the pre-tender estimate will be based on a formal cost plan. However, whether the pre-tender estimate is based on formal cost plan 1, 2 or 3 will depend on which stage of design development tenders are to be sought. For contract strategies such as design and build, the pre-tender estimate might be based on formal cost plan 1.
- **Post-tender estimate** – prepared after all the construction tenders have been received and evaluated. It is based on the outcome of any post-tender negotiations, including the resolution of any tender qualifications and tender price adjustments. The post-tender estimate will include the actual known construction costs and any residual risks. The aim of this estimate is to corroborate the funding level required by the employer to complete the building project, including cost updates of project and design team fees, as well as other development and project costs, where they form part of the costs being managed by the cost manager. When reporting the outcome of the tendering process to the employer, the cost manager should include a summary of the post-tender estimate(s). The post-tender estimate should be fairly accurate because the uncertainties of market conditions have been removed. Post-tender estimates are used as the control estimate during construction.

Cost-estimating and cost-planning processes in the context of the RIBA Work Stages and the OGC Gateways are set out in Tables 2.1 and 2.2, respectively.

2.4 Impact of procurement and contract strategies on cost planning

Different procurement and contract strategies will impact on the degree of cost planning undertaken by the cost manager and the use of cost plans. For contract strategies such as design and build or management contracting (chosen to produce earlier construction completion, made possible through overlapping activities), the principles of cost management throughout the design process remain identical, although the estimating stages cannot be separated so clearly.

In order to achieve an early start on site an employer might be content to procure the building project by means of a design and build strategy, simply using the drawings on which planning approval was granted and outline specification information as the basis of the employer's requirements. On the other hand, if an employer wishes to retain greater control over the final design of the building, then the building project might be tendered after completion of detailed design and formal cost plan 2. Under a design and build strategy, cost management during detailed design is carried out by the contractor since they are already committed to a firm

Table 2.1 Cost-estimating, cost-planning and cost control processes in the context of the RIBA Work Stages

RIBA Work Stage		Process	Function(s)	Method
0	Strategic Definition	Option costing – prepare order of cost estimate(s).	1. Establish estimated cost of proposed building project or comparative costs for a number of options. 2. Determine viability, affordability of building project. 3. Establish initial cost limit (i.e. authorised budget) for resultant building project.	1. Floor area method, functional unit method and/or elemental method. 2. Apply unit rates interpolated from historical cost data.
1	Preparation of Brief			
2	Concept Design	Cost planning – prepare initial cost plan (formal cost plan 1); or Cost estimating – prepare initial cost estimate.	1. Establish cost targets for each group element, element and sub-element. 2. Confirm cost limit (i.e. authorised budget) for building project.	1. Quantify elements, sub-elements and components (to be measured in accordance with the NRM 1 tabulated rules of measurement). 2. Apply unit rates interpolated from historical cost data and/or obtained via market testing. Note: (a) Where cost planning is not used, approximate estimating techniques can be used to establish the current estimated cost. This will help to identify if changes are required to the scope of works, design or specification prior to progressing to the next RIBA Work Stage.
3	Developed Design	Cost planning – prepare firm cost plan (formal cost plan 2); or Cost checking – prepare firm cost estimate.	1. Cost check of developing design against cost plan. 2. Reappraise cost targets for each group element, element and sub-element. 3. Confirm revisions to cost targets for each group element, element and sub-element. 4. Identify cost savings and/or cost concerns. 5. Confirm adequacy of cost limit for building project.	1. Check developing design drawings, specification information, as more design information about the site becomes available. 2. Quantify elements, sub-elements and components for which additional information has been received. 3. Check adequacy of unit rates used in formal cost plan 1. 4. Apply unit rates to additional items – interpolated from historical cost data and/or obtained via market testing. Notes: (a) Cost planning is a continuous process which includes comparative cost studies of

Table 2.1 *(Continued)*

RIBA Work Stage		Process	Function(s)	Method
				components, research and advice on the economic effect of proposed design details, including the use of alternative methods of construction and product/ material specifications. (b) Where cost planning is not used, approximate estimating techniques can be used to establish the current estimated cost. This will help to identify if changes are required to the scope of works, design or specification prior to progressing to the next RIBA Work Stage.
4	Technical Design	Cost planning – review and revise cost plan (formal cost plan 3)/ prepare pre-tender estimate; or Cost estimating – prepare pre-tender estimate.	1. Final cost checks of design against cost plan. 2. Reappraise cost targets for each group element, element and sub-element. 3. Confirm revisions to cost targets for each group element, element and sub-element. 4. Identify cost savings and/or cost concerns. 5. Confirm adequacy of cost limit for building project.	1. Check developing design drawings, specification information, as more design information about the site becomes available. 2. Quantify elements, sub-elements and components for which additional information has been received. 3. Check adequacy of unit rates used in formal cost plan 2. 4. Apply unit rates to additional items – interpolated from historical cost data and/or obtained via market testing. Notes: (a) Cost planning is a continuous process which includes comparative cost studies of components, research and advice on the economic effect of proposed design details, including the use of alternative methods of construction and product/ material specifications. (b) The result of formal cost plan 3 is a pre-tender estimate. (c) Where cost planning is not used, approximate estimating techniques can be used to establish the pre-tender estimate. It is common for an approximate estimate to

Table 2.1 *(Continued)*

RIBA Work Stage		*Process*	*Function(s)*	*Method*
				be based on the tender documents prior to tender action. This allows time for final scope, design or specification adjustments prior to the commencement of tender action.
		Cost estimating – prepare post-tender estimate.	Confirm cost limit.	1. Tender price for building works plus residual risks. Plus (if total costs to be estimated): 2. Cost estimate for project/design team fees. 3. Cost estimate for other development/project costs.
		Data collection – analyse tender prices.	Gather historical cost data.	Analysis of tender prices and storing data for future use.
Construction				
5	Construction	Cost control – manage expenditure of the risk allowance.	1. Check claims for extra costs and cost savings. 2. Confirm revisions to risk allowance. 3. Confirm adequacy of cost limit for building project.	1. Based on contract rates. 2. Actual cost scrutiny of accounts and contractor's cost proposals.
6	Handover and Closeout	Final account.	Agreement and settlement of final account.	Negotiation.
Use				
7	In Use	-		-

construction price. As a result, there will no requirement for a formal cost plan 2 or 3 to be prepared by the employer's cost manager, as formal cost plan 1 will form the basis of the pre- and post-tender estimates and, most likely, be used as the basis for cost control during the construction phase of the building project. However, during construction, the employer's cost manager will still be responsible for managing the expenditure of risk allowances, negotiating and agreeing costs with the main contractor brought about by changes and claims, and regularly advising the employer on the financial status of the building project.

Under a management contract it is even more important that continuous monitoring of the cost targets for the elements and the cost limit for the entire building project is established as the design proceeds rather than at set stages. Construction works will start on some parts of the building project before the detail design for other parts is complete. Commitment to proceed is therefore taken when the forecast final cost for the project is at a lower level of accuracy than would be in the case at construction commencement using 'traditional lump sum' or 'design and build' contract strategies.

Table 2.2 Cost-estimating, cost-planning and cost control processes in the context of the OGC Gateway Process

OGC Gateways		Process	Function(s)	Method
1	Business Justification	Option costing – prepare order of cost estimate(s).	1. Establish estimated cost of proposed building project or comparative costs for a number of options. 2. Determine viability, affordability of building project. 3. Establish initial cost limit (i.e. authorised budget) for resultant building project.	1. Floor area method, functional unit method and/or elemental method. 2. Apply unit rates interpolated from historical cost data.
2	Delivery Strategy			
3A	Design Brief and Concept Approval	Cost planning – prepare initial cost plan (formal cost plan 1); or Cost estimating – prepare initial cost estimate.	1. Establish cost targets for each group element, element and sub-element. 2. Confirm cost limit (i.e. authorised budget) for building project.	1. Quantify elements, sub-elements and components (to be measured in accordance with the NRM). 2. Apply unit rates interpolated from historical cost data and/or obtained via market testing. Note: Where cost planning has not been used, approximate estimating techniques will be used to establish the current estimated cost. This will help to identify if changes are required to the scope of works, design or specification prior to progressing to the next OGC Gateway.
3B	Detailed Design Approval	Cost planning – prepare firm cost plan (formal cost plan 2); or Cost estimating – prepare firm cost estimate.	1. Cost check of developing design against cost plan. 2. Reappraise cost targets for each group element, element and sub-element. 3. Confirm revisions to cost targets for each group element, element and sub-element. 4. Identify cost savings and/or cost concerns. 5. Confirm adequacy of cost limit for building project.	1. Check developing design drawings, specification information, as more design information about the site becomes available. 2. Quantify elements, sub-elements and components for which additional information has been received. 3. Check adequacy of unit rates used in formal cost plan 1. 4. Apply unit rates to additional items – interpolated from historical cost data and/or obtained via market testing. Notes: (a) Cost planning is a continuous process which includes comparative cost studies

Table 2.2 *(Continued)*

OGC Gateways		Process	Function(s)	Method
				of components, research and advice on the economic effect of proposed design details, including the use of alternative methods of construction and product/ material specifications. (b) Where cost planning is not used, approximate estimating techniques will be used to establish the current estimated cost. This will help to identify if changes are required to the scope of works, design or specification prior to progressing to the next OGC Gateway.
3C	Investment Decision	Cost planning – review and revise cost plan (formal cost plan 3)/ prepare pre-tender estimate; or Cost estimating – prepare pre-tender estimate.	1. Final cost checks of design against cost plan. 2. Reappraise cost targets for each group element, element and sub-element. 3. Confirm revisions to cost targets for each group element, element and sub-element. 4. Identify cost savings and/or cost concerns. 5. Confirm adequacy of cost limit for building project.	1. Check developing design drawings, specification information, as more design information about the site becomes available. 2. Quantify elements, sub-elements and components for which additional information has been received. 3. Check adequacy of unit rates used in formal cost plan 2. 4. Apply unit rates to additional items – interpolated from historical cost data and/or obtained via market testing. Notes: (a) Cost planning is a continuous process which includes comparative cost studies of components, research and advice on the economic effect of proposed design details, including the use of alternative methods of construction and product/ material specifications. (b) The result of formal cost plan 3 is a pre-tender estimate. (c) Where cost planning is not used, approximate estimating techniques will be used to establish the pre-tender

Table 2.2 *(Continued)*

OGC Gateways		Process	Function(s)	Method
				estimate. It is common for an approximate estimate to be based on the tender documents prior to tender action. This allows time for final scope, design or specification adjustments prior to the commencement of tender action.
		Cost estimating – prepare post-tender estimate.	Confirm cost limit.	1. Tender price for building works plus residual risks. Plus (if total costs to be estimated): 2. Cost estimate for project/design team fees. 3. Cost estimate for other development/project costs.
		Data collection – analyse tender prices.	Gather historical cost data.	Analysis of tender prices and storing data for future use.
		Cost control – manage expenditure of the risk allowance. Final account.	1. Check claims for extra costs and cost savings. 2. Confirm revisions to risk allowance. 3. Confirm adequacy of cost limit for building project. 4. Agreement and settlement of final account.	1. Based on contract rates. 2. Actual cost scrutiny of accounts and contractor's cost proposals. 3. Negotiation.
4	Readiness for Service	—	—	—
5	Benefits Evaluation	—	—	—

Notes:

(1) A prerequisite of OGC Gateway Review 3: Investment Decision is that the design brief, concept design and detailed design have been approved and signed off by the Senior Responsible Owner (SRO). For the purpose of comparing the OGC Gateway Process with the RIBA Work Stages. these two decision points are referred to as OGC Gateway 3A (Design Brief and Concept Approval) and OGC Gateway 3B (Detailed Design Approval), with OGC Gateway 3C representing the final OGC Gateway Review 3 (Investment Decision)

(2) In addition to certain pre-construction activities (i.e. pre-tender estimate, post-tender estimate and data collection), OGC Gateway 3C (Investment Decision) also comprises the construction phase of a building project

Box 2.1 Key definition

- **Employer's requirements** – one of the contract documents, used for a design and build contract, which describes the building which the employer requires the contractor to construct. The requirements are as comprehensive as necessary. It may range from little more than a description of the accommodation required to a full scheme design. If the employer wishes to keep tight control of the final product (i.e. the building), a very full and detailed performance specification will need to be produced. Despite what may be sometimes thought by the employer and other project team members, the preparation of a proper performance specification is a skilful and time-consuming task.

2.5 Review and approval process

Prior to the employer authorising commencement of the next RIBA Work Stage or OGC Gateway, the order of cost estimate or cost plan for the preceding RIBA Work Stage or OGC Gateway must be reviewed by the employer and the project team to ensure that:

- the building project is affordable;
- the cost target for each element of the project is reasonable and up to date; and
- the cost limit has not been exceeded.

Following the review, the employer will sign off the cost plan and give any necessary instructions and authorise commencement of the next RIBA Work Stage or OGC Gateway.

The employer is required to 'approve' each formal cost estimate, or cost plan, before authorising commencement of the next RIBA Work Stage or OGC Gateway. In order to avoid unnecessary conflict, it is essential that employers and other project team members are aware of what is included in each element of the cost plan. However, the extent to which the employer wishes to disclose cost information to the project team members is often a sensitive matter. Therefore, it is essential that the cost manager obtains clarification from the employer as to which project team member receives what cost information.

PART 2

Order of cost estimates

CHAPTER 3

Order of cost estimates – an overview

Introduction

This chapter:

- explains the purpose of an order of cost estimate;
- distinguishes between 'rough' order of cost estimates and order of cost estimates; and
- explains when to prepare an order of cost estimate.

3.1 Purpose of an order of cost estimate

Before deciding to build, an employer will want to establish if the proposed building project is affordable and, if so, to establish a realistic cost limit for the proposed building project. In practice, the employer's first approval is based on a broad brush estimate of cost of the building project, often on a global assessment of the cost per square metre (cost/m^2) of the building works, with appropriate additions for project and design team fees, risk allowances and inflation. This is called an 'order of cost estimate'. It is important to note that it is not necessarily based on actual design but on past experience of similar building projects.

The main purposes of an order of cost estimate are to:

- determine the possible construction costs of a prospective building project, based on the employer's fundamental requirements;
- establish if a prospective building project is viable and, if viable, to establish a realistic cost limit for the building project; and
- make the employer aware of the likely financial commitment as early as possible, but before design of the preferred option is developed, to avoid the waste of expensive resources.

3.2 Types of order of cost estimate

Order of cost estimates can be divided into two distinct types:

- rough order of cost estimate; and
- order of cost estimate.

It is not uncommon for an employer to request an order of cost estimate before a site has been acquired: often at short notice and with a quick response required (e.g. in half an hour). The aim of the order of cost estimate here is to provide the employer with an estimated construction cost for inclusion in his initial development, or investment, appraisal. 'Quick-and-dirty' cost estimates such as these are commonly referred to as 'rough' order of cost estimates. They are usually based on a single quantity (i.e. the required gross internal floor area – GIFA – or number of functional units) multiplied by a single 'all-in' unit rate, which includes all costs in connection with building works, external works, main contractor's preliminaries, and main contractor's overheads and profit. Such rates are derived through the analysis of the known total construction cost of a completed building project (called a cost analysis). Rates are calculated by dividing the total construction cost by the number of units – GIFA or number of functional units, whichever is required. Obviously, amongst other things, the rates need to be adjusted to take account of changes in price levels since the date of the cost analysis, differences in site conditions and location. The collection, adjustment and use of cost data are further discussed in Chapter 23 ('Deriving unit rates for building components, sub-elements and elements').

Sometimes, at the embryonic stage of a building project, an employer might simply ask the cost manager to provide a single 'all-in' unit rate for constructing the building (either the cost/m^2 of GIFA or a functional unit rate), which can be used in an initial appraisal. Alternatively, the employer might request a range of 'all-in' unit rates in order that different scenarios can be assessed (e.g. worst case, best case and realistic case scenarios). Equally, the employer might quantify the prospective building project and request the cost manager to advise the estimated total cost of construction (e.g. provide an area schedule of accommodation required). Should the appraisal prove positive (i.e. indicate that the desired profit margin is achievable), the employer might decide to acquire the site to develop in the future. Therefore, it is extremely important for the cost manager to obtain a clear understanding of what level of cost information the employer is after, and what the perceived accuracy is.

The problem that the cost manager has when asked for a 'rough' order of cost estimates is that not enough will be known about the building project to give an accurate budget price. Therefore, the cost manager's estimate will be based on very sketchy information only, e.g. without the benefits of surveys and designs. Order of cost estimates entail a more considered approach and take account of the main facets of the proposed building project.

Typically, the cost manager will prepare cost estimates for a number of alternative building types or development scenarios for a site. Such alternative cost estimates are often referred to as option costs, whereas the consideration and appraisal of alternative building types or development scenarios for a site is often referred to as option studies, which usually result from optioneering.

Box 3.1 Key definition

- **Optioneering** – the consideration and appraisal of alternative methods of achieving the project objectives. The process can be used throughout stages of a project. The term has become increasingly used in the construction industry when employers, or management, need to be confident of a course of action, particularly when regulatory or funding bodies seek demonstration of due process.

3.3 When is an order of cost estimate prepared?

Order of cost estimates are produced as an intrinsic part of RIBA Work Stages 0 (Strategic Definition) and 1 (Preparation of Brief), or OGC Gateways 1 (Business Justification) and 2 (Delivery Strategy), whichever process model is applicable. The requirements of RIBA Work Stages 0 and 1, as described in the *RIBA Outline Plan of Work*, are as follows:

- RIBA Work Stage 0: Strategic Definition:

  Identify client's Business Case and Strategic Brief and other core project requirements.

OGC Gateway 1 (Business Justification) is comparable with RIBA Work Stage 0.

- RIBA Work Stage 1: Preparation of Brief:

  Develop Project Objectives, including Quality Objectives and Project Outcomes, Sustainability Aspirations, Project Budget, other parameters or constraints and develop Initial Project Brief. Undertake Feasibility Studies and review of Site Information.

OGC Gateway 2 (Delivery Strategy) is comparable with RIBA Work Stage 1.

CHAPTER 4

Commencing an order of cost estimate

Introduction

This chapter:

- explains how to plan an order of cost estimate using a work breakdown structure and a cost breakdown structure;
- conveys the order of cost framework;
- outlines the basis for preparing order of cost estimates;
- looks at the elemental method of estimating; and
- defines the information that is required by the cost manager to prepare an order of cost estimate.

4.1 Planning an order of cost estimate

Before attempting to quantify and estimate the cost of a prospective building project, the composition of the building project needs to be determined and the structure of the order of cost estimate planned. The composition of an order of cost estimate can be viewed as a work breakdown structure. This is a tree structure that can be used to define and divide a building project into key facets. It is developed by starting with the end objective (i.e. WBS Level 0 – the entire building project), which is then successively subdivided into the main components and sub-components that make up the entire building project – providing a hierarchical breakdown. What is more, a work breakdown structure initiates the development of the cost breakdown structure. The cost breakdown structure is used to allocate costs to every facet of the building project at each level of the work breakdown structure. Together, the work breakdown structure and cost breakdown structure provide an initial frame of reference for the cost management of a building project.

Figure 4.1 illustrates the typical use of a work breakdown structure to organise an order of cost estimate. It shows a prospective building project comprising facilitating works, and building works consisting of a hotel building and an office building, together with the associated external works. The use of numerical codes and descriptors provides the basis for the cost breakdown structure.

WBS Level 0	WBS Level 1	WBS Level 2	WBS Level 3	WBS Level 4	WBS Level 5
Entire building project:	1. Building works:	1.1 Facilitating Works:	1.1.1 Facilitating works		
			1.1.2 Demolition works		
		1.2 Building works:	1.2.1 Residential (see note 4)		
			1.2.2 Retail (see note 4)		
			1.2.3 External works:	1.2.3.4.1 Landscaping	
				1.2.3.4.2. Mains connections:	1.2.3.4.2.1 Water
					1.2.3.4.2.2 Electricity
					1.2.3.4.2.3 Gas
					1.2.3.4.2.4 Telecommunications
	2. Main contractor's preliminaries				
	3. Main contractor's overheads and profit				
	4. Project/design team fees				
	5. Other development/ project costs				
	6. Risk:	6.1 Design development risk			
		6.2 Construction risk			
		6.3 Employer change risks			
		6.4 Employer other risks			
	7. Inflation:	7.1 Tender inflation			
		7.2 Construction inflation			
	8. Value added tax (VAT)				

Figure 4.1 Use of a work breakdown structure to organise an order of cost estimate

Notes:

(1) The end objective (WBS Level 0) is the resultant total cost of the building project. At WBS Level 1, the building project is subdivided into the main eight facets

(2) WBS Level 5 items can be further divided if required

(3) The numerical codes for each WBS level together with the descriptors also provide the basis for a cost breakdown structure

(4) Building works that are considered to be outside the norm should be identified and quantified as separate items against the relevant building – these are sometimes referred to as 'abnormals' (e.g. specialist piled foundation design to overcome railway tunnel). Failure to clearly and separately identify abnormal costs will result in distorted costs being reported by the cost manager to the employer and project team members. Facilitating works by their nature are also considered to be 'abnormal' items

4.2 Order of cost estimate framework

NRM 1 presents a framework for order of cost estimates. This framework comprises a number of separate estimates, which are as follows:

- building works estimate;
- main contractor's preliminaries estimate;
- main contractor's overheads and profit estimate;
- project and design team fees estimate;
- other development and project costs estimate;
- risk allowances estimate;
- inflation estimate; and
- VAT assessment.

The sum of these cost estimates gives the cost limit for the building project.

Considering each estimate in turn:

- **Building works estimate –** comprises cost estimates for the building activities required to construct a building or structure, including mechanical and electrical engineering services, external works and any facilitating works.
- **Main contractor's preliminaries estimate –** includes items that are required to construct the building project, but cannot be allocated to a specific building activity (e.g. constructing the foundations). They include the main contractor's costs associated with:
  - management and staff;
  - site establishment;
  - temporary services;
  - security, safety and environmental protection;
  - control and protection;
  - common user mechanical plant (i.e. mechanical plant used for more than one construction activity);
  - common user temporary works (i.e. temporary works used to facilitate more than one construction activity);
  - maintenance of site records;
  - completion and post-completion requirements;
  - cleaning;
  - fees and charges;
  - sites services; and
  - insurances, bonds, guarantees and warranties.
- **Main contractor's overheads and profit estimate –** this is the main contractor's charge associated with head office administration proportioned to each building contract plus the main contractor's return on capital investment.
- **Project and design team fees estimate –** covers project team and design team consultants' fees for pre-construction-, construction- and post-construction-related services, as well as for fees and charges for intrusive site investigations, and other specialist support consultants' fees.

- **Other development and project costs estimate** – comprises costs that are not necessarily directly associated with the cost of constructing the building, but form part of the total cost of the building project to the employer (e.g. land acquisition costs, marketing costs, decanting costs and temporary accommodation costs).
- **Risk allowances estimate** – deals with risks and uncertainties associated with design development (design development risks) and the nature of the site (construction risks, e.g. unexpected ground conditions). Risks associated with the employer changing either the quality of a building project or the scope of works are also contended with (employer change risks), together with other employer-related risks (other employer risks, e.g. financial and third party risks). The rules recommend that each category of risk be treated as an individual cost centre.
- **Inflation estimate** – an allowance for fluctuations in the basic prices of labour, plant and equipment, and materials during both the tender period (tender inflation) and the construction period (construction inflation). In the same way as risk, each category of inflation is to be treated as an individual cost centre.
- **Value added tax (VAT) assessment** – an assessment of the VAT liability for the building project.

Figure 4.2 shows how each of these separate cost estimates is integrated into the order of cost estimate framework. The cost limit is based on summing the cost estimates for the building project.

Initially, it is unlikely that the employer will require the cost manager to make allowances in an order of cost estimate for project and design team fees, other development and project costs, and certain categories or types of risk. Nonetheless, it is essential for the cost manager to discuss these aspects openly with the employer and ascertain how they are to be treated.

Box 4.1 Key definitions

- **Base cost estimate** – an evolving estimate of known factors without any allowances for risk and uncertainty, or element of inflation. The base cost estimate is the sum of the works cost estimate, the project/design team fees estimate and the other development/project costs estimate.
- **Building works estimate** – the estimated cost of building activities required to construct a building. It excludes main contractor's preliminaries and main contractor's overheads and profit.
- **Construction inflation estimate** – an allowance included in the order of cost estimate for fluctuations in the basic prices of labour, plant and equipment, and materials during the period from the date of tender return to the mid-point of the construction period.
- **Cost limit** – the maximum expenditure that the employer is prepared to make in relation to the completed building. Sometimes referred to as authorised budget or approved estimate.
- **Main contractor's preliminaries estimate** – the amount added to the building works estimate for main contractor's preliminaries to arrive at the works cost estimate.

- **Main contractor's overheads and profit estimate** – the amount added to the building works estimate for main contractor's overheads and profit to arrive at the works cost estimate.
- **Other development/project costs estimate** – the amount added to the works cost estimate for other development/project costs to arrive at the base cost estimate.
- **Project/design team fees estimate** – the amount added to the works cost estimate for fees in connection with employing the project/design team to arrive at the base cost estimate.
- **Risk allowance estimate** – the amount added to the base cost estimate for items that cannot be precisely predicted to arrive at the cost limit.
- **Tender inflation estimate** – an allowance included in the order of cost estimate for fluctuations in the basic prices of labour, plant and equipment, and materials during the period from the estimate base date to the date of tender return.
- **Works cost estimate** – the combined estimated cost of the building works, the main contractor's preliminaries, and the main contractor's overheads and profit.

Constituent
Building works estimate (1)
Main contractor's preliminaries estimate (2)
Subtotal (3) [(3) = (1) + (2)]
Main contractor's overheads and profit estimate (4)
Works cost estimate (5) [(5) = (3) + (4)]
Project/design team fees estimate (if required) (6)
Subtotal (7) [(7) = (5) + (6)]
Other development/project costs estimate (if required) (8)
Base cost estimate (9) [(9) = (7) + (8)]
Risk allowances estimate (10) [(10) = (10(a)) + (10(b)) + (10(c))+ (10(d))]
(a) Design development risks estimate (10(a))
(b) Construction risks estimate (10(b))
(c) Employer change risks estimate (10(c))
(d) Employer other risks estimate (10(d))

Figure 4.2 Order of cost estimate framework

Cost limit (excluding inflation) (11) [(11) = (9) + (10)]

Tender inflation estimate (12)

Cost limit (excluding construction inflation) (13) [(13) = (11) + (12)]

Construction inflation estimate (14)

Cost limit (including inflation) (15) [(15) = (13) + (14)]

VAT assessment

Figure 4.2 *(Continued)*

4.3 Methods of measurement

NRM 1 advocates the following methods of measurement as a basis for preparing an order of cost estimate:

- floor area method (or superficial method); and
- functional unit method (or unit method).

In practice, a combination of both methods is often employed to ascertain the estimated cost of a building project.

4.3.1 Floor area method (or superficial method)

This is a rough budget-setting method, which is based on the cost per square metre of gross internal floor area of a building. To determine the cost of a building, the total gross internal floor area of each building is measured and multiplied by an appropriate cost/m^2 of GIFA (i.e. a single unit rate).

The floor area method has the advantage that it is meaningful in its concept of measurement (the square metre), as most will be able to visualise a building (e.g. a building of 1,300m^2).

4.3.2 Measurement of gross internal floor area

Gross internal floor area (or gross internal area, GIA) is the principal method of measurement used in calculating building costs. The RICS Code of Measuring Practice defines the GIFA as being:

> the area of a building measured to the internal face of the perimeter walls at each floor level.

The term 'internal face' is defined by the Code as the internal perimeter surface of the brickwork, blockwork or plaster coat applied to the brickwork or blockwork, not the surface of the internal linings. Where modern construction products are used, such as curtain walling, the internal face can be construed as being the internal face of the

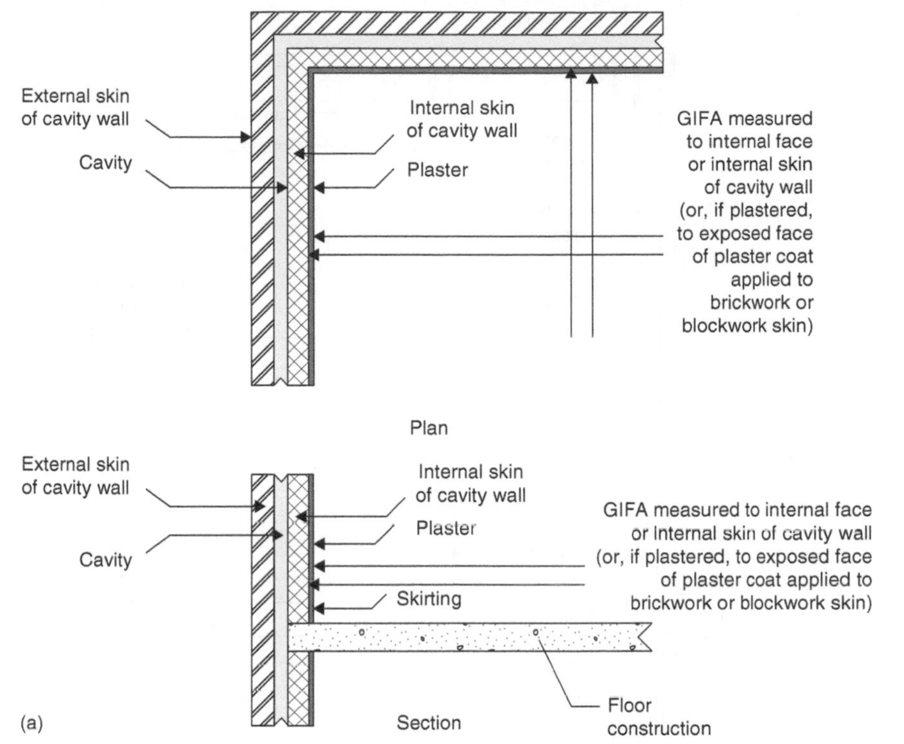

Figure 4.3A Interpretation of 'internal face'– (a) cavity wall construction

mullion or transom. However, where the construction entails an upstand edge beam or a backing wall to curtain walling or panel wall system (e.g. blockwork skin), the GIFA is to be measured from the internal face of the upstand edge beam or backing wall construction as appropriate. Figures 4.3A and 4.3B show how to interpret these different scenarios.

Table 4.1 sets out the method of measuring the GIFA.

Applying the rules for GIFA measurement in Table 4.1, Figure 4.4 and Figure 4.5 show how to interpret the appropriate dimensions for GIFA measurement in the context of an industrial (or warehouse unit) and a cinema, respectively.

4.3.3 Functional unit method (or unit method)

The functional unit method (or unit method) is another rough budget-setting technique which consists of selecting a suitable standard functional unit of use for the project and multiplying the projected number of units by an appropriate cost per functional unit (i.e. a single unit rate), or functional unit rate. Functional units are the units of occupancy or performance for which a building or part of a building is functionally designed (e.g. per bed space, per house type, per m^2 – based on net internal area (NIA), gross effective area and retail area). Each functional unit includes all associated circulation space. It is essential that the functional unit be clearly identified when measurements are expressed in this way. A list of commonly used functional units for buildings is provided in Appendix B of NRM 1.

Functional units based on NIA, retail areas and gross effective areas are measured in accordance with the appropriate definition within the RICS Code of Measuring Practice.

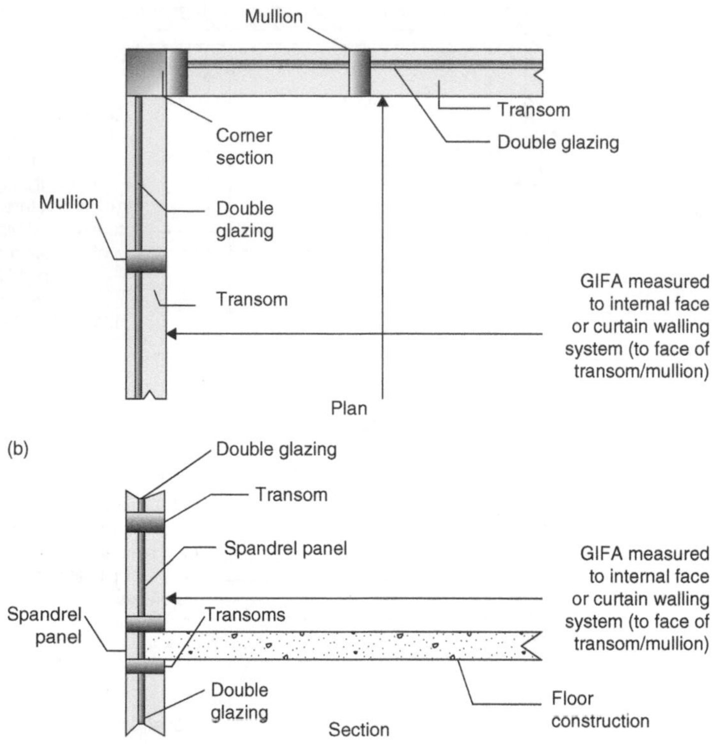

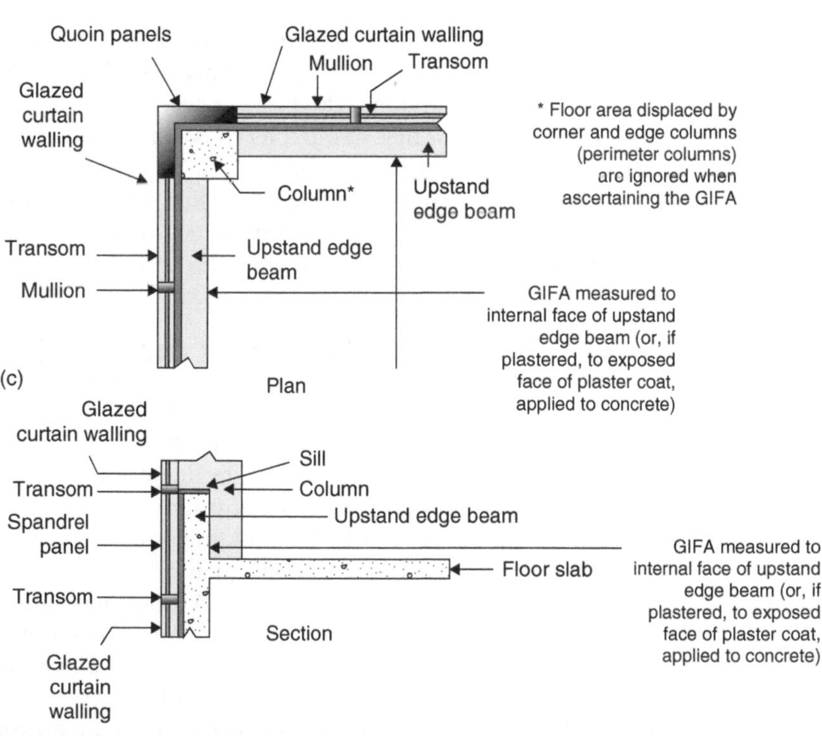

Figure 4.3B 'Interpretation of internal face' – (b) curtain walling arrangement (c) curtain walling arrangement with upstand edge beam

Table 4.1 GIFA – rules of measurement

Inclusions	*Exclusions*
Area of each floor level measured to the internal face of the perimeter walls, including: 1. Areas occupied by internal walls and partitions. Note: Internal party walls – in the absence of specific guidance within the Code on how to deal with areas occupied by internal party walls within a multifunction building (load bearing and non-load bearing), it is recommended that the centre line of the party wall be used to delineate the functions (refer to sub-paragraph 4.3.4 of this book). 2. Columns, piers, chimney breasts, stairwells, lift-wells, other internal projections, vertical ducts and the like. 3. Atria and entrance halls, with clear height above, measured at base level only. Note: Where an atrium-like space is formed to create an entrance feature and this also accommodates a staircase, this does not become a stairwell but remains an atrium measurable at base level only. 4. Internal open-sided balconies, walkways and the like. 5. Structural, raked or stepped floors are to be treated as a level floor measured horizontally. 6. Horizontal floors, with permanent access, below structural, raked or stepped floors. 7. Corridors of a permanent essential nature (e.g. fire corridors, smoke lobbies). 8. Mezzanine floor areas with permanent access. 9. Lift rooms, plant rooms, fuel stores, tank rooms and the like which are housed in a covered structure of a permanent nature, whether or not above the main roof level. Note: Lift rooms, plant rooms, fuel stores, tank rooms and the like should be included if housed in a roofed structure having the appearance of permanence (e.g. made of brick or similar building material). 10. Service accommodation such as toilets, toilet lobbies, bathrooms, showers, changing rooms, cleaners' rooms, and the like. 11. Projection rooms. 12. Voids over stairwells and lift shafts on upper floors. 13. Loading bays. 14. Areas with headroom of less than 1.5m. Note: Level changes – the presence of steps or a change in floor levels is to be noted. 15. Pavement vaults. 16. Garages. 17. Conservatories.	1. Perimeter wall thicknesses and external projections. 2. External open-sided balconies, covered ways and fire escapes. 3. Canopies. 4. Voids over or under structural, raked or stepped floors. 5. Greenhouses, garden stores, fuel stores and the like in residential property.

Source: adapted from the RICS Code of Measuring Practice: A Guide for Property Professionals (2007)

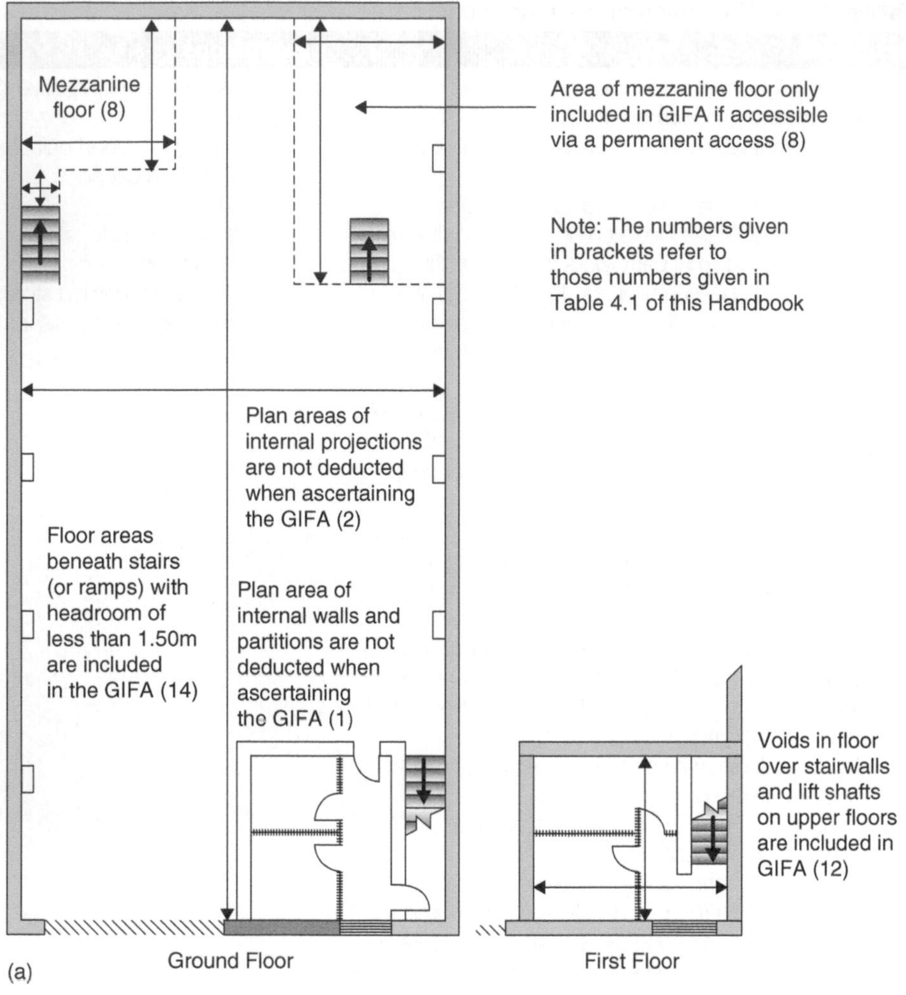
Mezzanine floor (8)
Area of mezzanine floor only included in GIFA if accessible via a permanent access (8)
Note: The numbers given in brackets refer to those numbers given in Table 4.1 of this Handbook
Plan areas of internal projections are not deducted when ascertaining the GIFA (2)
Floor areas beneath stairs (or ramps) with headroom of less than 1.50m are included in the GIFA (14)
Plan area of internal walls and partitions are not deducted when ascertaining the GIFA (1)
Voids in floor over stairwalls and lift shafts on upper floors are included in GIFA (12)
(a)
Ground Floor
First Floor

Figure 4.4A GIFA measurement – industrial warehouse unit, including offices at first floor level (adapted from the RICS Code of Measuring Practice: A Guide for Property Professionals, 2007)

4.3.4 Dealing with multiple buildings

Where measurement is for more than one building, irrespective of building use, the gross internal floor area, or functional unit, for each building is measured and shown separately for each separate building.

4.3.5 Dealing with single multifunction buildings

Where a single building comprises more than one user function (e.g. a single building comprising a combination of, say, office, retail and residential accommodation), the gross internal floor area, or functional unit, of each function is to be calculated and quantified separately. In the case of GIFA, the sum total of the GIFA for each separate function is to be equal to the GIFA for the whole building (refer to Figures 4.4A and 4.4B).

The GIFA excludes the thickness of perimeter walls, but includes the area occupied

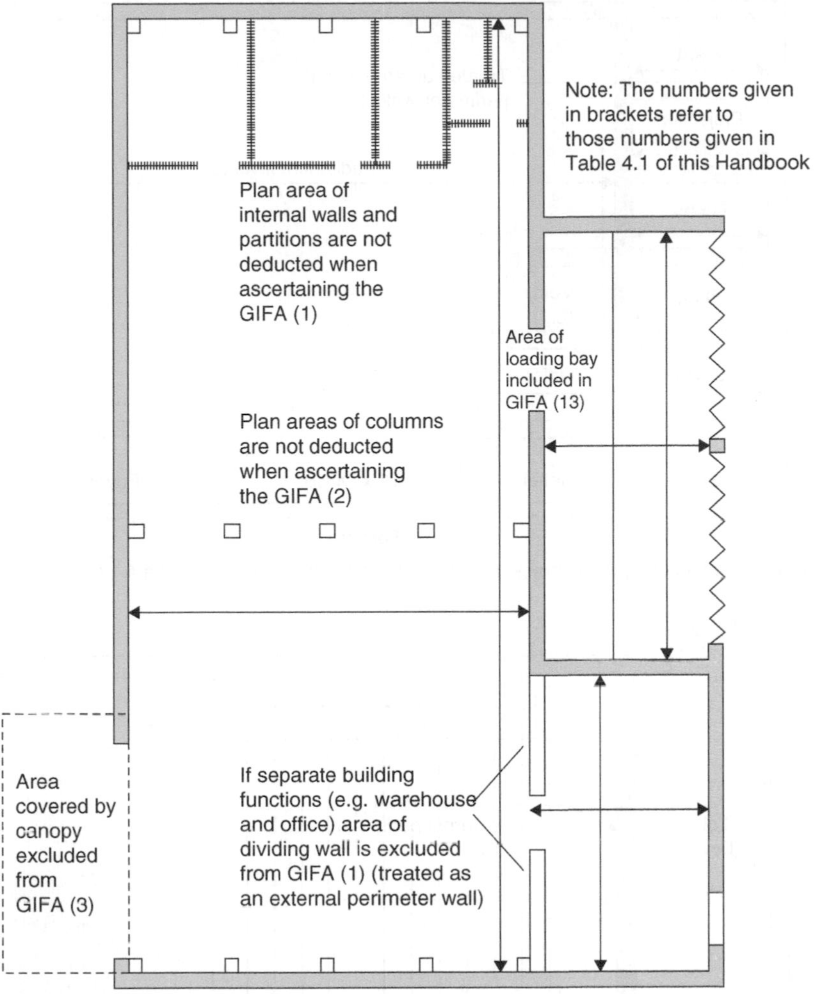
Note: The numbers given in brackets refer to those numbers given in Table 4.1 of this Handbook
Plan area of internal walls and partitions are not deducted when ascertaining the GIFA (1)
Area of loading bay included in GIFA (13)
Plan areas of columns are not deducted when ascertaining the GIFA (2)
Area covered by canopy excluded from GIFA (3)
If separate building functions (e.g. warehouse and office) area of dividing wall is excluded from GIFA (1) (treated as an external perimeter wall)

Figure 4.4B GIFA measurement – industrial/warehouse unit, including loading bay (adopted from the RICS Code of Measuring Practice: A Guide for Property Professionals, 2007)

by all internal walls (including internal walls acting as party walls). However, the RICS Code of Measuring Practice is vague on how to deal with areas occupied by internal walls acting as party walls (load bearing and non-load bearing) within a single multifunction building. Consequently, where the building comprises more than one user function, it is necessary for the cost manager to identify what area constitutes each separate function. In the absence of such guidance, it is recommended that the centre line of the party wall be used to delineate the functions (refer to Figure 4.6).

Using the dimensions given in Figure 4.6, the total GIFA of the building and for each function can be calculated as shown in Example 4.1.

4.4 Elemental method (or elemental estimate)

The elemental method of estimating (or elemental estimate) considers the major elements of a building and provides an order of cost estimate based on an elemental

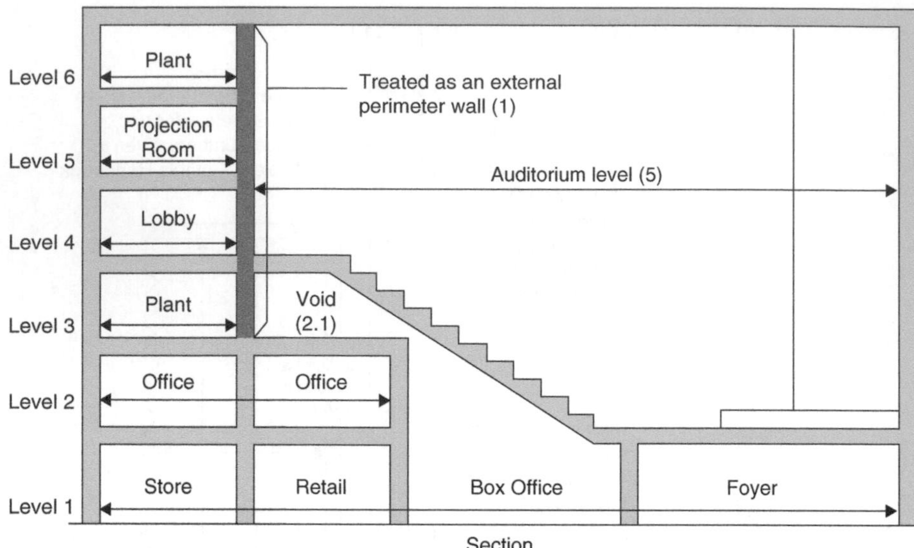

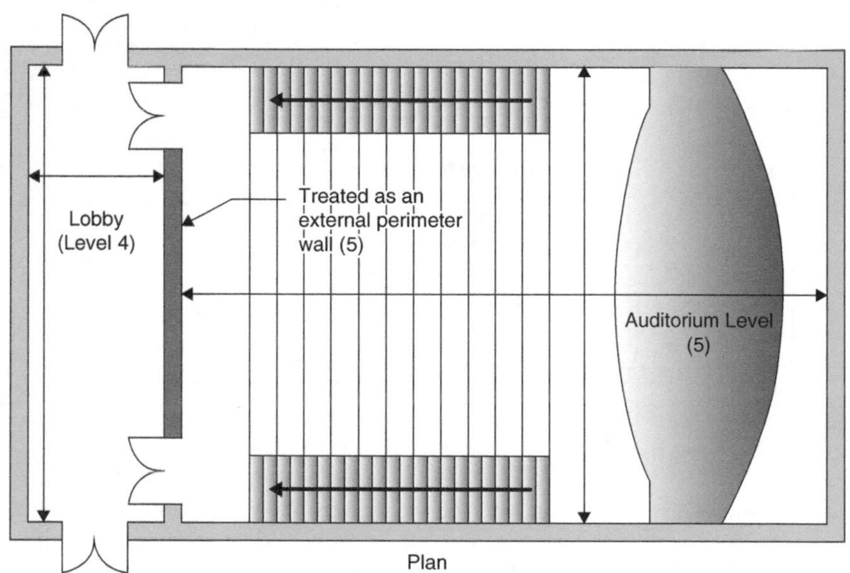

Figure 4.5 GIFA measurement – cinema (adapted from the RICS Code of Measuring Practice: A Guide for Property Professionals, 2007)

breakdown of a building project – this is often referred to as a preliminary cost plan or a cost model. Element unit quantities (EUQ) are measured for each element (i.e. group element or element, as appropriate) and multiplied by an appropriate element unit rate (EUR) to determine the cost target of each element. The cost targets for each element are added to provide the total estimated cost of the prospective building project.

The main purpose of an elemental estimate is to:

- reveal the distribution of the costs of a building among its elements in terms which are meaningful to the employer and the project team members; and
- develop an initial cost model as a prerequisite to developing a formal cost plan.

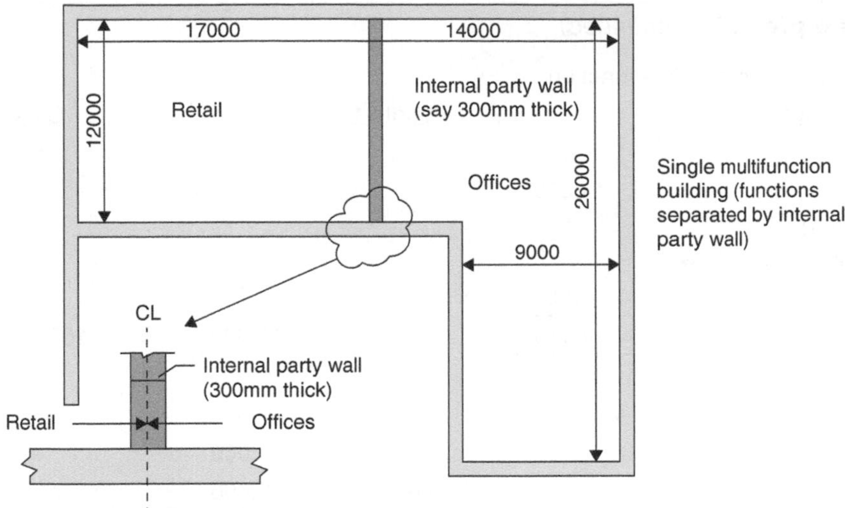

Figure 4.6 Delineation of functions in a multifunction building

Example 4.1 Calculation of total GIFA of a multifunction building

(a) Total GIFA of building:

Length 1			Width 1			GIFA
	m			m		m²
Retail	17.000					
Internal party wall	300					
Office	14.000					
	31.300	×		12.000	=	375.600
Length 2			Width 2			
	m			m		
			Office	26.000		
			Less			
			Retail	12.000		
	9.000	×		14.000	=	126.000
			Total GIFA of building		=	501.600

(b) Total GIFA of retail function:

Length			Width			GIFA
	m			m		m²
	17.000					
Add						
Internal party wall						
½/300	150					
	17.150	×		12.000	=	205.800
			Total GIFA of retail function		=	205.800

Example 4.1 *(Continued)*

(c) Total GIFA of office function:

Length 1			Width 1			GIFA
	m			m		m²
	14.000					
Add						
Internal party wall						
½/300	150					
	14.150	x		12.000	=	169.800
Length 2			Width 2			
	m			m		
				26.000		
			Less	12.000		
	9.000	x		14.000	=	126.000
			Total GIFA of office function		=	295.800

(d) Area check, to ensure that total GIFA and sum of GIFAs for each function correlate:

	m²
Total GIFA of building	501.600
Less	
Total GIFA of retail function:	205.800
	295.800
Less	
Total GIFA of office function:	295.800
Variance =	0

The method also provides a means of isolating the cost differences between the proposed building and the cost analyses of the historical buildings being used.

Guidance on how to prepare an initial cost model using the elemental method of estimating is given in Chapter 8 ('Preparation of an initial elemental cost model – using the elemental method of estimating').

Box 4.2 Key definitions

- **Element** – the secondary headings used to describe the key parts of a building, which make up each group element. The elements used are the same as those used for cost plans.
- **Elemental method** – a budget-setting technique that considers the major elements of a building and provides an order of cost estimate based on an elemental breakdown of a building project. The elemental method of cost estimating can also be used to develop an initial cost model as a prerequisite

to developing an elemental cost plan. The method involves the use of element unit quantities (EUQ) and element unit rates (EUR).

- **Element unit quantity (EUQ)** – a unit of measurement that relates solely to the quantity of the element or sub-element itself (e.g. the area of the external walls, the area of windows and external doors and the number of internal doors).
- **Functional unit** – a unit of measurement used to represent the prime use of a building or part of a building (e.g. per bed space, per house and per m^2 of retail area). It also includes all associated circulation space.
- **Functional unit method** – a rough budget-setting technique that consists of selecting a suitable standard functional unit of use for the project and multiplying the projected number of units by an appropriate cost per functional unit.
- **Gross internal floor area (GIFA) (or gross internal area – GIA)** – the area of a building measured to the internal face of the perimeter walls at each floor level. The rules of measurement of gross internal floor area are defined in the RICS Code of Measuring Practice.
- **Group element** – the primary headings used to describe the component a building. The group elements used are the same as those used for cost plans.

4.5 Information requirements at order of cost estimating stage

The accuracy of an order of cost estimate, as with all types of cost estimates and cost plans, is dependent on the quality of the information given to the cost manager. The more information provided, the more reliable the outcome will be. On the other hand, where little or no information is provided, the cost manager will need to heavily qualify the order of cost estimate.

It is important that the cost manager obtains as much information about the proposed building project as possible on which to base an order of cost estimate. This is important to producing an accurate cost estimate, on which decisions can be made. Nonetheless, cost managers are notoriously poor at advising the employer and designers of the information that they require to produce accurate cost estimates. This problem is recognised by NRM 1. As a result, the rules recommend the minimum information required by the cost manager at each stage of the cost estimating and cost planning process. The information required by the cost manager to compile an order of cost estimate is set out in Table 4.2. The schedule identifies the information that the cost manager should request from the employer and, where they have been appointed, the architect, structural engineer and mechanical and electrical (M&E) services engineer.

Table 4.2 Schedule of information required by the cost manager for the preparation of an order of cost estimate

Order of Cost Estimate			
Employer	***Architect***	***Structural engineer (if appointed)***	***M&E services engineer (if appointed)***
1. Location of the site and the availability of the site for commencement of the building project. 2. A statement of building use. 3. A statement of floor area (or number of functional units) and schedule of accommodation – in conjunction with the architect. 4. Requirements for refurbishment (if the project comprises rehabilitation of an existing building) – in conjunction with the architect. Details of the new use and any outstanding maintenance or repairs necessary to give the building fabric the required life expectancy are required. 5. Initial project/design brief, including statement of quality, sustainability requirements and 'fit-out' requirements – in conjunction with the architect. 6. Details of any enabling works, decanting or other specific requirements. 7. Indicative programme, including key dates (e.g. planning application and occupation dates).	1. Design study sketches or drawings for each alternative design/development option to a suitable scale, comprising: • floor plans (for each different floor plate configuration/shape and use); • roof plan(s); • elevations; and • sections. 2. Schedule of gross external areas, gross internal floor areas, net internal areas (i.e. usable area for shops, supermarkets and offices) and site area (SA). 3. Minimum storey heights. 4. Schedule of accommodation – in conjunction with the employer. 5. Number of car parking spaces and whether above ground or below ground.	1. Advice on probable ground conditions. 2. Indicative structural specification/design intent for building option(s). 3. Initial risk register/log.	1. Indicative services specification/design intent for building option(s). 2. Indicative environmental/ sustainability strategy – in conjunction with the architect. 3. Advice on availability and/or adequacy of utility services connections to the site. 4. Initial risk register/log.

Table 4.2 *(Continued)*

Order of Cost Estimate			
Employer	***Architect***	***Structural engineer (if appointed)***	***M&E services engineer (if appointed)***
8. Details of any particular restraints to be imposed by the employer, local planners or statutory undertakers – in conjunction with the architect (e.g. work in a secure area, limitations on building position, work in a conservation area, work to a historic or listed building, external appearance and number of storeys). 9. Details of any particular site conditions – in conjunction with the architect (e.g. sloping site, likelihood of contaminated ground, demolition of existing buildings, adequacy and condition of existing mains services). 10. Budget/cash flow constraints. 11. Initial views (if any) on construction procurement options and contract strategies. 12. Life span (e.g. 10-year, 25-year or 60-year target life span). 13. An indication of the proposed storey heights of the building – in conjunction with the architect. The introduction of raised access floors for IT cabling or deep suspended ceiling voids for mechanical and electrical services installations could significantly increase storey height, thus increasing estimated costs. Where such requirement is known, it is recommended that this be stated.	6. Indicative specification/design intent for building option(s). 7. Indicative environmental/ sustainability strategy – in conjunction with the mechanical and electrical services engineer. 8. Advice on likely site constraints. 9. Advice on likely planning constraints. 10. Definition of 'fit-out'. 11. Initial risk register/log.		

14. Particular requirements in respect of mechanical and electrical services installations – in conjunction with the architect (and mechanical and electrical services engineer – if appointed). 15. Requirements in respect of: • treatment of project/design team fees; • approach to other development/project costs; • treatment of inflation; and • treatment of value added tax (VAT). 16. Other considerations (e.g. approach to dealing with capital allowances, land remediation and grants).			

CHAPTER 5

Preparing an order of cost estimate

Introduction

This chapter:

- provides a process map which summarises the steps to be taken to compute an order of cost estimate;
- describes the steps involved in preparing an order of cost estimate; and
- summarises the composition of order of cost estimates.

5.1 Steps to preparing an order of cost estimate

There are 13 fundamental steps to preparing an order of cost estimate that are summarised in Figure 5.1.

The steps identified in Figure 5.1 are further explained below:

Step 1: Basic information gathering

Step 1.1 Establish employer's requirements:

1.1.1 Establish use of each building.

1.1.2 Establish quality requirements for each building.

1.1.3 Establish location of the site.

1.1.4 Establish accommodation requirements.

1.1.5 Obtain area schedules.

1.1.6 Determine probable procurement strategy.

1.1.7 Determine probable contract strategy.

1.1.8 Determine probable construction period.

1.1.9 Determine pre-construction programme, including target dates for tender action and commencement of works.

1.1.10 Determine probable phasing requirements.

Step 1.2 Appraise design information:

1.2.1 Review and interpret design information.

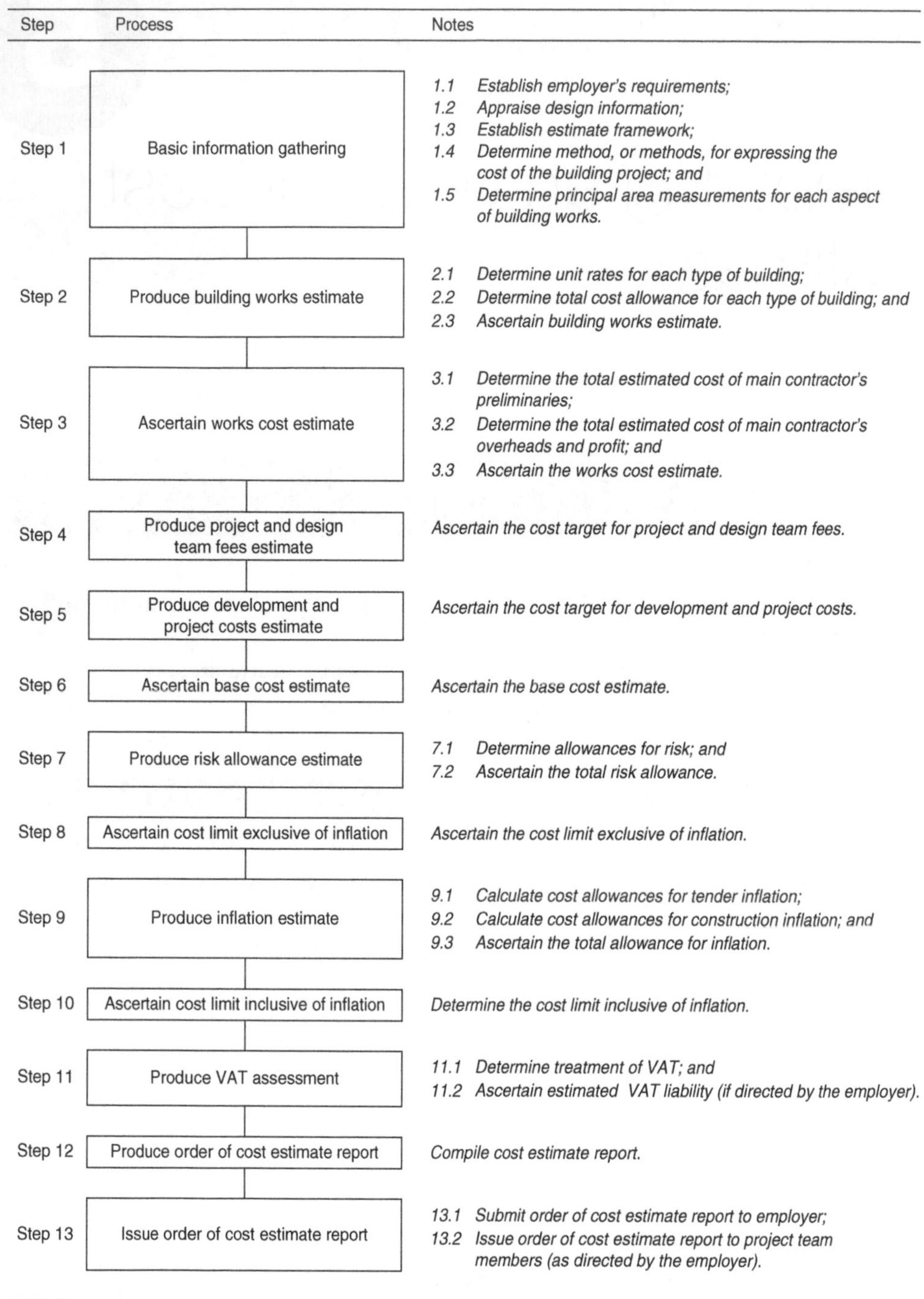

Step
Process
Notes
Step 1
Basic information gathering
1.1 Establish employer's requirements;
1.2 Appraise design information;
1.3 Establish estimate framework;
1.4 Determine method, or methods, for expressing the cost of the building project; and
1.5 Determine principal area measurements for each aspect of building works.
Step 2
Produce building works estimate
2.1 Determine unit rates for each type of building;
2.2 Determine total cost allowance for each type of building; and
2.3 Ascertain building works estimate.
Step 3
Ascertain works cost estimate
3.1 Determine the total estimated cost of main contractor's preliminaries;
3.2 Determine the total estimated cost of main contractor's overheads and profit; and
3.3 Ascertain the works cost estimate.
Step 4
Produce project and design team fees estimate
Ascertain the cost target for project and design team fees.
Step 5
Produce development and project costs estimate
Ascertain the cost target for development and project costs.
Step 6
Ascertain base cost estimate
Ascertain the base cost estimate.
Step 7
Produce risk allowance estimate
7.1 Determine allowances for risk; and
7.2 Ascertain the total risk allowance.
Step 8
Ascertain cost limit exclusive of inflation
Ascertain the cost limit exclusive of inflation.
Step 9
Produce inflation estimate
9.1 Calculate cost allowances for tender inflation;
9.2 Calculate cost allowances for construction inflation; and
9.3 Ascertain the total allowance for inflation.
Step 10
Ascertain cost limit inclusive of inflation
Determine the cost limit inclusive of inflation.
Step 11
Produce VAT assessment
11.1 Determine treatment of VAT; and
11.2 Ascertain estimated VAT liability (if directed by the employer).
Step 12
Produce order of cost estimate report
Compile cost estimate report.
Step 13
Issue order of cost estimate report
13.1 Submit order of cost estimate report to employer;
13.2 Issue order of cost estimate report to project team members (as directed by the employer).

Figure 5.1 Steps in preparing an order of cost estimate

1.2.2 Review and interpret reports, surveys and drawings relating to the nature of the site and the existing buildings.
1.2.3 Ascertain possible boundary issues.
1.2.4 Determine cost-significant risks.

1.2.5 Determine requirements for any facilitating works.
1.2.6 Record all assumptions that have been made.

Step 1.3 Establish estimate framework:
Determine work breakdown structure and cost breakdown structure.

Step 1.4 Determine method, or methods, for expressing the cost of the building project:
Establish method, or methods, for expressing the cost of the building project.

Notes:

(1) The cost/m² of GIFA is to be used to express the cost of a building project. When requested by the employer, costs can be expressed in different units and given in addition to the cost/m² of GIFA. For example, an employer might require the cost/ft² of GIFA, the cost/m² of functional area (e.g. cost/m² of NIA), or the cost per functional unit.

(2) Where an order of cost estimate is based solely on functional units, then the cost is to be expressed as the cost per functional unit.

Step 1.5 Determine principal area measurements:
Where order of cost estimate is based on the floor area method:

1.5.1 Determine the gross internal floor area of each building (including division of GIFA where building is divided into different uses).
1.5.2 Determine the gross external area (GEA) of each building, including existing buildings and buildings to be demolished within the curtilage of the site.
1.5.3 Determine the net internal area (NIA) of each building.
1.5.4 Determine the site area.
1.5.5 Determine area of external works.
1.5.6 Determine any other key measurements required.

Where order of cost estimate is based on the functional unit method:

1.5.7 Define functional unit.
1.5.8 Determine number of functional units for each building.

Step 2: Produce building works estimate

Step 2.1 Ascertain unit rates for building works:

Note:
Guidance on how to ascertain unit rates (i.e. cost/m² of GIFA and functional unit rates) for order of cost estimates is given in Chapter 23: 'Deriving unit rates for building components'.

2.1.1 Establish estimate base date (i.e. the date on which the cost limit, excluding inflation, is established as a basis for calculating inflation, changes or other related variances).
2.1.2 Select cost data (i.e. a selection of cost analyses – four to ten recommended – or benchmark analyses).
2.1.3 Interpolate cost data (i.e. identify and isolate major differences between initial brief and information contained within cost data, including quality, storey heights, different number of floors, specification, abnormal content, specific limitations and methods of working).
2.1.4 Adjust cost data for time and location (i.e. to be at estimate base date).
2.1.5 Determine unit rates.

Notes:

(1) Unit rates used are to be current at the time the cost estimate is produced. That is, they must exclude any allowances for future inflation or deflation.

(2) Unit rates applied to building works are to exclude allowances for main contractor's preliminaries, main contractor's overheads and profit, and inflation, as separate cost estimates are prepared for each of these facets. Allowances for all costs in connection with subcontractors' (or works contractors') preliminaries, overheads and profit, design fees and risk allowances should be made in the unit rates used. Similarly, allowances for suppliers' delivery charges should also be made within the unit rate.

2.1.6 Record all assumptions that have been made.

Step 2.2 Determine total cost allowance for each type of building and other facets of building works:

2.2.1 Assign unit rates to each type of building and other aspect of building works (i.e. external works, connections to statutory undertakers' mains services and facilitating works).

2.2.2 Determine estimated cost of each facet of building works by multiplying the quantity by the unit rate.

Step 2.3 Establish building works estimate:

Determine the total estimated cost of building works by adding together the estimated cost of each facet of building works.

Step 3: Ascertain works cost estimate

Step 3.1 Determine the cost target for main contractor's preliminaries:

3.1.1 Determine percentage addition to be applied for main contractor's preliminaries.

3.1.2 Determine cost target for main contractor's preliminaries by multiplying the total estimated cost of building works (i.e. the building works estimate) by the percentage addition for main contractor's preliminaries.

Step 3.2: Determine the cost target for main contractor's overheads and profit:

Where main contractor's overheads and main contractor's profit are to be treated as separate cost targets:

3.2.1 Ascertain combined total of estimated cost of building works (i.e. the building works estimate) and main contractor's preliminaries.

3.2.2 Determine percentage addition to be applied for main contractor's overheads.

3.2.3 Determine percentage addition to be applied for main contractor's profit.

3.2.4 Determine cost target for main contractor's overheads by multiplying the combined total estimated cost of building works (i.e. the building works estimate) and the cost target for main contractor's preliminaries by the percentage addition for main contractor's overheads.

3.2.5 Determine cost target for main contractor's profit by multiplying the combined total estimated cost of building works (i.e. the building works estimate) and the cost target for main contractor's preliminaries by the percentage addition for main contractor's profit.

3.2.6 Determine cost target for main contractor's overheads and profit by adding together the cost target for main contractor's overheads and the cost target for main contractor's profit.

Where main contractor's overheads and profit are to be treated as a single cost target:

3.2.7 Ascertain combined total of estimated cost of building works (i.e. the building works estimate) and main contractor's preliminaries.

3.2.8 Determine percentage addition to be applied for main contractor's overheads and profit.

3.2.9 Determine cost target for main contractor's overheads and profit by multiplying the combined total estimated cost of building works (i.e. the building works estimate) and cost target for main contractor's preliminaries by the percentage addition for main contractor's overheads and profit.

Step 3.3 Establish the works cost estimate:
Determine the works cost estimate by adding together the total estimated cost of building works (i.e. the building works estimate), the cost target for main contractor's preliminaries and cost target (or cost targets where treated as discrete cost targets) for main contractor's overheads and profit.

Step 4: Produce project/design team fees estimate
Determine the cost target for project and design team fees (if requested by the employer).

Step 5: Produce development/project costs estimate
Determine the cost target for development and project costs (if requested by the employer).

Step 6: Ascertain base cost estimate
Determine the base cost estimate by adding together the works cost estimate, the cost target for project and design team fees and the cost target for development and project costs.

Step 7: Produce risk allowance estimate

Step 7.1 Determine allowances for risk:

7.1.1 Determine allowance for design development risks.

7.1.2 Determine allowance for construction risks.

7.1.3 Determine allowance for employer change risk (if requested by the employer).

7.1.4 Determine allowance for employer other risks (if requested by the employer).

Step 7.2 Establish the total risk allowance:
Determine the total risk allowance by adding together the allowances for design development risks, construction risks, employer's change risk and employer's other risks.

Step 8: Ascertain cost limit exclusive of inflation
Ascertain the cost limit exclusive of inflation by adding together the base cost estimate and the total risk allowance.

Step 9: Produce inflation estimate

Step 9.1 Calculate cost allowances for tender inflation:

9.1.1 Determine percentage addition to be applied for tender inflation (i.e. estimated cost effect of inflation for period from the estimate base and the date of tender return).

9.1.2 Ascertain cost allowance for tender inflation by multiplying the cost limit exclusive of inflation by the percentage addition for tender inflation.

Step 9.2 Calculate cost allowances for construction inflation:

9.2.1 Ascertain combined total of the base cost estimate and the cost allowance for tender inflation.

9.2.2 Determine percentage addition to be applied for construction inflation (i.e. estimated cost effect of inflation for period from date of tender return to the mid-point of the construction phase).

9.1.3 Ascertain cost allowance for construction inflation by multiplying the combined total of the cost limit exclusive of inflation and the cost allowance for tender inflation.

Step 9.3 Establish the total allowance for inflation:
Determine the total allowance for inflation by adding together the cost allowances for tender inflation and construction inflation.

Step 10: Ascertain cost limit inclusive of inflation
Determine the cost limit inclusive of inflation by adding together the cost limit exclusive of inflation, the total allowance for risk allowance and the total allowance for inflation.

Step 11: Produce VAT assessment

Step 11.1 Determine treatment of VAT:
Determine whether or not the employer requires VAT to be included in the order of cost estimate.

> Note:
> Where VAT is to be excluded from the cost estimate, it is essential that this is clearly pointed out to the employer in the order of cost estimate report submitted to the employer; otherwise there is a risk that the employer might think that the cost estimate represents their total liability.

If VAT assessment required:

Step 11.2 Ascertain estimated VAT liability:

11.2.1 Determine the rate of VAT applicable.

11.2.2 Ascertain the VAT liability.

Step 12: Produce order of cost estimate report
Compile order of cost estimate report.

> Note:
> The characteristics of a good order of cost report are set out in Chapter 34 ('Reporting of cost estimate and cost plans').

Step 13: Issue order of cost estimate report

Step 13.1 Submit order of cost estimate report to employer.

Step 13.2 Issue order of cost estimate report to project team members (as directed by the employer).

5.2 Composition of an order of cost estimate

The composition of an order of cost estimate is shown in Figure 5.2.

Estimate	Process
Estimate 1	Works cost estimate
	+
Estimate 2	Project and design team fees estimate
	+
Estimate 3	Other development and project costs estimate
	+
Estimate 4	Risk allowances estimate
	+
Estimate 5	Inflation estimate
	=
	COST LIMIT
	VAT assessment

Figure 5.2 Composition of an order of cost estimate

5.3 Checklist for order of cost estimates

A typical example of a checklist of the principal tasks that need to be considered when compiling an order of cost estimate is set out in Example 5.1:

Example 5.1 Checklist for order of cost estimates

Bright Kewess Partnership		**Form DPB03**
Order of cost estimate: checklist		
Project no:		
Project title:		
Date:		
Confirm that the following tasks have been carried out before issue of order of cost report:		
1.	**Employer's requirements:**	☐
1.1	Location of the site established?	☐
1.2	Use of each building established?	☐
1.3	Accommodation requirements established?	☐
1.4	Area schedules obtained and checked?	☐
1.5	All information, reports, surveys and drawings relating to existing site requested and reviewed?	☐

Example 5.1 *(Continued)*

Bright Kewess Partnership		**Form DPB03**
	Order of cost estimate: checklist	
1.6	Extent of work to existing buildings and structures established (e.g. employer's requirements for de-rating of building; advanced removal of asbestos; advanced strip-out (and whether in readiness for demolition or refurbishment), and extent of demolition works)?	☐
1.7	Potential boundary issues identified (and action plan in place)?	☐
1.8	Sufficiency of statutory undertakers' mains services determined?	☐
1.9	All design information relating to proposed building, or development, requested and reviewed?	☐
1.10	Quality requirements for each building established, including services?	☐
1.11	Minimum storey heights determined?	☐
1.12	Fit-out specification determined (if applicable)?	☐
1.13	Key dates identified?	☐
1.14	Details of any works required to be completed early ascertained (i.e. any enabling works)?	☐
1.15	Phasing requirements determined?	☐
1.16	Likely construction period established?	☐
1.17	Initial programme established (i.e. both pre-construction (including design) and construction)?	☐
1.18	Likely overall procurement strategy determined?	☐
1.19	Considered tender and contract strategies taken into account?	☐
1.20	Treatment of project and design team fees established?	☐
1.21	Treatment of other development project costs established?	☐
1.22	(If included) extent and nature of other development project costs determined?	☐
1.23	Nature of risk allowances to be considered agreed with employer (i.e. design development risk, construction risks, employer's change risks, employer's other risks)?	☐
1.24	Method of calculating risk allowances agreed?	☐
1.25	Work breakdown structure for order of cost estimate determined?	☐
1.26	Method(s) of expressing costs agreed?	☐
1.27	GIFA(s) determined?	☐
1.28	NIA(s) determined (if required)?	☐
1.29	Type(s) and number(s) of functional units determined (if required)?	☐
1.30	GEA of existing buildings to be demolished determined?	☐

Example 5.1 *(Continued)*

Bright Kewess Partnership		**Form DPB03**
	Order of cost estimate: checklist	
1.31	Site area determined?	☐
1.32	Area of external works determined?	☐
1.33	Estimate base date established?	☐
1.34	Treatment of VAT ascertained?	☐
1.35	Recipients of 'Order of cost estimate report' agreed?	☐
2.	**Works cost estimate:**	
2.1	Unit rates used applicable, appropriate and adjusted for time and location?	☐
2.2	Building works estimate completed?	☐
2.3	Main contractor's preliminaries estimate completed?	☐
2.4	Percentage addition used for main contractor's preliminaries applicable and appropriate?	☐
2.5	Phasing requirements factored into cost estimate?	☐
2.6	Main contractor's overheads and profit estimate completed?	☐
2.7	Percentage addition used for main contractor's overheads and profit applicable and appropriate?	☐
3.	**Project and design team fees estimate:**	
3.1	(If included) percentage rates used for consultants' fees applicable and appropriate?	☐
3.2	Project and design team fees estimate completed?	☐
4.	**Development and project costs estimate:**	
4.1	Development and project costs estimate completed?	☐
5.	**Risk allowance estimate:**	
5.1	Percentage addition used for design development risk appropriate?	☐
5.2	Percentage addition used for construction risk appropriate?	☐
5.3	Percentage addition used for employer's change risk appropriate (and agreed with employer)?	☐
5.4	Percentage addition used for employer's other risk appropriate (and agreed with employer)?	☐
5.5	Risk allowance estimate completed?	☐
6.	**Inflation estimate:**	
6.1	Percentage addition used for tender inflation appropriate?	☐
6.2	Percentage addition used for construction inflation appropriate?	☐
6.3	Inflation estimate completed?	☐
7.	**VAT assessment:**	
7.1	VAT assessment completed?	☐

Example 5.1 *(Continued)*

Bright Kewess Partnership		**Form DPB03**
	Order of cost estimate: checklist	
8.	**Assumptions and checks:**	
8.1	All assumptions clearly recorded (i.e. inclusions and exclusions from order of cost estimate)?	☐
8.2	All calculations independently checked?	☐

OCE Prepared by:	Verified by:
Signed: Position: Date:	Signed: Position: Date:

CHAPTER 6

Computing the works cost estimate

Introduction

This chapter:

- provides step-by-step guidance on how to compute each potential facet of a building works estimate;
- gives advice on alternative methods of quantification for building works;
- provides step-by-step guidance on how to quantify:
  - main contractor's preliminaries; and
  - main contractor's overheads and profit; and
- explains how to compute the works cost estimate.

Figure 6.1 highlights where this part of the order of cost estimate fits within the order of cost estimate framework and provides a quick cross-reference to the other relevant chapters.

6.1 Computing the works cost estimate

The works cost estimate is made up of the following:

- building works estimate;
- main contractor's preliminaries estimate; and
- main contractor's overheads and profit estimate.

Figure 6.2 lists the steps to producing the works cost estimate.

6.2 Estimating the cost of building works (estimate 1A)

The building works estimate deals with the following:

- facilitating works;

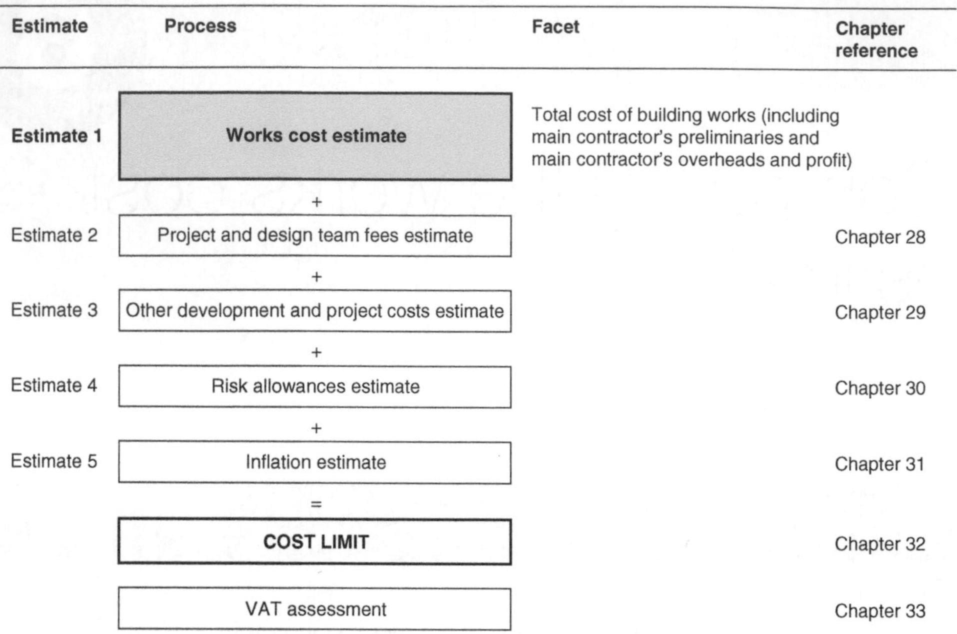

Estimate
Process
Facet
Chapter reference
Estimate 1
Works cost estimate
Total cost of building works (including main contractor's preliminaries and main contractor's overheads and profit)
+
Estimate 2
Project and design team fees estimate
Chapter 28
+
Estimate 3
Other development and project costs estimate
Chapter 29
+
Estimate 4
Risk allowances estimate
Chapter 30
+
Estimate 5
Inflation estimate
Chapter 31
=
COST LIMIT
Chapter 32
VAT assessment
Chapter 33

Figure 6.1 Works cost estimate in the context of the order of cost estimate framework

Estimate	Process	Facet
Estimate 1A	Produce building works estimate	Total cost of building works (excluding main contractor's preliminaries and main contractor's overheads and profit)
	+	
Estimate 1B	Estimate cost of main contractor's preliminaries	Main contractor's preliminaries
	+	
Estimate 1C	Estimate cost of main contractor's overheads and profit	Main contractor's overheads and profit
	=	
Estimate 1	**Produce works cost estimate**	Total cost of building works (including main contractor's preliminaries and main contractor's overheads and profit)

Figure 6.2 Works cost estimate process map

- demolition works;
- basements, including underground car parking;
- new buildings;
- extensions to existing buildings;
- refurbishment works to existing buildings; and
- external works, including site clearance, preparatory earthworks, soft and hard landscaping, site-surface-level car parking and connections to statutory undertakers' mains services.

Estimate	Process	Facet
Estimate 1A1	Estimate cost of facilitating works	Facilitating works
	+	
Estimate 1A2	Estimate cost of demolition works	Facilitating works
	+	
Estimate 1A3	Estimate cost of new basements	Building works
	+	
Estimate 1A4	Estimate cost of new buildings	Building works
	+	
Estimate 1A5	Estimate cost of new extensions to existing buildings	Building works
	+	
Estimate 1A6	Estimate cost of refurbishment works to existing buildings	Building works
	+	
Estimate 1A7	Estimate cost of external works	Building works
	+	
Estimate 1A8	Estimate cost of connections to statutory undertakers' mains services	Building works
	=	
Estimate 1A	**Produce building works estimate**	Total cost of building works (excluding main contractor's preliminaries and main contractor's overheads and profit)

Figure 6.3 Composition of building works estimate

However, the actual make-up of the building works estimate will be dependent on the composition of the building project and the resulting work breakdown and cost breakdown structures. For example, a building project might simply comprise the construction of a new single-function building on a greenfield site. Here, the building works estimate will be made up of the estimated cost of the new building, together with the estimated cost of external works and connections to statutory undertakers' mains services. On the other hand, a more complex building project might comprise the construction of a number of new buildings with different functions on a brownfield site, with extensions to, and refurbishment of, existing buildings, new basement car parks, major demolition works and site clearance works. In this case, a multifaceted work breakdown structure will ensue.

The estimated cost of building works is ascertained for each facet, determined by the work breakdown structure, of the building project, with the total cost of building works (i.e. the building works estimate) being the sum total of the estimated cost of each facet of the building project. Figure 6.3 illustrates the composition of a building works estimate.

6.2.1 Enabling works or facilitating works?

The terms, enabling works and facilitating works, should not be confused. Enabling works is a common term used by cost managers and others to describe the works that are to be executed before the main construction works commence. On the other hand, facilitating works is a term used by NRM 1 to describe specialist works that, normally, need to be completed before any construction of the building(s) can commence (e.g. works involving the removal of asbestos).

Enabling works might consist solely of facilitating works items as defined by the rules; equally, they might comprise the early construction of certain permanent works as well as the provision of some temporary works. Examples are the early construction of access roads, pathways, foul and surface water sewers and associated drain runs, alterations to an existing building, the construction of retaining walls on a sloping site and the provision of temporary accommodation for the purposes of decanting occupants from a building which is to be refurbished. What is more, enabling works are often procured separately from the main construction works as a discrete contract. For this reason, it is recommended that enabling works be treated as a separate works package and identified accordingly in the order of cost estimate. Therefore, the cost manager will need to decide how enabling works are to be dealt with when formulating the work breakdown structure for the order of cost estimate.

6.2.2 Estimating the cost of facilitating works (estimate 1A1)

By and large, facilitating works can be divided into two main categories:

- **Facilitating works** – works involving the removal of toxic and hazardous material, specialist groundworks required to stabilise or improve the bearing capacity of the site before construction can commence, temporary diversion of services, extraordinary site investigation works and reptile or wildlife mitigation measures. These are sometimes referred to as 'abnormal' construction costs; and
- **Demolition works** – works comprising demolition of entire, or major parts of, buildings and structures.

While the rules consider demolition works under facilitating works, they are most commonly dealt with as a separate cost centre under facilitating works (refer to sub-paragraph 6.2.3).

Examples of facilitating works items include:

- the eradication or treatment of Japanese knotweed or other invasive plant species;
- the removal of, or on-site treatment of, combustible or carbonaceous fills from beneath the footprints of proposed buildings;
- the on-site treatment of highly contaminated materials by specialist techniques such as encapsulation or entombment, bio-remediation or thermal desorption;
- the provision of a capillary break layer to prevent recontamination of near-surface soils as a result of recharging of potentially contaminated ground water;
- the provision of an engineered cap layer to protect end-users and the building fabric from contaminants;
- probe drilling and pressure grouting of cavities and voids associated with former mine workings and geological faulting beneath the footprints of buildings;
- diversion of existing services, sewers, culverted watercourses and overhead power lines;
- removal or treatment of underground obstructions, cellars, basements and storage tanks;

- protection measures to foundations and drainage systems to safeguard against very aggressive ground conditions (e.g. sacrificial materials and protective coatings and treatments); and
- provision of active gas protection measures and certain aspects of passive gas protection measures to safeguard occupants of proposed buildings from elevated levels of ground gas (e.g. gas-proof membranes).

Where the nature or extent of any facilitating works is unknown, the cost manager will need to ascertain whether or not an allowance ought to be made within the construction risk allowance for dealing with eventualities, should they materialise. On receipt of more detailed information (i.e. environmental surveys, geotechnical investigations and other site surveys), the risk allowances should be reviewed and reappraised by the cost manager.

For the purpose of an order of cost estimate, the estimated cost of a facilitating works item is determined as follows:

Step 1: Determine the nature of facilitating works item.

Step 2: Determine the area to be used to quantify facilitating works items.

Notes:

(1) The total area of the site within the site title boundaries, or the area within site title boundaries denoted as being the site area, whichever is applicable, measured on a horizontal plane, with the footprint (i.e. gross external area at ground level) of all existing buildings deducted. No deduction is made for the footprint of any new proposed buildings within the curtilage of the site.

(2) Where the extent of the area to be treated is known, then measurement of the applicable facilitating works item can be based on the actual area to be treated. However, the cost manager must make sure that a unit rate applicable to the method of measurement is used to estimate the cost of the works.

(3) The method of measurement used must be clearly stated in the order of cost estimate by the cost manager.

Step 3: Determine the unit rate to be applied in respect of the facilitating works item.

Notes:

(1) Unit rates used are to be current at the time the cost estimate is produced. That is, they must exclude any allowances for future inflation or deflation.

(2) Unit rates applied to facilitating works items are to exclude allowances for main contractor's preliminaries and main contractor's overheads and profit, as separate cost estimates are prepared for each of these aspects and added to the total building works estimate to give the 'works cost estimate' (see paragraphs 6.3 and 6.4 respectively). Moreover, allowances for all costs in connection with subcontractors' (or works contractors') preliminaries, overheads and profit, design fees and risk allowances are to be made in the unit rates used. Similarly, allowances for suppliers' delivery charges should also be made within the unit rate.

Step 4: Ascertain the estimated cost of facilitating works item by multiplying the area of facilitating works by the appropriate unit rate (i.e. cost/m^2 of facilitating works area).

Notes:
The equation for calculating the estimated cost of each facilitating works item is as follows:

d = (a – b) × c

Where:

a = total area of the site within the site title boundaries, or the area within site title boundaries denoted as being the site area, whichever is applicable
b = footprint of all existing buildings
c = cost/m^2 of facilitating works area
d = estimated cost of facilitating works item.

The estimated cost of each individual facilitating works item is calculated in the same way, with the total estimated cost of facilitating works being the sum of the estimated cost of each facilitating works item.

Alternatively, facilitating works items can be measured as spot items (i.e. itemised), with the estimated costs based on individual lump sum allowances. To avoid any misunderstanding, it is essential that the cost manager clearly defines the method of measurement used in the order of cost estimate.

6.2.3 Estimating the cost of demolition works (estimate 1A2)

The estimated cost of demolition works is calculated as follows:

Step 1: Determine the GIFA of each building to be demolished.

Note:
Whether measured from plans or taken from an area schedule provided by the employer, or the architect, the GIFA is deemed to have been measured strictly in accordance with the 'Core definition: gross internal area (GIA)' of the RICS Code of Measuring Practice, which is reproduced in Appendix A of NRM 1.

Step 2: Determine the unit rate to be applied to each building to be demolished.

Notes:

(1) Unit rates used are to be current at the time the cost estimate is produced. That is, they must exclude any allowances for future inflation or deflation.
(2) Unit rates applied to demolition works are to exclude allowances for main contractor's preliminaries and main contractor's overheads and profit, as separate cost estimates are prepared for each of these aspects and added to the total building works estimate to give the 'works cost estimate' (see paragraphs 6.3 and 6.4 respectively). Moreover, allowances for all costs in connection with subcontractors' (or works contractors') preliminaries, overheads and profit, design fees and risk allowances should be made in the unit rates used.

Step 3: Ascertain the estimated cost of demolition works by multiplying the GIFA of each building to be demolished by the appropriate unit rate (i.e. cost/m^2 of GIFA).

Note:
The equation for calculating the estimated cost of demolishing a single building is as follows:

$c = a \times b$

Where:

a = GIFA of the building to be demolished
b = cost/m^2 of GIFA applicable to the building to be demolished
c = estimated cost of the building to be demolished.

Each building or major part of a building to be demolished is to be measured and described separately. The same method is used to calculate the estimated cost of each building to be demolished, and the estimated cost of demolishing each building is added to give the total estimated cost of demolition works.

Where the GIFA of each building to be demolished is unknown, the area used to calculate the estimated cost of demolition works is the total area of the site within the site title boundaries. Where this alternative method of measurement is employed, care must be taken to make sure that the unit rate applied to the area correlates with the method of quantification. Additionally, in order to avoid any confusion over the method used, it is essential that the cost manager clearly states the method of measurement used in the order of cost estimate.

6.2.4 Estimating the cost of basements (estimate 1A3)

The estimated cost of a basement is determined as follows:

Step 1: Determine the GIFA of basement.

Note:
Whether measured from plans or from an area schedule provided by the employer, or the architect, the GIFA is deemed to have been measured strictly in accordance with the 'Core definition: gross internal area (GIA)' of the RICS Code of Measuring Practice, which is reproduced in Appendix A of NRM 1.

Step 2: Determine the unit rate to be applied to basement.

Notes:

(1) Unit rates used are to be current at the time the cost estimate is produced. That is, they must exclude any allowances for future inflation or deflation.
(2) Unit rates applied to building works are to exclude allowances for main contractor's preliminaries and main contractor's overheads and profit, as separate cost estimates are prepared for each of these aspects and added to the total building works estimate to give the 'works cost estimate' (see paragraphs 6.3 and 6.4 respectively). Moreover, allowances for all costs in connection with subcontractors' (or works contractors') preliminaries, overheads and profit, design fees and risk allowances should be made in the unit rates used. Similarly, allowances for suppliers' delivery charges should also be made within the unit rate.

Step 3: Ascertain the estimated cost of basement by multiplying the GIFA of each basement by the appropriate unit rate (i.e. cost/m^2 of GIFA).

Note:
The equation for calculating the estimated cost of constructing a basement is:

$c = a \times b$

Where:

a = GIFA of the basement
b = cost/m² of GIFA applicable to the basement
c = estimated cost of the basement.

The same method is used to calculate the estimated cost of each basement.

Sometimes the GIFA of a basement will be unknown (i.e. it will not be possible to measure GIFA of a basement from plans). Consider, for example, a requirement for basement car parking. The expected number of car parking spaces is known, but the basement layout has not yet been considered. In this situation the functional unit (i.e. car parking spaces) can be used to calculate the estimated cost of the basement car park. Alternatively, the functional area of a car space can be ascertained and multiplied by the proposed number of car parking spaces to calculate the expected GIFA. When using functional areas to estimate the GIFA of a basement car park, sufficient allowances for basement plant rooms, building cores and ancillary rooms must be included in the area calculation.

Again, where alternative methods of measurement are used, it is imperative that the cost manager clearly explains the method of measurement adopted in the order of cost estimate so as to avoid any misunderstanding.

6.2.5 Estimating the cost of new buildings (estimate 1A4)

Quantities for new building can be determined by measuring the total gross internal floor area (GIFA) of the building or buildings (using the floor area method) or by projecting the number of functional units (using the functional unit method), although a combination of both methods is usually employed. As a result, the estimated cost of new buildings can be calculated in the following ways:

Using the floor area method:

Step 1: Determine the GIFA of each building.

Notes:

(1) The GIFA of new buildings, refurbished buildings and extensions can normally be obtained by the cost manager from either the employer's brief or an area schedule prepared by the architect. Alternatively, where plans exist of a suitable scale, the gross internal floor areas or functional areas, whichever is applicable, can be measured from these.

(2) Some single buildings comprise multiple functions (e.g. a mixture of retail, leisure and offices). Here, the GIFA is ascertained for each function and, when added together, give the total GIFA of the building. The cost manager is responsible for checking that the GIFAs ascertained for each individual function correlates with the total GIFA of the building.

(3) Whether measured from plans or taken from an area schedule provided by the employer, or the architect, the GIFA is deemed to have been measured strictly in accordance with the 'Core definition: gross internal area (GIA)' of

the RICS Code of Measuring Practice, which is reproduced in Appendix A of NRM 1.

Step 2: Determine the unit rate or rates to be applied to each building.

Notes:

(1) Unit rates used are to be current at the time the cost estimate is produced. That is, they must exclude any allowances for future inflation or deflation.

(2) Estimating the cost of single multiple functioned buildings will involve the use of more than one unit rate. Unit rates applicable to each function are applied to each function.

(3) Unit rates applied to building works are to exclude allowances for main contractor's preliminaries and main contractor's overheads and profit, as separate cost estimates are prepared for each of these aspects and added to the total building works estimate to give the 'works cost estimate' (see paragraphs 6.3 and 6.4 respectively). Moreover, allowances for all costs in connection with subcontractors' (or works contractors') preliminaries, overheads and profit, design fees and risk allowances should be made in the unit rates used. Similarly, allowances for suppliers' delivery charges should also be made within the unit rate.

Step 3: Ascertain the estimated cost of each building by multiplying the GIFA of each building by the appropriate unit rate (i.e. cost/m² of GIFA).

Notes:

(1) The equation for calculating the estimated cost of a single-use building is:

$c = a \times b$

Where:

a = GIFA of the building
b = cost/m² of GIFA applicable to the building
c = estimated cost of the building.

(2) The equation for calculating the estimated cost of a single multifunction building using the floor area method is:

$c = \Sigma\ (a1 \times b1) + (a2 \times b2) + (a3 \times b3) + \text{etc.} \ldots$

Where:

a1 = GIFA applicable to building use type 1
a2 = GIFA applicable to building use type 2
a3 = GIFA applicable to building use type 3
b1 = cost/m² of GIFA applicable to building use type 1
b2 = cost/m² of GIFA applicable to building use type 2
b3 = cost/m² of GIFA applicable to building use type 3
c = estimated cost of the building.

Using the functional unit method:

Step 1: Determine the number of functional units.

Notes:

(1) Where measurement for the functional unit is to be 'net internal area', the NIA is to be measured in accordance with the 'Core definition: net

internal area (NIA)' of the RICS Code of Measuring Practice, which is reproduced in Appendix C of NRM 1.

(2) Where measurement for the functional unit is to be expressed as 'retail area', the retail area of the shop is to be measured in accordance with the 'Special use definition: shops' of the RICS Code of Measuring Practice, which is reproduced in Appendix D of NRM 1.

(3) A functional unit includes all circulation necessary.

Step 2: Determine the functional unit rate to be applied to each building.

Notes:

(1) Functional unit rates used are to be current at the time the cost estimate is produced. That is, they must exclude any allowances for future inflation or deflation.

(2) Functional unit rates are to exclude allowances for main contractor's preliminaries and main contractor's overheads and profit, as separate cost estimates are prepared for each of these aspects and added to the total building works estimate to give the 'works cost estimate' (see paragraphs 6.3 and 6.4 respectively). Moreover, allowances for all costs in connection with subcontractors' (or works contractors') preliminaries, overheads and profit, design fees and risk allowances should be made in the unit rates used. Similarly, allowances for suppliers' delivery charges should also be made within the unit rate.

Step 3: Ascertain the estimated cost of each building by multiplying the number of functional units by the appropriate functional unit rate.

Notes:

(1) The equation for calculating the estimated cost of a building based on a single functional unit rate is:

$$c = a \times b$$

Where:

a = number of functional units
b = functional unit rate
c = estimated cost of the building.

(2) The equation for calculating the estimated cost of a single multifunction building using the functional unit method is:

$$c = \Sigma\ (a1 \times b1) + (a2 \times b2) + (a3 \times b3) + \text{etc.} \ldots$$

Where:

a1 = number of functional units applicable to building use type 1
a2 = number of functional units applicable to building use type 2
a3 = number of functional units applicable to building use type 3
b1 = functional unit rate applicable to building use type 1
b2 = functional unit rate applicable to building use type 2
b3 = functional unit rate applicable to building use type 3
c = estimated cost of the building.

6.2.6 Estimating the cost of new extensions to existing buildings (estimate 1A5)

The estimated cost of new extensions (e.g. additional rooms, annexes or wings) is calculated in the same way as the estimated cost of new buildings, with the GIFA of the new extension measured.

Preparatory works to enable the extension to connect with the existing building can be separately itemised (i.e. measured as spot items), with the estimated costs based on discrete lump sum allowances.

6.2.7 Estimating the cost of refurbishment works to existing buildings (estimate 1A6)

Estimated costs for internal refurbishment works to existing buildings (e.g. redecoration, realigning internal walls, structural alterations, revamping and the like) can be calculated in the same way as the estimated cost of new buildings. Alternatively, refurbishment works can be measured as spot items (i.e. itemised), with the estimated costs based on discrete lump sum allowances.

6.2.8 Estimating the cost of external works (estimate 1A7)

The estimated cost of external works is ascertained as follows:

Step 1: Determine the total area of the site within the site title boundaries, or the area within site title boundaries denoted as being the site area, whichever is applicable.

Note: The total area of the site is measured on a horizontal plane.

Step 2: Determine the footprint of each new building or structure (i.e. the gross external area of new buildings and structures at ground level).

Step 3: Determine the footprint of each existing building or structure (i.e. the GEA of existing buildings and structures at ground level) where encircled by the external works scheme.

Notes:

(1) The footprint of a building (i.e. the GEA of building at ground level) is measured on a horizontal plane.

(2) Whether measured from plans or taken from an area schedule provided by the employer, or the architect, the GEA is deemed to have been measured strictly in accordance with the 'Core definition: gross external area (GEA)' of the RICS Code of Measuring Practice, which is reproduced in Appendix A of NRM 1.

Step 4: Determine the area of external works by subtracting the footprint of each new building, and existing building encircled by the external works scheme, from the total area of the site.

Note:
Areas used temporarily for building works that do not form part of a new or extended building are not deducted from the total area of the site.

Step 5: Determine the unit rate to be applied in respect of external works.

Step 6: Ascertain the estimated cost of external works by multiplying the area of external works by the appropriate unit rate (i.e. cost/m² of external works area).

Note:
The equation for calculating the estimated cost of external works is:

$$e = (a - (b + c)) \times d$$

Where:

a = total area of the site within the site title boundaries, or the area within site title boundaries denoted as being the site area, whichever is applicable
b = footprint of new buildings
c = footprint of existing buildings (i.e. encircled by the external works scheme)
d = cost/m² to be applied for external works
e = estimated cost of external works.

Where different unit rates are to be applied to different parts of the external works (e.g. to hard and soft landscaping), the total area of external works is apportioned accordingly. Apportionment of the area of external works can be achieved by either a percentage split or, if practicable, by measuring the actual areas from plans. In both cases, the total area of the parts must equal the total area of external works.

6.2.9 Estimating the cost of connections to statutory undertakers' mains services (estimate 1A8)

Connections to statutory undertakers' mains services (e.g. surface water and foul water sewers, electricity, water, gas and telecommunications) are measured as enumerated items. Alternatively, connections to statutory undertakers' mains services can be measured superficially (in square metres), in the same way as external works.

The estimated cost of connections to statutory undertakers' mains services are derived as follows:

Step 1: Determine the nature of each type of mains services connections.

Step 2: Determine the number of each type of mains services connections.

Step 3: Determine the unit rate to be applied in respect of each type of mains services connections.

Step 4: Ascertain estimated cost of connections to statutory undertakers' mains services by multiplying the number of mains services connections by the appropriate unit rate.

Note:
The equation for calculating the estimated cost of each type of mains services connections is as follows:

$$c = a \times b$$

Where:

a = number of mains services connections
b = unit rate applicable to the mains services connection
c = estimated cost of mains services connection (or connections).

The estimated cost of each mains services connection is calculated using the same method. They are added to give the total estimated cost of connections to statutory undertakers' mains services.

6.3 Estimating the cost of main contractor's preliminaries (estimate 1B)

For the purpose of order of cost estimates, the estimated cost of main contractor's preliminaries is calculated as a percentage of the total cost of building works (i.e. the building works estimate). Accordingly, the estimated cost of main contractor's preliminaries is calculated as follows:

Step 1: Determine the building works estimate.

Note:
The building works estimate is the sum of the estimated costs for each facet of building works (i.e. the sum of cost estimates 1A1, 1A2, 1A3, 1A4, 1A5, 1A6, 1A7 and 1A8).

Step 2: Determine the percentage addition to be applied in respect of main contractor's preliminaries.

Step 3: Ascertain the estimated cost of main contractor's preliminaries by multiplying the (total) building works estimate by the selected percentage addition for main contractor's preliminaries.

Note:
The equation for calculating the estimated cost of main contractor's preliminaries is therefore:

$$c = a \times p$$

Where:

a = building works estimate (i.e. total estimated cost of the building works)
p = percentage addition for main contractor's preliminaries
c = estimated cost of main contractor's preliminaries.

Where known at this early stage of the building project, costs relating to known site constraints, special construction methods, sequencing of works or other non-standard requirements should be assessed and identified separately. If unknown, the cost manager will need to ascertain whether or not allowances ought to be made within the construction risk allowance for such eventualities arising. The cost manager can reappraise these risk allowances as more information about the site and the employer's phasing requirements become available.

Main contractor's preliminaries are a cost-significant cost item, often equating to 10% to 30% of the works cost estimate. Although it is considered acceptable practice to estimate the cost of main contractor's preliminaries based on a percentage for the purpose of an order of cost estimate, the cost target should be reviewed and validated as more information becomes available about the nature of the site, site constraints and construction method. Chapter 25 ('Main contractor's preliminaries') considers the estimation of main contractor's preliminaries in more detail.

6.4 Estimating the cost of main contractor's overheads and profit (estimate 1C)

The estimated cost of main contractor's overheads and profit is calculated as a percentage of the combined total cost of building works (i.e. the building works estimate) and main contractor's preliminaries. Main contractor's overheads and profit are considered in more detail in Chapter 26 ('Main contractor's overheads and profit').

6.5 Compilation of the works cost estimate (estimate 1)

The works cost estimate is simply the sum of the building works estimate (estimate 1A), the main contractor's preliminaries estimate (estimate 1B) and the main contractor's overheads and profit estimate (estimate 1C).

Note:
The equation for calculating the estimated cost of works cost estimate is:

$$d = a + b + c$$

Where:

a = building works estimate (i.e. estimate 1A)
b = main contractor's preliminaries estimate (estimate 1B)
c = main contractor's overheads and profit estimate (estimate 1C)
d = works cost estimate.

CHAPTER 7

Order of cost estimate – a worked example

Introduction

This chapter:

- provides step-by-step guidance, with worked examples, on how to develop an order of cost estimate based on indicative sketched floor plans and area schedule, prepared by an architect, and an indicative programme;
- considers issues that need to be contemplated by the cost manager when calculating an order of cost estimate; and
- illustrates how to present an order of cost estimate.

7.1 Information gathering

The worked example of an order of cost estimate provided in this chapter is based on the initial indicative sketched elevations and floor plans (refer to Figure 7.2) and area schedule (Figure 7.3) prepared by the architect for a proposed 12-storey hotel building and a nine-storey office building. Details of the site and the existing building are shown in Figure 7.1. The drawings and area schedule has been developed from the employer's initial brief.

It is unlikely that the architect will provide dimensioned drawings at this early stage in the design development of the proposed building project. Drawings provided will normally be design study sketches illustrating floor plans of the proposed building, together with a simple section through the building. The architect might also provide sketches to illustrate how the proposed building might look in its environment once completed. It is usual, therefore, for the architect to prepare the area schedules and the cost manager to obtain area measurements on which to ascertain the order of cost of the proposed building project from the architect's area schedule.

Quality attributes and indicative programme information will normally be ascertained from the employer's initial brief. If the employer's initial brief is lacking sufficient information on which to prepare an order of cost estimate, the cost manager will need to ascertain any further information requirements from the employer (or the employer's representative). In the event that the information required is not available, or as yet to be decided, the cost manager will need to make any necessary assumptions. Such assumptions must be clearly conveyed to the employer by the cost manager in any cost estimate report (refer to Chapter 34: 'Reporting of cost estimate and cost plans').

The information required by a cost manager to compile an order of cost estimate is set out in Table 4.1 (refer to Chapter 4: 'Commencing an order of cost estimate').

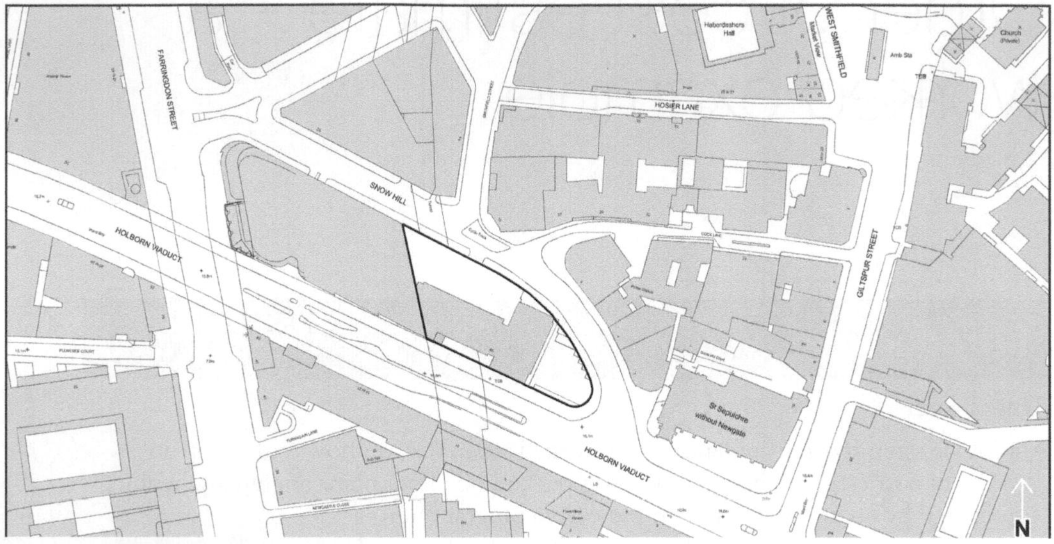

Figure 7.1 (a) Existing site plan showing the existing office building

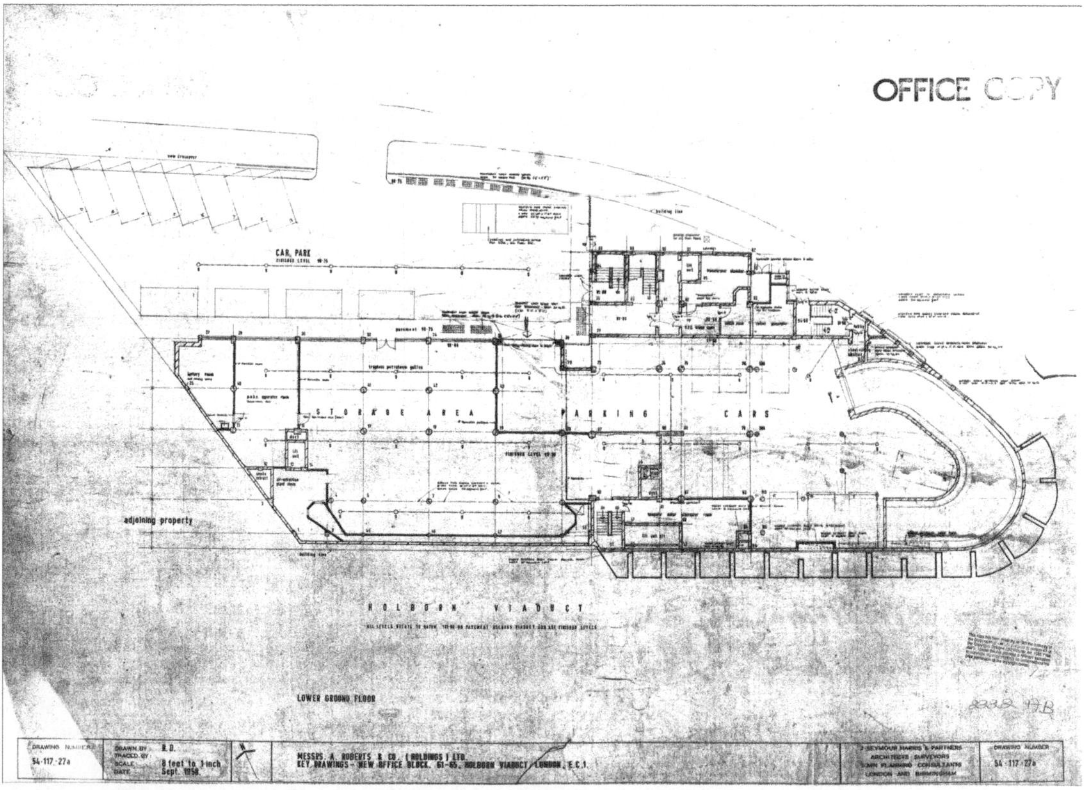

Figure 7.1 (b) Existing office building – basement floor plan

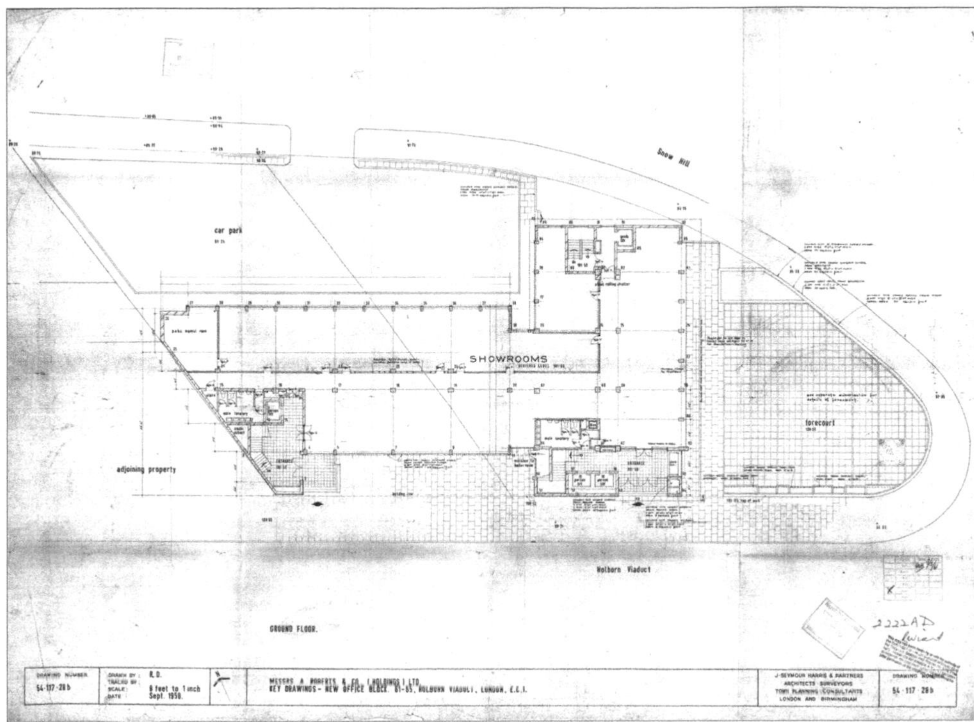

Figure 7.1 (c) Existing office building – ground floor plan

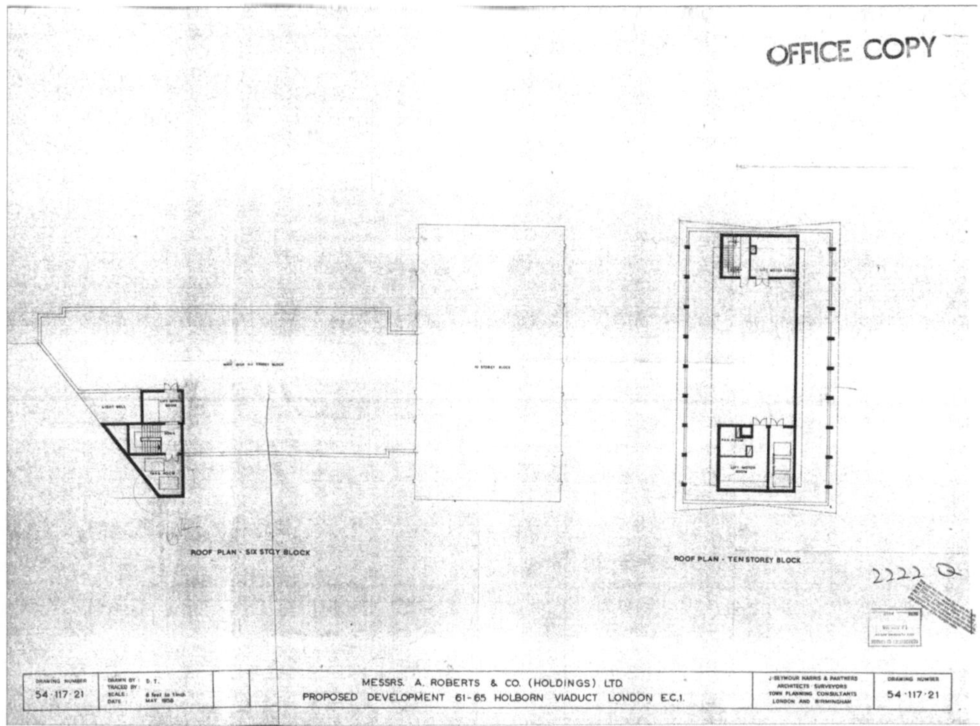

Figure 7.1 (d) Existing office building – roof plan

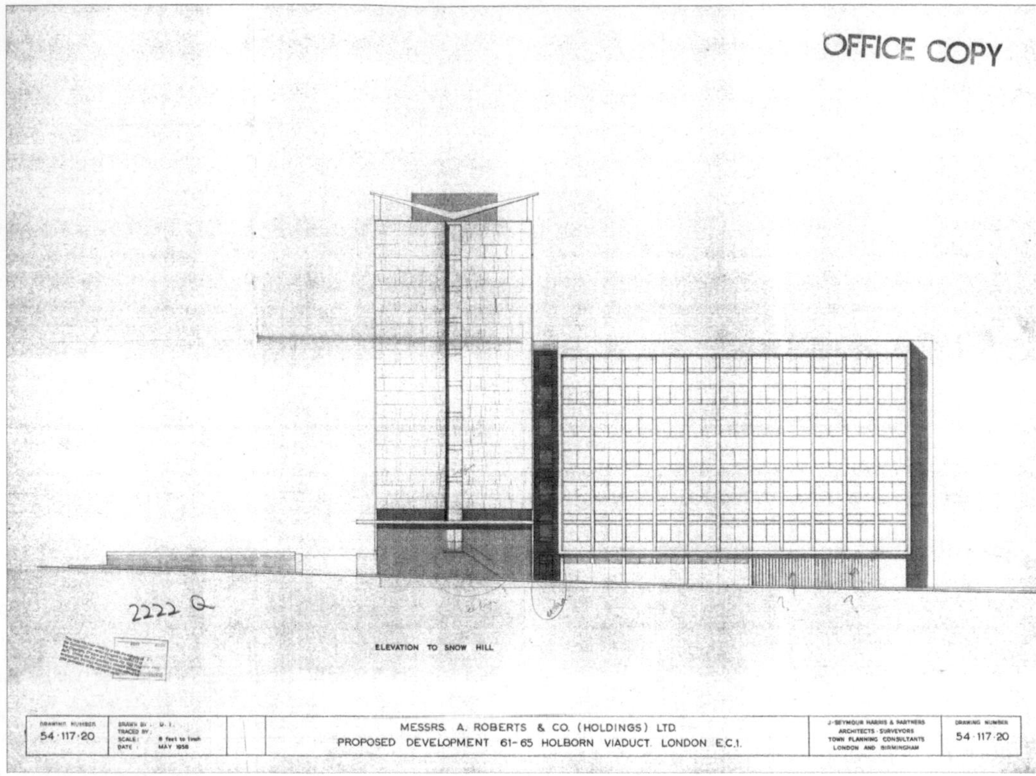

Figure 7.1 (e) Existing office building – north elevation

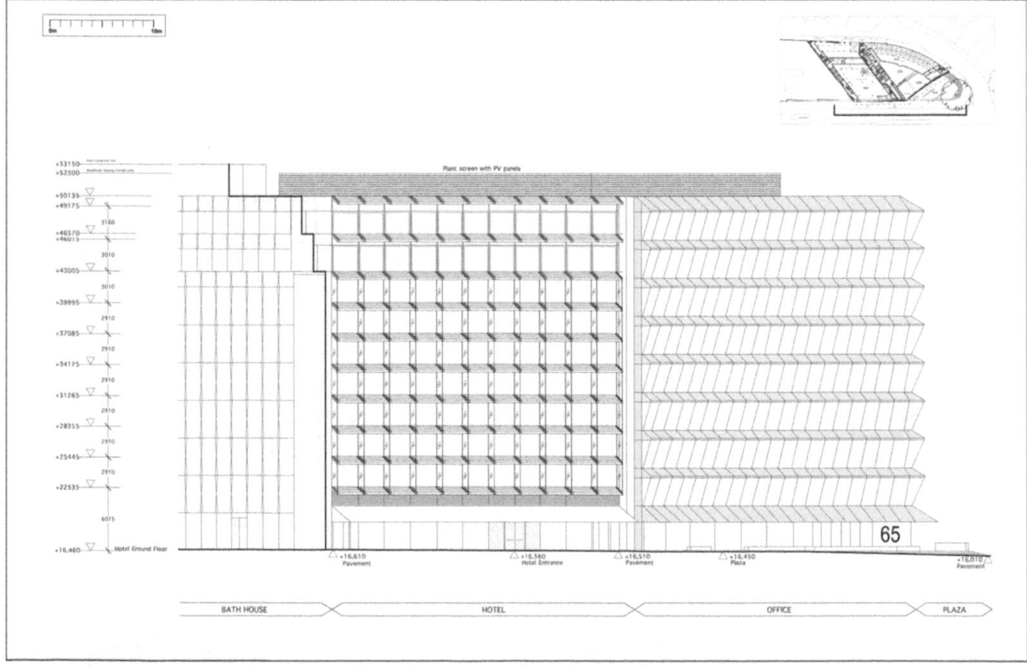

Figure 7.2 (a) Proposed hotel and office development – south elevation

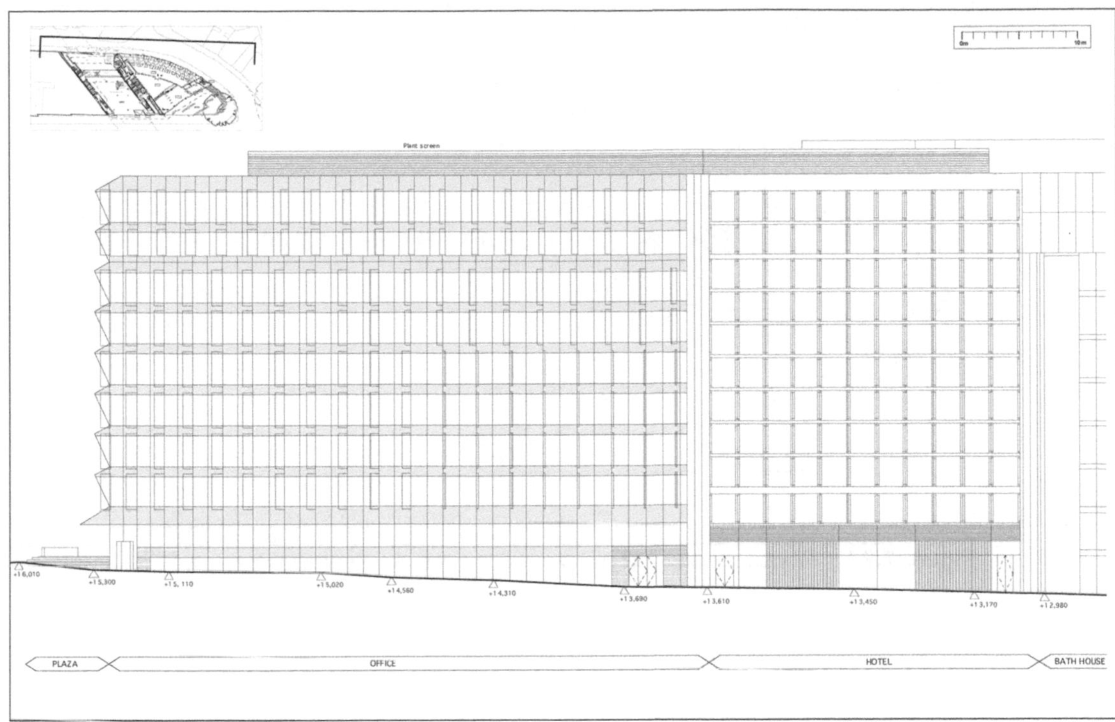

Figure 7.2 (b) Proposed hotel and office development – north elevation

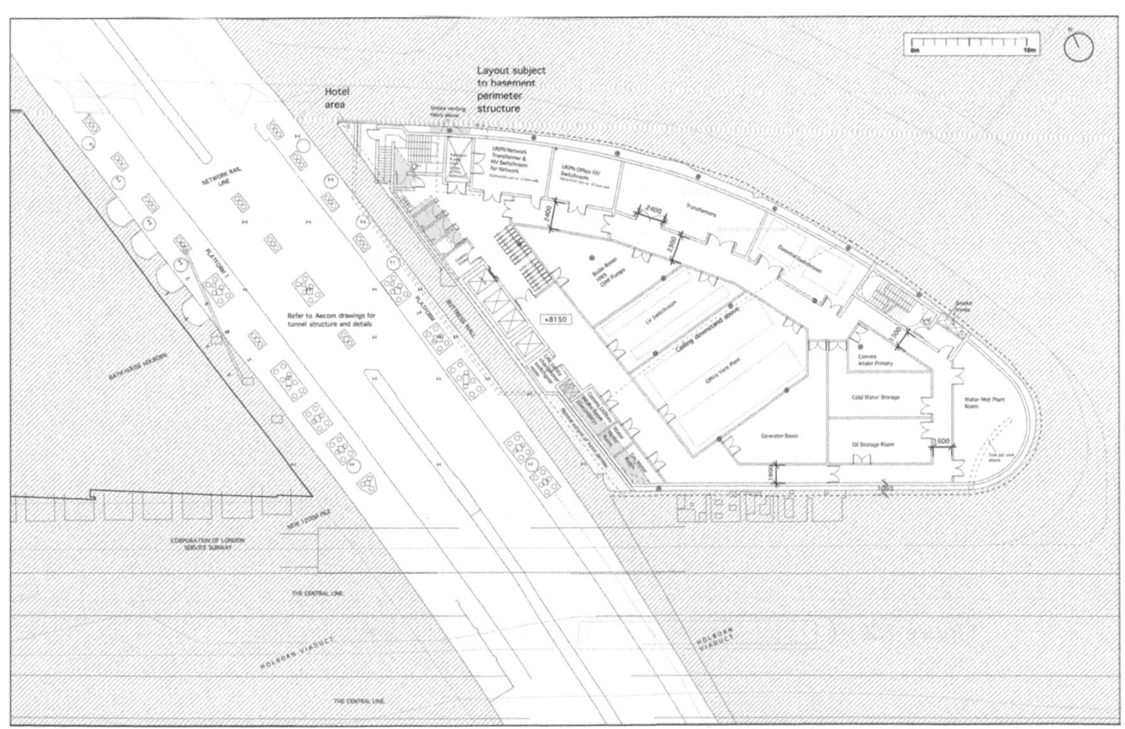

Figure 7.2 (c) Proposed hotel and office development – basement floor plan

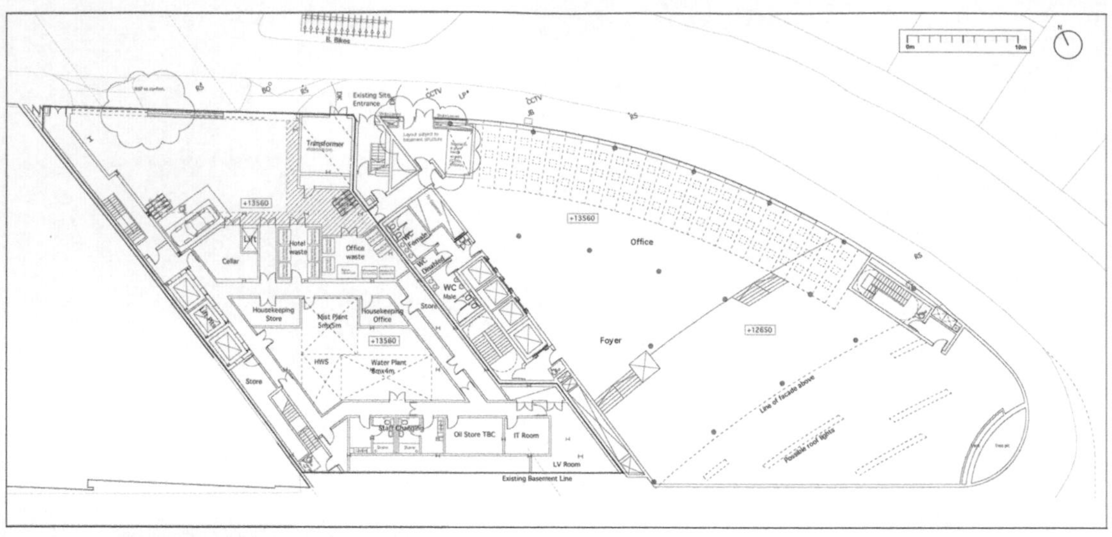

Figure 7.2 (d) Proposed hotel and office development – lower ground floor plan

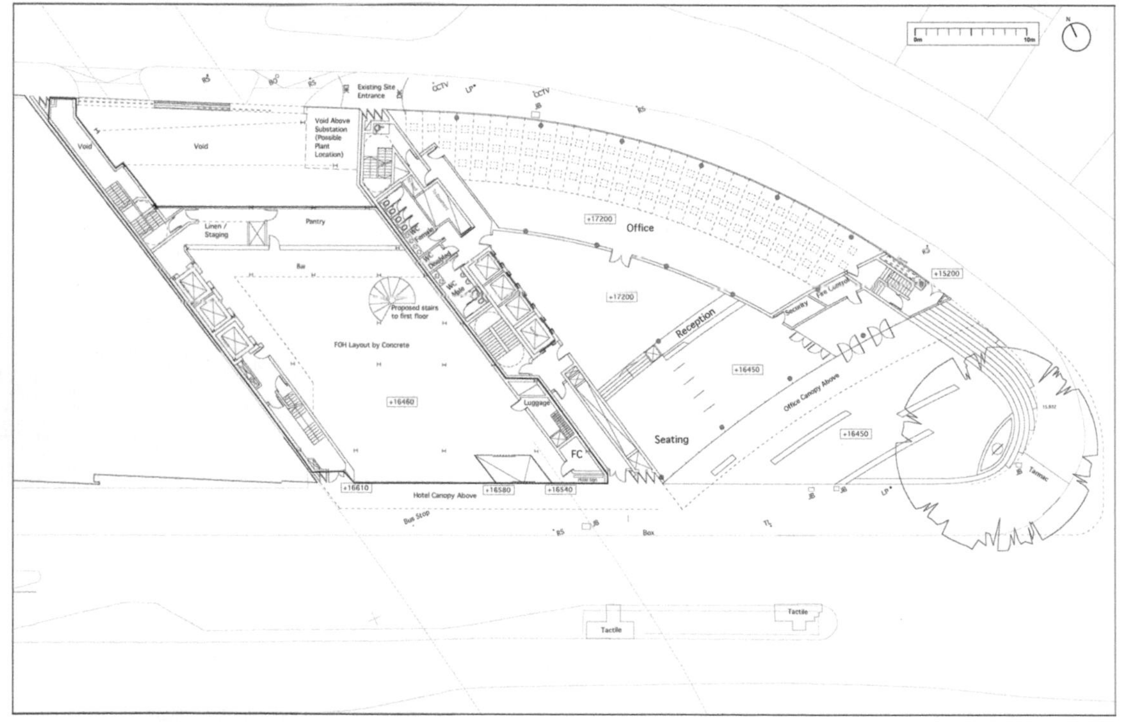

Figure 7.2 (e) Proposed hotel and office development – ground floor plan

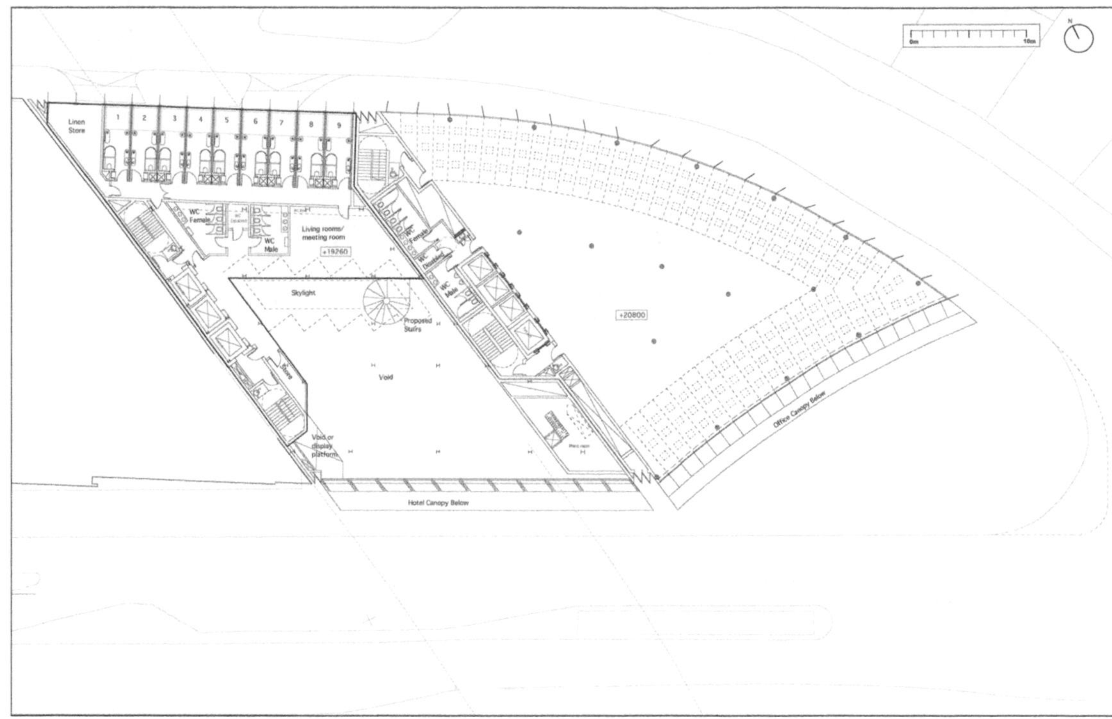

Figure 7.2 (f) Proposed hotel and office development – first floor plan

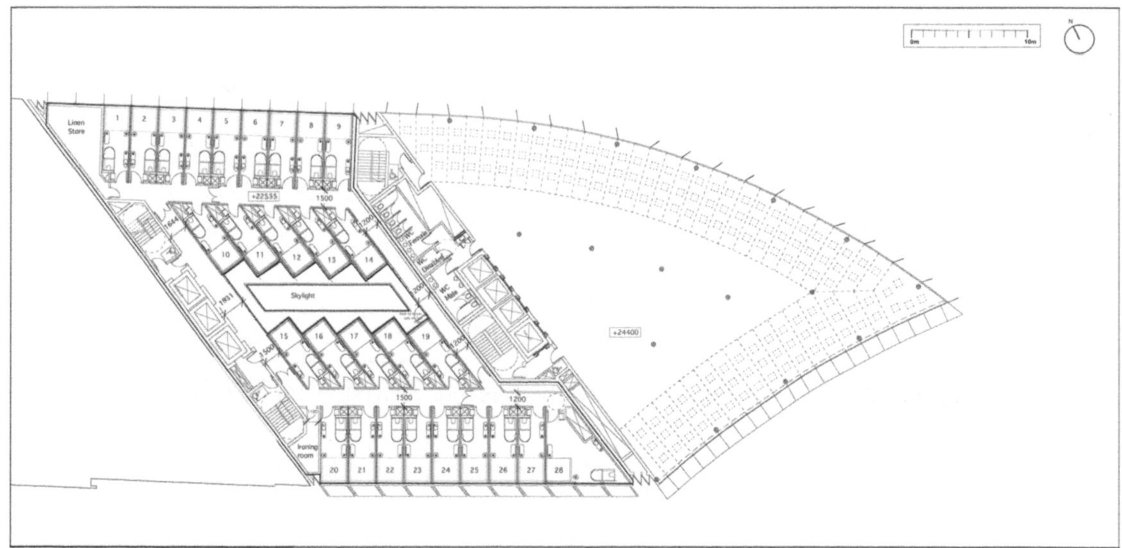

Figure 7.2 (g) Proposed hotel and office development – typical upper floor plan

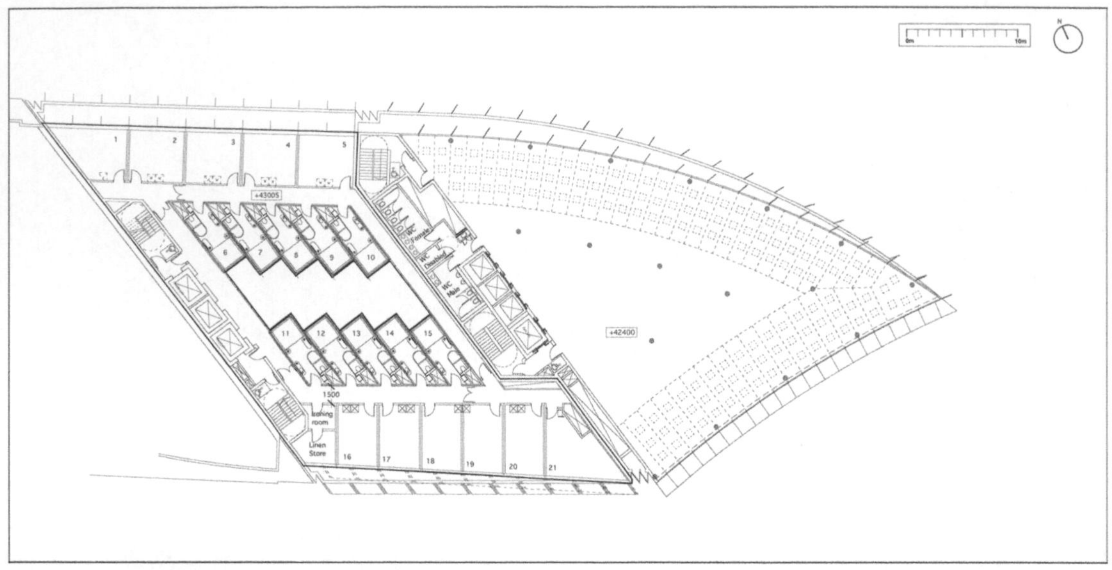

Figure 7.2 (h) Proposed hotel and office development – 7th office/9th hotel floor plan

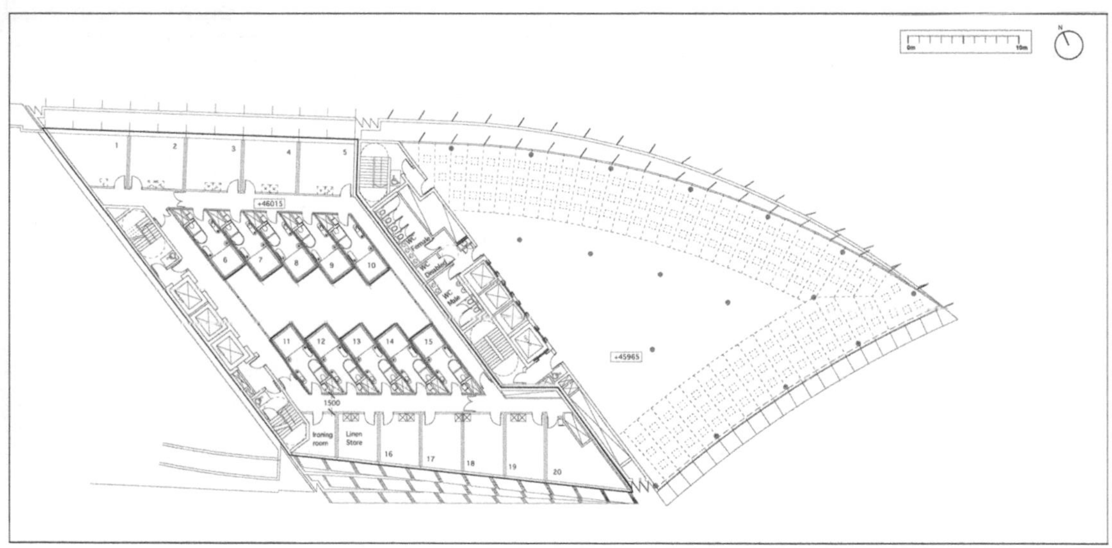

Figure 7.2 (i) Proposed hotel and office development – 8th office/10th hotel floor plan

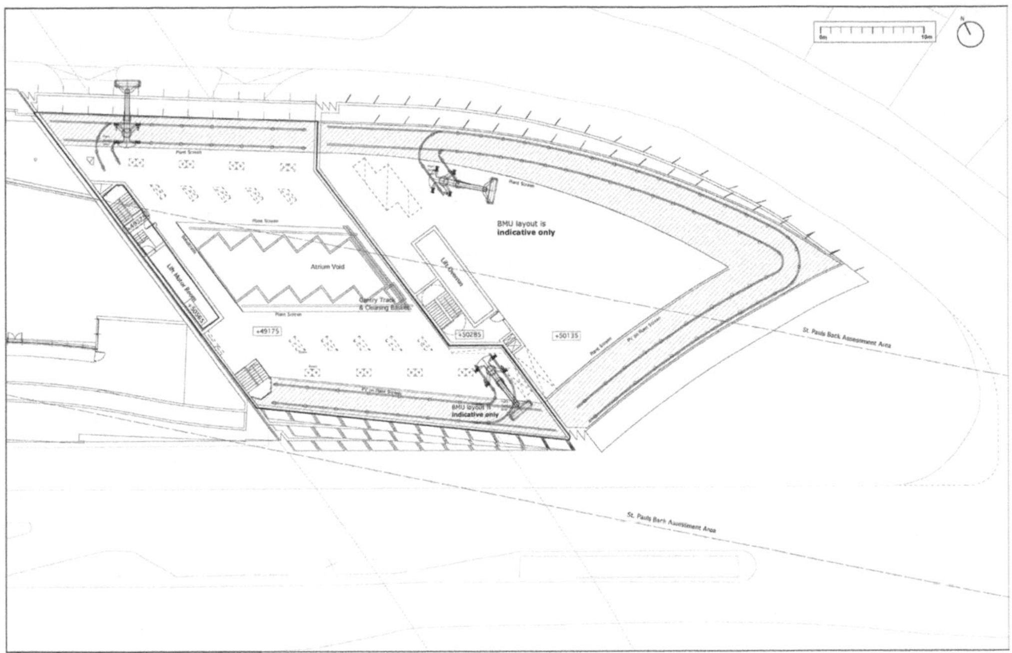

Figure 7.2 (j) Proposed hotel and office development – roof plan

7.2 Development of a work breakdown structure

Based on a review of the sketches, area schedule and employer's initial brief for the proposed building project, an initial work breakdown structure can be prepared by the cost manager to organise the order of cost estimate. The method of developing a WBS is explained in Chapter 4 ('Commencing an order of cost estimate').

The WBS developed for the proposed building project is shown in Figure 7.4.

AREA SCHEDULE
Prepared by: RK Tex Limited
Project no. RKT-1106
Project: Woollard Hotel and City Rise House, Holborn Viaduct, St Leonards-on-Sea, East Sussex
Date: 6 August 2013

Hotel					
Floor Level	Keys	Gross Internal Floor Area (GIFA)			
		Guest Floors	Front of House	Back of House	Total
	nr	m²	m²	m²	m²
Roof (Lift Motor Room)	–	–	–	43	43
Level 10	20	644	–	–	644
Level 9	21	668	–	–	668

Figure 7.3 Architect's area schedule for proposed hotel and office development

Hotel					
Floor Level	Keys	Gross Internal Floor Area (GIFA)			
		Guest Floors	Front of House	Back of House	Total
	nr	m²	m²	m²	m²
Level 8	28	731	–	–	731
Level 7	28	731	–	–	731
Level 6	28	731	–	–	731
Level 5	28	731	–	–	731
Level 4	28	731	–	–	731
Level 3	28	731	–	–	731
Level 2	28	731	–	–	731
Level 1	9	216	314	47	577
Level 0 (Ground Floor)	–	–	438	126	564
Level -1 (Lower Ground Floor)	–	–	–	786	786
Totals	246	6,645	752	1,002	8,399

Offices				
Floor Level	Gross Internal Floor Area (GIFA)		Net Internal Area (NIA)	
	m²	ft²	m²	ft²
Roof (Plant Rooms)	–	–	–	–
Level 8	845	9,096	678	7,298
Level 7	845	9,096	678	7,298
Level 6	920	9,903	748	8,051
Level 5	920	9,903	748	8,051
Level 4	920	9,903	748	8,051
Level 3	920	9,903	748	8,051
Level 2	920	9,903	748	8,051
Level 1	920	9,903	748	8,051
Level 0 (Ground Floor)	940	10,118	467	5,027
Level -1 (Lower Ground Floor)	1,224	13,175	975	10,495
Level -2 (Basement – Plant Rooms)	1,069	11,507	0	0
Totals	10,443	112,410	7,286	78,424

Efficiency: NIA to GIFA = 70.00%

Site Area	
	m²
Site area	2,144
Totals	2,144

Figure 7.3 *(Continued)*

PROJECT WORK BREAKDOWN STRUCTURE
Prepared by: Bright Kewess Partnership
Project no. BKP-0693
Project: Woollard Hotel and City Rise House, Holborn Viaduct, St Leonards-on-Sea, East Sussex

WBS Level 0	WBS Level 1	WBS Level 2	WBS Level 3	WBS Level 4	WBS Level 5
Entire building project:	1. Building works:	1.1 Facilitating Works:	1.1.1 Facilitating works		
			1.1.2 Demolition works		
		1.2 Building works:	1.2.1 Hotel building		
			1.2.2 Office building		
			1.2.3 External works:	1.2.3.1 Landscaping	
				1 2.3.2. Mains connections:	1.2.3.2.1 Sewer (Drainage)
					1.2.3.2.2 Water
					1.2.3.2.3 Electricity
					1.2.3.2.4 Gas
					1.2.3.2.5 Telecommunications
	2. Main contractor's preliminaries				
	3. Main contractor's overheads and profit				
	4. Project/design team fees				
	5. Other development/ project costs				
	6. Risk:	6.1 Design development risk			
		6.2 Construction risk			
		6.3 Employer change risks			
		6.4 Employer other risks			
	7. Inflation:	7.1 Tender inflation			
		7.2 Construction inflation			
	8. Value added tax (VAT)				

Figure 7.4 Work breakdown structure for order of cost estimate for proposed hotel and office development

7.3 Cost manager's assumptions and data gathered

The assumptions made by the cost manager are as follows:

- Quality indicators:

  Basement: plant areas.

  Hotel: luxury boutique quality; to achieve a BREEAM [Building Research Establishment Environmental Assessment Method] rating of 'very good'.

  > Note:
  > In practice, the cost manager will need to ascertain the employer's interpretation of 'luxury boutique' quality – as it means different things to different people. Without definition, there is a significant risk of over- or under-estimating the cost by the cost manager.

  Offices: fit-out to Cat. A; building to achieve a BREEAM rating of 'excellent'.

  > Note:
  > Cat. A fit-out as defined by the latest edition of the British Council for Offices Guide: Best Practice in the Specification for Offices. Fit-out to include services, life safety elements, and basic fittings and finishes for the operation of the lettable workspace.

- Work breakdown structure for the proposed new building development: refer to Figure 7.4.
- Quantification of building works:

  Gross internal floor areas for the site and the building have been taken from the area schedule prepared by RK Tex Limited, dated 6 August 2013. The schedule has been verified by the Bright Kewess Partnership.

  The area measured for landscaping works is area of the site to be landscaped (i.e. site area less the GEA of the ground floor of the building).

  Quantities for mains services connections are based on the GIFA of the proposed new building.

- Unit rates for building works:

  (a) Facilitating works: Stripping-out building, including removal of asbestos and other toxic substances, in readiness for full demolition. Estimate also includes the removal of some staircases and decommissioning of lifts for the purpose of de-rating the existing building for taxation purposes.

  > Note:
  > An allowance for potential works arising out of party wall awards has been included in initial design development and construction risk allowances.

  (b) Stripping-out works: £15/m^2 of GEA (existing building)

  (c) Demolition works: £70/m^2 of GEA (existing building)

  (d) Basement: £2,000/m^2 of GIFA

  (e) Hotel: £2,270/m^2 of GIFA, plus £1,500,000 allowance for piling through railway tunnel

(f) Offices: £1,475/m² of GIFA – for shell, core and envelope
£460/m² of GIFA – for fit-out to Cat A.

(g) External works: £350/m² of actual area

(h) Mains services connections:

Based on an all-in rate of £36/m², which includes for the following mains services connections:

- mains electricity connections;
- mains water connections;
- mains gas connections; and
- telecommunications connections.

Note:
All selected unit rates have been adjusted by the cost manager for time (to be at September 2013 price levels), regional variation and quality (the method of selecting and adjusting unit rates is explained in Chapter 23: 'Deriving unit rates for building components').

Main contractor's preliminaries: 13.60%
Main contractor's overheads and profit: 4.00%
Project and design team fees: 11.50%
Other development and project costs: Excluded

Notes:

(1) At the direction of the employer, no allowance for other development and project costs are to be made in the cost estimate.
(2) Further discussions to be held with the employer regarding other development and project costs development of the project.

Risk allowances:

(a) for design development risks: 1.50%
(b) for construction risks: 3.50%
(c) for employer's change risk: Excluded
(d) for employer's other risks: Excluded

Notes:

(1) At the direction of the employer, no allowance for employer's change risk and employer's other risks is to be made in the cost estimate.
(2) Further discussions to be held with the employer regarding these categories of risk during the development of the project.
(3) Risk allowances to be reviewed as more information about the site and design information becomes available.

Inflation:

Estimate base date is May 2013

Programme assumptions:

- Estimate base date: May 2013
- Tender issue: 1 October 2013
- Tender period: 9 weeks
- Tender return date: 1 December 2013

- Tender evaluation, reporting, post-tender negotiations: 6 weeks (includes Christmas and New Year Holidays)
- Contract award: 1 February 2014
- Mobilisation period: 4 weeks
- Construction phase commencement: 1 March 2014 (Commencement Date)
- Construction period: 65 weeks
- Construction phase completion date: 30 June 2015

VAT:
Excluded.

> Note:
> At the direction of the employer, no allowance for VAT is to be made in the cost estimate.

7.4 Generation of quantities for order of cost estimate

From the data gathered the cost manager can calculate the quantities for the order of cost estimate as follows (Figure 7.5):

SCHEDULE OF QUANTITIES: ORDER OF COST ESTIMATE
Prepared by: Bright Kewess Partnership
Project no. BKP-0693
Project: Woollard Hotel and City Rise House, Holborn Viaduct, St Leonards-on-Sea, East Sussex

Component	Basis of quantity		Quantity
Site area:			
Site area:	Site area; measured within title boundaries =		2,144m²
Existing building:			
Demolition works:	Total GEA of existing building =		7,710m²
Building:			
Basement and lower ground floors:	Hotel =	786m²	
	Offices =	2,293m²	
	GIFA of basement and lower ground floors:		3,079m²
Ground and upper floors:	Hotel =	7,613m²	
	Offices =	8,150m²	
	GIFA of ground and upper floors =		15,763m²
		Total GIFA =	18,842m²

Figure 7.5 Calculation of quantities for order of cost estimate for proposed hotel and office development

Component	Basis of quantity			Quantity
External works:				
	526m²			
	Site area =		2,144m²	
	Less GEA of the ground floor of the proposed building:			
	Hotel =	564m²		
	Offices =	1,124m²		
			(1,688)m²	
	Area of landscaping works =			456m²
Statutory connections:				
(1) Sewer (mains drainage) connections:	GIFA of entire building =			18,842m²
(2) Mains water connections:	GIFA of entire building =			18,842m²
(3) Mains electricity connections:	GIFA of entire building =			18,842m²
(4) Mains gas connections:	GIFA of entire building =			18,842m²
(5) Telecommunications:	GIFA of entire building =			18,842m²

Figure 7.5 *(Continued)*

7.5 Computing the order of cost estimate

7.5.1 Produce the building works estimate (estimate 1A)

The building works estimate is calculated as follows:

> Note:
> Facilitating works and building works are calculated using either of the following formulae, whichever is appropriate:
>
> - GIFA × cost/m² of GIFA; or
> - Number of functional units × functional unit rate

- The estimated cost of facilitating works is:

Facilitating works:	7,710m² at £15/m² of GIFA =	£115,650
Demolition works:	7,710m² at £56/m² of GIFA =	£539,700
Total – Facilitating works:		£655,350

- The estimated cost of building works is:

Basement:	Total GIFA = 3,079m²	
(1) Hotel building:	786m² at £2,000/m² of GIFA =	£1,572,000
(2) Office building:	2,293m² at £2,000/m² of GIFA =	£4,586,000

Hotel:		
(1) Hotel building:	7,613m² at £1,650/m² of GIFA =	£17,281,510
(2) Abnormal piled foundation design:	Say:	£1,500,000
Offices (to Cat. A fit-out):		
(1) Shell, core and envelope:	8,150m² at £1,475/m² of GIFA =	£12,021,250
(2) Fit-out to Cat A:	8,150m² at £460/m² of GIFA =	£3,749,000
Landscaping:	526m² at £456/m² of GIFA =	£159,600
Statutory connections:	18,842m² at £36/m² of GIFA =	£678,312
Total – Building works:		£41,574,672

Therefore, the building works estimate is:

Facilitating works:	£655,350
Building works:	£41,574,672
Building works estimate:	£42,203,022

7.5.2 Produce main contractor's preliminaries estimate (estimate 1B)

The main contractor's preliminaries estimate is calculated as follows:

Main contractor's preliminaries estimate =

£42,203,022 × 13.60% = £5,739,611

7.5.3 Produce main contractor's overheads and profit estimate (estimate 1C)

The main contractor's overheads and profit estimate is calculated as follows:

Building works estimate:	£42,203,022
Main contractor's preliminaries estimate:	£5,739,611
Combined total of building works estimate and main contractor's preliminaries estimate:	£47,942,633

Therefore:

Main contractor's overheads and profit estimate =

£47,942,633 × 4.00% = £1,917,705

7.5.4 Produce the works cost estimate (estimate 1)

The works cost estimate is calculated as follows:

Building works estimate:	£42,203,022
Main contractor's preliminaries estimate:	£5,739,611

Main contractor's overheads and profit estimate:	£1,917,705
Works cost estimate:	£49,860,338

7.5.5 Produce the project and design fees estimate (estimate 2)

The project and design fees estimate is calculated as follows:

Project and design fees estimate =

£49,860,338 × 11.50% = £5,733,939

7.5.6 Produce the other development and project costs estimate (estimate 3)

No allowance for other development and project costs are to be made in the cost estimate.

7.5.7 Produce the base cost estimate

The base cost estimate is calculated as follows:

Works cost estimate:	£49,860,338
Project and design fees estimate:	£5,733,939
Other development and project costs estimate:	Excluded
Base cost estimate:	£55,594,277

7.5.8 Risk allowance estimate (estimate 4)

The risk allowance estimate is calculated as follows:

Design development risk:	£55,594,277 × 1.50% =	£833,914
Construction risk:	£55,594,277 × 3.50% =	£1,945,800
Employer's change risk:		Excluded
Employer's other risk:		Excluded
Risk allowance estimate:		£2,779,714

7.5.9 Cost limit (excluding inflation)

The cost limit (excluding inflation) is calculated as follows:

Works cost estimate:	£49,860,338
Project and design fees estimate:	£5,733,939
Other development and project costs estimate:	Excluded
Risk allowance estimate:	£2,779,714
Cost limit (excluding inflation):	£58,373,991

7.5.10 Produce the inflation estimate (estimate 5)

- Tender inflation:
  Estimate base date: May 2013
  Tender return date: December 2013

Therefore, tender inflation is measured for the period from May 2013 to December 2013.

Tender price indices (TPIs – using BCIS all-in TPIs):

May 2013 =	228 (forecast)
December 2013 =	231 (forecast)

Note:
The equation for determining the percentage change is:

$$\% \text{ Change} = \frac{\text{The index at the later date} - \text{the index at the earlier date}}{\text{the index at the earlier date}} \times 100$$

Thus, the percentage applicable for tender inflation is:

$$\frac{(231 - 228)}{228} \times 100 = 1.32\%$$

Note:
In this case it is forecast that tender prices will increase during this period by 1.32% (i.e. inflation).

The allowance for tender inflation is calculated by multiplying the cost limit (excluding inflation, at estimate base date) by the percentage allowance for tender inflation.

Thus, the tender inflation estimate is:

£58,373,991 × 1.32% = £770,537

i.e. a forecast increase of £770,537.

- Construction inflation:
  Tender return date: December 2013
  Construction phase commencement: March 2014
  Construction phase completion date: June 2015

  Construction period: 65 weeks
  Mid-point construction phase occurs at, say, end of week 33 (33 weeks = approximately 8 months)

Tender evaluation, reporting, post-tender negotiations and contract award:	6 weeks
Mobilisation period:	4 weeks
Mid-point construction phase occurs after:	33 weeks
Period for which construction inflation is measured:	43 weeks

Therefore, mid-point of construction period occurs 43 weeks after the tender return date. Thus, the mid-point of construction period is November 2014.

Therefore, construction inflation is measured for the period from December 2013 to November 2014.

Tender price indices:

December 2013 = 231 (forecast)
November 2014 = 238 (forecast)

Thus, the percentage applicable for construction inflation is:

$$\frac{(238 - 231)}{238} \times 100 = 3.03\%$$

Note:
In this case it is forecast that tender prices will increase during this period by 3.03%.

The allowance for construction inflation is calculated by multiplying the sum of the cost limit (excluding inflation, at estimate base date) and the allowance for tender inflation the percentage allowance for construction inflation. That is:

Cost limit (excluding inflation):	£58,373,9911
Tender inflation:	770,537
Cost limit (including tender inflation):	£59,144,528

Thus, the construction inflation estimate is:

£59,144,528 × 3.03% =

£1,792,079; a forecast increase of £1,792,079.

The total inflation estimate is calculated by adding together the cost allowances for tender inflation and construction.

The total inflation estimate is therefore:

Tender inflation:	£770,537
Construction inflation:	£1,792,079
Inflation estimate:	£2,562,616

7.5.11 Cost limit (excluding inflation):

The cost limit (including inflation) is calculated as follows:

Cost limit (excluding inflation):	£58,373,991
Inflation estimate:	£2,562,616
Cost limit (including inflation):	£60,936,607

7.5.12 VAT assessment

No allowance for VAT is to be made in the cost estimate.

7.6 Presenting the order of cost estimate

A standard form should be used to present the order of cost estimate. The form shown in Figure 7.6 is an example of how an order of cost estimate should be set out based on the framework advocated by NRM 1. The make-up of the building cost estimate will vary according to the work breakdown structure for the proposed building project, but the overarching framework should remain constant, whether or not all facets are used. The phrase 'excluded' should be simply inserted against facets when not used.

ORDER OF COST ESTIMATE – DETAILED SUMMARY
Prepared by: Bright Kewess Partnership
Project no. BKP-0693
Project: Woollard Hotel and City Rise House, Holborn Viaduct, St Leonards-on-Sea, East Sussex
Estimate base date: March 2013

Code	Element	Qty	Unit	Rate	Element Cost	Total Cost	Total Cost
				£	£	£	£
1.	**Building Works**						
1.1	Facilitating works					655,350	655,350
1.1.1	Facilitating works	7,710	m²	15	115,650		
1.1.2	Demolition works	7,710	m²	70	539,700		
1.2	Building works						41,547,672
1.2.1	Basement (including lower ground floors)	3,079	m²			6,158,000	
1.2.1.1	Hotel building	786	m²	2,000	1,572,000		
1.2.1.2	Office building	2,293	m²	2,000	4,586,000		
1.2.2	Hotel					18,781,510	
1.2.2.1	Hotel building	7,613	m²	2,270	17,281,510		
1.2.2.2	Abnormal – piling through railway tunnel		item		1,500,000		
1.2.3	Office building (to Cat. A fit-out)					15,770,250	
1.2.3.1	Office building – shell, core and envelope	8,150	m²	1,475	12,021,250		
1.2.3.2	Cat. A fit-out	8,150	m²	460	3,749,000		
1.2.4	External works					837,912	
1.2.4.1	Landscaping	526	m²	200	105,200		
1.2.4.2	Mains connections						

Figure 7.6 Order of cost estimate – form of presentation

Code	Element	Qty	Unit	Rate	Element Cost	Total Cost	Total Cost
				£	£	£	£
1.2.4.2.1	Sewer (mains drainage) connection:	18,842	m^2				
1.2.4.2.2	Mains water connection	18,842	m^2				
1.2.4.2.3	Mains electricity connection	18,842	m^2	30	565,260		
1.2.4.2.4	Mains gas connection	18,842	m^2				
1.2.4.2.5	Telecommunications	18,842	m^2				
1.	Building works estimate				42,203,022	42,203,022	42,203,022
2.	Main contractor's preliminaries	13.60%	of	42,203,022	5,739,611	5,739,611	5,739,611
	Subtotal				47,942,633	47,942,633	47,942,633
3.	Main contractor's overheads and profit	4.00%	of	47,942,633	1,917,705	1,917,705	1,917,705
	Works cost estimate				49,860,338	49,860,338	49,860,338
4.	Project/design team fees estimate	11.50%	of	49,860,338	5,733,939	5,733,939	5,733,939
5.	Other development/project costs estimate				Excluded	Excluded	Excluded
	Base cost estimate				55,594,277	55,594,277	55,594,277
6.	Risk allowance estimate					2,779,714	2,779,714
6.1	Design development risks	1.50%	of	55,594,277	833,914		
6.2	Construction risks	3.50%	of	55,594,277	1,945,800		
6.3	Design development risks				Excluded		
6.4	Construction risks				Excluded		
	Cost limit (excluding inflation @ estimate base date)				58,373,991	58,373,991	58,373,991
7.	Inflation estimate					2,562,616	2,562,616
7.1	Tender inflation	1.32%	of	58,373,991	770,537		
7.2	Construction inflation	3.03%	of	59,144,528	1,792,079		
	Cost limit (including inflation)						**60,936,607**
	Cost limit (including inflation) to 3 significant figures						**60,937,000**
8.	**VAT assessment**						**Excluded**

Figure 7.6 *(Continued)*

CHAPTER 8

Preparation of an initial elemental cost model – using the elemental method of estimating

Introduction

This chapter:

- provides an overview of the elemental method of estimating;
- describes the steps to preparing an initial elemental cost model;
- describes the elemental breakdown for building works;
- defines element unit quantities; and
- provides step-by-step guidance on how to measure EUQs for each principal building element.

Figure 8.1 highlights where this part of the order of cost estimate fits within the order of cost estimate framework and provides a quick cross-reference to the other relevant chapters.

8.1 The elemental method of estimating

The elemental method of estimating (or elemental estimate) involves subdividing an order of cost estimate into a number of defined group elements (each group element being subdivided into as many defined elements as might be needed). A key benefit of the elemental approach is that it reveals the distribution of the costs of a building among its principal elements in terms that are meaningful to the employer and the project team members. Essentially, the method is used to generate an initial elemental cost model (often referred to as either a preliminary cost plan or a cost model) at the commencement of RIBA Work Stage 3: Concept Design or OGC Gateway 3A: Design Brief and Concept Approval, whichever is applicable – as a prerequisite to developing a formal cost plan. Breaking down the building costs in this way facilitates the comparison of different design options (i.e. optioneering –

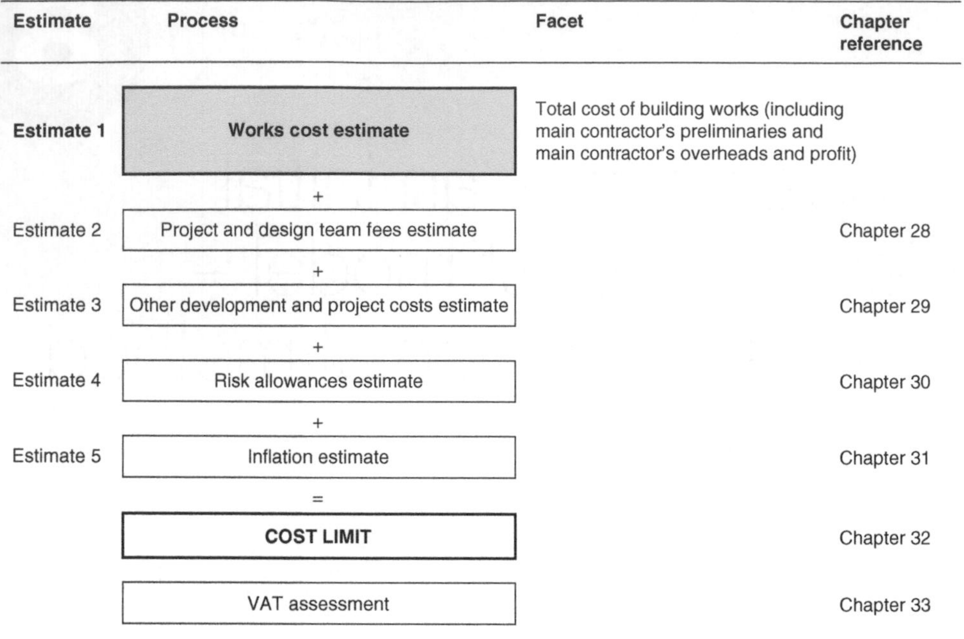

Estimate
Process
Facet
Chapter reference
Estimate 1
Works cost estimate
Total cost of building works (including main contractor's preliminaries and main contractor's overheads and profit)
+
Estimate 2
Project and design team fees estimate
Chapter 28
+
Estimate 3
Other development and project costs estimate
Chapter 29
+
Estimate 4
Risk allowances estimate
Chapter 30
+
Estimate 5
Inflation estimate
Chapter 31
=
COST LIMIT
Chapter 32
VAT assessment
Chapter 33

Figure 8.1 Works cost estimate in the context of the order of cost estimate framework

the consideration and appraisal of alternative methods of achieving the project objectives). Moreover, this elemental breakdown provides a frame of reference from which the first formal cost plan can be developed (see Part 3: Cost planning). The initial element unit quantities and element unit rates will eventually be superseded by more detailed measurement of elements (sub-elements and components) and unit rates once further design information has been prepared and the cost plan evolves.

Box 8.1 Key definitions

- **Element** – a subdivision of a group element.
- **Element unit quantity (EUQ)** – a unit of measurement that relates solely to the quantity of the element.
- **Element unit rate (EUR)** – the unit rate applied to an element unit quantity. EURs are derived from cost information relating to one or more historical building projects. They are ascertained from a historical building project by calculating the total cost of an element and dividing it by the EUQ for the applicable element.
- **Group element** – used to describe the main facets of a building project.

8.2 Steps to preparing an initial elemental cost model

Except for the way in which the building works estimate is produced (step 2), the approach to preparing an initial elemental cost model using the elemental method of estimating is the same as that for preparing an order of cost estimate (refer to Figure 8.2).

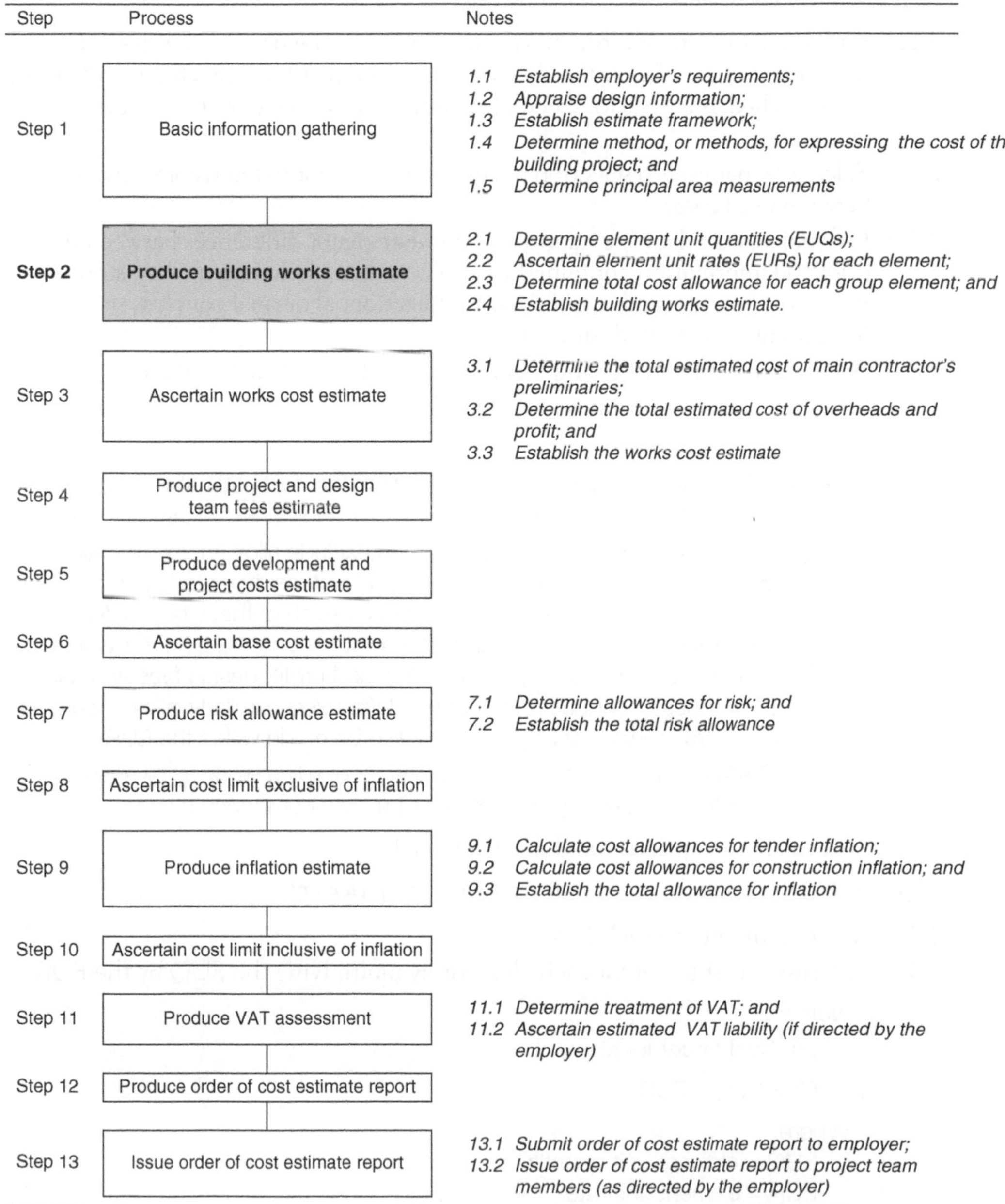
Step
Process
Notes
Step 1
Basic information gathering
1.1 Establish employer's requirements;
1.2 Appraise design information;
1.3 Establish estimate framework;
1.4 Determine method, or methods, for expressing the cost of the building project; and
1.5 Determine principal area measurements
Step 2
Produce building works estimate
2.1 Determine element unit quantities (EUQs);
2.2 Ascertain element unit rates (EURs) for each element;
2.3 Determine total cost allowance for each group element; and
2.4 Establish building works estimate.
Step 3
Ascertain works cost estimate
3.1 Determine the total estimated cost of main contractor's preliminaries;
3.2 Determine the total estimated cost of overheads and profit; and
3.3 Establish the works cost estimate
Step 4
Produce project and design team fees estimate
Step 5
Produce development and project costs estimate
Step 6
Ascertain base cost estimate
Step 7
Produce risk allowance estimate
7.1 Determine allowances for risk; and
7.2 Establish the total risk allowance
Step 8
Ascertain cost limit exclusive of inflation
Step 9
Produce inflation estimate
9.1 Calculate cost allowances for tender inflation;
9.2 Calculate cost allowances for construction inflation; and
9.3 Establish the total allowance for inflation
Step 10
Ascertain cost limit inclusive of inflation
Step 11
Produce VAT assessment
11.1 Determine treatment of VAT; and
11.2 Ascertain estimated VAT liability (if directed by the employer)
Step 12
Produce order of cost estimate report
Step 13
Issue order of cost estimate report
13.1 Submit order of cost estimate report to employer;
13.2 Issue order of cost estimate report to project team members (as directed by the employer)

Figure 8.2 Steps in preparing an initial elemental cost model

The revised step 2 for producing the building works estimate is as follows:

Step 2: Produce building works estimate

Step 2.1 Determine element unit quantities:

2.1.1 Determine group elements and elements to be quantified (refer to paragraph 8.3, Elemental breakdown for building works, below).

2.1.2 Determine EUQs for each group element and element, as required.

2.1.3 Record all assumptions that have been made.

Step 2.2 Ascertain element unit rates for each element:

2.2.1 Establish base date of estimate (i.e. the estimate base date – the rules define the estimate base date as the date on which the cost limit (excluding inflation) is established as a basis for calculating inflation, changes or other related variances).

2.2.2 Select cost data (i.e. a selection of cost analyses – four to ten recommended – or benchmarked data).

2.2.3 Interpolate cost data (i.e. identify and isolate major differences between initial brief and information contained within cost data, including quality, storey heights, different number of floors, specification, abnormal content, specifics limitations and methods of working).

2.2.4 Adjust cost data for time and location (i.e. to be at estimate base date).

2.2.5 Establish element unit rates.

Notes:

(1) EURs used are to be current at the time the cost estimate is produced. That is, they must exclude any allowances for future inflation or deflation.

(2) EURs applied to building works are to exclude allowances for main contractor's preliminaries and main contractor's overheads and profit, as separate cost estimates are prepared for each of these facets. Moreover, allowances for all costs in connection with subcontractors' (or works contractors') preliminaries, overheads and profit, design fees and risk allowances should be made in the EURs used. Similarly, allowances for suppliers' delivery charges should also be made within the EURs.

(3) Guidance on how to ascertain EURs for building works is given in Chapter 23: 'Deriving unit rates for building components'.

2.2.6 Record all assumptions that have been made.

Step 2.3 Determine total cost allowance for each group element:

2.3.1 Assign unit rates to each EUQ.

2.3.2 Determine cost target for each element by multiplying the EUQ by the EUR.

Notes:

(1) Cost target for element =

Ct = EUQ × EUR

Where:

EUQ = element unit quantity
EUR = element unit rate
Ct = cost target for element.

(2) To illustrate cost target for element as a cost/m² of GIFA:

$$\text{Cost/m}^2 \text{ of GIFA} = \frac{Ct}{GIFA}$$

Where:

Ct = cost target for element
GIFA = gross internal floor area.

2.3.3 Determine cost target for group elements by adding the cost targets for the elements that comprise the group element.

2.3.4 Repeat step 2.3.3 for each group element.

Step 2.4 Establish building works estimate:
Determine the total estimated cost of building works by adding together the cost target for each group element.

8.3 Elemental breakdown for building works

For the purpose of producing an initial elemental cost model, building works are subdivided into nine group elements, together with their applicable elements – as shown in Figure 8.3. These are the same group elements and major elements that are used to produce a formal cost plan (see Part 3: Cost planning). This is sometimes referred to as a condensed list of elements.

Group element	Element	
0 Facilitating works	0.1	Toxic/hazardous/contaminated material treatment
	0.2	Major demolition works
	0.3	Temporary support to adjacent structures
	0.4	Specialist groundworks
	0.5	Temporary diversion works
	0.6	Extraordinary site investigation works
1 Substructure	1.1	Substructure
2 Superstructure	2.1	Frame
	2.2	Upper floors
	2.3	Roof
	2.4	Stairs and ramps
	2.5	External walls
	2.6	Windows and external doors
	2.7	Internal walls and partitions
	2.8	Internal doors

Figure 8.3 Group elements and elements for building works

Group element	Element
3 Internal finishes	3.1 Wall finishes 3.2 Floor finishes 3.3 Ceiling finishes
4 Fittings, furnishings and equipment	4.1 Fittings, furnishings and equipment
5 Services	5.1 Sanitary installations 5.2 Services equipment 5.3 Disposal installations 5.4 Water installations 5.5 Heat source 5.6 Space heating and air conditioning 5.7 Ventilation 5.8 Electrical installations 5.9 Fuel installations 5.10 Lift and conveyor installations 5.11 Fire and lightning protection 5.12 Communication, security and control systems 5.13 Special installations 5.14 Builder's work in connection with services
6 Prefabricated buildings and building units	
7 Works to existing buildings	
8 External works	

Figure 8.3 *(Continued)*

8.4 Measurement of building elements

Quantification of building elements for the initial elemental cost model is through the use of element unit quantities. The element unit quantity is measured for each element (i.e. group element or element, as appropriate) and multiplied by an appropriate element unit rate to determine the cost target of each element. The cost targets for each element are then added to determine the total estimated cost of the building project.

The rules define an EUQ as:

> a unit of measurement which relates solely to the quantity of the element . . . (e.g. the area of the external walls, the area of windows and external doors, and the number of internal doors).

However, regardless of this definition, the method of quantifying many EUQs for elements will be dictated by the information available at the time the initial cost model is prepared. In practice, the EUQ for many elements will be based on the gross internal floor area. This is because very little design information will be available, with design information still only likely comprising the architect's indicative floor and roof plans, area schedule and a statement about the employer's expectations regarding quality. In actual fact, it is not unusual for a cost model to be developed with all EUQs based on the GIFA, as this will provide an appropriate initial assessment of cost targets for the main elements. The cost manager can begin to measure bespoke EUQs for specific elements (e.g. upper floors, roof, staircases and ramps, external walls, windows and external doors, and internal finishes) as more information about the shape, height and specification of the building becomes available. In practice, it is most likely that a combination of both methods will be used by the cost manager.

The method of measuring the EUQ for each element is defined in Table 2.1 Rules of measurement for elemental method of estimating, which can be found in Part 2 (Measurement rules for order of cost estimating) of NRM 1. The GIFA is to be measured in accordance with the 'Core definition: gross internal floor area (GIA)' of the RICS Code of Measuring Practice (refer to sub-paragraph 4.3.2 in Chapter 4: 'Commencing an order of cost estimate').

Note:
Whether measured from plans or taken from an area schedule provided by the employer, or the architect, unless stated otherwise, the GIFA should be deemed to have been measured strictly in accordance with the 'Core definition: gross internal area (GIA)' of the RICS Code of Measuring Practice. It is essential that the employer and all project team members understand the definition of GIFA and quantity areas in the same way.

8.5 Production of the building works estimate (estimate 1A)

The steps to preparing the building works estimate for an initial elemental cost model are illustrated in Figure 8.4.

The estimated cost of building works (estimate 1) is simply the sum of the estimated cost of each facet (i.e. sum of cost targets for each group element).

Although Figure 8.4 shows the process for dealing with a building project comprising a single building, the same approach is also applicable when dealing with a building project consisting of more than one building. In this case, a separate cost model is prepared for each building, culminating with an overall cost model summary. The breakdown of the overall elemental cost model will depend on the chosen work breakdown structure (WBS) – see paragraph 4.1 (Planning an order of cost estimate) in Chapter 4 ('Commencing an order of cost estimate').

8.6 Calculation of element unit quantities

8.6.1 Facilitating works (estimate 1A1)

The rules divide facilitating works into six elements as follows:

0.1 Toxic/hazardous/contaminated material removal
0.2 Major demolition works
0.3 Temporary support to adjacent structures
0.4 Specialist groundworks
0.5 Temporary diversion works
0.6 Extraordinary site investigation works.

For the purpose of order of cost estimating, with the exception of major demolition works (element 0.2 – see paragraph 8.6.2), and preparatory soft-strip works, quantities for all other types of facilitating works are generally based on the site area.

Element unit quantities for facilitating works items are measured superficially, in square metres. The area measured is the total area of the site within the site title boundaries, or the area within site title boundaries denoted as being the site area, whichever is applicable, measured on a horizontal plane, with the footprint of all

Estimate	Process	Facet
Estimate 1A1	Estimate cost of facilitating works	Facilitating works – Group element 0
	+	
Estimate 1A2	Estimate cost of demolition works	Facilitating works – Group element 0
	+	
Estimate 1A3	Estimate cost of substructure	Building works (including basement structures) – Group element 1
	+	
Estimate 1A4	Estimate cost of superstructure	Building works – Group element 2
	+	
Estimate 1A5	Estimate cost of internal finishes	Building works – Group element 3
	+	
Estimate 1A6	Estimate cost of fittings, furnishings and equipment	Building works – Group element 4
	+	
Estimate 1A7	Estimate cost of services	Building works (building engineering services) – Group element 5
	+	
Estimate 1A8	Estimate cost of prefabricated buildings and building units	Building works – Group element 6
	+	
Estimate 1A9	Estimate cost of works to existing buildings	Building works (including building engineering services within existing buildings) – Group element 7
	+	
Estimate 1A10	Estimate cost of external works	Building works (including building engineering services located beyond the face of the external envelop of the building) – Group element 8
	+	
Estimate 1A11	Estimate cost of connections to statutory undertakers' mains services	Building works – Group element 8
	=	
Estimate 1A	**Produce building works estimate**	Total cost of building works (excluding main contractor's preliminaries and main contractor's overheads and profit)

Figure 8.4 Steps in computing the building works estimate

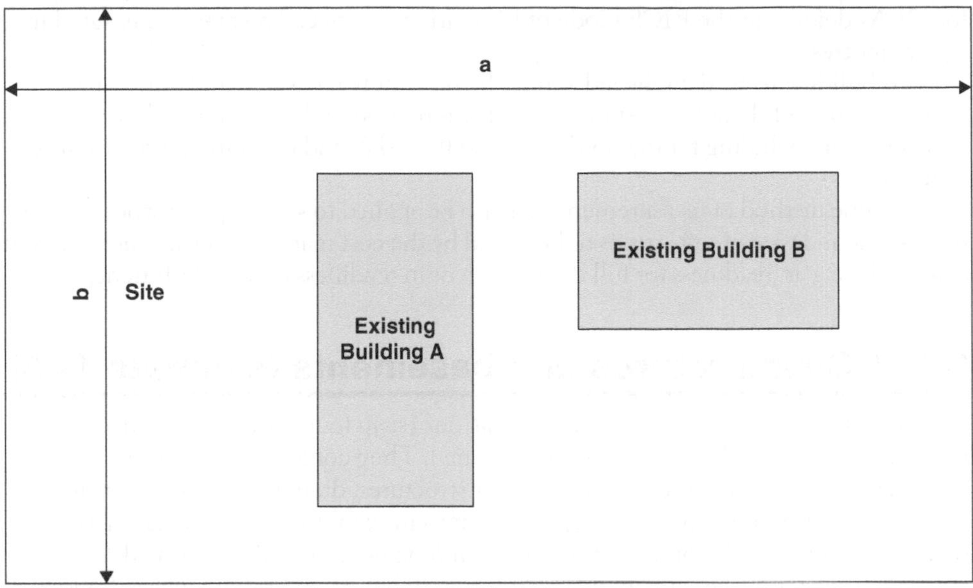

Figure 8.5 Calculating the EUQ for facilitating works

existing buildings deducted (see Figure 8.5). No deduction is made for the footprint of any new proposed buildings within the curtilage of the site. Each facilitating works item should be separately measured and described.

Where the extent of the area to be treated is known, then measurement of the applicable facilitating works item can be based on the actual area to be treated. However, the cost manager must make sure that a unit rate applicable to the method of measurement is used to estimate the cost of the works. Alternatively, facilitating works items can be measured as spot items (i.e. itemised), with the estimated costs based on discrete lump sum allowances. Whichever method of measurement is adopted, the cost manager must clearly state the method used in the cost model.

The EUQ for facilitating works items is:

(a × b) − (footprint area of existing building A + footprint area of existing building B)

Where:

(a × b) = the total area of the site within the site title boundaries (or the total area within the site title boundaries defined by the employer as the site for the building).

> Note:
> The footprint area of a building is the gross external area of the building measured at ground floor level only. The GEA is the area of a building measured externally (i.e. to the external face of the perimeter walls). The rules of measurement of GEA are defined in the RICS Code of Measuring Practice.

8.6.2 Major demolition works (estimate 1A2)

The element unit quantity for demolition works is the gross external floor area of the building to be demolished, measured using the principles of the rules of measurement for GEAs defined in the RICS Code of Measuring Practice. The area is measured in square metres.

Each building to be demolished should be separately measured and annotated, and the extent of demolition stated (e.g. down to ground floor level; or down to ground level, including taking up the ground floor slab and grubbing up the existing foundations).

The same method of measurement can also be applied to soft strip-out works. Again the extent and type of soft strip is to be stated by the cost manager, as there are different standards (e.g. in readiness for full demolition or in readiness for refurbishment).

8.6.3 Substructures and basements (estimate 1A3)

Substructures include conventional foundations (strip foundations) and piled foundations (piles, pile caps and ground beams). They comprise foundation and basement excavations, ground-bearing floor structures, drainage below or within the floor construction, entrance steps and ramps integral to the floor construction, structural screeds and toppings, internal swimming pools and all other work in constructing ground-bearing floor structures up to but excluding the lowest floor finishes. Basement retaining walls are also included.

Excluded are:

- site preparation (see element 8:1: site preparation works);
- columns, beams and integral structural walls within a basement (see element 2.2: frame);
- intermediate suspended floors which form part of a basement structure (see element 2.3: upper floors);
- podium slabs which form part of the basement structure (i.e. the roof to the basement – see element 2.3: upper floors);
- internal stairs and ramps (see element 2.4: stairs and ramps);
- non-retaining basement walls (see element 2.5: external walls);
- internal walls and partitions (see element 2.7: internal walls and partitions);
- internal doors (see element 2.8: internal doors)
- internal finishes (see element 3: internal finishes);
- fittings, furnishings and equipment (FF&E – see element 4: fittings, furnishings and equipment); and
- mechanical and electrical services installations (see element 5: services).

By way of different scenarios, Figures 8.6, 8.7 and 8.8 illustrate the main elements which make up the substructure element.

Substructures and basements are to be measured and annotated separately. Element unit quantities for both substructures and basements are measured superficially, in square metres. The area measured is the area of the lowest floor construction measured to the internal face of the external perimeter walls. Both EUQs for substructures and basements are measured using the principles of the rules of measurement for gross internal floor areas (GIFAs).

It is advocated that basement retaining walls be measured and annotated separately. Again, the EUQ is measured superficially, in square metres. The EUQ for basement retaining walls is derived by multiplying the length of the wall, measured on the internal face, by the height of the wall. The height is measured from the

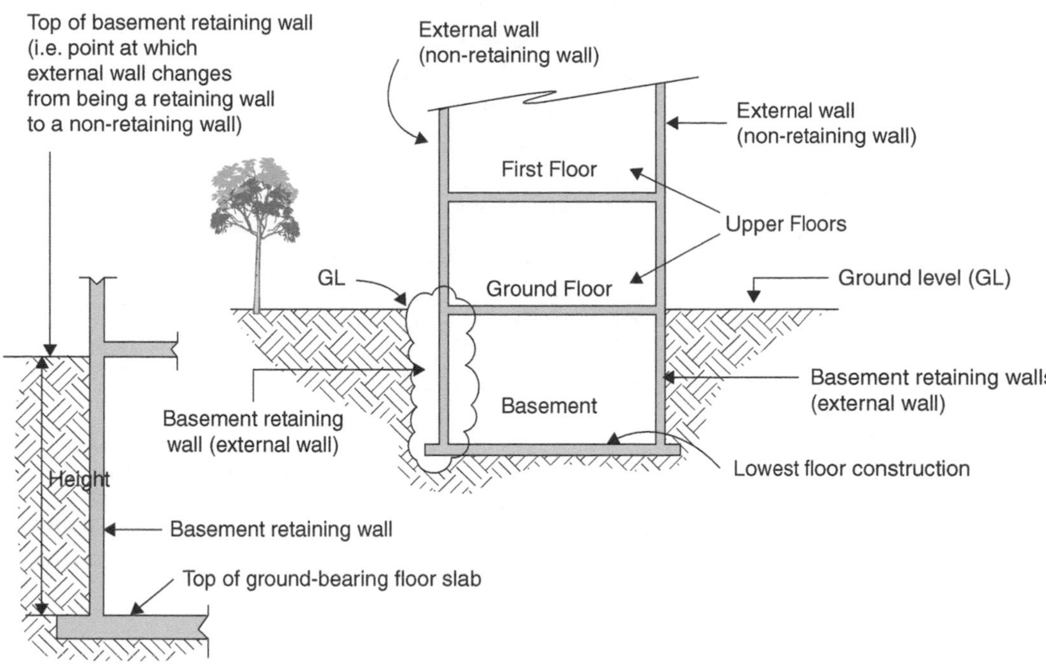

Figure 8.6 Substructure elements – scenario 1

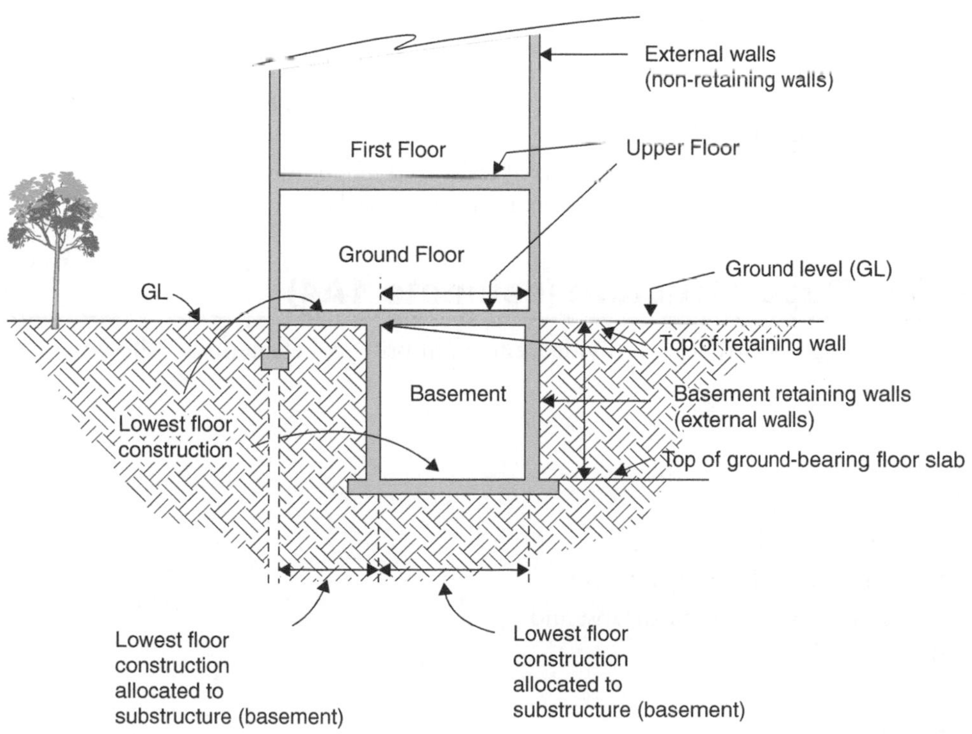

Figure 8.7 Substructure elements – scenario 2

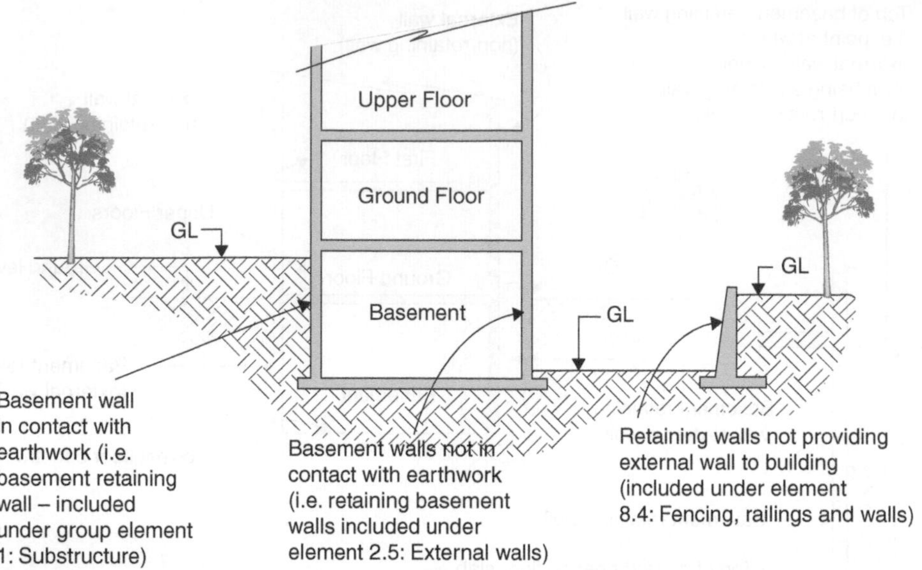

Figure 8.8 Substructure elements – scenario 3

top of the ground-bearing floor slab to the level at which the basement retaining wall connects with the external wall above ground (i.e. the point at which the basement retaining wall ceases being in contact with earthwork – refer to Figure 8.6). When measuring the height of basement retaining walls, any intersections with intermediate suspended floors are ignored.

Excluded are:

- basement walls not in contact with earthwork (see Figure 8.8 – measured with element 2.5: external walls); and
- retaining walls not providing external walls to the building (see Figure 8.8 – measured with element 8.4: fencing, railings and walls).

8.6.4 Superstructure (estimate 1A4)

The building superstructure is divided into eight principal elements as follows:

2.1 Frame;
2.2 Upper floors;
2.3 Roof;
2.4 Stairs and ramps;
2.5 External walls;
2.6 Windows and external doors;
2.7 Internal walls and partitions; and
2.8 Internal doors.

Frame

Frames comprise the columns, beams and integral structural walls (i.e. walls constructed from the same material as, and that form an integral part of, the frame) which support the upper floors and roof. Columns, beams and integral structural walls below the lowest floor, but above ground, are included in this element because of the impracticality of splitting components into two separate elements (e.g. where the design incorporates ground level car parking beneath the building – resulting in the frame being exposed).

Excluded are:

- stairs and ramps between floor levels (see element 2.4: stairs and ramps).

Element unit quantities for frames are measured superficially, in square metres. The area measured is the area of the floors related to the frame, and the area is measured using the principles of the rules of measurement for gross internal floor areas.

To enable costs in connection with the building superstructure and basement to be readily identifiable, it is suggested that columns, beams and integral structural walls relating to each of these facets be measured and annotated separately. Where required for the purpose of distributing costs, columns, beams and integral structural walls within a basement, which either act as foundations to the building's superstructure or form part of the basement support structure itself, can also be measured and annotated separately.

If a building has no frame (e.g. a bungalow or a two-storey house), then the element is not applicable as the supporting structure will be the external walls, and only the external wall element need be measured.

Upper floors

Upper floors include suspended floors, galleries, tiered terraces, service floors, internal walkways, internal bridges and external links. They also include balconies, roofs to internal buildings and external walkways which form an integral part of the upper floor construction.

Excluded are:

- suspended slabs forming the roof structure ramps (see element 2.3: roof); and
- landings between floor levels (see element 2.4: stairs and ramps).

Element unit quantities for upper floors are measured superficially, in square metres. The area measured is the total area of upper floors, and the area is measured in accordance with the principles for measuring gross internal floor area. Sloping surfaces such as galleries and tiered terraces are measured flat on plan. Each category of floor, as well as intermediate suspended floors and podium slabs which form part of the basement structure, are to be measured and annotated separately (refer to Figure 8.9). See also Figures 8.6, 8.7 and 8.8.

As a rule, the sum of the upper floors and the lowest floor will equal the gross internal floor area of the building. However, where balconies or external walkways are included as part of the element unit quantity, the sum of the upper floors and the lowest floor will exceed the GIFA of the building.

Roof

This element includes the roof construction, roof coverings, thermal insulation, rooflights and dormers, roof features and rainwater disposal systems.

Excluded are:

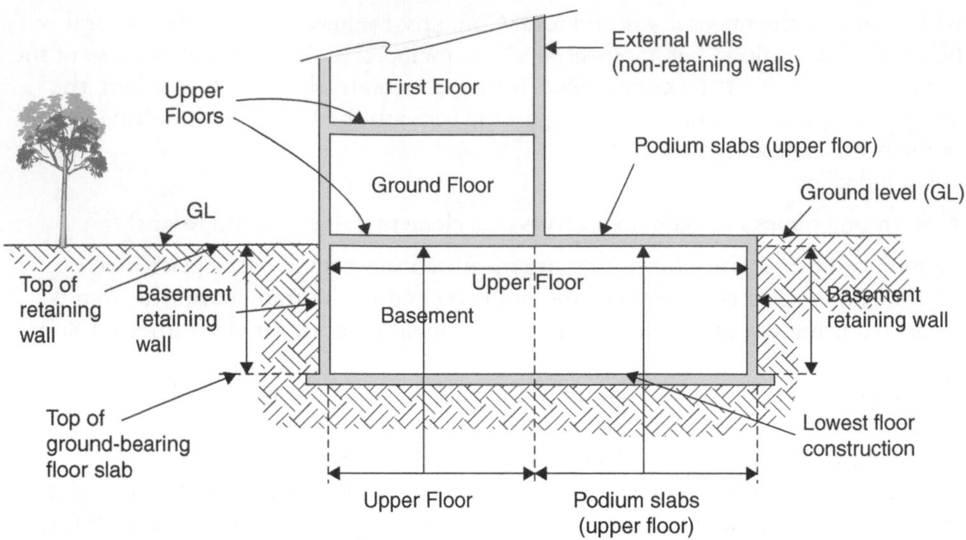

Figure 8.9 Categories of upper floor

- parapet walls, and roof perimeter balustrades and guards (see element 2.5: external walls); and
- windows to dormers (see element 2.6: windows and external doors).

Element unit quantities for roofs are measured superficially, in square metres, based on the area of the roof on plan measured to the inside face of the external walls, irrespective of the overhang at the eaves. See Figure 8.10.

Stairs and ramps

This element consists of the structural connections between floor levels, including connections to the roof and roof-top plant rooms, together with their associated finishings, balustrades and handrails. It covers staircases, including fire escape staircases, landings and ramps.

Stairs and ramps are enumerated (nr), giving the total number of storey flights (i.e. the number of staircases or ramps multiplied by the number of floors served) – see Figure 8.11. The total vertical rise of each staircase or ramp is also to be stated in metres (m). This is measured from top of structural floor level to top of structural floor level.

External walls

External walls comprise the vertical enclosure around the building: from the substructure to the roof. They include structural walls, non-retaining basement walls, curtain and window walls, external shop fronts, glazed screen walls, balcony walls and balustrades, solar screen walls, plant room airflow screens and louvres, all insulation to external walls and all external applied finishings to walls. External false ceilings and demountable suspended ceilings which form an integral part of the building envelope are also included under external walls.

Excluded are:

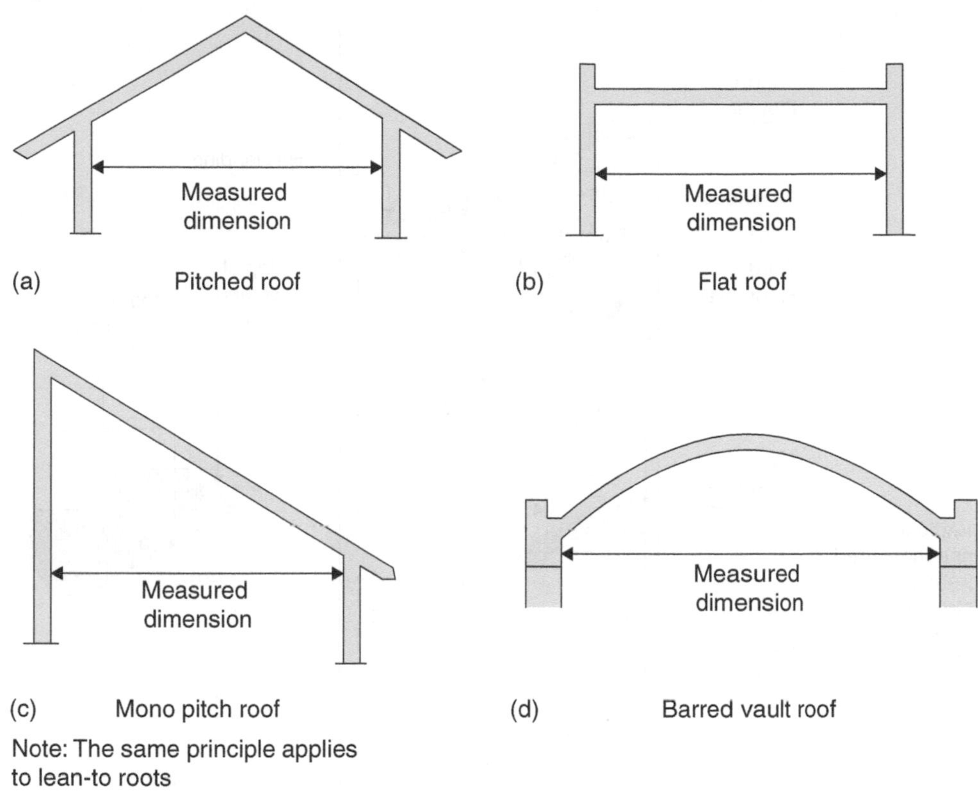

Figure 8.10 Measurement of roofs

- basement retaining walls (see group element 1: substructures);
- walls and retaining walls not providing external walls to the building (see element 8.4: fencing, railings and walls);
- windows and external doors (see element 2.5: windows and external doors); and
- all finishes to internal face of external walls (see element 3: internal finishes);

Element unit quantities for external walls are measured superficially, in square metres. The EUQ for external walls is derived by multiplying the length of the wall, measured on the internal face, by the height of the wall. The height is measured from the top of the lowest ground-bearing floor slab, or top of basement retaining wall, to the underside of the roof, or top of parapet wall, whichever is applicable, ignoring any intersections with suspended floors (see Figures 8.12A, 8.12B and 8.12C). The area is measured gross, with no deductions made for the area of windows and external doors.

Forming openings for windows and external doors in external walls will normally be carried out by the works contractor who constructs the external walls. As a consequence, historical cost data gathered for external walls (element 2.5) will include costs in connection with forming openings for windows and external doors. Therefore, costs associated with forming openings are deemed to be included in the external wall element. Cost data collected in respect of windows and external doors will only include the cost of works in connection with supplying and installing windows and external doors.

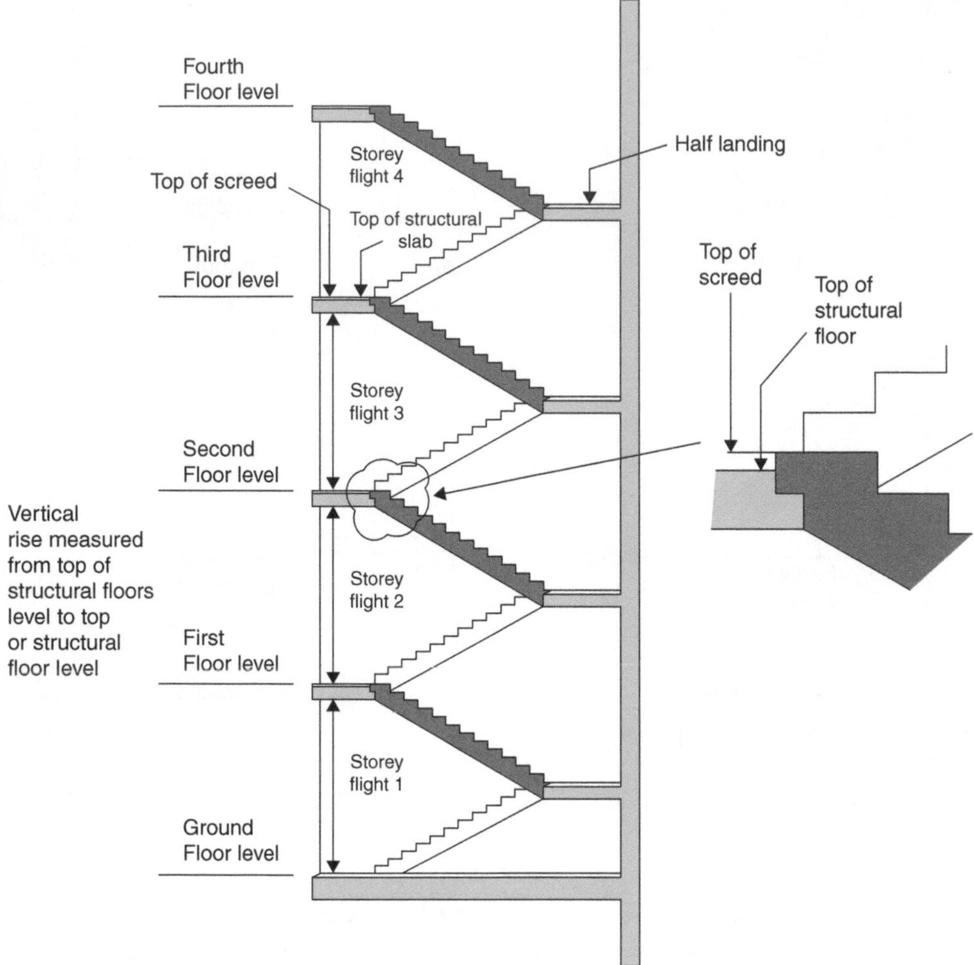

Figure 8.11 Measurement of stairs

External false ceilings and demountable suspended ceilings should be measured and annotated separately. The area is measured superficially, in square metres, based on the surface area of the external ceiling. Although the cost of external ceilings forms part of the cost target for external walls, the overall EUQ for external walls excludes the area of external ceilings. The overall EUQ for the external wall element is simply the EUQ for external walls.

Windows and external doors

This element covers windows and louvres, as well as doors providing access into the building. Windows include glazed windows, louvre windows, ironmongery, integral blinds, solid infill panels, protection film, decoration and sealants. Clerestory windows and dormer windows are also included under this element. External doors comprise doors for both pedestrian and vehicular access and include entrance doors, fire escape doors, garage doors, roller shutters, insect screen doors, gates, frames, linings, thresholds, architraves, glazing, ironmongery, sidelights and over panels, protection film, decoration and sealants.

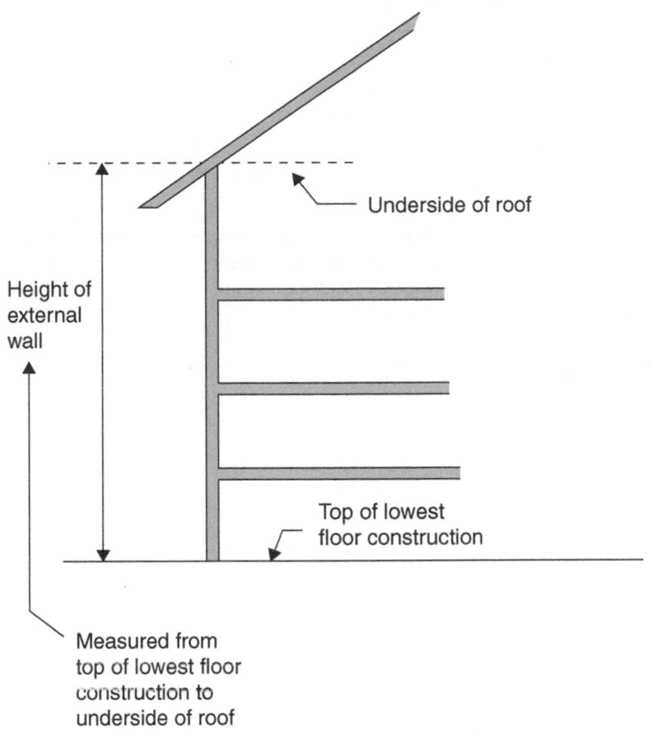

Figure 8.12A Measurement of external walls – pitched roof scenario

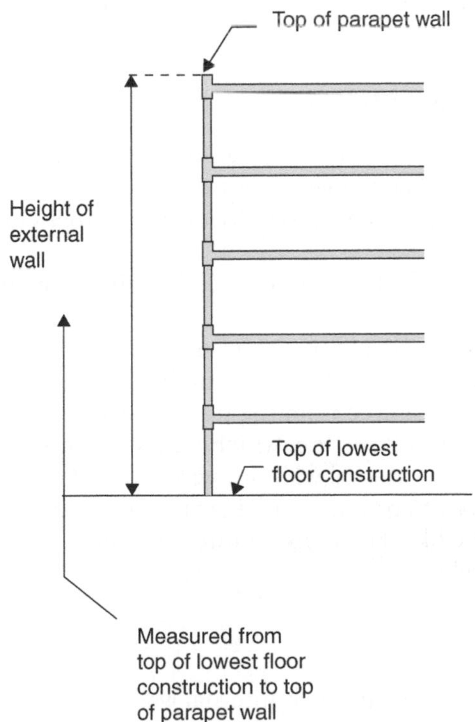

Figure 8.12B Measurement of external walls – flat roof scenario

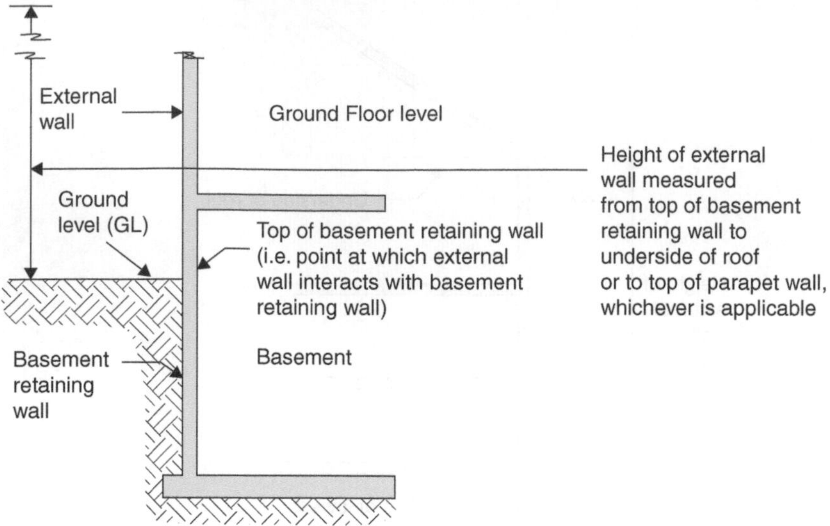

Figure 8.12C Measurement of external walls – basement scenario

Excluded are:

- construction of and coverings to dormers (see group element 2.4: roof); and
- curtain walling and other glazed walling systems (see group element 2.5: external walls).

Element unit quantities for windows and external doors are measured superficially, in square metres, based on the area of windows and external doors measured over all frames.

Internal walls and partitions

This element deals with the division of internal spaces within the building. It includes internal walls and permanent partitions, glazed screens, internal shop fronts, balustrades and handrails, movable room dividers and proprietary pre-finished toilet, bath and shower cubicles.

Element unit quantities for internal walls and partitions are measured superficially, in square metres.

The EUQ for internal walls and partitions is ascertained by multiplying the length of the wall or partition, measured on the centre line, by the height of the wall or partition. The height is measured from the top of the floor structure to the underside of the floor structure, or roof structure, whichever is applicable. For the same reason as for forming openings in external walls for windows and external doors, the area of internal walls and partitions is measured gross, with no deduction made for door openings, screens or the like. Each type of internal wall or partition is to be measured and annotated separately.

Internal doors

Internal doors provide passage through internal walls and partitions, including doors providing access to service cupboards and ducts. They include standard internal doors, fire-resisting doors, door sets and composite door units with sidelights and over panels, roller shutters, glazing, linings, frames, architraves, ironmongery and decoration.

Internal doors are enumerated (nr), irrespective of door size or type, with the total number of internal doors stated.

8.6.5 Internal finishes (estimate 1A5)

Group element 3 deals with internal finishes and is divided into the following three elements:

3.1 Wall finishes;

3.2 Floor finishes; and

3.3 Ceiling finishes.

Wall finishes

Wall finishes comprise finishes applied to all internal faces of external walls, internal walls and partitions (both faces), and of internal columns (all exposed faces). Element unit quantities are measured superficially, in square metres, with the area measured being the surface area of walls to which finishes are to be applied.

Finishes to walls include in situ coatings, sheet linings, acoustic wall linings, decorative sheet coverings, tiling (including splash-backs), painting and decorating. They also include picture rails, dado rails, bumper guards and the like.

Floor finishes

Element unit quantities for floor finishes are measured superficially, in square metres, with the area measured being the surface area of floors to which finishes are to be applied. Floor finishes include non-structural screeds, latex screeds, floating floors, resilient layers, in situ floor finishes, tiling, flexible and semi-flexible tile and sheet coverings, and carpeting. Raised access floor systems are also covered by this element.

Excluded are:

- structural screed and toppings (see group element 2.2: upper floors, or element 2.3: roof, whichever is applicable); and
- floor finishes to half-landings, treads and risers to staircases (see group element 2.4: stairs and ramps).

Ceiling finishes

Ceiling finishes comprise finishes applied to all internal soffits, soffits and sides of beams, and vertical faces of bulkheads formed within in ceilings. They comprise linings to ceilings, in situ coatings, and painting and decorating. They also comprise both false ceilings and demountable suspended ceilings.

Element unit quantities for ceiling finishes are measured superficially, in square metres, with the area measured being the surface area of ceilings to which finishes are to be applied.

Excluded are:

- external soffits (see group element 2.5: external walls); and

- finishes to soffits of staircases, including soffits of landings between floors (see group element 2.4: stairs and ramps).

8.6.6 Fittings, furnishings and equipment (estimate 1A6)

This group element includes fittings, furnishings and equipment that are either to be fixed to the building fabric or to be provided loose within the building. Included in this element are items such as counters, benches, blinds, mirrors, storage racks, domestic kitchen units and appliances, signs and notices, works of art and internal planting. Specialist fittings, furnishings and equipment are also dealt with under this element. These are items that are designed for the particular purpose of a building (e.g. hospitals, dental surgeries, sports buildings and repositories).

The element unit quantity for fittings, furnishings and equipment are measured superficially, in square metres; based on the gross internal floor area of the building.

8.6.7 Services (estimate 1A7)

This group element deals with mechanical and electrical services installations. It is divided into 14 principal elements as follows:

5.1 Sanitary installations;
5.2 Services equipment;
5.3 Disposal installations;
5.4 Water installations;
5.5 Heat source;
5.6 Space heating and air conditioning;
5.7 Ventilation;
5.8 Electrical installations;
5.9 Fuel installations;
5.10 Lift and conveyor installations;
5.11 Fire and lightning protection;
5.12 Communication, security and control systems;
5.13 Specialist installations; and
5.14 Builder's work in connection with services.

When preparing an initial elemental cost model, information about the mechanical and electrical services installations will be vague and limited to the employer's brief. Therefore, the cost manager will have to use the gross internal floor area as the basis for measuring some or all of the initial elemental unit quantities for mechanical and electrical services installations. However, as more information about the services becomes available, then the cost manager can begin to measure elements in accordance with methodology described in the rules of measurement for EUQs. The following describes how each services element is measured.

Costs in connection with the testing and commissioning of services installations are to be included in the element unit rates used.

Sanitary installations

This element deals with the supply and fixing of sanitary appliances (e.g. WC suites, urinals, basins, sinks, baths, bidets and shower trays) and bathroom, toilet and shower room pods). Fittings and ancillary items associated with the provision of sanitary appliances are also covered under this element.

Excluded are:

- sinks which are supplied as an integral part of the kitchen fittings (see group element 4: fittings, furnishings and equipment);
- sinks which are supplied as an integral part of catering equipment (see element 5.2: services equipment);
- bathroom pods (see sub-element 6.1.3: pods);
- toilet pods (see sub-element 6.1.3: pods);
- shower room pods (sub-element 6.1.3: pods);
- waste installations (see element 6.3: disposal installations); and
- water supply (see element 5.4: water installations).

Element unit quantities for sanitary appliances are based on the number of items (i.e. enumerated (nr)), and measured under the following categories:

- domestic sanitary appliances;
- specialist sanitary appliances.

The total number of appliances for each category is to be stated.

Services equipment

Services equipment deals with catering equipment designed for use on a communal or commercial scale. Element unit quantities for services equipment are based on the number of items (i.e. enumerated (nr)), and measured under the following categories:

- commercial catering equipment;
- sinks supplied as an integral part of catering equipment;
- food storage equipment; and
- specialist equipment (e.g. dental chair and equipment in dental surgery).

The total number of items for each category is to be stated.

Disposal installations

Disposal installations deal with the disposal of all waste and soiled water from sanitary appliances, pods and services equipment out to the external face of the external walls, or ground-bearing floor structure, whichever is applicable. That is, above-ground waste installations to sanitary appliances and services equipment, and entry chutes to refuse disposal installation. Excluded are:

- Rainwater disposal systems (see element 2.3: roof).

Element unit quantities for disposal installations are based on the number of waste disposal points to sanitary appliances and services equipment. Waste points are enumerated (nr), and measured under the following categories:

- waste points to sanitary appliances (e.g. WC suites, urinals, basins, sinks, baths, bidets and shower trays);
- waste points to services equipment (e.g. sinks);
- waste points to laboratory and industrial liquid waste;
- entry points to refuse disposal installations; and
- entry points to chemical and industrial refuse disposal installations.

The total number of waste disposal, or entry, points for each category is to be stated.

Water installations

This element deals with the supply of water from the point of entry into a building to the points of consumption. It includes storage, cold and hot water distribution pipelines, pumps, taps, valves, water heaters and coolers, thermal insulation, as well as internal rainwater-harvesting systems and grey water collection pipe systems.

Excluded are:

- connections to statutory undertaker's mains (see element 8.7: external services); and
- water supply from statutory undertaker's mains to point of entry into the building (see element 8.7: external services).

Element unit quantities for water installations are based on the number of draw-off points. Draw-off points are enumerated (nr), and measured under the following categories:

- mains supply draw-off points;
- cold water draw-off points;
- hot water draw-off points; and
- steam and condensate draw-off points.

The total number of draw-off points for each category is to be stated.

Heat source

Heat sources include boilers, package steam generators, central heat and power boiler plant, heat pumps, ground source heating or any other component which provides the source of heat to the building.

Element unit quantities are based on the number of heat sources. Heat sources are enumerated (nr), and the rating in kilowatts (kW) is to be stated for each heat source. Each type of heat source is to be measured and annotated separately, with the total number of each type stated.

Space heating and air conditioning; ventilation systems; electrical installations; and gas and other fuel installations

Element unit quantities for heating and air conditioning; ventilation systems; electrical installations; and gas and other fuel installations are measured superficially, in square metres, based on the area serviced by the system or installation. The way in which the area is measured is described in Chapter 19 ('Services (group element 5)').

Each type of system or installation is to be measured and annotated separately.

Lift and conveyor installations

Element unit quantities for lift and conveyor installations are enumerated (nr), with the total number of each type of installation stated. They are measured under the following categories, with the additional information requested given in the description:

- lifts (passenger, goods, fire fighting, etc) – stating the number of levels served (nr);
- enclosed hoists – stating the number of levels served (nr);
- escalators – stating the number of levels served (nr), the rise (m) and length of travel (m);
- moving pavements – stating the length of travel (m);
- powered stairlifts;
- conveyors (passenger or goods) – stating the length of travel (m);
- dock levellers and scissor lifts – stating the total rise (m) and the designed load (in kilonewtons, kN)
- cranes and unenclosed hoists – stating the total rise (m) and the designed load (kN);
- car lifts – stating the number of levels served;
- car-stacking systems – stating the capacity (i.e. the number of cars (nr));
- car and lorry turntables and the like; and
- document handling systems.

Each type of lift and conveyor installation, or different composition of the same type of lift or conveyor installation, is to be measured and annotated separately.

Fire and lightning protection; communication, security and control systems; and specialist installations

Element unit quantities for fire and lightning protection, communication, security and control systems, and specialist installations are measured superficially, in square meters. The method of calculating the area is described in Chapter 19: (Services (group element 5)).

Each type of system or installation is to be measured and annotated separately.

Builder's work in connection with services

Element unit quantities for builder's work in connection with services, and testing and commissioning of services, are measured superficially, in square metres, based on the gross internal floor area of the building.

Builder's work in connection with services, and testing and commissioning of services, are to be measured and annotated separately.

8.6.8 Complete buildings and building units (estimate 1A8)

Prefabricated buildings and prefabricated room units are measured superficially, in square metres, with the element unit quantity based on the gross internal floor area of the building or room units, whichever is applicable. The area is measured in accordance with the principles for measuring the GIFA. Complete prefabricated buildings and prefabricated room units are to be measured and described separately. The EUQ for complete buildings and building units is the total GIFA of all prefabricated buildings and room units.

The type of prefabricated buildings and prefabricated room units is to be given in the description. In the case of prefabricated room units, the description is to include the number of each type of room unit.

8.6.9 Works to existing buildings (estimate 1A9)

Works to existing buildings comprise six elements as follows:

7.1 Minor demolition and alteration works;
7.2 Repairs to existing services;
7.3 Damp-proof courses/fungus and beetle eradication;
7.4 Facade retention;
7.5 Cleaning existing surfaces; and
7.6 Renovation works.

Minor demolition and alteration works, and repairs to existing services

Element unit quantities for minor demolition works, minor alteration works and repairs to existing services are measured superficially, in square metres. The EUQ is based on the gross internal floor area of the building concerned and is measured in accordance with the principles for measuring the GIFA. Minor demolition and alteration works, and repairs to existing services, are to be dealt with as two distinct facets.

Damp-proof courses/fungus and beetle eradication

The element unit quantity for damp-proof courses, fungus eradication and beetle eradication is measured superficially, in square metres, based on the gross internal floor area of the room or rooms to be treated. The area is measured in accordance with the principles for measuring the GIFA. Where there is more than one type of treatment to be undertaken, each type of treatment is to be measured and described separately. However, the EUQ is the total GIFA of all rooms to be treated, irrespective of the number of different types of treatment.

Facade retention

Facade retention works are measured superficially, in square metres, with the element unit quantity being the gross area of facade to be retained, with no deductions for window or door openings, or other voids.

Cleaning existing surfaces

Cleaning existing surfaces is measured superficially, in square metres. The element unit quantity is the area of the surface to be cleaned. The gross area of the surface to be cleaned is measured, with no deductions for voids.

Where there is more than one type of surface to be cleaned, each type of surface is to be measured and described separately. Irrespective of the number of different surfaces to be cleaned, the element unit quantity is the total area of all surfaces that are to be cleaned.

Renovation works

Renovation works are measured superficially, in square metres, with the element unit quantity being the actual surface area or areas to be renovated. Where there is more than one type of surface to be renovated, each type is to be measured and annotated separately. However, irrespective of the number of different surfaces to be renovated, the EUQ is the total area of all surfaces that are to be renovated.

8.6.10 External works (estimate 1A10)

External works are divided into eight elements as follows:

8.1 Site preparation works;
8.2 Roads, paths, pavings and surfacings;
8.3 Soft landscaping, planting and irrigation systems;
8.4 Fencing, railings and walls;
8.5 External fixtures;
8.6 External drainage;
8.7 External services; and
8.8 Minor building works and ancillary buildings.

Site preparation works; roads, paths, pavings and surfacings; soft landscaping, planting and irrigation systems; fencing, railings and walls; external fixtures; external drainage; and external services

Element unit quantities for these items are measured superficially, in square metres, and based on the site area. The site area is the total area of the site within the site title boundaries (or the total area within the site title boundaries defined by the employer as the site for the building), measured on a horizontal plane, excluding the area of the footprint of all buildings (new and existing) within the site boundaries (refer to Figure 8.13). Any areas used temporarily for the building works that will not form part of the delivered building project are to be excluded from the area calculated. Each element is to be measured and annotated separately (i.e. site preparation works; roads, paths, pavings and surfacings; soft landscaping, planting and irrigation systems; fencing, railings and walls; external fixtures; external drainage; and external services).

In Figure 8.13, the EUQ for site preparation works; roads, paths and pavings; planting; fencing, railings and walls; site and street furniture and equipment; external drainage; and external services equals:

(a × b) – footprint area of proposed new building

Where:

a × b = the total area of the site within the site title boundaries (or the total area within the site title boundaries defined by the employer as the site for the building).

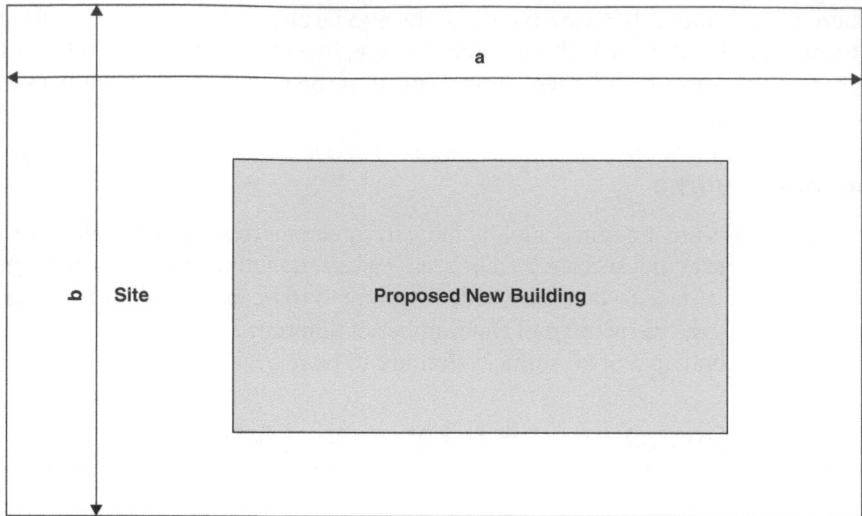

Figure 8.13 Calculating the EUQ for external works

Note:
The footprint area of a building is the gross external area of the building measured at ground floor level only. The GEA is the area of a building measured externally (i.e. to the external face of the perimeter walls). The rules of measurement of GEA are defined in the RICS Code of Measuring Practice.

Minor building works and ancillary buildings

Element unit quantities for minor building works and ancillary buildings are measured superficially, in square metres, with the EUQ based on the gross internal floor area of the buildings which make up the element. The EUQ is measured in accordance with the principles for measuring the GIFA.

Alternatively, spot items can be measured; but the EUQ is to be expressed as the GIFA of the buildings which make up the element.

8.6.11 Connections to statutory undertakers' mains services (estimate 1A11)

For the purpose of preparing an initial elemental cost model, connections to statutory undertakers' mains services are measured in the same way as for an order of cost estimate (see paragraph 6.2.9 (Estimating the cost of connections to statutory undertakers' mains services (estimate 1A8) of Chapter 6: 'Computing the works cost estimate').

8.7 Calculating elemental cost targets and the building works estimate

Cost targets for each element are simply calculated by multiplying the element quantity for the element by the element unit rate. Accordingly, the equation for calculating the cost target for an element is therefore:

Ct = EUQ × EUR

Where:

EUQ = element unit quantity
EUR = element unit rate
Ct = cost target (for element).

The cost target for individual group elements is the sum total of the cost targets for the elements applicable to the group element, while the total cost of building works (i.e. the building works estimate) is the sum of the cost targets of each group element (refer to Figure 8.4).

8.8 Computing the works cost estimate

The method of calculating the works cost estimate is shown in Figure 8.14:

Estimate	Process	Facet	Chapter reference
Estimate 1A	**Produce building works estimate**	Total cost of building works (excluding main contractor's preliminaries and main contractor's overheads and profit)	
	+		
Estimate 1B	Estimate cost of main contractor's preliminaries	Main contractor's preliminaries	Chapter 6, paragraph 6.3
	+		
Estimate 1C	Estimate cost of main contractor's overheads and profit	Main contractor's overheads and profit	Chapter 6, paragraph 6.4
	=		
Estimate 1	**Produce works cost estimate**	Total cost of building works (including main contractor's preliminaries and main contractor's overheads and profit)	Chapter 6, paragraph 6.5

Figure 8.14 Works cost estimate process map

CHAPTER 9

Initial elemental cost model – a worked example

Introduction

This chapter:

- provides step-by-step guidance on how to generate element unit quantities to provide an initial elemental breakdown of the cost of a building project; and
- explains how to bring together the various aspects of an elemental cost model.

9.1 Basis of elemental cost model

The following worked example of an initial elemental cost model is based on the architect's indicative sketched floor plans and area schedule, together with the employer's required quality attributes and indicative programme information, for the proposed hotel development referred to in Chapter 7 ('Order of cost estimate – a worked example'). Although to be constructed on a single site, the hotel and office buildings will be two separate buildings, as they will be separate legal entities when they have been completed. Moreover, GMB Developments Limited has a hotel operator on board, but no potential taker for the offices. In view of this, construction of the hotel will commence approximately 6 months ahead of the offices.

Since preparation of the order of cost estimate, it has been assumed that the only additional information provided is a drawing by the architect, RK Tex Limited, showing an indicative section through the proposed hotel building (Figure 9.1). However, at this early stage of the design process, it has been assumed that the architect has not yet considered the internal layout of the building or construction methods in detail.

The schedule of floor areas is the same as for the original order of cost estimate (refer to Figure 7.2 in Chapter 7).

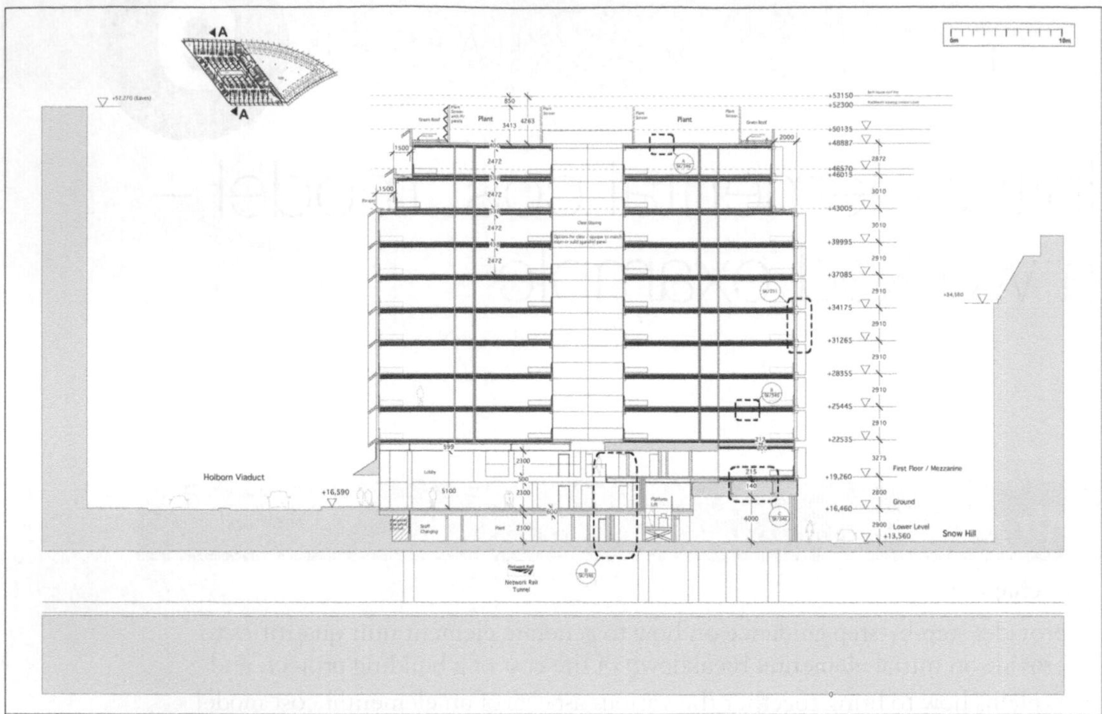

Figure 9.1 Drawing showing section through proposed hotel building (not to scale)

9.2 Assumptions and data gathered

With the exception of the quantification of building works and unit rates for building works, the assumptions made by the cost manager in respect of the initial elemental cost model are the same as those used for the order of cost estimate – refer to paragraph 7.3 of Chapter 7 ('Order of cost estimate – a worked example').

Element unit quantities have been generated for each element of the proposed building (see paragraph 9.3), while cost targets and element unit rates are derived from previous cost analysis (see paragraph 9.4).

9.3 Generation of element unit quantities

From the assembled data the cost manager can calculate the element unit quantities for the cost model as shown in Example 9.1.

9.4 Compiling an initial elemental cost model

Using the element unit quantities calculated, the initial elemental cost model can be compiled.

The purpose of the elemental approach is to distribute the cost limit for building works (set in the order of cost estimate approved by the employer) to establish initial

Example 9.1 Calculation of element unit quantities for the proposed hotel development

Element	EUQ Calculation	EUQ
Demolition:	GEA of existing office building obtained from architect.	7,710m²
Substructure:	Note: GIFA of level –1 taken from area schedule.	786m²
Basement:	Note: No new basement needed to be constructed, but retaining wall required to south elevation only – abutting existing highway.	Not applicable
Basement walls:	South elevation — Height Datum (ground floor level) = 16.590m Datum (lower ground floor level) = 13.560m Height = 3.030m Area Perimeter: 28.330m Height: × 3.030m EUQ = 85.840m²	86m²
Frame:	Note: Total GIFA taken from area schedule.	8,399m²
Upper floors:	Note: GIFAs of floor levels taken from area schedule. Roof (area occupied by Lift Motor Room only, remainder is classified as Roof) 43m² Level 10 644m² Level 9 668m² Level 8 731m² Level 7 731m² Level 6 731m² Level 5 731m² Level 4 731m² Level 3 731m² Level 2 731m² Level 1 577m² Level 0 (ground floor) – (sits above lower ground floor) 564m² EUQ =	7,613m²
Roof:	As for Level 10	644m²
Stairs and ramps:	2 no. staircases – 11 flights; total rise 32.455m + 1 nr spiral staircase – 1 flight; total rise 6.075m. Datum 46.015m 13.560m 32.455m	23 nr
External walls:	Gross external wall area: (1) External walls – south elevation — Height Datum (roof level) = 48.887m Datum (ground floor level) = 16.590m Height = 32.297m	

Example 9.1 *(Continued)*

Element	EUQ Calculation			EUQ
		Area		
	Perimeter:	28.330m		
	Height:	× 35.297m		
	Area =		999.96m²	
	(2) External walls – All other elevations	Height		
	Datum (level 10 – roof level) =	48.887m		
	Datum (lower ground floor level) =	13.560m		
	Height =	35.327m		
		Perimeter		
	North elevation	29.600m		
	East elevation	44.050m		
	West elevation	42.850m		
	Perimeter:	116.500m		
	Height:	× 35.327m		
	Area =		4,115.60m²	
	(3) External walls – area adjustments to east and west elevations (i.e. for walkways/ balconies):			
	North elevation adjustment:	Height		
	Datum (roof level) =	48.887m		
	Datum (level 9) =	43.005m		
	Height =	5.882m		
	Less	Perimeter		
	East elevation	2.000m		
	West elevation	2.000m		
	Perimeter:	4.000m		
	Height:	× 5.882m		
	Area =		(23.53)m²	
	South elevation adjustment (1):	Height		
	Datum (roof level) =	48.887m		
	Datum (level 9) =	43.005m		
	Height =	5.882m		
	Less	Perimeter		
	East elevation	1.500m		
	West elevation	1.500m		
	Perimeter:	3.000m		
	Height:	× 5.882m		
	Area =		(17.65)m²	
	South elevation adjustment (2):	Height		
	Datum (roof level) =	48.887m		
	Datum (level 10) =	46.015m		
	Height =	2.872m		

Example 9.1 *(Continued)*

Element	EUQ Calculation	EUQ
	Less — Perimeter East elevation 1.500m West elevation 1.500m Perimeter: 3.000m Height: × 2.872m Area = (8.62)m² (4) Roof (Lift Motor Room): Height Datum (Lift Motor Room roof level) = 52.300m Datum (roof level) 48.887m Height = 3.413m Perimeter: 27.423m Height: × 3.413m Area = 93.59m² EUQ: 5,159.35m²	5,159m²
Windows and external doors:	Assumptions: (1) No allowance made for windows, as external walls to be constructed from glazed curtain walling system. (2) Area of external doors based on 2.50% of EUQ for external walls. 5,159.35m² × 3.50% EUQ: 180.58m²	181m²
Internal walls and partitions:	Based on GIFA Note: The internal layout and extent of internal walls and partitions have not been fully considered by the architect at this stage of design development. Consequently, the EUQ is based on the GIFA.	8,399m²
Internal doors:	Based on GIFA Note: Ditto internal walls and partitions.	8,399m²
Internal finishes:	Based on GIFA Note: Ditto internal walls and partitions.	8,399m²
Fittings, furnishings and equipment:	Based on GIFA Note: Ditto internal walls and partitions.	8,399m²
Services:	Based on GIFA Note: Ditto internal walls and partitions.	8,399m²
External works:	Note: No external works associated with hotel; offices only.	Not applicable

cost targets for each group element and element of the building works. To accomplish this, EUQs are simply inputted into the cost model, cost targets determined and element unit rates (EURs) ascertained.

Cost targets for each element are calculated by applying an appropriate percentage to each element derived from a range of cost analyses (e.g. the cost target for substructure equates to 6.10% of the total cost of building works); and EURs are calculated by dividing the cost target ascertained for the element by the EUQ for the element. This method of deriving initial cost targets and EURs for the purpose of distributing the budget (i.e. cost limit) amongst the group elements and elements is explained in Chapter 23 ('Deriving unit rates for building components, sub-elements and elements').

9.5 Presenting the initial elemental cost model

A standard form should be used to present the order of cost estimate. The form shown in Figure 9.2 is an example of how an initial elemental cost model should be set out based on the framework advocated by NRM 1. The make-up of the cost model will vary according to the work breakdown structure for the proposed building project, but the overarching framework should remain constant.

INITIAL ELEMENTAL COST MODEL
Prepared by: Bright Kewess Partnership
Project no. BKP-0693
Project: Woollard Hotel, Holborn Viaduct, St Leonards-on-Sea, East Sussex
Estimate base date: May 2013
Gross internal floor area (GIFA): 8,399m²

Code	Group element/element	EUQ		EUR	Element cost target	Cost/m² of GIFA
				£	£	£
1	**Building works:**					
1.0	Facilitating works					
1.0.1	Facilitating works	7,710	m²	15.00	115,650	13.77
1.0.2	Major demolition	7,710	m²	70.00	539,700	64.26
1.1	Substructure					
1.1.1	Substructure	786	m²	1,950.00	1,532,700	182.49
1.1.2	Basement	–	m²	–	–	–
1.1.3	Basement walls	86	m²	–	–	–
1.2	Superstructure					
1.2.1	Frame	8,399	m²	136.91	1,149,907	136.91
1.2.2	Upper floors	7,613	m²	102.00	776,526	92.45
1.2.3	Roof	644	m²	641.55	413,158	49.19
1.2.4	Stairs and ramps	23	nr	12,217.39	281,000	33.46

Figure 9.2 Initial elemental cost model – form of presentation

Code	Group element/element	EUQ		EUR	Element cost target	Cost/m² of GIFA
				£	£	£
1.2.5	External walls	5,159	m²	509.65	2,629,284	313.05
1.2.6	Windows and external doors	181	m²	472.50	85,523	10.18
1.2.7	Internal walls and partitions	8,399	m²	20.98	176,211	20.98
1.2.8	Internal doors	8,399	m²	39.25	329,661	39.25
1.3	Internal finishes					
1.3.1	Wall finishes	8,399	m²	241.32	2,026,847	241.32
1.3.2	Floor finishes	8,399	m²	49.43	415,163	49.43
1.3.3	Ceiling finishes	8,399	m²	19.81	166,384	19.81
1.4	Fittings, furnishings and equipment	8,399	m²	70.42	591,458	70.42
1.5	Services					
1.5.1	Sanitary installations	8,399	m²	3.06	25,701	3.06
1.5.2	Services equipment	8,399	m²	28.00	235,172	28.00
1.5.3	Disposal installations	8,399	m²	8.76	73,575	8.76
1.5.4	Water installations	8,399	m²	44.76	375,939	44.76
1.5.5	Heat source	8,399	m²	27.39	230,049	27.39
1.5.6	Space heating and air conditioning	8,399	m²	34.35	288,506	34.35
1.5.7	Ventilation	8,399	m²	55.60	466,984	55.60
1.5.8	Electrical installations	8,399	m²	59.39	498,817	59.39
1.5.9	Fuel installations	8,399	m²	53.29	447,583	53.29
1.5.10	Lift and conveyor installations	8,399	m²	40.79	342,595	40.79
1.5.11	Fire and lightning protection	8,399	m²	63.39	532,413	63.39
1.5.12	Communication, security and control systems	8,399	m²	31.17	261,797	31.17
1.5.13	Special installations	8,399	m²	17.86	150,006	17.86
1.5.14	Builder's work in connection with services	8,399	m²	13.06	109,691	13.06
1.6	Proprietary buildings and building units	8,399	m²	898.74	7,548,517	898.74
1.7	Works to existing buildings	0		0.00	0	0.00
1.8	External works					
1.8.1	Landscaping	0	m²	0.00	0	0.00
1.8.2	Mains connections	8,399	m²	36.00	302,364	36.00
1	Building works estimate				23,118,879	2,752.58
2	Main contractor's preliminaries	13.60%			3,144,168	374.35
	Subtotal				26,263,047	3,126.93

Figure 9.2 *(Continued)*

Code	Group element/element	EUQ	EUR	Element cost target	Cost/m² of GIFA
			£	£	£
3	Main contractor's overheads and profit	4.00%		1,052,522	125.32
	Works cost estimate			27,315,569	3,252.25
4	Project/design team fees estimate	11.50%		3,141,290	374.00
5	Other development/project costs estimate			0	0.00
	Base estimate			30,456,859	3,626.25
6	Risk allowance estimate				
6.1	Design development risks	1.50%	30,456,859	456,853	54.39
6.2	Construction risks	3.50%	30,456,859	1,065,990	126.92
6.3	Design development risks	0.00%	Excluded	0	0.00
6.4	Construction risks	0.00%	Excluded	0	0.00
	Cost limit (excluding inflation)			31,979,702	3,807.56
7	Inflation estimate				
7.1	Tender inflation	1.32%	31,979,702	422,132	50.26
7.2	Construction inflation	3.03%	32,401,834	981,756	116.89
	Cost limit (including inflation)			**33,383,590**	**3,974.71**
	Cost limit (including inflation) to 3 significant figures			**32,384,000**	**3,823.00**
8	VAT assessment			0	0.00

Figure 9.2 *(Continued)*

PART 3

Cost planning

CHAPTER 10

Cost planning – an overview

Introduction

This chapter:

- explains the concept of cost planning;
- points out the main objectives of cost planning;
- distinguishes between the different types of cost plan; and
- defines the information that is required by the cost manager to prepare cost plans during the pre-construction phases of a building project.

10.1 What is the purpose of cost planning?

Cost planning can be described as a system of procedures and techniques used by quantity surveyors and cost managers to manage costs in connection with a building project. The purpose of cost planning is:

> to ensure that [employers] are provided with value for money on their [building] projects; that [employers] and designers are aware of the cost [implications] of their proposals; that if they do so choose, [employers] may establish budgets for their [building] projects; and that designers are given advice which enables them to arrive at practical and balanced designs within budget.
>
> Property Services Agency (1981), *Cost Planning and Computers*, HMSO, London, UK

According to Seeley, cost planning aims at:

> ascertaining costs before many of the decisions are made relating to the design of a building. It provides a statement of the main issues, identifies the various courses of action, determines the cost implications of each course [of action] and provides a comprehensive economic picture of the whole [building project]. The [quantity surveyor/cost manager and designers] should be continually questioning whether a specific item of cost is really necessary, whether it is giving value for money or whether there is not a better way of performing a particular function [or achieving certain quality characteristics required by the employer].
>
> Seeley, I. H. (1996), *Building Economics*, 4th edition, McMillan, London, UK

Cost planning is an iterative process, which is performed in steps of increasing detail as more design information becomes available. It is essentially a budget distribution technique that is implemented during the design stages of a building project. It involves a critical breakdown of the cost limit (i.e. the employer's authorised budget) for the building project into cost targets for each element of the buildings that constitute the building project. Cost targets are the recommended expenditure for each element (e.g. substructure, frame, upper floors, roof, staircases and ramps, and external walls). The cost plan that results is a statement of how the project team proposes to distribute the available budget among the elements of the building. It provides a frame of reference from which to develop the design and maintain cost control. What is more, the cost plan provides both a work breakdown structure and a cost breakdown structure that, through codification, can be used to redistribute works in elements to construction works packages for the purpose of procurement and cost control during the construction phase of the building project.

The main objectives of cost planning are to:

- ensure that employers are provided with value for money;
- make employers and designers aware of the cost consequences of their desires and/ or proposals;
- provide advice to designers that enables them to arrive at practical and balanced designs within budget;
- integrate costs with time and performance (quality);
- keep expenditure within the cost limit approved by the employer; and
- provide robust cost information upon which the employer can make informed decisions.

The cost plan should also be used as a tool to promote effective communication between the project's team members and to ensure that risks associated with the building project are identified, analysed and responded to.

10.2 Cost plans

Three formal cost plans are identified by NRM 1:

- Cost plan 1;
- Cost plan 2: and
- Cost plan 3.

The time at which formal cost plans are produced is correlated with both the RIBA Work Stages and OGC Gateways. For the purpose of comparing the OGC Gateway Process with the RIBA Work Stages, RIBA has divided OGC Gateway 3 (Investment Decision) into three discrete gateways. These are:

- OGC Gateway 3A (Design Brief and Concept Approval), which can be compared with RIBA Work Stages 1 and 2;
- OGC Gateway 3B (Detailed Design Approval), which can be compared with RIBA Work Stages 3 and 4; and
- OGC Gateway 3C representing the final OGC Gateway Review 3 (Investment Decision), which can be compared with RIBA Work Stages 4 and 5.

Whether or not a formal cost plan is prepared at each RIBA Work Stage or OGC Gateway will be dependent on the procurement and contract strategy selected (see Chapter 2, paragraph 2.4).

The cost targets within each formal cost plan approved by the employer will be used as the baseline for future cost comparisons. Each subsequent cost plan will require reconciliation with the preceding cost plan and explanations relating to changes made. In view of this, it is essential that records of any transfers made to or from the risk allowances and any adjustments made to cost targets are maintained, so that explanations concerning changes can be provided to both the employer and the project team.

10.2.1 Cost plan 1 – based on concept design

Cost plan 1 is the first formal cost plan and is produced as an intrinsic part of RIBA Work Stage 2 (Concept Design) or OGC Gateway 3A (Design Brief and Concept Approval), whichever process model is applicable. It is concerned with isolating the best means of satisfying the requirements of the brief and is prepared at a point where the scope of work is fully defined and key criteria are specified but no detailed design has commenced.

The requirements of RIBA Work Stage 2, as described in the *RIBA Outline Plan of Work*, are as follows:

RIBA Work Stage 2: Concept Design:

Prepare Concept Design, including outline proposals for structural design, building services systems, outline specifications and preliminary Cost Information along with relevant Project Strategies in accordance with Design Programme. Agree alterations to brief and issue Final Project Brief.

NRM 1 recognises that cost planning is an iterative process, and acknowledges that limited information will be available to the cost manager when preparing cost plan 1. Should there be insufficient information available on which to quantify certain aspects of building works, the rules stipulate that quantification is to be based on the gross internal floor area. Where information is not available, the cost manager will need to record any assumptions made when compiling the cost plan.

Cost plan 1 provides the frame of reference for cost plan 2. However, if cost plan 1 is to be used as a basis for procuring all or any part of the building project, then it will become the frame of reference for appraising tenders.

10.2.2 Cost plan 2 – based on developed design

Cost plan 2 is concerned with checking that the design is effective, confirming that costs based on detailed design are within the cost targets and confirming that the cost limit is achievable. Cost plan 2 is produced as an integral part of RIBA Work Stage 3 (Developed Design) or OGC Gateway 3B (Detailed Design Approval), whichever process model is applicable.

The requirements of RIBA Work Stage 3, as described in the *RIBA Outline Plan of Work*, are as follows:

RIBA Work Stage 3: Developed Design:

Prepare Developed Design, including coordinated and updated proposals for structural design, building services systems, outline specifications, Cost Information and Project Strategies in accordance with Design Programme.

Note:
Planning applications are typically made using the Stage 3 output. A bespoke *RIBA Plan of Work 2013* will identify when the planning application is to be made.

> RIBA Work Stage 4: Technical Design:
>
> Prepare Technical Design in accordance with Design Responsibility Matrix and Project Strategies to include all architectural, structural and building services information, specialist subcontractor design and specifications, in accordance with Design Programme.

This is the second formal cost plan, which corresponds with the completion of the design development. Cost plan 2 is a progression of cost plan 1. It is developed by cost checking cost-significant items as more detailed design information is made available. As part of this process, elements previously quantified using the gross internal floor area should be revisited by the cost manager. The aim should be, wherever possible, to measure the cost-significant components which make up the elements.

Cost plan 2 will provide the frame of reference for cost plan 3. However, if cost plan 2 is to be used as a basis for procuring all or any part of the building project, it will become the frame of reference for appraising any tenders in the same way as cost plan 1.

10.2.3 Cost plan 3 – based on technical design

Cost plan 3 is concerned with ensuring that the cost of the building project, based on the finished documentation, is within cost limit by continual cost checks throughout the documentation process (sometimes referred to as a pre-tender cost plan). It is produced as part of RIBA Work Stage 4 (Technical Design) or OGC Gateway 3B (Detailed Design Approval), whichever process model is applicable.

The requirements of RIBA Work Stage 4, as described in the *RIBA Outline Plan of Work*, are as follows:

> RIBA Work Stage 4: Technical Design:
>
> Prepare Technical Design in accordance with Design Responsibility Matrix and Project Strategies to include all architectural, structural and building services information, specialist subcontractor design and specifications, in accordance with Design Programme.

This third formal cost plan stage is based on technical designs, specifications and detailed information for construction. Cost plan 3 is a progression of cost plan 2. In the same way as cost plan 2, it is developed by cost checking of cost-significant cost targets for elements as more detailed design information is made available from the design team. Cost plan 3 will provide the frame of reference for appraising tenders.

10.3 Information requirements

To enable a meaningful cost plan to be prepared, on which decisions can be made, the cost manager will require certain information from the employer and the designers about the proposed building project. Therefore, to assist the cost manager, NRM 1 sets out the information that is typically required from the employer and the designers at each stage of the cost-planning process (see Tables 10.1, 10.2 and 10.3 – Schedule of information required for the preparation of formal cost plan 1, 2 and 3, respectively).

In practice, the cost manager would have done exceptionally well to obtain all the information stipulated by the rules. However, without this information, the cost manager will undoubtedly need to caveat any cost plan report that he/she produces. Failure to obtain the information will also affect the accuracy of the cost plan.

Table 10.1 Cost plan 1 – schedule of information required by cost manager at RIBA Work Stage 2 (Concept Design)

Formal cost plan 1				
Employer	***Architect***	***Structural engineer***	***M&E services engineer***	***Specialist consultants***
1. Confirmation of the cost limit (i.e. the authorised budget). Where alternative cost options were reported to the employer, confirmation of the preferred design/ development option and cost limit is required. 2. Confirmation of the project/design brief, including statement of quality and 'fit-out' requirements. 3. Confirmation of programme, including timetable of critical events (including timetable for design, construction start date, construction time, construction completion date and required occupation dates).	1. Concept design drawings to a suitable scale, comprising: • general arrangement plans (for all floors, including basement levels, and roofs); • general elevations (with materials clearly annotated); • general sections; • external landscaping – general arrangement plan(s); • plans of key building functions; • detailed elevations showing construction of external walls, roofs, ground floor; • construction and upper floor construction; • sketches showing key details/interfaces (e.g.	1. Reports based on desktop studies, including: • environmental contamination (Phase 1 audit – i.e. to establish the nature of any subsurface contaminated soil and/or ground water); • geotechnical properties; and • bombs. 2. Reports based on fieldwork, sampling and analysis (where commissioned by the employer), including: • environmental contamination (Phase 2 audit): and	1. Concept design drawings to a suitable scale, comprising: • general arrangement for each main system; • schematic diagrams for each major system; • plant room layouts, including roof plant layout; • single line diagrams showing primary service routes; and • typical layouts of landlord's areas, service areas and cores. 2. Outline specification information, including: • mechanical services; • electrical services; • transportation systems (e.g.	

Table 10.1 *(Continued)*

Formal cost plan 1				
Employer	**Architect**	**Structural engineer**	**M&E services engineer**	**Specialist consultants**
4. Confirmation of requirements in respect of: • procurement strategy, including phasing of construction works, temporary access requirements and the like; • contract strategy; • phasing, including requirements relating to decanting, temporary access and the like; • facilitating works, including demolition, preparatory site works, and early infrastructure works (e.g. mains services connections and roadworks); • treatment of project/ design team fees; • insurances; • approach to dealing with other development/project	interface between curtain walling system and structure, balconies and the like); • concept design for rooms and common areas; and • site constraints plan. 2. Schedule of gross external areas (GEA), gross internal floor areas (GIFA), net internal areas (NIA – i.e. usable area for shops, supermarkets and offices) and site area (SA). 3. Outline specification information, including: • specification/design intent for all main elements; • statement of required quality; • outline specification for components, materials and finishes; • acoustics/vibration requirements; • outline performance criteria for main element;	• geotechnical properties. Note: Fieldwork comprises trial pits, auger holes, window samplers, boreholes, probing and the like. 3. Environmental risk assessment. 4. Advice on ground conditions. 5. Concept design drawings to a suitable scale, comprising: • general arrangement; • frame configuration; • layout of shear walls, core walls, columns and beams; • sections; • foundation layouts, including pile (and pile cap and ground beam) layouts; • sections, showing ground slab construction, basement wall construction, pile	lifts, hoists and escalators); • protective installations; • communication, security and control systems; • special installations; • plant/equipment schedule (for primary plant/equipment); • approximate duties, output, and sizes of primary plant/ equipment; • schedule of cost-significant builder's work in connection with mechanical and electrical engineering services installations/ systems; and • details of alternative specifications. 3. Strategies, including: • environmental/ sustainability (in conjunction with the architect), including:	

costs (e.g. Section 106; • contributions, party wall works and decanting costs; • planning and Section 106 and 278 requirements required to be incorporated in the building design and/or building works contract(s); • treatment of employer's risks; • treatment of inflation; • treatment of value added tax (VAT); and • other considerations (e.g. approach to dealing with capital allowances, land remediation allowances and grants). 5. Post-completion requirements. 6. Authority to commence the next RIBA Work Stage or proceed to the next OGC VGateway.	• schedule of finishes; and • details of alternative specifications. 4. Room data sheets. 5. Schedules of key fittings, furnishings and equipment. 6. Strategies, including: • environmental/ sustainability (in conjunction with the mechanical and electrical services engineer), including: – measures to achieve BREEAM or Code for Sustainable Homes; – Building Regulations requirements; – sustainability requirements and assumptions; – renewable energy requirements and assumptions; – employer's specific requirements. • car parking, including motorcycles and bicycles;	caps construction and the like; and • indicative drainage solution. 6. Formation and excavation levels. 7. Outline specification information, including: • specification/design intent for all main elements; • outline specification for components and materials; • structural performance criteria (e.g. design loadings); • pile sizes, including indicative lengths; • statement on strategy for integration of mechanical and electrical engineering; • services with structural components; and	– measures to achieve BREEAM or Code for Sustainable Homes; – Building Regulations requirements; – sustainability requirements and assumptions; – renewable energy requirements and assumptions; – employer's specific requirements. • vertical movement (in conjunction with the architect); and • removal/ decommissioning of existing plant and or equipment. 4. Reports, including: • survey of underground services. 5. Identification of requirements for any abnormal mechanical and electrical engineering services installations/systems.	

Table 10.1 *(Continued)*

Formal cost plan 1				
Employer	***Architect***	***Structural engineer***	***M&E services engineer***	***Specialist consultants***
	• vertical movement (in conjunction with the mechanical and electrical services engineer); • information technology (IT); • fire; • acoustics; • security; • DDA (Disability Discrimination Act); • window cleaning; • refuse/waste disposal; • public art; • conservation/listed buildings and the like (if applicable); and • other important aspects of the building project. 7. Reports, including: • archaeological assessment/report (desktop study); • measured survey (i.e. topographical survey). 8. Phasing and outline construction methodology. 9. Definition of 'fit-out'. 10. Risk register/log.	• details of alternative specifications. 8. Estimates of reinforcement content for all reinforced concrete components. 9. Mass of steelwork in steel-framed structures. 10. Methodologies for: • facilitating works, including demolition, preparatory site works, and • early infrastructure works (e.g. roadworks and drainage); • temporary works; • alterations; and • drainage (indicative solution). 11. Risk register/log.	6. Details of utilities services connections, including: • connections; • upgrading requirements; and • diversions. 7. Methodology for facilitating works (i.e. early provision of mains services to site). 8. Risk register/log.	

Table 10.2 Cost plan 2 – schedule of information required by cost manager at RIBA Work Stage 3 (Developed Design)

Formal cost plan 2				
Employer	**Architect**	**Structural engineer**	**M&E services engineer**	**Specialist consultants**
1. Confirmation that Formal Cost Plan 1, prepared at RIBA Work Stage C: Concept, or OGC Gateway 3A: Design Brief and Concept Approval, is acceptable. 2. Confirmation of any preferred alternatives given in cost report for Cost Plan 1. 3. Confirmation of the project/design brief, including statement of quality and 'fit-out' requirements. 4. Confirmation of programme, including timetable of critical events (including timetable for design, construction start date, construction time, construction completion date and required occupation dates). 5. Confirmation of requirements in respect of:	1. Detailed design drawings to a suitable scale, comprising: • general arrangement plans (for all floors, including basement levels, and roofs); • general elevations (with materials clearly annotated); • general sections; • external landscaping – general arrangement plan(s); • plans of key building functions; • detailed elevations; • detailed sections, showing construction of external walls, roofs, ground floor construction and upper floor construction; • drawings showing key details/interfaces (e.g. interface between curtain walling system and structure, balconies and the like); and	1. Reports based on fieldwork, sampling and analysis (where commissioned by the employer), including: • environmental contamination (Phase 2 audit). • geotechnical properties. Note: Fieldwork comprises trial pits, auger holes, window samplers, boreholes, probing and the like. 2. Environmental risk assessment. 3. Updated advice on ground conditions. 4. Detailed design drawings to a suitable scale, comprising: • general arrangement; • layout of shear walls, core walls, columns and beams; • frame configuration.	1. Detailed design drawings to a suitable scale, comprising: • general arrangement for each main system; • schematic diagrams for each major system; • plant room layouts, including roof plant layout; • single line diagrams showing primary service routes; and • typical layouts of landlord's areas, service areas and cores. 2. Updated outline specification information, including: • mechanical services; • electrical services; • transportation systems (e.g. lifts, hoists and escalators); • protective installations;	1. Design development drawings. 2. Outline specification information. 3. Reports.

Table 10.2 *(Continued)*

Formal cost plan 2				
Employer	**Architect**	**Structural engineer**	**M&E services engineer**	**Specialist consultants**
• procurement strategy, including phasing of construction works, temporary access requirements and the like; • contract strategy; • phasing, including requirements relating to decanting, temporary access and the like; • facilitating works, including demolition, preparatory site works, and early infrastructure works (e.g. mains services connections and roadworks); • treatment of project/ design team fees; • insurances; • approach to dealing with other development/project costs (e.g. Section 106 contributions,	• detailed floor plans, showing the layout of rooms and common areas. 2. Updated site constraints plan. 3. Schedule of gross external areas (GEA), gross internal floor areas (GIFA), net internal areas (NIA – i.e. usable area for shops, supermarkets and offices) and site area (SA). 4. Updated outline specification information, including: • specification/design intent for all main elements; • statement of required quality; • outline specification for components, materials and finishes; • acoustics/vibration requirements; • outline performance criteria for main element;	• foundation layouts, including pile (and pile cap and ground beam) layouts; • sections, showing ground slab construction, basement wall construction, pile caps construction and the like; and • indicative drainage solution. 5. Formation and excavation levels. 6. Updated outline specification information, including: • specification/design intent for all main elements; • outline specification for components and materials; • structural performance criteria (e.g. design loadings);	• communication, security and control systems; • specialist installations; • plant/equipment schedule (for primary plant/equipment); • approximate duties, output, and sizes of primary plant/ equipment; • schedule of cost-significant builder's work in connection with mechanical and electrical engineering services installations/ systems. 3. Updated strategies, including: • environmental/ sustainability (in conjunction with the architect), including: – measures to achieve required environmental rating (e.g. via BREEAM, Code for	

party wall works and decanting costs); • planning and Section 106 and 278 requirements required to be incorporated in the building design and/or building works contract(s); • treatment of employer's risks; • treatment of inflation; • treatment of value added tax (VAT); and • other considerations (e.g. approach to dealing with capital allowances, land remediation allowances and grants). 6. Post-completion requirements. 7. Acceptance of any other matters within the cost report for Formal Cost Plan 1. 8. Authority to commence the next RIBA Work Stage or proceed to the next OGC Gateway.	• schedule of finishes; and • details of alternative specifications. 5. Updated room data sheets. 6. Updated schedules of key fittings, furnishings and equipment. 7. Updated strategies, including: • environmental/ sustainability (in conjunction with the mechanical and electrical services engineer and environmental consultant, if appointed), including: – measures to achieve required environmental rating (e.g. via BREEAM, Code for Sustainable Homes or other environmental assessment method); – Building Regulations requirements; – sustainability requirements and assumptions; – renewable energy requirements and assumptions;	• pile sizes, including indicative lengths; • statement on strategy for integration of mechanical and electrical engineering services with structural components; • details of alternative specifications. 7. Updated estimates of reinforcement content for all reinforced concrete components. 8. Mass of steelwork in steel-framed structures. 9. Updated methodologies for: • facilitating works, including demolition, preparatory site works, and early infrastructure works (e.g. roadworks and drainage); • temporary works; • alterations; and • drainage (indicative solution).	Sustainable Homes or other environmental assessment method); – Building Regulations requirements; – sustainability requirements and assumptions; – renewable energy requirements and assumptions; – employer's specific requirements. • vertical movement (in conjunction with the architect); and • removal/ decommissioning of existing plant and or equipment. 4. Identification of requirements for any abnormal mechanical and electrical engineering services installations/systems. 5. Details of utilities services connections, including: • connections; • upgrading requirements;	

Table 10.2 *(Continued)*

Formal cost plan 2				
Employer	**Architect**	**Structural engineer**	**M&E services engineer**	**Specialist consultants**
	– employer's specific requirements. • car parking, including motorcycles and bicycles; • vertical movement (in conjunction with the mechanical and electrical services engineer); • information technology (IT); • fire; • acoustics; • security; • DDA (Disability Discrimination Act); • facade access and window cleaning; • refuse/waste disposal; • public art; • conservation/listed buildings and the like (if applicable); and • other important aspects of the building project. 8. Updated reports, including: • archaeological assessment/report (desktop study); and • measured survey (i.e. topographical survey). 9. Updated phasing and outline construction methodology. 10. Updated risk register/log.	10. Risk register/log.	• diversions; and • quotations from statutory undertakers. 6. Updated methodology for facilitating works (i.e. early provision of mains services to site). 7. Updated risk register/log.	

Table 10.3 Cost plan 3 – schedule of information required by cost manager at RIBA Work Stage 4 (Technical Design)

Formal cost plan 3				
Employer	***Architect***	***Structural engineer***	***M&E services engineer***	***Specialist consultants***
1. Confirmation that Formal Cost Plan 2, prepared at RIBA Work Stage D: Design Development, or OGC Gateway 3B: Detailed Design Approval, is acceptable. 2. Confirmation of any preferred alternatives given in cost report for Cost Plan 2. 3. Confirmation of the project/design brief, including statement of quality and 'fit-out' requirements. 4. Confirmation of programme, including timetable of critical events (including timetable for design, construction start date, construction time, construction completion date and required occupation dates). 5. Confirmation of requirements in respect of:	1. Final design drawings to a suitable scale, including: • final plans/layouts; • elevations; • sections; • location drawings; • assembly drawings; and • component drawings. 2. Updated site constraints plan. 3. Schedule of gross external areas (GEA), gross internal floor areas (GIFA), net internal areas (NIA – i.e. usable area for shops, supermarkets and offices) and site area (SA). 4. Final specification information, including: • specification/design for all main elements; • statement of required quality; • final specification for components, materials and finishes; • acoustics/vibration requirements;	1. Reports based on fieldwork, sampling and analysis (where commissioned by the employer), including: • environmental contamination (Phase 2 audit); and • geotechnical properties. Note: Fieldwork comprises trial pits, auger holes, window samplers, boreholes, probing and the like. 2. Updated environmental risk assessment. 3. Updated advice on ground conditions. 4. Final design drawings to a suitable scale, including: • general arrangement; • layout of shear walls, core walls, columns and beams; • frame configuration;	1. Detailed design drawings to a suitable scale. 2. Final specification information, including: • mechanical services; • electrical services; • transportation systems (e.g. lifts, hoists and escalators); • protective installations; • communication, security and control systems; • specialist installations; • plant/equipment schedule (for primary plant/equipment); • duties, output, and sizes of primary plant/equipment; and • schedule of cost-significant builder's work in connection with mechanical and electrical engineering	1. Final design drawings to a suitable scale. 2. Final specification information. 3. Final reports.

Table 10.3 Cost plan 3 – schedule of information required by cost manager at RIBA Work Stage 4 (Technical Design)

Formal cost plan 3				
Employer	***Architect***	***Structural engineer***	***M&E services engineer***	***Specialist consultants***
• procurement strategy, including phasing of construction works, temporary access requirements and the like; • contract strategy; • phasing, including requirements relating to decanting, temporary access and the like; • facilitating works, including demolition, preparatory site works, and early infrastructure works (e.g. mains services connections and roadworks); • treatment of project/ design team fees; • insurances; • approach to dealing with other development/project costs (e.g. planning contributions, party wall works and decanting costs); • planning and Section 106 and 278	• final performance criteria for main element; • schedule of finishes; and • details of alternative specifications. 5. Final room data sheets. 6. Final schedules of key fittings, furnishings and equipment. 7. Final strategies, including: • environmental/ sustainability (in conjunction with the mechanical and electrical services engineer and, if appointed, the environmental consultant), including: – measures to achieve required environmental rating (e.g. via BREEAM, Code for Sustainable Homes or other environmental assessment method) – Building Regulations requirements; – sustainability requirements and assumptions;	• foundation layouts, including pile (and pile cap and ground beam) layouts; • sections, showing ground slab construction, basement wall construction, pile caps construction and the like; and • final drainage solution. 5. Formation and excavation levels. 6. Final specification information, including: • specification/ design for all main elements; • final specification for components and materials; • structural performance criteria (e.g. design loadings); • pile sizes, including indicative lengths; and	services installations/ systems. 3. Final strategies, including: • environmental/ sustainability (in conjunction with the structural engineer, mechanical and electrical services engineer and, if appointed, the environmental consultant), including: – measures to achieve required environmental rating (e.g. via BREEAM, Code for Sustainable Homes or other environmental assessment method) – Building Regulations requirements; – sustainability requirements and assumptions; – renewable energy requirements and assumptions;	

requirements required to be incorporated in the building design and/or building works contract(s); • treatment of employer's risks; • treatment of inflation; • treatment of value added tax (VAT); and • other considerations (e.g. approach to dealing with capital allowances, land remediation allowances and grants). 6. Post-completion requirements. 7. Acceptance of any other matters within the cost report for Cost Plan 2. 8. Authority to commence the next RIBA Work Stage or proceed to the next OGC Gateway.	– renewable energy requirements and assumptions; – employer's specific requirements. • car parking, including motorcycles and bicycles; • vertical movement (in conjunction with the mechanical and electrical services engineer); • information technology (IT); • fire; • acoustics; • security; • DDA (Disability Discrimination Act); • facade access and window cleaning; • refuse/waste disposal; • public art; • conservation/listed buildings and the like (if applicable); and • other important aspects of the building project. 8. Updated reports, including: • archaeological assessment/report (desktop study); and • measured survey (i.e. topographical survey). 9. Final phasing and outline construction methodology. 10. Updated risk register/log.	• statement on strategy for integration of mechanical and electrical engineering services with structural components. 7. Updated estimates of reinforcement content for all reinforced concrete components. 8. Mass of steelwork in stee-framed structures. 9. Final methodologies for: • facilitating works, including demolition, preparatory site works, and early infrastructure works (e.g. roadworks and drainage); • temporary works; • alterations; and • drainage. 10. Updated risk register/log.	– employer's specific requirements. • vertical movement (in conjunction with the architect); and • removal/decommissioning of existing plant and or equipment. 4. Details of utilities services connections, including: • connections; • upgrading requirements; • diversions; and • quotation from statutory undertakers. 5. Final methodology for facilitating works (i.e. early provision of mains services to site). 6. Updated risk register/log.	

CHAPTER 11

Format, structure and codification of cost plans

Introduction

This chapter:

- describes the hierarchical structure of elements and components used in cost planning;
- conveys the framework for cost plans;
- explains how to codify elemental cost plans;
- contains step-by-step guidance on how to develop codes for elements and components; and
- describes how cost plans can be restructured for the purpose of cost control during the construction phase of a building project.

11.1 Hierarchy of elements and components

NRM 1 provides a hierarchy of standard elements and cost-significant components. The purpose of this hierarchical structure is to subdivide the composition of a building into logical and manageable elements for which cost targets can be estimated. This hierarchical structure can be viewed as both a work breakdown structure and a cost breakdown structure.

The hierarchical structure consists of four primary levels, namely:

- Group elements
- Elements;
- Sub-elements; and
- Components.

Group elements are used to describe the main facets of a building project. The rules divide a building project into the following group elements. Together, the combined cost of all these group elements equals the 'total' capital cost of a building project.

0: Facilitating works;
1: Substructure;
2: Superstructure;

3: Internal finishes;
4: Fittings, furnishings and equipment;
5: Services;
6: Prefabricated buildings and building units;
7: Work to existing buildings;
8: External works;
9: Main contractor's preliminaries;
10: Main contractor's overheads and profit;
11: Project and design team fees;
12: Other development and project costs;
13: Risks; and
14: Inflation.

Elements are a subdivision of a group element (e.g. elements that form group element 2: Superstructure comprises frame, upper floors, roof, stairs and ramps, external walls, windows and external doors, internal walls and partitions, and internal doors).

Sub-elements are a subdivision of an element (e.g. sub-elements that create element 2.1: Frames are steel frames, space frames/decks, concrete casings to steel frames, concrete frames, timber frames and specialist frames).

Components are cost-significant items which form part of a sub-element. The quantity of more than one component will be measured and the cost estimated to ascertain the cost target for a sub-element.

11.2 Cost plan framework

The overarching framework for a cost plan (Figure 11.1) is simply an expansion of the order of cost estimate framework (see Chapter 4, paragraph 4.2). In fact, the only difference is the subdivision of the building works estimate into group elements 0 to 8 (see paragraph 11.1).

This framework represents a financial breakdown of the building project into cost

Constituent
0: Facilitating works
1: Substructure
2: Superstructure
3: Internal finishes
4: Fittings, furnishings and equipment
5: Services
6: Prefabricated buildings and building units
7: Work to existing buildings
8: External works
Building works estimate [(1)]
9: Main contractor's preliminaries estimate [(2)]

Figure 11.1 Cost plan framework

Subtotal (3) [(3) = (1) + (2)]
10: Main contractor's overheads and profit estimate (4)
Works cost estimate (5) [(5) = (3) + (4)]
11: **Project/design team fees estimate** (if required) (6)
Subtotal (7) [(7) = (5) + (6)]
12: **Other development/project costs estimate** (if required) (8)
Base cost estimate (9) [(9) = (7) + (8)]
13: **Risk allowances estimate** (10) [(10) = (10(a)) + (10(b)) + (10(c))+ (10(d))]
(a) Design development risks estimate (10(a))
(b) Construction risks estimate (10(b))
(c) Employer change risks estimate (10(c))
(d) Employer other risks estimate (10(d))
Cost limit (excluding inflation) (11) [(11) = (9) + (10)]
14A: **Tender inflation estimate** (12)
Cost limit (excluding construction inflation) (13) [(13) = (11) + (12)]
14B: **Construction inflation estimate** (14)
Cost limit (including inflation) (15) [(15) = (13) + (14)]
VAT assessment

Figure 11.1 *(Continued)*

targets (i.e. budgets) for each group element – i.e. a cost breakdown structure (CBS). This facilitates top-down cost targets and bottom-up estimating (see Figure 11.2).

Through cost planning, cost targets can be established for group elements, elements and sub-elements. The cost breakdown structure for group element 1 (substructure) is shown in Figure 11.3.

Box 11.1 Key definitions

- **Component** – a measured item which forms part of a sub-element.
- **Cost breakdown structure (CBS)** – in the context of estimating and cost planning, represents the financial breakdown of a building project into cost targets for elements or work packages.
- **Cost limit** – the maximum expenditure that the employer is prepared to make in relation to the completed building. Sometimes referred to as authorised budget or approved estimate.
- **Cost target** – the recommended total expenditure for an element. The cost target for each group element is likely to be derived from a number of elements, sub-elements and components.
- **Element** – a subdivision of a group element.
- **Group element** – used to describe the main components of a building project.
- **Sub-element** – a subdivision of an element.
- **Work breakdown structure (WBS)** – in the context of estimating and cost planning, used to subdivide a building project into meaningful elements or work packages.

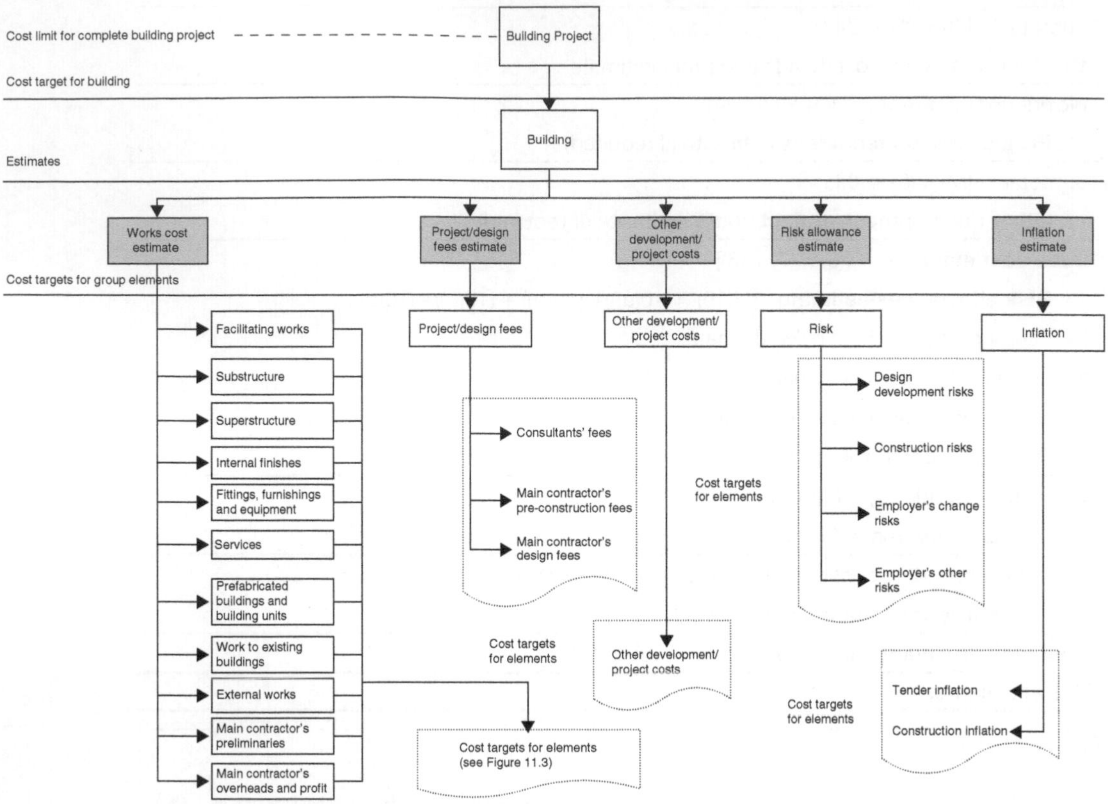

Cost limit for complete building project
Building Project
Cost target for building
Building
Estimates
Works cost estimate
Project/design fees estimate
Other development/ project costs
Risk allowance estimate
Inflation estimate
Cost targets for group elements
Facilitating works
Substructure
Superstructure
Internal finishes
Fittings, furnishings and equipment
Services
Prefabricated buildings and building units
Work to existing buildings
External works
Main contractor's preliminaries
Main contractor's overheads and profit
Project/design fees
Consultants' fees
Main contractor's pre-construction fees
Main contractor's design fees
Cost targets for elements
Cost targets for elements (see Figure 11.3)
Other development/ project costs
Cost targets for elements
Other development/ project costs
Risk
Design development risks
Construction risks
Employer's change risks
Employer's other risks
Cost targets for elements
Inflation
Tender inflation
Construction inflation

Figure 11.2 Cost breakdown structure for a building project

Note: WBS is replicated for each building or principle facet of the building project. Refer to paragraph 11.3.2

Cost target for group element
Substructure
Cost targets for elements
Substructure
Cost targets for sub-elements
Standard foundations
Specialist foundations
Lowest floor construction
Basement excavation
Basement retaining walls
Composition of cost targets for sub-elements
Components
Components
Components
Components
Components

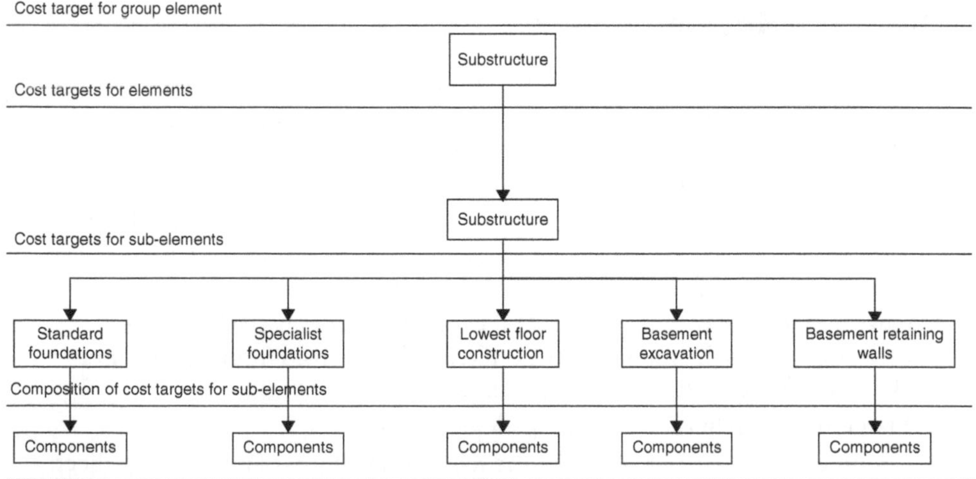

Figure 11.3 Typical cost breakdown structure for group element – substructure

11.3 Codification of cost plans

One of the beneficial features of elemental cost planning is the ability to uniquely identify by a code all group elements, elements, sub-elements and components in a numerical and logical manner. With a unique code, all components can be linked to the sub-elements, sub-elements to elements, elements to group elements, and group elements to the cost limit (i.e. the total estimated cost of the building project). This makes it easier to retrieve, manage and restructure information (costs and building works items).

The coding system advocated by NRM 1 is numeric. However, alphanumeric codes can also be used (letters and numbers).

11.3.1 The coding system

Under the rules, the group elements, elements and sub-elements are uniquely identified. For example, the reference for piled foundations is: 1.1.2. The first number represents the group element (Substructures), the second number represents the element (Foundations), and the third number represents the sub-element (Piled foundations). These are predefined codes. Although components have been given numeric references, they cannot be used as part of the code. This is because of the sheer number of variable components that could be generated for any one sub-element. Consider the following components:

- Component 1:
  Pile caps; excavate pit, partial backfill, partial disposal, earthwork support, compact base of pit; reinforced concrete 35N/mm^2 – 20mm aggregate reinforced concrete (reinforcement rate 100kg/m^3), formwork; pile cap size 2.10m × 2.10m × 1.25m deep.
- Component 2:
  Pile caps; excavate pit, partial backfill, partial disposal, earthwork support, compact base of pit; reinforced concrete 35N/mm^2 – 20mm aggregate reinforced concrete (reinforcement rate 100kg/m^3), formwork; pile cap size 4.00m × 2.10m × 1.25m deep.

As can be seen, the only difference between the two measured components is the size of the pile cap. Consequently, using the numeric component references given in the rules would result in both components having the same codes. Because of this, the rules advocate that the number used to differentiate between similar components be user defined. One method is to continuously and sequentially number the components which form a sub-element. This approach is illustrated in Example 11.1 and is used in the examples throughout this book. The reference for the first pile cap is 1.1.2.1 and the second pile cap is 1.1.2.2; with both being unique identification numbers.

The method described for codifying components is also flexible enough to allow other cost-significant components to be codified (i.e. cost-significant components unspecified by the rules).

For practical purposes, five to six levels of code should be sufficient to achieve the desired level of cost control. The main levels to be considered are as follows:

Level 0: **Project number** – most building projects will be given a project number, together with a project title or name, to distinguish them from all other projects the company might be working on.

Example 11.1 Codification of cost plans

SUBSTRUCTURE					
Code	**Element/Sub-elements/Components**	**Qty**	**Unit**	**Rate**	**Total**
				£p	**£**
1	**SUBSTRUCTURE**				
1.1	**Substructure**				
1.1.2	*Specialist foundations*				
1.1.2.1	Pile caps; excavate pit, partial backfill, partial disposal, earthwork support, compact base of pit; reinforced concrete 35N/mm² – 20mm aggregate reinforced concrete (reinforcement rate 100kg/m³), formwork; pile cap size 2.10m × 2.10m × 1.25m deep	26	nr		
1.1.2.2	Pile caps; excavate pit, partial backfill, partial disposal, earthwork support, compact base of pit; reinforced concrete 35N/mm² – 20mm aggregate reinforced concrete (reinforcement rate 100kg/m³), formwork; pile cap size 4.00m × 2.10m × 1.25m deep	3	nr		
1.1.2	**TOTAL – Specialist foundations: To element 1.1 summary**				

Level 1: **Cost plan number** – where a building project comprises more than one building or facet, a discrete cost plan will most likely be prepared for each building and key facet, culminating in a 'summary cost plan' (see paragraph 11.3.2). Therefore, an identification number will be required to distinguish cost plans. This code will not be required for a single cost plan.

Level 2 **Group element** – identification number predefined by the rules.

Level 3 **Element** – identification number predefined by the rules.

Level 4 **Sub-element** – identification number predefined by the rules.

Level 5 **Component** – user defined.

Further levels of code can be added as necessary to suit user requirements. For example, a user-defined level 6 code can be introduced if there is a need to further break down components into sub-components. The way in which a code can be expanded to include a level 6 code for sub-components is shown in Example 11.2.

11.3.2 Dealing with building projects that comprise multiple buildings

Where a building project comprises more than one building or facet, it is advisable that a separate cost plan be prepared for each building and key facet, culminating in a 'summary cost plan' for the entire building project. Therefore, before embarking on the cost plan, the composition of the building project needs to be determined and

Example 11.2 Additional code level (level 6)

Level	Description	Item	Identification no.	Resultant code
Level 1	Cost plan number	Cost plan no. 3	3	
Level 2	Group element	Substructure	1	
Level 3	Element	Substructure	1	
Level 4	Sub-element	Specialist foundations	2	
Level 5	Component	Pile cap (Type PC1)	1	
Level 6	Sub-component	Excavation	1	3.1.1.2.1.1
Level 6	Sub-component	Disposal	2	3.1.1.2.1.2
Level 6	Sub-component	Earthwork support	3	3.1.1.2.1.3
Level 6	Sub-component	Concrete	4	3.1.1.2.1.4
Level 6	Sub-component	Reinforcement	5	3.1.1.2.1.5
Level 6	Sub-component	Formwork	6	3.1.1.2.1.6

Note:
The prefix to the code will be the project number

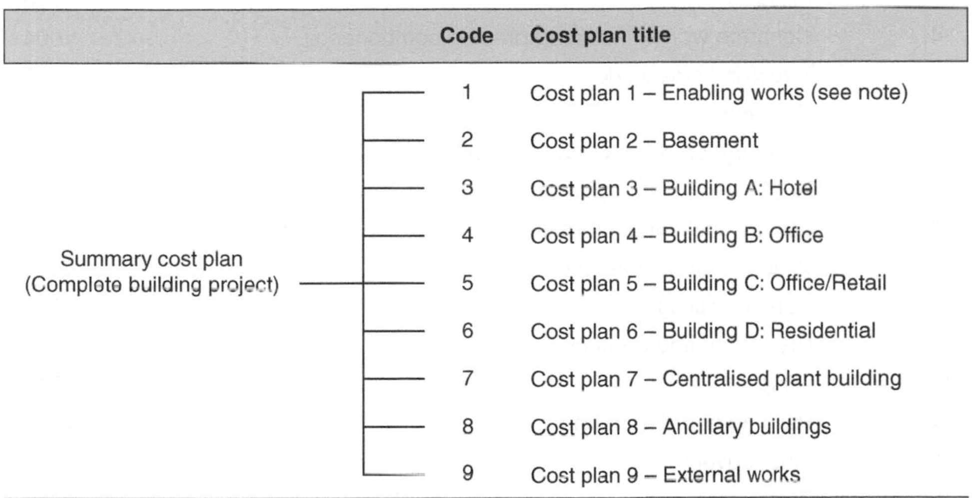

Code
Cost plan title
Summary cost plan
(Complete building project)
1
Cost plan 1 – Enabling works (see note)
2
Cost plan 2 – Basement
3
Cost plan 3 – Building A: Hotel
4
Cost plan 4 – Building B: Office
5
Cost plan 5 – Building C: Office/Retail
6
Cost plan 6 – Building D: Residential
7
Cost plan 7 – Centralised plant building
8
Cost plan 8 – Ancillary buildings
9
Cost plan 9 – External works

Figure 11.4 Typical cost breakdown structure for a complete building project

Note: The terms enabling works and facilitating works should not be confused (refer to Chapter 6, paragraph 6.2.1)

the overall structure of the cost plan agreed on. A typical cost breakdown structure of a cost plan for a complete building project is shown in Figure 11.4.

11.3.3 Restructuring the cost plan for the purpose of procurement

It is unlikely that either the procurement strategy or contract strategy would have been formulated at the start of the cost-planning process. However, more often than not, the method of procurement, or contract strategy, will result in a need for the cost plan to be restructured into either contract or work packages. For example, on the advice of the project team, the employer might decide to procure the building project through a number of separate contracts, such as:

- asbestos removal;
- demolition works;
- site clearance and other preparatory works;
- basement structure:
- shell and core, including provision of mains services;
- fit-out; and
- external works.

Then again, it might be decided that the main building contract should be procured using a number of discrete work packages (see Figure 11.5). These scenarios

Serial No.	Work package	Suffix
1.	Main contractor's preliminaries	/001
2.	Groundworks	/002
3.	Piling	/003
4.	Concrete works (including precast components)	/004
5.	Structural steelwork	/005
6.	Carpentry	/006
7.	Masonry (brickwork and blockwork)	/007
8.	Roof systems and rainwater goods	/008
9.	Joinery (including internal doors, toilet cubicles and vanity units)	/009
10.	Windows and external doors	/010
11.	Curtain walling	/011
12.	Dry linings and partitions	/012
13.	Tiling	/013
14.	Painting and decorating	/014
15.	Floor coverings	/015
16.	Suspended ceilings	/016
17.	Mechanical and electrical services installations (including sanitary appliances)	/017
18.	Pods	/018
19.	Lifts	/019
20.	Fittings, furnishings and equipment	/020
21.	Architectural metalwork	/021
22.	External drainage	/022
23.	External works – soft landscape works	/023
24.	External works – hard landscape works	/024
25.	Main contractor's overheads and profit	/025

Figure 11.5 Contract breakdown structure – works packages

emphasise the importance to the cost manager of codifying cost plans at the beginning of the cost-planning process.

NRM 1 recognises that cost plans will need to be restructured from elements to contract or works packages for the purposes of procurement. However, it makes no attempt at standardising works packages. This is because the content of works packages is likely to be different from one building contract to another – with the content of works packages often based on the perception of risk of those ultimately liable for the construction works. For example, for one building contract it might be deemed appropriate to have all concrete work carried out by a single subcontractor. However, for another building contract, because of the perceived risks associated with the drainage passing through the ground floor construction, it is considered more appropriate to include the construction of pile caps, ground beams and base slab in the works package for groundworks.

How the cost plan is to be restructured for the purposes of procurement will most likely be decided as the cost plan is being developed. Therefore, a method of allocating elements and components to contract packages or works packages needs to be pre-empted. The method advocated by the rules involves the provision of a secondary code that acts as a suffix to the primary code described in paragraph 11.3.2. Figure 11.5 illustrates typical suffix codes for work packages. Example 11.3 illustrates the use of suffix codes. In this example, the code for bathroom pods is 6.1.3.1/018, which will result in the component being allocated to work package 18 (Pods).

Example 11.3 Codification of components for the purposes of procurement

SERVICES						
Code	**Suffix**	**Element/Sub-elements/Components**	**Qty**	**Unit**	**Rate**	**Total**
					£p	**£**
6		**PREFABRICATED BUILDINGS AND BUILDING UNITS**				
6.1		**Prefabricated Buildings and Building Units**				
6.1.3		Pods				
6.1.3.1	/018	Bathroom pods; comprising WC, wash handbasin, bath (including bath panel and trim) and shower over bath; fully fitted out, including coloured glazed ceramic tiles to all walls and sanitary fittings; installed (Type A)	12	nr	4,360.00	52,320
5.1.2		**TOTAL – Pods: To element 5.1 summary**				**52,320**

CHAPTER 12

Preparing a cost plan

Introduction

This chapter:

- contains a process map which summarises the steps to be taken to prepare a cost plan;
- describes the steps involved in preparing a cost plan;
- summarises the composition of a cost plan; and
- explains the composition of the works cost estimate for a cost plan.

12.1 Steps to preparing a cost plan

With the exception of the building works estimate (step 2) and producing and issuing the cost plan report (steps 12 and 13), the steps in preparing a cost plan are the same as for preparing an order of cost estimate (refer to Chapter 5: 'Preparing an order of cost estimate'). See Figure 12.1.

The revised steps 2, 12 and 13 applicable to the cost-planning process are as follows:

Step 2: Produce building works estimate

Step 2.1 Determine quantities:

2.1.1 Determine elements and components to be quantified.

2.1.2 Describe elements and components to be quantified.

> Note:
> Elements mean group elements, elements or sub-elements, whichever are applicable.

2.1.3 Determine quantities for each element and component, as required.

> Note:
> Measured in accordance with NRM 1.

2.1.4 Record any cost-significant design development and construction risks identified during the quantification process (suitable cost allowances to be made in risk allowance estimate).

2.1.5 Record all assumptions that have been made.

Step 2.2 Ascertain unit rates:

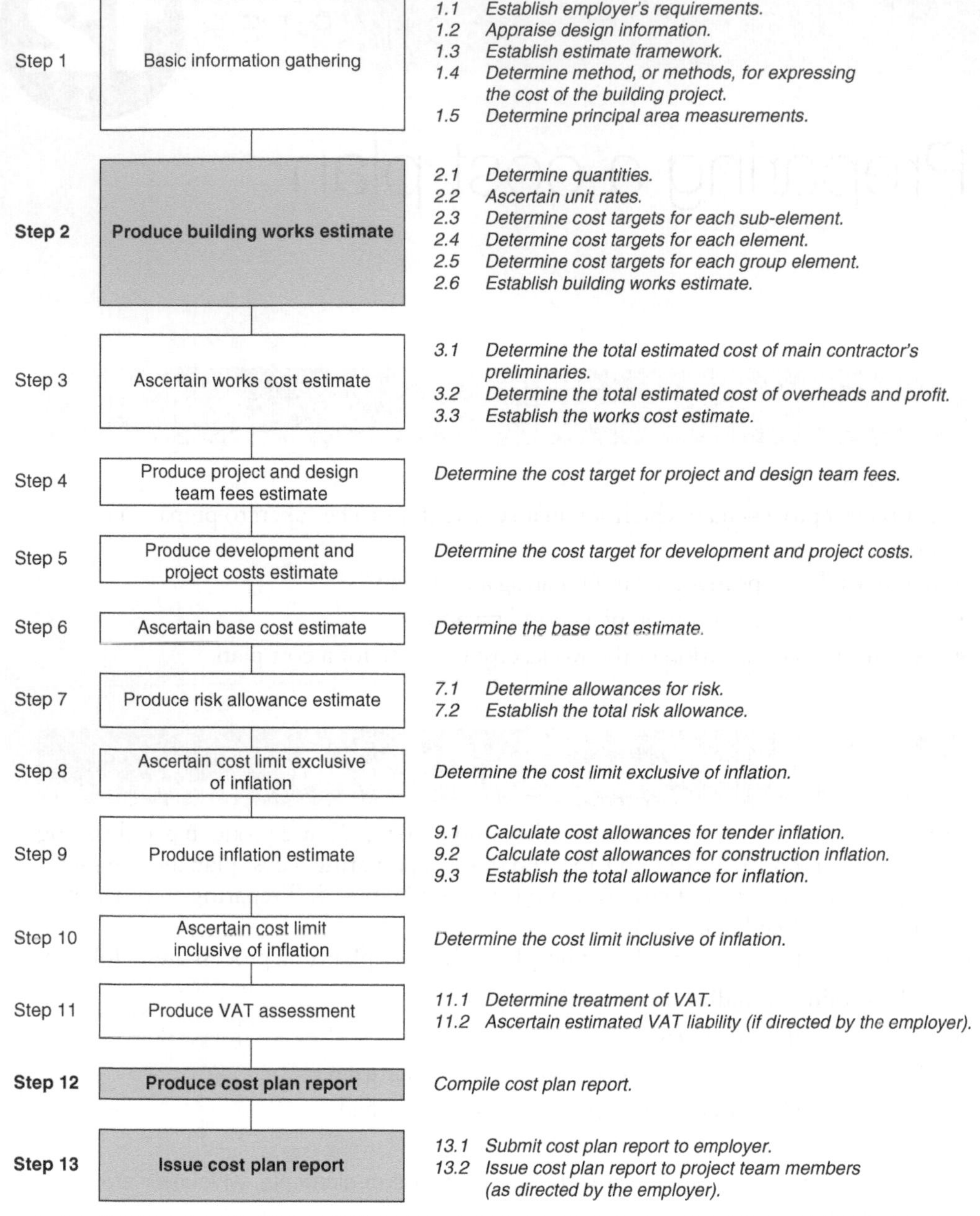
Step 1
Basic information gathering
1.1 Establish employer's requirements.
1.2 Appraise design information.
1.3 Establish estimate framework.
1.4 Determine method, or methods, for expressing the cost of the building project.
1.5 Determine principal area measurements.
Step 2
Produce building works estimate
2.1 Determine quantities.
2.2 Ascertain unit rates.
2.3 Determine cost targets for each sub-element.
2.4 Determine cost targets for each element.
2.5 Determine cost targets for each group element.
2.6 Establish building works estimate.
Step 3
Ascertain works cost estimate
3.1 Determine the total estimated cost of main contractor's preliminaries.
3.2 Determine the total estimated cost of overheads and profit.
3.3 Establish the works cost estimate.
Step 4
Produce project and design team fees estimate
Determine the cost target for project and design team fees.
Step 5
Produce development and project costs estimate
Determine the cost target for development and project costs.
Step 6
Ascertain base cost estimate
Determine the base cost estimate.
Step 7
Produce risk allowance estimate
7.1 Determine allowances for risk.
7.2 Establish the total risk allowance.
Step 8
Ascertain cost limit exclusive of inflation
Determine the cost limit exclusive of inflation.
Step 9
Produce inflation estimate
9.1 Calculate cost allowances for tender inflation.
9.2 Calculate cost allowances for construction inflation.
9.3 Establish the total allowance for inflation.
Step 10
Ascertain cost limit inclusive of inflation
Determine the cost limit inclusive of inflation.
Step 11
Produce VAT assessment
11.1 Determine treatment of VAT.
11.2 Ascertain estimated VAT liability (if directed by the employer).
Step 12
Produce cost plan report
Compile cost plan report.
Step 13
Issue cost plan report
13.1 Submit cost plan report to employer.
13.2 Issue cost plan report to project team members (as directed by the employer).

Figure 12.1 Steps in preparing a cost plan

Note:
Guidance on how to ascertain unit rates for elements, sub-elements and components is given in Chapter 23: 'Deriving unit rates for building compenents, sub-elements and elements'.

2.2.1 Establish estimate base date (i.e. the date on which the cost limit (excluding inflation) is established as a basis for calculating inflation, changes or other related variances).

2.2.2 Determine unit rates.

Notes:

(1) Unit rates used are to be current at the time the cost estimate is produced. That is, they must exclude any allowances for future inflation or deflation.

(2) Unit rates applied to building works are to exclude allowances for main contractor's preliminaries, main contractor's overheads and profit, and inflation, as separate cost estimates are prepared for each of these facets. Allowances for all costs in connection with subcontractors' (or works contractors') preliminaries, overheads and profit, design fees and risk allowances are to be made in the unit rates used. Similarly, allowances for suppliers' delivery charges should also be made within the unit rate.

2.2.3 Record all assumptions that have been made.

Step 2.3 Determine cost targets for each sub-element:

2.3.1 Assign unit rate to component (or sub-element where no components measured).

2.3.2 Determine total estimated cost of component (or sub-element where no components measured) by multiplying the quantity by the unit rate.

Estimated total cost of component =

$$c = q \times r$$

Where

q = quantity
r = rate
c = estimated total cost of component.

2.3.3 Repeat step 2.3.2 for each component.

2.3.4 Determine cost target for sub-element by adding together the estimated total cost of all components which comprise the sub-element.

2.3.5 Repeat step 2.3.4 for each sub-element.

Step 2.4 Determine cost targets for each element:

2.4.1 Determine the cost target for each element by adding together the cost target for each sub-element which forms part of the relevant element.

2.4.2 Repeat step 2.4.1 for each element.

Step 2.5 Determine cost targets for each group element:

2.5.1 Determine the cost target for each group element by adding together the cost target for each element which forms part of the relevant group element.

2.5.2 Repeat step 2.5.1 for each group element.

Step 2.6 Establish building works estimate:
Determine the total estimated cost of building works by adding together the cost target for each group element.

Step 12 Produce cost plan report

Compile cost plan report.

Note:
The characteristics of a good order of cost report are set out in Chapter 34: 'Reporting of cost estimate and cost plans'.

Step 13: Issue cost plan report

Step 13.1: Submit cost plan report to employer.

Step 13.2: Issue cost plan report to project team members (as directed by the employer).

12.2 Composition of a cost plan

The principal constituents of a cost plan are shown in Figure 12.2. Cross-references to the chapter or other location in the book where each aspect of a cost plan is addressed are also provided.

The works cost estimate is the sum of the building works estimate (estimate 1A), the cost target for main contractor's preliminaries (estimate 1B) and the cost target for main contractor's overheads and profit (estimate 1C). Figure 12.3 shows the make-up of a works cost estimate for a cost plan of a project comprising a single building. The building works estimate (estimate 1A) is the sum of the cost targets for each group element applicable to the building.

Chapter 13 ('Quantification and the use of the tabulated rules of measurement for cost planning') provides guidance on the general approach to measuring building works, whereas subsequent Chapters 14 to 22 explain how to interpret the tabulated rules of measurement in NRM 1 in order to quantify and describe the elements and building works items which make up each group element (refer to Figure 12.3). Chapter 24 shows how to calculate the building works estimate, whereas the approach to determining cost targets for main contractor's preliminaries and main contractor's overheads and profit are described in Chapters 25 and 26, respectively. Lastly, the method of calculating the works cost estimate is explained in Chapter 27.

Estimate	Process	Facet	Chapter/reference
Estimate 1	Works cost estimate	*Building works estimate plus the cost targets for main contractor's preliminaries and main contractor's overheads and profit.*	Refer to Figure 12.3 Chapter 27
	+		
Estimate 2	Project and design team fees estimate	*Cost target for project and design team fees.*	Chapter 28
	+		
Estimate 3	Other development and project costs estimate	*Cost target for other development and project costs.*	Chapter 29
	+		
Estimate 4	Risk allowances estimate	*Cost target for risk.*	Chapter 30
	+		
Estimate 5	Inflation estimate	*Cost target for inflation.*	Chapter 31
	=		
	COST LIMIT		Chapter 32
	VAT assessment		Chapter 33

Figure 12.2 Composition of a cost plan

Estimate	Group element	Facet	Chapter/reference
Estimate 1A1	0: Facilitating works	*Cost target. Includes major demolition works.*	Chapter 14
	+		
Estimate 1A2	1: Substructure	*Cost target.*	Chapter 15
	+		
Estimate 1A3	2: Superstructure	*Cost target.*	Chapter 16
	+		
Estimate 1A4	3: Internal finishes	*Cost target.*	Chapter 17
	+		
Estimate 1A5	4: Fittings, furnishings and equipment	*Cost target.*	Chapter 18
	+		
Estimate 1A6	5: Services	*Cost target.*	Chapter 19
	+		
Estimate 1A7	6: Prefabricated buildings and building units	*Cost target.*	Chapter 20
	+		
Estimate 1A8	7: Work to existing buildings	*Cost target.*	Chapter 21
	+		
Estimate 1A9	8: External works	*Cost target. Includes cost of connections to statutory undertakers' mains services.*	Chapter 22
	=		
Estimate 1A	**Building works estimate**	*Total cost of building works (excluding main contractor's preliminaries and main contractor's overheads and profit).*	Chapter 24
	+		
Estimate 1B	**9: Main contractor's preliminaries**	*Cost target.*	Chapter 25
	+		
Estimate 1C	**10: Main contractor's overheads and profit**	*Cost target.*	Chapter 26
	=		
Estimate 1	**Works cost estimate**	*Total cost of building works (including main contractor's preliminaries and main contractor's overheads and profit).*	Chapter 27

Figure 12.3 Composition of the works cost estimate

12.3 Checklist for cost plan

A typical checklist for preparing a cost plan is set out in Example 12.1:

Example 12.1 Cost plan checklist

Bright Kewess Partnership	**Form DPB03**
Cost plan: checklist	

Project no:

Project title:

Date:

Confirm that the following tasks have been carried out before issue of cost plan report:

1.	**Employer's requirements:**	
1.1	Cost limit established?	☐
1.2	Project/design brief obtained?	☐
1.3	Programme established (i.e. for both pre-construction, including design, and construction activities)?	☐
1.4	Key dates established?	☐
1.5	Details of any works required to be completed early ascertained (e.g. any enabling works)?	☐
1.6	Phasing requirements determined, including any requirements for decanting, temporary access or similar?	☐
1.7	Likely construction period established?	☐
1.8	Procurement strategy, including contractor selection and tender strategies, determined?	☐
1.9	Enabling works contract or contracts required?	☐
1.10	Considered tender and contract strategies taken into account?	☐
1.11	Extent of facilitating works determined?	☐
1.12	Location of the site established?	☐
1.13	Use of each building established?	☐
1.14	Accommodation requirements established?	☐
1.15	Area schedule(s) obtained and checked for accuracy?	☐
1.16	Site area determined?	☐
1.17	All information, reports, surveys and drawings relating to existing site requested, obtained and reviewed? • Existing buildings and structures? • Above-ground and below-ground services? • Ground conditions? • Existing foundations? • Measured survey? • Party wall and/or boundary issues?	☐
1.18	Extent of work to existing buildings and structures established (e.g. employer's requirements for de-rating of building; advanced removal of asbestos; advanced strip-out (and whether in readiness for demolition or refurbishment), and extent of demolition works)?	☐

Example 12.1 *(Continued)*

Bright Kewess Partnership		**Form DPB03**
	Cost plan: checklist (cont'd)	
Project no:		
Project title:		
Date:		
Confirm that the following tasks have been carried out before issue of cost plan report:		
1.19	Potential boundary issues identified (and action plan in place)?	☐
1.20	Temporary works requirements determined?	☐
1.21	Sufficiency of statutory undertakers' mains services determined?	☐
1.22	All design information relating to proposed building, or development, requested and reviewed? • Architectural? • Structural? • Building engineering services? • Public health? • Archaeological? • Specialist consultants? (Use tables for group element 11 as a checklist.)	☐
1.23	Environmental assessment method(s) applicable determined?	☐
1.24	Environmental assessment ratings to be achieved determined?	☐
1.25	Details of measures to be taken to achieve required environmental assessment rating(s) determined?	☐
1.26	Estimated mass of steelwork for steel-framed buildings obtained from structural engineer?	☐
1.27	Estimates of reinforcement for reinforced concrete components obtained from structural engineer?	☐
1.28	Quality requirements for each building established, including services?	☐
1.29	Room data sheets obtained?	☐
1.30	Fit-out specification determined (if applicable)?	☐
1.31	Method of providing building 'operating and maintenance (O&M) manual(s)' and health and safety file(s) determined (e.g. paper-based, electronic or web-based information management system)?	☐
1.32	Post-completion requirements determined for employer (i.e. any maintenance requirements)?	☐
1.33	Treatment of project and design team fees established?	☐
1.34	Treatment of other development project costs established?	☐
1.35	(If included) extent and nature of other development project costs determined?	☐

Example 12.1 *(Continued)*

Bright Kewess Partnership		Form DPB03
Cost plan: checklist (cont'd)		
Project no:		
Project title:		
Date:		
Confirm that the following tasks have been carried out before issue of cost plan report:		
1.36	Nature of risk allowances to be considered agreed with employer (i.e. design development risk, construction risks, employer's change risks, employer's other risks)?	☐
1.37	Up-to-date risk register or log obtained?	☐
1.38	Method of calculating risk allowances agreed?	☐
1.39	Insurance requirements determined?	☐
1.40	Work breakdown structure (WBS) determined?	☐
1.41	Method(s) of expressing costs agreed (i.e. has the employer any specific requirements)?	☐
1.42	Treatment of VAT ascertained?	☐
1.43	Recipients of 'Order of cost estimate report' agreed?	☐
2.	**Works cost estimate:**	
2.1	GIFA(s) determined?	☐
2.2	NIA(s) determined (if required)?	☐
2.3	Type(s) and number(s) of functional units determined (if required)?	☐
2.4	GEA of existing buildings to be demolished determined?	☐
2.5	Area of external works determined?	☐
2.6	Estimate base date established?	☐
2.7	Unit rates used applicable, appropriate and adjusted for time and location?	☐
2.8	Works required to complete each element relevant to the building project included in cost plan?	☐
2.9	NRM 1 tabulated rules for the measurement of elements used as checklist?	☐
2.10	Building works estimate completed?	☐
2.11	Main contractor's preliminaries estimate completed?	☐
2.12	Percentage addition used for main contractor's preliminaries applicable and appropriate?	☐
2.13	Phasing requirements factored into cost estimate?	☐
2.14	Main contractor's overheads and profit estimate completed?	☐
2.15	Percentage addition used for main contractor's overheads and profit applicable and appropriate?	☐

Example 12.1 *(Continued)*

Bright Kewess Partnership		**Form DPB03**
	Cost plan: checklist (cont'd)	
Project no:		
Project title:		
Date:		
Confirm that the following tasks have been carried out before issue of cost plan report:		
3.	**Project and design team fees estimate:**	
3.1	(If included) percentage rates used for consultants' fees applicable and appropriate?	☐
3.2	Project and design team fees estimate completed?	☐
4.	**Development and project costs estimate:**	
4.1	Development and project costs estimate completed?	☐
5.	**Risk allowance estimate:**	
5.1	Percentage addition used for design development risk appropriate?	☐
5.2	Percentage addition used for construction risk appropriate?	☐
5.3	Percentage addition used for employer's change risk appropriate (and agreed with employer)?	☐
5.4	Percentage addition used for employer's other risk appropriate (and agreed with employer)?	☐
5.5	Risk allowance estimate completed?	☐
6.	**Inflation estimate:**	
6.1	Percentage addition used for tender inflation appropriate?	☐
6.2	Percentage addition used for construction inflation appropriate?	☐
6.3	Inflation estimate completed?	☐
7.	**VAT assessment:**	
7.1	VAT assessment completed?	☐
8.	**Assumptions and checks:**	
8.1	All assumptions clearly recorded (i.e. inclusions and exclusions from order of cost estimate)?	☐
8.2	All calculations independently checked?	☐
OCE Prepared by: Signed: Position: Date:	Verified by: Signed: Position: Date:	

CHAPTER 13

Quantification and the use of the tabulated rules of measurement for cost planning

Introduction

This chapter:

- gives guidance on the general approach to measuring building works;
- explains how to use the tabulated rules for each group element;
- explains how to deal with components excluded by NRM 1;
- describes how to record dimensions and quantities;
- considers the pitfalls of recording dimensions and quantities ascertained using electronic measuring devices;
- defines and distinguishes between prime cost (PC) sums and provisional sums;
- explains the use of PC sums and provisional sums in cost planning;
- looks at accuracy in measurement;
- discusses the importance of annotating dimensions and quantities;
- explains how to quantify and describe components for cost plans;
- advises how to formulate descriptions for components;
- makes clear how work to and within existing buildings is to be measured under the rules;
- explains how to measure components which are not covered by the rules; and
- describes how to measure components when there is inadequate design information.

13.1 General approach to measuring building works

The measurement and description of building works items can be divided into two distinct processes:

- initial measurement and abbreviated description of components (i.e. 'taking off'); and
- compiling the cost plan from the measured data, including writing concise descriptions.

13.1.1 Measurement by elements

The elemental structure set out in NRM 1 provides the framework for measuring building works. This elemental structure conveys a systematic approach to measurement by providing a checklist of the elements and components that need to be considered by the cost manager – the rules provide a ready made 'taking-off' list of cost-significant items.

This approach also aids the breaking down of large building projects, where measurement is normally undertaken by a team of cost managers, with each member taking responsibility for measuring one or more element of the building.

The measurement of building works items, elements and components forming each group element is decribed in the chapters that ensue. For ease of reference, these chapters are written in the same sequence as Part 4 (Tabulated rules of measurement for elemental cost planning) of NRM 1.

13.1.2 Initial review of design information

Before any building works are measured, the cost manager should review the design information provided by all design team members to get a general understanding of the proposed construction of the building. Drawings should be checked to ensure that all floor plans, roof plans, elevations and sections have been provided, and that they align with each other, as well as making sure drawings provided by different design disciplines correlate. These checks will help prevent the possibility of inconsistency in dimensions and measurements.

This initial review will also aid the cost manager in interpreting the design information and identifying any initial concerns about the information.

13.1.3 Site visit

Whenever possible, the cost manager should visit, and undertake a visual inspection of, the site to be developed before commencing measurement. This will help the cost manager visualise references to the site on the drawings.

For commercial reasons, it is not uncommon for cost managers to have had to produce cost plans without having actually seen the site or, at best, having only been permitted a walk around the perimeter of the site. This situation might arise where an employer does not wish to openly be known to be involved with a site, or the use of the proposed building is considered to be sensitive.

13.1.4 Resolving queries with design team members

Initial review of the drawings, specification and other design information by the cost manager will normally result in a number of queries (e.g. information requirements, points of clarification and discrepancies). The early communication, and resolution, of queries with design team members is important and simplifies the cost-planning process. However, there is nothing more infuriating than sporadically receiving a large number of questions. In view of this, it is recommended that, whenever possible, the cost manager prepare a comprehensive list of questions before approaching design team members.

13.2 Use of the tabulated rules for building works

The rules of measurement for building works are set out in tabular form, and the order and layout of the pages direct how the rules are to be interpreted and used (see Figure 13.1).

Group element number and heading (e.g. Group Element 3: Internal finishes)
Element number and heading (e.g. Element 3.3: Ceiling finishes)

Sub-element	Component	Unit	Measurement rules for components	Included	Excluded
1	1 2 3	m^3	C1 C2 C3	1 2 3	1 2 3
2	1 2	m^2	C1 C2 C3	1 2 3	1 2 3
3	1 2 3	m	C1 C2 C3	1 2 3	1
4	1	nr	C1 C2	1 2 3	1 2
5	1	item	C1 C2	1 2 3	1 2 3
6	1	m^2/nr	C1 C2	1 2	1 2 3

Figure 13.1 Structure and format of the tabulated rules for the measurement of building works

Notes:
(1) The left hand column lists the sub-elements and contains the definition of the sub-element
(2) Where exclusions are stated, cross-references to the appropriate element or sub-element are given
(3) Horizontal lines divide the tables to denote the end of a sub-element
(4) The symbol '/' (i.e. a backward slash) used between two or more units of measurement in the third column (unit of measurement) means 'or'
(5) The rules are written in the present tense

Tabulated rules for group elements 0 to 8 comprise the rules of measurement for building works:

0: Facilitating works;
1: Substructure;
2: Superstructure;
3: Internal finishes;
4: Fittings, furnishings and equipment;
5: Services;
6: Prefabricated buildings and building units;
7: Work to existing buildings; and
8: External works.

Although NRM 1 provides a comprehensive list of components, there is always going to be the probability that a certain component has been overlooked. This is where common sense needs to be applied by the cost manager. It is recommended that missing components are simply included within the most appropriate sub-element and measured using the most applicable unit of measurement in accordance with the rules for 'Other cost significant components'.

13.3 Use of the tabulated rules for other group elements

As for building works, the rules of measurement for main contractor's preliminaries; main contractor's overheads and profit; project/design team fees; other development/project costs; and inflation are set out in tabular form (Figure 13.2). Separate tables are provided for each sub-element.

13.4 Use of the tabulated rules for risks

Lists of potential causes of risk are given in a single-column table of rules, with a separate table provided for each category of risk (Figure 13.3).

13.5 Degree of measurement

NRM 1 adopts the *Pareto* theory. That is, 80% of the cost is in 20% of the items. Therefore, only the cost-significant components (i.e. the 20%) are measured, with allowances being made for minor sundry items, labour and ancillary items in the unit rates applied to the item (refer to Chapter 23: 'Deriving unit rates for building components, sub-elements and elements').

Group element number and heading (e.g. Group Element 9: Main contractor's preliminaries)
Element number and heading (e.g. Element 9.2: Main contractor's cost items)
Sub-element number and heading (e.g. Element 9.2.2: Site establishment)

Component	Included	Unit	Excluded
1	1 2 3	item	1 2 3
2	1 2	per week	1 2 3
3	1 2 3	per person	1
5	1	nr	1 2
6	1	m^2/nr	1 2 3

Figure 13.2 Structure and format of the tabulated rules for the measurement of other group elements (main contractor's preliminaries, main contractor's overheads and profit, project/design team fees, other development/project costs)

Notes:

(1) The left-hand column contains descriptors of the components that comprise the sub-element (e.g. Site accommodation, Temporary works in connection with site establishment, Furniture and equipment and IT systems)
(2) The second column identifies the items that are to be included within the component (i.e. the sub-components)
(3) The third column identifies the unit of measurement for each sub-component.
(4) The right-hand column identifies specific exclusions from a component and cross-references to the appropriate component where the item is to be measured
(5) Horizontal lines divide the tables to denote the end of a component
(6) The symbol '/' (i.e. a backward slash) used between two or more units of measurement in the third column (unit of measurement) means 'or'
(7) The rules are written in the present tense

1. 2. 3. 4.

Figure 13.3 Structure and format of the tabulated rules for the quantification of risks

13.6 Measurement notation

13.6.1 Recording dimensions and quantities

Dimensions and quantities of elements and components are measured from the drawings and recorded by the cost manager using traditional dimension paper or estimating paper (refer to Figures 13.4 and 13.5).

Group element number and heading (e.g. Group element 13: Risks)
Element number and heading (e.g. Element 13.1: Design development risks)

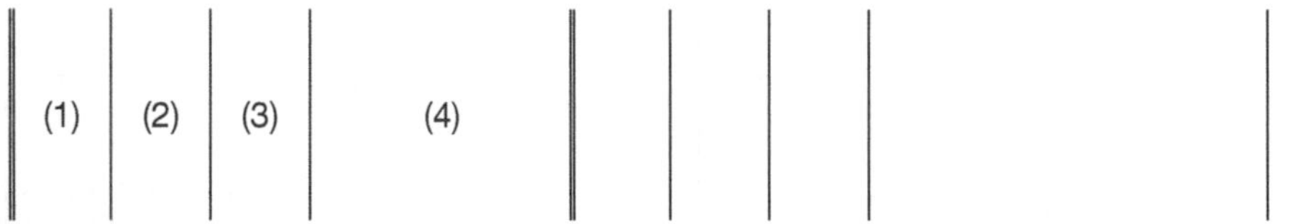

Figure 13.4 Traditional dimension paper

Notes:
(1) Timesing column – used for timesing and dotting on (see Figure 13.6)
(2) Dimension column, in which measurements are set down as taken from the drawings
(3) Squaring column, in which are set out the calculated quantities
(4) Description column, in which is written the description of the component
(5) The next four columns are a repeat of the first four columns

Figure 13.5 Traditional estimating paper

Notes:
(1, 2, 3 and 4) The same as traditional dimension paper (refer to Figure 13.4)
(5) Unit rate column, in which the rate for the component per unit of measurement is inserted
(6 and 7) The extension column, in which is written the total cost of component measured
(8) Columns (5), (6) and (7) are not normally used, as the estimating paper is generally only used as a means of documenting quantities and outline descriptions. Unit rates are normally calculated separately and inputted directly into the cost plan, which is usually generated using a spreadsheet or a bespoke computerised cost-planning program. Component costs and cost targets are automatically generated

Traditional taking-off techniques are normally employed by the cost manager to quantify and briefly describe components for inclusion in the cost plan. Figure 13.6 illustrates the traditional form of writing dimensions. A brief description of the component measured is jotted down opposite the measurement in the description column.

	19.65 *1.80* *3.20*	*Brief description of component measured*	Cubic measurement (m^3).
	7.60 *2.95*	*Brief description of component measured*	Superficial measurement (m^2).
	6.90	*Brief description of component measured*	Linear measurement (m).
	18	*Brief description of component measured*	Enumerated item (nr).
	Item	*Brief description of component measured*	Itemised item (item).
6/	*6.90*	*Brief description of component measured*	Timesing – multiplying dimensions (in this example, the linear measurement (6.00m) is multiplied by 6).
3/6/	*6.90*	*Brief description of component measured*	Timesing – multiplying dimensions (in this example, the linear measurement (6.00m) is multiplied by 18 (i.e. 3 × 6)).
6/ · *2*	*3*	*Brief description of component measured*	Dotting on – repeating or adding dimensions (in this example, the enumerated item is multiplied by 8 (i.e. 2 + 6)). The dot is placed below the top figure to avoid confusion with decimals.
3/6/ · *2*	*7.60* *2.95*	*Brief description of component measured*	Timesing and dotting on used collectively (in this example, the superficial measurement is multiplied by 24, i.e. 3 × (2 + 6)). The technique of timesing and dotting on can be used in conjunction with any form of dimension.

Figure 13.6 Form of dimensions

13.6.2 Distinguishing between dimensions and quantities

With the aid of digitisers and other electronic measuring devices, it is possible to ascertain quantities from drawings without the need to measure each dimension and calculate the quantities from the measured dimensions. However, use of such measuring devices can result in the misinterpretation of dimensions and measurements. This is because of the way in which dimensions are usually recorded. For example, a single

Dimension

	9.00 *3.00*	*Brief description* *m²*	Superficial measurement (two dimensions)

Quantity

	27.00	*Brief description* *m²*	Superficial measurement (quantity) measured using a digitiser or similar measuring device. Where single measurements are used to represent an area, it is essential that the cost manager annotates the applicable measurements accordingly.

Quantity and dimension

	27.00 *3.90*	*Brief description* *m³*	The same equally applies to situations where an area measurement (quantity) and a height or depth measurement (dimension) are used to represent a cubic measurement.

Figure 13.7 'Dimensions' or 'quantities'?

dimension representing a superficial measurement, or a cubic measurement represented by two dimensions, might be recorded. Therefore, to avoid unintentional misuse of measured data, it is essential that the cost manager ensures that dimensions and quantities are clear to others. In this instance, the cost manager should make a note of the unit of measurement adjacent to the recorded dimensions (refer to Figure 13.7).

13.6.3 Supporting calculations

Dimensions should not be calculated mentally, as this can result in errors. The risk of errors will be reduced if supporting calculations showing how dimensions have been ascertained are written down. This will also enable others to readily see how the dimension has been derived, as well as allow the calculation of the dimension to be checked. Supporting calculations are made on the right-hand side of the description column and must be written clearly (refer to Figure 13.8). These supporting calculations are sometimes referred to as 'waste calculations'.

13.6.4 Gross measurements

Measuring gross with adjustment later is preferable to piecemeal measuring. When dealing with sub-elements and components that are measured superficially, it is usual in the first instance to measure the overall area of the item. Different sub-components or component specifications can be dealt with by adjustment later.

Measurement of external walls is a good example of where measuring gross with adjustment later is important. It is not uncommon for an architect to incorporate a number of different external wall systems in a single building. If the external wall systems were measured piecemeal, there is a real risk that the cost manager might

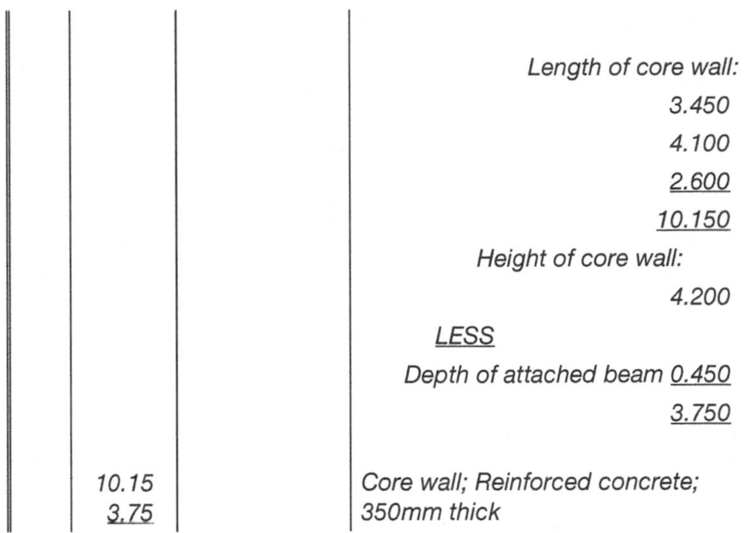

			Length of core wall: *3.450* *4.100* *2.600* *10.150* *Height of core wall:* *4.200* *LESS* *Depth of attached beam 0.450* *3.750*	
	10.15 *3.75*		*Core wall; Reinforced concrete; 350mm thick*	

Figure 13.8 Supporting calculations

neglect to measure a system. As a consequence, not only will the external wall system not have been measured, it will also result in the external wall element being undermeasured – potentially, a cost-significant mistake. This risk is significantly reduced when the area is measured gross, and the resulting quantity adjusted to compensate for the quantity of the other external wall systems as they are measured. In this case, should the cost manager fail to measure a system, the risk is limited to the difference in cost between the common system and the system not measured.

13.6.5 Quantities resulting from measurement

For the purposes of measurement, quantities are calculated to two decimal places (e.g. 39.60m^3). It is only when they are transferred to the cost plan that they are rounded to the nearest whole unit (refer to paragraph 13.7.1 below).

13.6.6 Degree of accuracy

Clearly, the cost manager must aim to measure the building works components as accurately as possible; however, the level of accuracy of measurement will be dependent on the quality of drawings and other design information available. Where figured dimensions given on drawings are the basis of measurement the measurement taken should be exact. However, where dimensions are scaled using a scale ruler there is some excuse for slight over-measure. This is because there might be some shrinkage in drawings caused through the printing process.

The cost manager can also use electronic measuring tools such as a digitiser, which is a highly accurate measuring device that interfaces with a computer. However, since building details are often supplied to the cost manager in electronic form, the use of computer-aided measurement software is increasing (e.g. AutoCAD or IntelliCAD for Quantity Surveyors). The software allows the cost manager to extract measurement data directly from CAD (computer aided design) data using a computer, rather than the traditional manual methods of extraction off paper drawings. When using digitisers, or computer-aided measurement programs, the overall accuracy achieved will clearly depend on the care taken by the cost manager in outlining a given area or length. The more points used to define an area or length, the better. With sufficient care, accuracy is usually better than +/− 1%.

Obviously, there is a sensible limit to the degree of accuracy to which the cost manager should measure; experienced cost managers will often take the cost significance of a component into account when deciding on the degree of accuracy required. The cost manager owes a duty of care to the employer to prepare a realistic cost plan, which the employer will use to decide whether or not to progress the building project.

13.6.7 Annotating dimensions and quantities

It must be remembered that the measured data will need to go forward to the cost plan, so it is important that all relevant dimensions and quantities can be identified. It is essential, therefore, that the cost manager annotates drawings and keeps notes to show how dimensions and quantities were derived. Failure to do this might result in substantial remeasurement at later cost-planning stages. Clear annotation and cross-referencing will make it easier to undertake cost comparisons of different component specifications, as well as facilitate the cost-checking process during subsequent cost-planning stages.

13.6.8 Initial descriptions

Descriptions used for the measurement process need only be scant, with sufficient information to enable a fuller description for the cost plan to be developed.

13.7 Quantifying and describing components for cost plans

13.7.1 Quantities for components

Measured building work items are deemed to be supply and fix – they include all subcontractors' preliminaries, overheads and profit, design fees and risk allowances. Quantities are calculated to the nearest whole unit (e.g. 23.60m^3 is taken as 24m^3, and 19.36m is taken as 19m). The two exceptions to this rule are:

- any quantity less than one whole unit, which is to be given as one unit; and
- structural steelwork, which is measured in tonnes and given to two decimal places (i.e. 79.63t). This is because structural steel is a high-cost component.

13.7.2 Descriptions of components

Descriptions for building work items within cost plans need not be comprehensive, but must contain sufficient information so that the cost manager and all project team members can clearly see what has been allowed for in the cost plan – in particular, the quality of specification. The cost manager must, therefore, ensure that descriptions for components are clear and concise. Without adequate descriptions, the cost manager will be unable to carry out proper cost checks during the later design and cost-planning stages, which becomes even more difficult when there is a long gap in time between the stages (e.g. where the building project is put on hold by the employer), or a different cost manager is tasked with cost checking.

When framing descriptions, the requirements of the rules should be followed carefully. While NRM 1 sets out certain information to be included in descriptions, additional information should be given where necessary to convey the nature of the component.

It should be remembered that, in addition to the employer and the project team members, the cost plan is likely to be seen by other organisations involved in the building project (e.g. funders, solicitors and planning departments). Therefore, consistency in language and spelling is important. Minor differences in phraseology might confuse readers (i.e. is the meaning intended to be different?). Obviously, spell checking in computer programs, coupled with common sense, is an easy solution to the spelling problem.

Examples of descriptions that have been formulated using the rules are provided in the ensuing chapters, which give guidance on how to measure and describe building works associated with each group element.

13.7.3 Dimensions in descriptions

Where the rules call for dimensions to be included in descriptions, all dimensions should be given to enable the cost of the component to be fully appraised. Dimensions are used in descriptions to provide clarity as to what the cost manager has allowed for in the cost plan. Dimensions are written: length × width × height (depth or thickness). Therefore, 6.00 × 2.00 × 1.25m would indicate an item 6.00m long × 2.00m wide × 1.25m high (deep or thick).

13.7.4 Prime cost sums

The phrase, 'prime cost sum' is often confused with the phrase, 'provisional sum'. Moreover, they are often erroneously used indiscriminately by cost managers. In the context of cost planning, a PC sum is a sum of money included in a measured component to confirm the price allowed in the unit rate for a specific sub-component (e.g. the price allowed for facing bricks or allowance for ironmongery to an internal door). Prime cost means the price of supplying the sub-component. A PC sum is sometimes referred to as a PC price.

Typical examples of PC sums are as follows:

- supply only of facing bricks at £360.00 per 1,000; and
- supply only ironmongery at £130.00 per door.

13.7.5 Provisional sums

The rules do not define the term 'provisional sum'. Customarily, the term has been used to cover the cost of something which cannot be entirely foreseen (i.e. unknown or uncertain in extent). For this reason, the use of provisional sums for building works items are to be avoided as their use implies uncertainty. Such items should possibly be treated as either a potential design development or construction risk (see Chapter 30: 'Setting and managing risk allowances').

However, if it is decided to include a provisional sum, the cost manager will need to consider the risk of the provisional sum being insufficient.

Box 13.1 Key definitions

- **Prime cost** – the price of supplying the sub-component (e.g. materials).
- **Prime cost (PC) sum (or price)** – (in the context of cost planning) a sum of money included in a measured component to confirm the price allowed in the unit rate for a specific sub-component.
- **Provisional sum** – a sum of money set aside to carry out work not sufficiently identified.

13.8 Curved work

Where required by NRM 1, curved work is to be described and identified separately as 'curved work'. This is because of the difficulties involved in setting out and constructing components to a curve.

13.9 Work to and within existing buildings

NRM 1 contains specific rules associated with work to or within existing buildings. Work in altering, adapting or repairing the existing structure and fabric of existing buildings is measured under group element 7 (Work to existing buildings). Such work includes taking out existing components (i.e. soft strip), forming new openings in the existing structure, taking out windows and doors, filling in openings, the refurbishment of existing services installations and equipment, damp-proof treatment, and the eradication of infestation and facade improvement works.

New works to or within existing buildings are measured in accordance with the rules of measurement applicable to the element or component to be measured. However, the rules require such work to be described and identified separately. This can be done by simply inserting the relevant measured items under a sub-heading in the cost plan – 'Work to existing buildings'. Examples of new work to or within existing buildings include replacement windows, 'fit-out' works, new services installations, and new internal walls, partitions, internal doors and internal finishes to existing surfaces.

Extensions to existing buildings constitute new works. Therefore, they are measured and described in the same way as new work to or within existing buildings.

13.10 Work not covered by the rules

Some cost-significant components might not be specifically covered by the rules. In these cases, rules of measurement for similar components should be used. However, for purposes of clarity, the method of measurement adopted for components not covered by the rules should be stated in the cost plan by the cost manager.

13.11 Inadequate design information

Cost planning is an iterative process, which is performed in steps of increasing detail as more design information becomes available. Correspondingly, measurement is performed in steps of increasing detail, and this is recognised by NRM 1.

Where the cost manager has inadequate design information to enable the measurement of components, the rules advocate that the basis of quantification is to be the gross internal floor area (GIFA – refer to Section 3.11 (Measurement rules for building works) of NRM 1).

13.12 Information required by the cost manager

The information required by the cost manager from the employer and the designers at each stage of the cost-planning process (refer to Tables 10.1, 10.2 and 10.3 – Schedule of information required for the preparation of formal cost plan 1, 2 and 3 – in Chapter 10: 'Cost planning – an overview').

CHAPTER 14

Facilitating works (group element 0)

Introduction

This chapter:

- provides step-by-step guidance on how to measure and describe facilitating works; and
- contains examples illustrating how to formulate descriptions for components.

14.1 Method of measurement

The rules for measuring facilitating works are found in group element 0, which is divided into six elements as follows:

0.1 Toxic/hazardous/contaminated material treatment;
0.2 Major demolition works;
0.3 Temporary support to adjacent structures;
0.4 Specialist groundworks;
0.5 Temporary diversion works; and
0.6 Extraordinary site investigation works.

Works associated with general site preparation and groundworks, minor demolition works, and permanent roads, paths and pavings are measured under group element 8 (External works). The provision of temporary roads and services required as part of the main contractor's site establishment and logistics are included in group element 9 (Main contractor's preliminaries).

14.2 Toxic/hazardous material removal (element 0.1)

This element is further subdivided into the following three sub-elements:

0.1.1 Toxic/hazardous material removal;

0.1.2 Contaminated land; and
0.1.3 Eradication of plant growth.

The UK government encourages the clean-up of contaminated land by offering tax relief for the costs incurred by companies in cleaning up land they acquire in a contaminated state. See Chapter 33 ('Taxes and incentives').

14.2.1 Toxic/hazardous material treatment (sub-element 0.1.1)

Works comprising the treatment or removal of toxic or hazardous materials that form part of the existing fabric (e.g. asbestos linings or fire barriers) or insulating materials or components from existing mechanical and electrical services installations are separately itemised (item). Descriptions are to include details of any key components, together with any key quantities (see Example 14.1).

Further advice about asbestos-containing materials (ACMs) and the reporting thereof can be obtained in the United Kingdom (UK) from the Health and Safety Executive (HSE).

Example 14.1 Formulation of descriptions for toxic and hazardous material removal

FACILITATING WORKS					
Code	**Element/sub-elements/components**	**Qty**	**Unit**	**Rate**	**Total**
				£p	**£**
0	**FACILITATING WORKS**				
0.1	**Toxic/hazardous material removal**				
0.1.1	*Toxic and hazardous material removal*				
0.1.1.1	Toxic and hazardous material removal; asbestos; all in accordance with Refurbishment/Demolition Survey Report (Reference: 123/XYZ, dated 7 December 2013); by Licensed Asbestos Removal Specialist		item		
0.1.1.2	Toxic and hazardous chemical removal; decommission underground petrol storage tank; by Specialist		item		
0.1.1.3	Toxic and hazardous chemical removal; decommission existing chillers; by Specialist		item		
0.1.1	**TOTAL – Toxic and hazardous material removal: to element 0.1 summary**				

14.2.2 Contaminated land (sub-element 0.1.2)

The removal of contaminated ground material using dig and dump strategy can either be measured superficially, in square metres, or by volume, in cubic metres (m^3). Where measured superficially, the surface area of the contaminated land is measured.

Example 14.2 Formulation of descriptions for contaminated land

FACILITATING WORKS					
Code	**Element/sub-elements/components**	**Qty**	**Unit**	**Rate £p**	**Total £**
0	**FACILITATING WORKS**				
0.1	**Toxic/hazardous material removal**				
0.1.2	*Contaminated land*				
0.1.2.1	Contaminated ground material removal; excavation, disposal off site; non-hazardous material		m^3		
0.1.2.2	Contaminated ground material; excavation, disposal off site; hazardous material		m^3		
0.1.2.3	Contaminated ground material treatment; on-site encapsulation		m^2		
0.1.2	**TOTAL – Contaminated land: to element 0.1 summary**				

Where the volume is measured, the volume is the surface area of the contaminated land multiplied by the average depth of the contaminated material. A quantity given for the disposal of contaminated ground material is the bulk before excavating, and no allowance is made for subsequent variations to bulk. Therefore, it is important that adequate allowance is made for bulking when computing the unit rate for removing contaminated ground material.

The classification of material removed (e.g. non-hazardous or hazardous) is to be stated in the description. Unit rates are to include the safe disposal of excavated material to a licensed tip, tipping charges and associated landfill tax. (See Example 14.2.)

Contaminated ground material treatments, using in situ methods, are measured superficially, with the method to be used stated in the description. Methods include:

- dilution;
- clean cover;
- on-site encapsulation;
- bio-remediation;
- soil washing;
- soil flushing;
- thermal treatment;
- vacuum extraction; and
- stabilisation.

14.2.3 Eradication of plant growth (sub-element 0.1.3)

Invasive plants are introduced plant species that can thrive in areas beyond their natural range of dispersal. These plants are characteristically adaptable and aggressive, and have a high reproductive capacity. Their vigour combined with a lack of natural enemies often leads to outbreak populations.

Japanese knotweed is the most widespread and troublesome bankside species in the United Kingdom, followed closely by hybrid knotweed, which has a similarly high regeneration capacity. It is a highly invasive plant that is not easy to control due to its extensive underground rhizome system, which enables it to survive when all above-ground parts of the plant have been removed. It has been known to grow through materials such as concrete, bituminous bound materials and masonry.

Only female plants are present in the UK. Japanese knotweed forms dense clumps with fleshy, red/green shoots, 2–3m tall, which have hollow green stems with red/purple flecks. Leaves are green, heart or shield shaped, with a flat base, up to 120 millimetres (mm) long. Creamy clusters of flowers are borne on the tips of most stems in late summer. The root system consists of rhizomes that are orange/yellow when cut. The underground rhizome system can extend at least 7m from the parent plant and reach a depth of 3m or more. A piece of rhizome the size of a little fingernail can grow into a new plant. The crown, located at the base of the stem, will produce new plants. The stems die back in winter and take up to three years to decompose. Japanese knotweed must not be removed from site without a waste licence.

Unsurprisingly, in the UK there is legislation that covers Japanese knotweed and other invasive species. The two main pieces of legislation are:

- Wildlife and Countryside Act 1981 – under the Act, it is an offence to plant or otherwise cause the species to grow in the wild.
- Environmental Protection Act 1990 – Japanese knotweed is classed as 'controlled waste' and as such must be disposed of safely at a licensed landfill site according to the Environmental Protection Act (Duty of Care) Regulations 1991. Soil containing rhizome material can be regarded as contaminated and, if taken off site, must be disposed of at a suitably licensed landfill site and buried to a depth of at least 5m. Japanese knotweed must not be removed from site without a waste licence.

An offence under the Wildlife and Countryside Act can result in a criminal prosecution. An infringement under the Environmental Protection Act can result in enforcement action being taken by the Environment Agency, which can result in an unlimited fine. Property owners can also be held liable for costs incurred from the spread of knotweed into adjacent properties and for the disposal of infested soil off site during development which later leads to the spread of Japanese knotweed onto another site.

Equally vilified is giant hogweed, as it not only creates serious environmental problems but also possesses the ability to create very unpleasant reoccurring skin conditions in humans, which can persist for many years. It is, therefore, an offence to cause growth of Japanese knotweed, giant hogweed and other invasive plants, as well as allowing it to spread onto neighbouring properties.

Invasive plant growth can be eradicated by either a 'dig and dump' strategy or by chemical treatment. However, eradication of Japanese knotweed by chemical treatment (i.e. the application of herbicides) is not a one-off treatment and can take up to three years to complete. The most common and sustainable method of treatment for Japanese knotweed is through the use of herbicides.

Eradication of Japanese knotweed, giant hogweed or other invasive plant by 'dig and dump' strategy, including inserting root barrier membrane systems, can either be measured superficially or by volume. Both areas and volumes are measured in the same way as contaminated ground material removal. Similarly, the effects of bulking must be considered when compiling the unit rate.

While not specifically stated in the rules, it is probably best to treat root barrier membrane systems as other cost-significant items in accordance with measurement rule C5, measured in linear in metres. Root barrier membrane systems are designed to prevent the further intrusion of Japanese knotweed or other invasive plant. The root barrier consists of a membrane which must be installed vertically along the boundary of the areas treated, particularly at boundaries to adjacent properties.

Unit rates are to include the safe disposal of excavated material to a licensed tip, tipping charges and associated landfill tax. It will also be necessary to make provision in the unit rate for backfilling voids with inert material following the removal of invasive plants. Moreover, it is important that the eradication of invasive plants by 'dig and dump' strategy is continuously monitored by a specialist in order to make sure that rhizomes (i.e. the reproductive structures) are not inadvertently transferred to other locations. This is crucial if the employer requires the works guaranteed. Should this method of eradication be used, then the cost manager ought to make allowance in element 11.1 (Consultants' fees) for the fees in connection with the employment of a specialist.

Eradication by chemical treatment can either be enumerated or measured superficially, in square metres, with the nature and period over which the treatment is to be carried out (or the number of applications) stated in the description. The area designated as infected by the invasive plant is measured. (See Example 14.3.)

Example 14.3 Formulation of descriptions for eradication of plant growth

FACILITATING WORKS					
Code	**Element/sub-elements/components**	**Qty**	**Unit**	**Rate**	**Total**
				£p	**£**
0	**FACILITATING WORKS**				
0.1	**Toxic/hazardous material removal**				
	Note:				
	Specialist supervision: eradication works to be continuously monitored by Japanese knotweed specialist. Costs in connection with specialist attendance included in element 12.1 (Consultants' fees)				
0.1.3	*Eradication of plant growth*				
0.1.3.1	Eradication by dig and dump strategy; removal of Japanese knotweed, including excavation and disposal of hazardous material		m^3		
0.1.3.2	Root barrier; to site boundaries		m		
0.1.3.3	Eradication by chemical treatment; Japanese knotweed eradication by herbicides		m^2		
0.1.3	**TOTAL – Eradication of plant growth: to element 0.1 summary**				

14.3 Major demolition works (element 0.2)

This element comprises a single sub-element as follows:

0.2.1 Demolition works.

This deals with the demolition of entire buildings and structures, as well as the demolition of major parts of existing buildings, down to ground level. Demolition works are measured superficially, in square metres, with the nature of the demolition works stated in the description. The area measured is the GIFA of the building, or part of building, to be demolished. Grubbing up existing foundations and piles is measured under sub-element 8.1.2 (Site preparation works).

Demolition works are often carried out in two operations:

1 soft-strip works, which comprise the removal of existing mechanical and electrical services installations, non-structural building components such as partitions and doors, and finishes; and

2 demolition of the external building fabric and structural elements.

Typical examples of descriptions for demolition activities are shown in Example 14.4.

Example 14.4 Formulation of descriptions for major demolition works

FACILITATING WORKS					
Code	**Element/sub-elements/components**	**Qty**	**Unit**	**Rate**	**Total**
				£p	**£**
0	**FACILITATING WORKS**				
0.2	**Major demolition works**				
0.2.1	*Demolition works*				
	Note:				
	Removal and disposal of asbestos within building and remaining chemicals/fuels to old storage tanks are measured under sub-element 0.1.1 (Toxic and hazardous material removal)				
0.2.1.1	Soft strip; (GEA: 1,995m²); comprising:		item		
	(1) removal of all existing non-structural walls and partitions				
	(2) removal of all existing internal doors, screens, balustrades, handrails and the like				
	(3) removal of all existing wall, floor and ceiling finishes				
	(4) removal of all existing fixtures and fittings, balustrades, handrails and the like				
	(5) surveying existing buildings to locate, identify and isolate existing mechanical services, electrical services and public health systems and installations				
	(6) isolating and disconnecting all statutory mains connections (with the exception of sewer connections) and leaving safe				

Example 14.4 *(Continued)*

FACILITATING WORKS					
Code	**Element/sub-elements/components**	**Qty**	**Unit**	**Rate**	**Total**
				£p	**£**
	(7) decommissioning, draining down and stripping out of all mechanical services, electrical services and public health systems and installations, including all plant and equipment (including lifts, roof-top plant and equipment, and plant screens)				
	(8) preparation of Health and Safety File				
	Existing warehouse (old printers)				
0.2.1.2	Demolition of entire building, comprising robust two-storey reinforced concrete-framed superstructure with deep haunched concrete beams at first floor level; pitched roof construction of steel roof trusses with metal cladding; wall construction of brick/block cavity wall construction; taking down to ground floor slab level		m^2		
	Existing office building				
0.2.1.3	Demolition of major parts of existing building; existing cores, including reinforced concrete shear walls, adjacent floor slabs and staircases; height of core: 6 floors; (as shown on drawing nr SE/2953/DEM/009)		m^2		
0.2.1.4	Demolition of major parts of existing building; existing roof slab; prestressed concrete; including parapet walls and insulated asphalt roof coverings; approximately 360m^2 on plan; (as shown on drawing nr. SE/2610/DEM/004)		item		
0.2.1	**TOTAL – Demolition works: to element 0.2 summary**				

14.4 Temporary support to adjacent structures (element 0.3)

Element 0.3 comprises a single sub-element as follows:

0.3.1 Temporary support to adjacent structures.

This sub-element deals with the provision of temporary supports to unstable structures that are to be retained (e.g. a series of raking shores upholding a single wall or flying shores between two buildings). Support structures are measured as enumerated (nr) items, with the type of support stated in the description. Components measured are deemed to include cutting holes in existing structures to take the support structure, as well as all foundations. Although not specifically stated by the rules, the removal of support structures is best measured under this sub-element.

Costs in connection with the inspection and maintenance of temporary supports will also need to be considered and allowed for by the cost manager. Although not specifically mentioned by the rules, such costs can be significant, especially where they are likely to be in place for a lengthy period. Furthermore, considerable additional costs can also arise where there is a risk of the temporary works impacting on adjoining properties, particularly where road, rail networks or other operational environments are involved. The cost manager will need to consider how to address such costs in the cost plan. It is suggested that the inspection and maintenance of temporary support measures are initially treated as an 'other significant cost item' and measured under sub-element 0.3 (Temporary support to adjacent structures). Example 14.5 provides typical examples of descriptions for temporary support structures.

Example 14.5 Formulation of descriptions for temporary support works

WORK TO EXISTING BUILDINGS					
Code	**Element/sub-elements/components**	**Qty**	**Unit**	**Rate £p**	**Total £**
0	**FACILITATING WORKS**				
0.3	**Temporary supports to adjacent structures**				
0.3.1	*Temporary supports to adjacent structures*				
	Design and installation of temporary supports measures:				
0.3.1.1	Raking shores; triple raker		nr		
0.3.1.2	Flying shores; single; approximately 9.85m span		nr		
0.3.1.3	Flying shores; double; approximately 14.60m span		nr		
0.3.1.4	Unsymmetrical flying shores; approximately 11.50m span		nr		
0.3.1.5	Dead shore; approximately 3.60m high		nr		
0.3.1.6	Inspection and maintenance of temporary supports (number of inspections measured)		nr		
	Note:				
	Remedial works/making good to existing structures following removal of temporary supports measured and described elsewhere under sub-element 7.1 (Minor demolition and alteration works).				
	Removal of temporary supports, including temporary bases:				
0.3.1.7	Raking shores; triple raker		nr		
0.3.1.8	Flying shores; single; approximately 9.85m span		nr		

Example 14.5 *(Continued)*

WORK TO EXISTING BUILDINGS					
Code	**Element/sub-elements/components**	**Qty**	**Unit**	**Rate**	**Total**
				£p	**£**
0.3.1.9	Flying shores; double; approximately 14.60m span		nr		
0.3.1.10	Unsymmetrical flying shores; approximately 11.50m span		nr		
0.3.1.11	Dead shore; approximately 3.60m high		nr		
0.3.1	**TOTAL – Temporary supports to adjacent structures: to element 0.3 summary**				

14.5 Specialist groundworks (element 0.4)

Element 0.4 is divided into three sub-elements as follows:

0.4.1 Site dewatering and pumping;
0.4.2 Soil stabilisation measures; and
0.4.3 Ground gas-venting measures.

14.5.1 Site dewatering and pumping (sub-element 0.4.1)

Temporary works in connection with lowering the ground water level over the whole of the site to facilitate the construction of the building can be either itemised or measured superficially, with the method of dewatering stated in the description. Where measured superficially, in square metres, the area to be dewatered is measured. Examples of typical descriptions for site dewatering and pumping operations are shown in Example 14.6.

Example 14.6 Formulation of descriptions for site dewatering and pumping

FACILITATING WORKS					
Code	**Element/sub-elements/components**	**Qty**	**Unit**	**Rate**	**Total**
				£p	**£**
0	**FACILITATING WORKS**				
0.4	**Specialist groundworks**				
0.4.1	*Site dewatering and pumping*				
0.4.1.1	Site dewatering; (allowance)		item		
0.4.1	**TOTAL – Site dewatering and pumping: to element 0.4 summary**				

14.5.2 Soil stabilisation measures (sub-element 0.4.2)

Measures to stabilise or improve the bearing capacity or slip resistance of existing ground to facilitate construction (e.g. by injecting or otherwise introducing stabilising materials, by power vibrating, by soil nailing or by ground anchors) are measured superficially, in square metres, with the method used stated in the description. The area measured is the area affected by the soil stabilisation measures.

Soil stabilisation measures might include:

- cement or chemical grouting;
- electrochemical stabilisation;
- sand stowing;
- forming regular pattern of holes, compacting surrounding soil and filling with aggregates or hard fill, all by means of power vibrators;
- soil nailing;
- ground anchors;
- pressure grouting;
- compacting;
- freezing of ground water and subsoil; and
- stabilising soil in situ by incorporating cement with a rotavator.

Example 14.7 illustrates descriptions for typical soil stabilisation measures.

Example 14.7 Formulation of descriptions for soil stabilisation measures

FACILITATING WORKS					
Code	**Element/sub-elements/components**	**Qty**	**Unit**	**Rate**	**Total**
				£p	**£**
0	**FACILITATING WORKS**				
0.4	**Specialist groundworks**				
0.4.2	*Soil stabilisation measures*				
0.4.2.1	Soil stabilisation measures; bank stabilisation, using erosion control mats		m²		
0.4.2.2	Soil stabilisation measures; in situ by incorporating cement in substrate using rotavator		m²		
0.4.2	**TOTAL – Soil stabilisation measures: to element 0.4 summary**				

14.5.3 Ground gas-venting measures (sub-element 0.4.3)

Ground gas-venting measures (i.e. systems employed to prevent accumulation of radon or landfill gases) are measured and described in the same way as soil stabilisation measures. The area measured is the area affected by the ground gas-venting measures. Such measures might include:

- gas-proof membranes;
- perforated collection pipes;
- proprietary gas dispersal fin layers;
- radon sumps; and
- vent pipes, including vertical risers to vent at high level.

Typical descriptions for ground gas-venting measures are shown in Example 14.8.

Example 14.8 Formulation of descriptions for ground gas-venting measures

FACILITATING WORKS					
Code	Element/sub-elements/components	Qty	Unit	Rate	Total
				£p	£
0	**FACILITATING WORKS**				
0.4	**Specialist groundworks**				
0.4.3	*Ground gas-venting measures*				
0.4.3.1	Ground gas venting; using gas proof membrane		m^2		
0.4.3.2	Ground gas venting; using perforated collection pies		m^2		
0.4.3.1	Ground gas venting; using proprietary gas dispersal fin layers		m^2		
0.4.3	**TOTAL – Ground gas-venting measures: to element 0.4 summary**				

14.6 Temporary diversion works (element 0.5)

Element 0.5 comprises a single sub-element as follows:

0.5.1 Temporary diversion works.

Temporary diversion works in connection with existing drainage systems, existing services installations and systems, rivers, streams and other waterways required to facilitate construction are itemised. Descriptions for temporary diversion works are to include the nature of the works.

Example 14.19 provides typical descriptions for temporary diversion works.

Example 14.9 Formulation of descriptions for temporary diversion works

FACILITATING WORKS					
Code	Element/sub-elements/components	Qty	Unit	Rate	Total
				£p	£
0	**FACILITATING WORKS**				
0.5	**Temporary diversion works**				
0.5.1	*Temporary diversion works*				

Example 14.9 *(Continued)*

FACILITATING WORKS					
Code	**Element/sub-elements/components**	**Qty**	**Unit**	**Rate**	**Total**
				£p	**£**
0.5.1.1	Temporary diversion of drains; surface water sewer; reconnecting on completion		item		
0.5.1.2	Temporary diversion of services; buried HV cable, 6350/11,000 volts; including reconnecting on completion		item		
0.5.1.3	Temporary diversion of stream; by use of small diversion ditch; reinstating stream on completion		item		
0.5.1	**TOTAL – Temporary diversion works: to element 0.5 summary**				

14.7 Extraordinary site investigation works (element 0.6)

Element 0.6 is divided into three sub-elements as follows:

0.6.1 Archaeological investigation;

0.6.2 Reptile/wildlife mitigation measures; and

0.6.3 Other extraordinary site investigation works.

14.7.1 Archaeological investigation (sub-element 0.6.1)

Physical archaeological investigation works (e.g. site-based excavation works in search of artefacts and the like), which encompass attendance by the main contractor or the archaeologist(s), are itemised (item). Details of the work required to be executed by the main contractor are to be given in the description. Where information about the site archaeology is unknown, an appropriate risk allowance should be made for carrying out both desk-based and physical archaeological investigation works.

14.7.2 Reptile and wildlife mitigation measures (sub-element 0.6.2)

Works in connection with relocating of reptiles and wildlife, including the provision of fences and barriers to cordon off the working area, are itemised (item), with the nature of the work described (see Example 14.10).

Example 14.10 Formulation of descriptions for reptile and wildlife mitigation measures

FACILITATING WORKS					
Code	**Element/sub-elements/components**	**Qty**	**Unit**	**Rate**	**Total**
				£p	**£**
0	**FACILITATING WORKS**				
0.6	**Extraordinary site investigation works**				
0.6.2	*Reptile and wildlife mitigation measures*				
0.6.2.1	Trapping and relocating slow worms; including slow worm fencing (approximately 300m)		item		
0.6.2	**TOTAL – Reptile and wildlife mitigation measures: to element 0.6 summary**				

14.7.3 Other extraordinary site investigation works (sub-element 0.6.3)

Works associated with other extraordinary site investigation works are also itemised (item). The type of work should be stated in the description.

CHAPTER 15

Substructure (group element 1)

Introduction

This chapter:

- provides step-by-step guidance on how to measure and describe components relating to substructure works;
- describes the principal risks associated with excavation and earthworks and how to deal with them;
- contains worked examples showing how to measure key dimensions; and
- gives examples illustrating how to formulate descriptions for components.

15.1 Method of measurement

The rules for measuring the substructure works are found in group element 1 of NRM 1, which comprises a single element:

1.1 Substructure.

This element covers works associated with:

- strip foundations;
- isolated pad foundations;
- piled foundations;
- caissons;
- underpinning;
- lowest floor construction, including under-floor drainage installations;
- basement excavation; and
- retaining walls enclosing basements.

15.2 Risks associated with substructure works

When measuring substructures, there are a number of common, but cost-significant, risks that need to be considered by the cost manager. Such risks include:

- ground water;
- tidal conditions;
- underground obstructions, including bands of hard material;
- contaminated material; and
- other risks associated with unknown ground conditions.

The presence and extent of risks in the ground will either be confirmed or otherwise, as more detailed geotechnical, environmental and archaeological investigations are completed.

15.2.1 Ground water

It is important to appreciate the various categories into which the problem of water can fall. There is water in the form of rain or other water falling directly into excavations or flowing from a nearby surface into the excavation. This is known as 'surface water'. There is also the problem of water naturally running through the ground at levels above the bottom of excavations. This is known as 'ground water'.

Encountering ground water when excavating can be a significant risk, as it can have a considerable impact on the contractor's method of working and can result in substantial additional costs as specialist dewatering techniques are needed. Notwithstanding this, the rules do not require a separate item to be measured for excavating below ground water, which infers that an allowance for such working conditions should be included in the unit rate. However, the rules permit an 'extra over' item to be measured and described if required (by treating the disposal of ground water as an 'other significant cost component' in accordance with the measurement rules for components). Alternatively, an allowance can be made within the construction risk allowance for excavating below ground water level.

Care must be taken when quantifying and costing works associated with the removal of ground water, as the level at which ground water will be encountered will change depending on what time of year the excavation work is to be undertaken.

15.2.2 Tidal conditions

Construction works carried out adjacent to the sea, the coast, estuaries, rivers and the like will undoubtedly be affected by tidal conditions. Not only can tidal conditions impact on when works can be undertaken (e.g. only working at low tide), they can also have a significant impact on the way in which the contractor can execute the works (e.g. plant restrictions due to weight, to avoid becoming stuck in sand). Therefore, the cost manager will need to make adequate allowance in the cost plan for the extra cost of the working restrictions imposed by tidal conditions. Although the rules are silent on this matter, construction works to be executed in tidal conditions should be separately identified and annotated accordingly.

15.2.3 Underground obstructions, bands of hard material and the like

Where information about the ground strata is unknown, the rules recommend that an allowance be made within the construction risk allowance for breaking through obstructions.

However, if details of underground obstructions, bands of hard material and the like are known, the rules permit an 'extra over' item to be measured and described should it be required (i.e. by treating the removal of underground obstructions, bands of hard material and the like as 'other significant cost components' in accordance with measurement rules for components).

15.2.4 Contaminated material

Similarly, where it is unknown if the ground contains any contaminated material (i.e. non-hazardous or hazardous material), the rules recommend that an allowance be made within the construction risk allowance for the extra cost associated with the disposal of contaminated material found during excavation. Where investigations and tests for environmental contamination and geotechnical properties have been completed, and the extent of any potential contamination has been identified, categorised and quantified by the environmental consultant, the rules permit items to be measured and described for the disposal of contaminated materials. The volume of contaminated material to be disposed off site is measured as 'extra over' the volume of disposing of excavated material. This allows the extra costs associated with the disposal of contaminated material to be calculated and identified separately.

15.2.5 Other risks associated with unknown ground conditions

There are many other unforeseeable risks that might be encountered during excavation and earthworks, including the discovery of archaeological remains or unexploded devices.

Risks associated with the discovery of protected species or invasive plant growths are discussed in Chapter 14 ('Facilitating works – group element 0').

15.2.6 Risk allowances

It is vital that the cost manager considers the cost implications of potential ground conditions. The cost manager will need to include sufficient allowances in the cost plan to cover the extra costs of dealing with construction risks should they arise.

15.3 Substructures (element 1.1)

The rules for element 1 (Substructures) are divided into the following five sub-elements:

1.1.1 Standard foundations;

1.1.2 Specialist foundations, which covers piled foundations, caissons and underpinning;
1.1.3 Lowest floor construction, which includes drainage under or within the construction;
1.1.4 Basement excavation; and
1.1.5 Basement retaining walls.

Raft foundations (i.e. base slabs with integral foundations) are treated as a lowest ground floor assembly and are, therefore, measured under sub-element 1.1.3 (Lowest floor construction).

15.4 Standard foundations (sub-element 1.1.1)

This sub-element comprises strip foundations and isolated pad foundations.

Strip foundations are measured linear, in metres, and measured on the centre line of the wall construction, irrespective of whether or not the projections to the concrete footing are equal. Descriptions for strip foundations are to include reference to excavation and earthworks and the make-up (i.e. the sub-components) of the strip foundation. Should the concrete footing be reinforced concrete, the reinforcement content in kilograms per cubic metre (kg/m^3) should be stated. Moreover, the principal dimensions that quantify the sub-components should also be stated in the description, namely: width and depth of concrete footing, height of wall construction and total depth of foundation. The total depth of the foundation is measured from ground level to the underside of the concrete foundation. The sub-components and the principal dimensions for typical strip foundation types are shown in Figure 15.1.

Isolated pad foundations (sometimes referred to as 'pad' foundations) are measured and described as enumerated items (nr). The construction and dimensions of key sub-components are to be stated in the description, including the content of reinforcement in the concrete pad (in kg/m^3) and overall size (i.e. length, width and depth). The total depth of the pad foundation is measured from ground level to the underside of the concrete foundation. Figure 15.2 illustrates the sub-components and the principal dimensions to be stated for a typical isolated pad foundation.

Pile caps, ground beams between pile caps and isolated ground beams are measured and described under sub-element 1.1.2 (Specialist foundations).

It is not uncommon for beams to be used to connect different foundation types, for example, used to connect strip foundations to pad foundations or to link isolated pad foundations to one another (see Figure 15.3). It is recommended that beams used in such situations are measured and described in the same way as ground beams (refer to paragraph 15.5.1 – Piled foundations and caissons), but included in the cost plan as a standard foundation component.

Where measured under this sub-element, the disposal of contaminated excavated material is dealt with as an 'extra over' item, with the volume measured in cubic metres. Quantities given for disposal of contaminated excavated material are the bulk before excavating, with no allowance made for subsequent variations to bulk or for extra space to accommodate earthwork support.

Typical component descriptions for standard foundations are shown in Example 15.1.

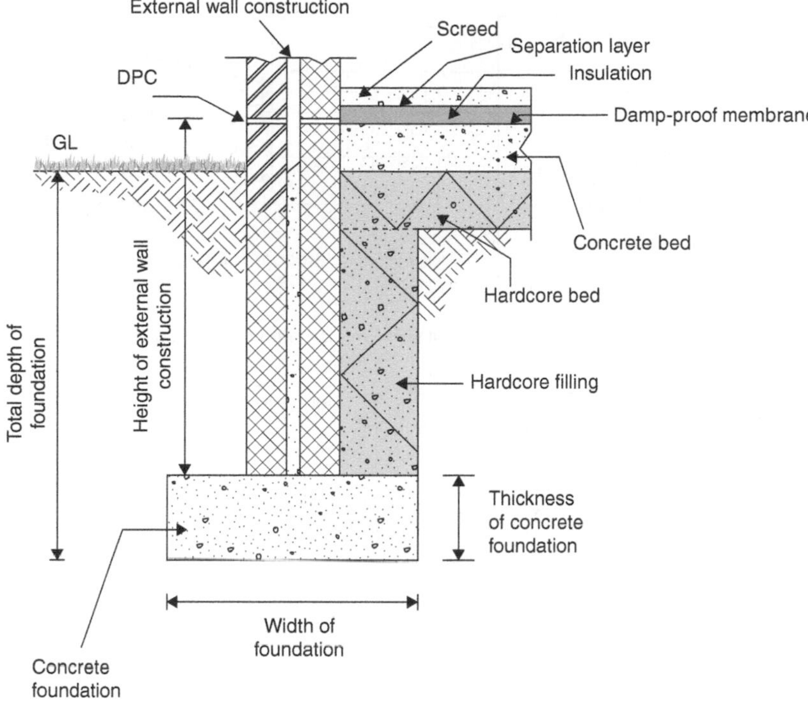

(a) Traditional strip foundation

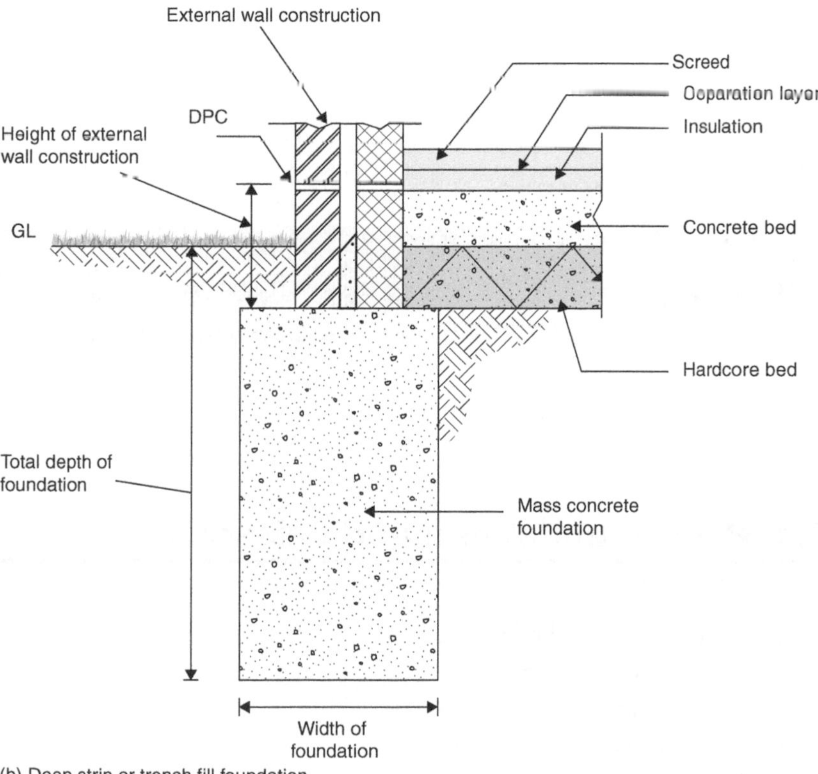

(b) Deep strip or trench fill foundation

Figure 15.1 Sub-components and the principal dimensions for typical strip foundation.

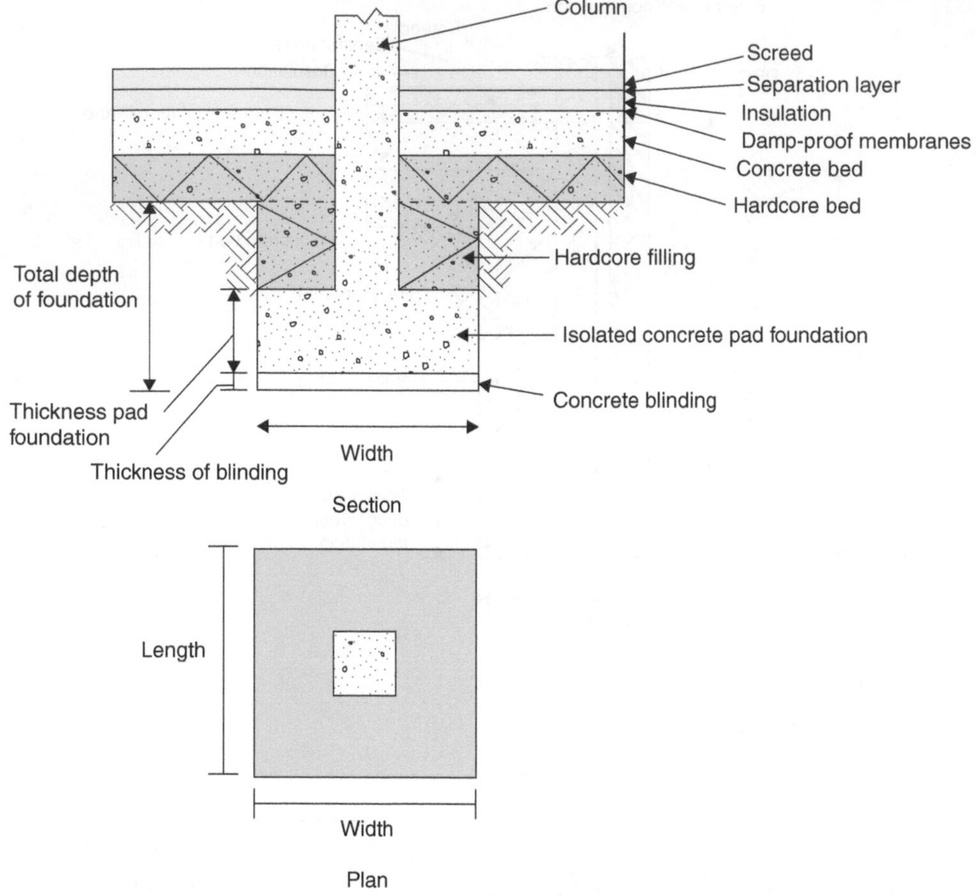

Figure 15.2 Sub-components and the principal dimensions for isolated pad foundations

15.5 Specialist foundations (sub-element 1.1.2)

NRM 1 divides sub-element 1.1.2 (Specialist foundations) into two principal categories, namely:

- piled foundations, including caissons; and
- underpinning.

15.5.1 Piled foundations and caissons

Piled foundations include:

- precast concrete reinforced piles;
- precast prestressed concrete piles;
- precast reinforced segmental concrete piles;
- bored cast-in-place concrete piles;
- driven cast-in-place concrete piles;
- steel bearing piles;

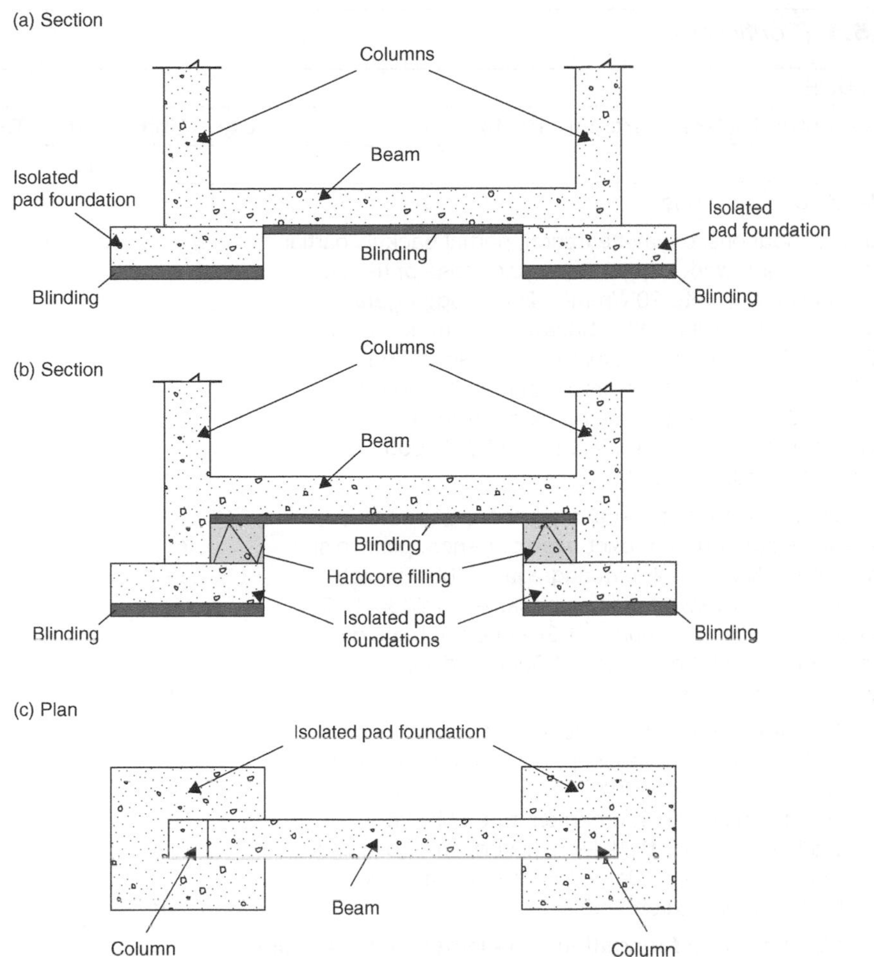

Figure 15.3 Beams connecting isolated pad foundations

Example 15.1 Formulation of descriptions for standard foundations

SUBSTRUCTURE					
Code	**Element/sub-elements/components**	**Qty**	**Unit**	**Rate**	**Total**
				£p	**£**
1	**SUBSTRUCTURE**				
1.1	**Substructure** Notes: (1) Breaking through obstructions: an allowance for the extra cost of breaking through or excavating obstructions is included within the construction risk allowance (2) Disposal of contaminated material: an allowance for the extra cost of disposing of contaminated material arising from excavations is included within the construction risk allowance				

Example 15.1 *(Continued)*

SUBSTRUCTURE					
Code	**Element/sub-elements/components**	**Qty**	**Unit**	**Rate**	**Total**
				£p	**£**
1.1.1	*Standard foundations*				
1.1.1.2	Strip foundations; excavate trench, partial backfill, partial disposal, earthwork support, compact base of trench, plain in situ concrete 20N/mm^2 – 20mm aggregate – 300mm thick, cavity wall in brickwork (common bricks – PC Sum £200/1,000)/blockwork in cement mortar (1:3) – 1.30m high, damp-proof courses; 1,000mm wide × 400mm deep; including extra for three course of facing bricks to outer skin (PC Sum £330/1,000); total depth of foundation 1.40m		m		
1.1.1.3	Trench fill foundations; excavate trench, disposal, earthwork support, compact base of trench, plain in situ concrete 20N/mm^2 – 20mm aggregate, 300mm high cavity wall in brickwork (facing bricks – PC £330/1,000)/ blockwork in cement mortar (1:3) – 400mm high, damp-proof courses; 450mm wide × 1,300mm deep; total depth of foundation 1.40m		m		
1.1.1.4	Isolated pad foundations; excavate pit, partial backfill, partial disposal, earthwork support, compact base of pit, reinforced concrete 30N/mm^2 – 20mm aggregate reinforced concrete (reinforcement rate 50kg/m^3), formwork; on 100mm thick lean mix concrete blinding; pad foundation size 1.50m × 1.50m × 600mm thick, total depth of pad foundation 1.60m		nr		
1.1.1	**TOTAL – Standard foundations: to element 1.1 summary**				

- timber bearing piles;
- mini piles;
- permanent caissons;
- vibro-compacted columns;
- pile caps; and
- ground beams.

Piles that are an integral part of an external retaining wall structure (e.g. a site boundary wall external to the building itself) are dealt with under element 8.4 (Retaining walls – i.e. those comprising part of the external works).

There are a number of cost-significant activities and components associated with the installation of piled foundations. Such items that need to be considered by the cost manager include the following:

- piling mats, or piling platforms;
- piling rigs and other equipment used specifically for the purpose of piling;
- piles and caissons;

- disposal of excavated material;
- cutting off the tops of piles;
- testing piles; and
- pile caps and ground beams.

Piling mats and piling platforms are temporary foundations for piling rigs and other heavy ground improvement machinery used to transfer the load as well as aid movement, particularly when working over very weak ground. It is common for piling mats and piling platforms to be constructed by a groundworks contractor. However, a very small soft spot can be sufficient to unbalance a piling rig and cause it to topple over. Therefore, it has become increasingly common for a piling specialist to insist on a 'Working Platform Certificate (WPC)' being provided by either the main contractor or the groundworks contractor. The WPC has been developed by the Federation of Piling Specialists in conjunction with the Health and Safety Executive in order to raise awareness of the importance of providing a proper platform for the plant that will work on it – and the importance of maintaining it during the course of the piling works.

Piling mats and piling platforms are measured superficially, in square metres, with the nature and thickness of the material to be used stated in the description. The description should also include details about the removal and the disposal of the piling mat or piling platforms on completion of piling operations. Whether or not the piling mat or piling platform is to be taken up and repositioned also needs to be considered by the cost manager. The structural engineer should be able to advise on the nature and thickness of the piling mat or platform, as well as methodology to be adopted. Alternatively, advice could be sought from a piling contractor. The area measured for piling mats or platforms is the surface area of the piling mat or platform.

Care must be taken to avoid duplicating the provision for piling mats or piling platforms. When a building comprises both ground-bearing piles and, say, a secant or contiguous piled walls, it will need to be decided under which sub-element the piling mat or piling platform is to measured – either sub-element 1.1.2 (Specialist foundations) or sub-element 1.1.5 (Basement retaining walls).

Piling rigs and other equipment specific to the piling contractor are itemised. Where known, the type and capacity of the piling rig should be given in the description. The unit rate used by the cost manager for the provision of piling rigs is to include bringing the rig to site, erection of the rig, operation of the rig (including the cost of the operator), maintenance of the rig, and dismantling and removing the rig on completion of the piling works. Moving the piling rig to each pile position is treated as a separate item, with the number of movements enumerated. The number of piles served by each movement of the piling rig should also be stated in the description.

If there is a likelihood of rig standing time, the cost manager will need to make sufficient allowance in the cost plan within construction risks.

Piles are measured as enumerated items, with the type (e.g. bored cast-in-place concrete pile or vibro-compacted columns), diameter and depth of pile stated in the description. The disposal of excavated material arising from piling is measured by volume, in cubic metres. Quantities given for disposal of excavated material are the bulk before excavating, with no allowance made for subsequent variations to bulk. The cost manager will need to make allowance for variations to bulk in the unit rate disposal of excavated material arising from piling.

Figure 15.4 illustrates the sub-components and the principal dimensions to be stated for components of a piled foundation.

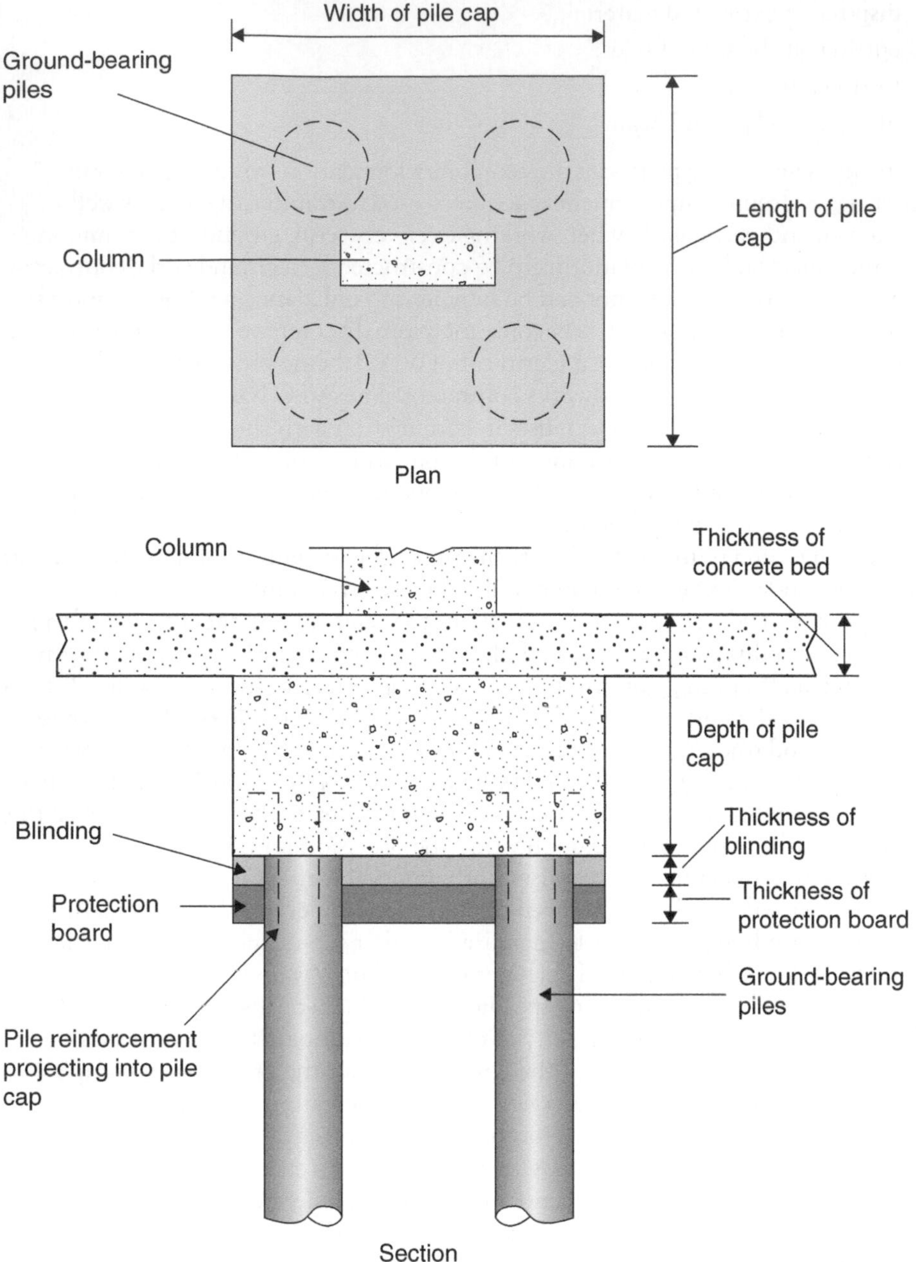

Figure 15.4 Sub-components and the principal dimensions for piled foundations

Cutting off the tops of concrete piles is enumerated (nr), with the diameter of the pile stated in the description. Such work includes cutting off excess lengths of piles, cutting out concrete tops to piles and preparing the pile heads (including reinforcement) in readiness to receive pile caps, ground beams or the like.

Where there is a risk of obstructions, bands of hard material or the like being encountered within the ground, then sufficient allowance should be made by the cost manager within the construction risk allowance for the extra cost of breaking through obstructions. In the same way, if there is a risk that contaminated material will be discovered within the ground, the cost manager should include a risk allowance for the extra cost of disposing of such hazardous material.

Pile tests can be itemised or measured as enumerated (nr) items, with the nature of the testing given in the description (e.g. load tests).

Caissons are measured in the same way as piles. Caissons are hollow cylindrical or boxlike structures, usually of reinforced concrete or precast concrete, sunk into a riverbed or the like to form the foundations of a structure. Once positioned, the hollow structures are normally filled with concrete.

Pile caps are measured and described as enumerated items (nr), with their composition, including the reinforcement content in kg/m^3, and overall size (i.e. length, width and depth) stated in the description. Depths of pile caps are measured from the top of the concrete base slab (i.e. the top of the pile cap) to the bottom of the pile cap. Details of construction are to include any blinding material and any protection boarding (e.g. heave protection board). Each different size of pile cap is to be measured and described separately.

Ground beams connecting pile caps and isolated ground beams are measured separately as linear items (m), measured on the centre line of the component. The description for ground beams is to include their composition, including the reinforcement content in kg/m^3, overall size (i.e. width and depth – measured in the same way as for pile caps) and details of any blinding material and any protection boarding. Each different size of ground beam is to be measured and described separately.

When measuring the ground floor construction, the area must be adjusted to compensate for the area of the ground floor construction displaced by pile caps and ground beams – as pile caps and ground beams pass through the concrete base slab and are measured and priced accordingly. Significant over-measurement and over-pricing of the sub-element might occur should this not be considered and addressed by the cost manager.

Figure 15.5 shows a typical piled foundation arrangement.

Based on the piled foundation arrangement in Figure 15.5, the initial measurement and supporting calculations are as shown in Example 15.2.

Using the initial measurements above, read in conjunction with all relevant design information, the cost manager can determine a cost target for piled foundations; by identifying the components to be addressed, formulating component descriptions, quantifying components and applying applicable unit rates. The computation of the cost target for the piled foundation arrangement (see Figure 15.6) is shown in Example 15.3. The example also includes notes that explain the approach taken by the cost manager.

The importance of including notes about assumptions, and cross-referencing to other elements, in the cost plan should not be underestimated. As previously mentioned, such notes will be invaluable to the cost manager when communicating the content of the cost plan to others, as well as make future cost-checking processes more straightforward.

15.5.2 Underpinning

Underpinning is the work of deepening or strengthening the foundations to prevent or rectify fractures, damage or settlement of existing walls. The work comprises inserting additional foundation support under and around the existing foundations.

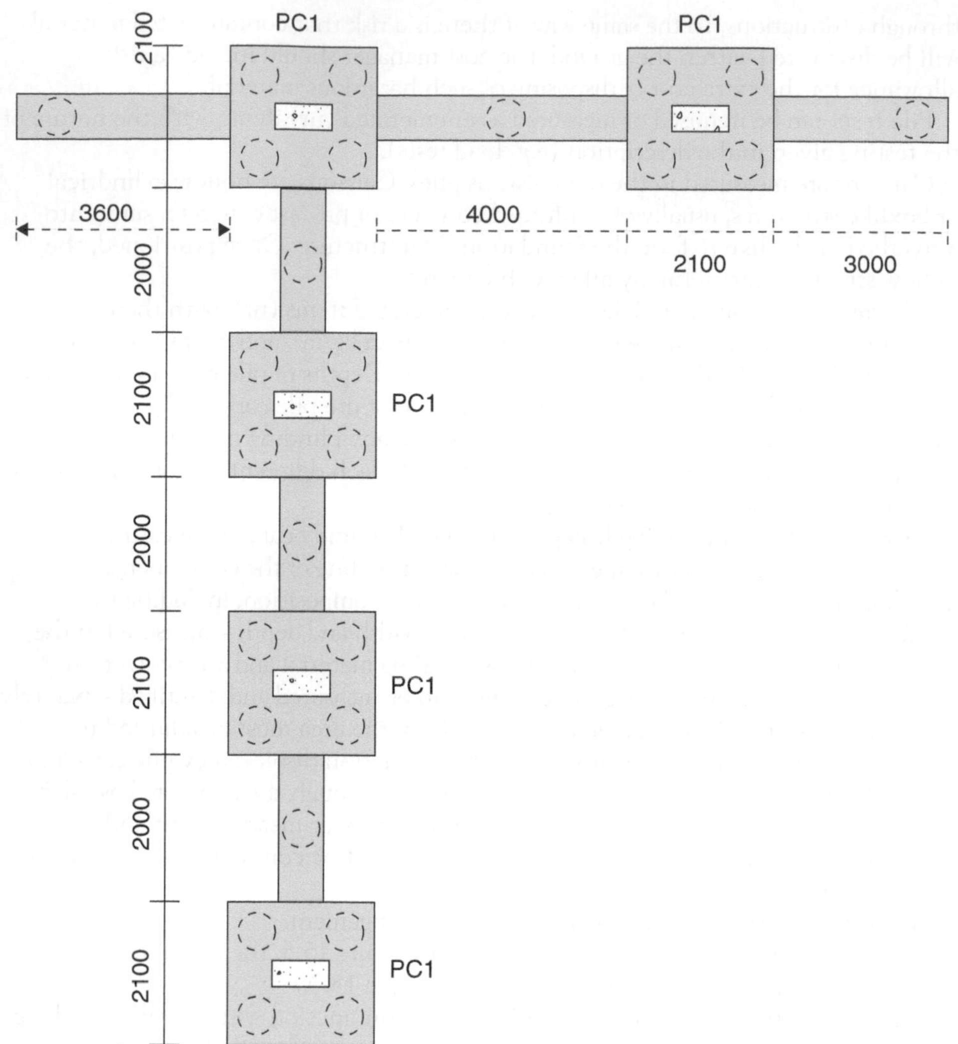

Figure 15.5 Piled foundation arrangement

Work in underpinning is measured linear, in metres, with the extreme length measured (i.e. the actual length of the wall to be underpinned – see Figure 15.6). Descriptions for underpinning are to state the extent of work and the methodology envisaged. The structural engineer should be able to advise the cost manager on the underpinning methodology to be adopted.

If it is uncertain as to whether or not underpinning is required, an allowance should be made by the cost manager within either the design development or construction risk allowance for underpinning of existing foundations.

Underpinning to walls within existing buildings are measured under group element 7 (Works in existing buildings), and underpinning to walls which are an integral part of the external works, including boundary walls, is measured under group element 8 (External works).

A typical description for underpinning works is shown in Example 15.4. Again, notes have been included to explain and amplify the cost manager's approach.

Example 15.2 Record of initial measurement and quantification of piled foundation components

			Piled foundations	
	30.00 10.00		Allowance for pile mat	
		300.00	m²	
5/	4 5	20 5	Piles; 650mm diameter; total length 15.00m	
		25	nr	
25/∏/	0.33 0.33 15.00		Disposal of excavated material arising from piling; off site	Volume of material arising from piling which is to be disposed of is calculated by multiplying the cross-sectional area of pile by total length of pile
		136.87	m³	
	5		Pile caps; 2.10 × 2.10 × 1.25m	
	3.00 4.00 3.60		Ground beams; 1.80 × 1.25m	
2/	2.00	4.00		
		15.60	m	

Example 15.3 Formulation of descriptions for piled foundations

Code	Element/sub-elements/components	Qty	Unit	Rate	Total
SUBSTRUCTURE					
				£p	£
1	**SUBSTRUCTURE**				
1.1	**Substructure**				
	Notes: (1) Piling mat: it has been assumed that a piling mat will be required over the entire site (2) Breaking through obstructions: an allowance for the extra cost of breaking through or excavating obstructions is included within the construction risk allowance (3) Disposal of contaminated material: an allowance for the extra cost of disposing of contaminated material arising from excavations is included within the construction risk allowance (4) Ground-bearing piles: it has been assumed that piling will be completed in a continuous operation. An allowance for the extra cost of rig standing time is included within the construction risk allowance				

Example 15.3 *(Continued)*

SUBSTRUCTURE					
Code	**Element/sub-elements/components**	**Qty**	**Unit**	**Rate**	**Total**
				£p	**£**
	(5) Ground-bearing piles: it has been assumed that rotary bored piles will be used. An allowance for the extra cost of employing CFA (continuous flight auger) piles or cased flight auger piles (CFP) is included within the construction risk allowance.				
	(6) Pile caps and ground beams: in accordance with the rules, all excavation, disposal of excavated material, earthwork support and excavating and filling of working space are included within the items. Costs in connection with these activities will need to be shown separately if groundworks and concrete works items are to be executed by different works contractors.				
	(7) Pile caps and ground beams: laid on 50mm thick sand blinding, 150mm heave protection board, 100mm thick lean mix concrete blinding – sub-components measured with ground floor construction (sub-element 1.1.3).				
1.1.2	*Specialist foundations*				
1.1.2.1	Piling mat/platforms; imported granular material; 450mm thick; taking up and disposing on completion of piling operations		m²		
1.1.2.2	Piling plant; rig		item		
1.1.2.3	Moving piling rig to pile position		nr		
1.1.2.4	Piles; auger bored; 650mm diameter; total depth 15.00m		nr		
1.1.2.4	Disposal of excavated material arising from piling; off site		m³		
1.1.2.5	Cutting off tops of concrete piles: 650mm diameter		nr		
1.1.2.6	Pile tests; load tests on 650mm diameter pile, including reaction test		item		
1.1.2.7	Pile tests; integrity tests; minimum 20 piles per visit		nr		
1.1.2.8	Pile caps; type PC1; excavate pit, partial backfill, partial disposal, earthwork support, compact base of pit; reinforced concrete 35N/mm² – 20mm aggregate reinforced concrete (reinforcement rate 100kg/m³), formwork; pile cap size 2.10m × 2.10m × 1.25m deep		nr		
1.1.2.9	Ground beams; excavate trench, partial backfill, partial disposal, earthwork support, compact base of trench, reinforced concrete 35N/mm² – 20mm aggregate reinforced concrete (reinforcement rate 75kg/m³), formwork; ground beam size 1.80m wide × 1.25m deep		m		
1.1.2	**TOTAL – Specialist foundations: to element 1.1 summary**				

Note:
The method used to calculate the unit rates for the pile caps (Code 1.1.2.8) is shown in Chapter 23 ('Deriving unit rates for building components, sub-elements and elements')

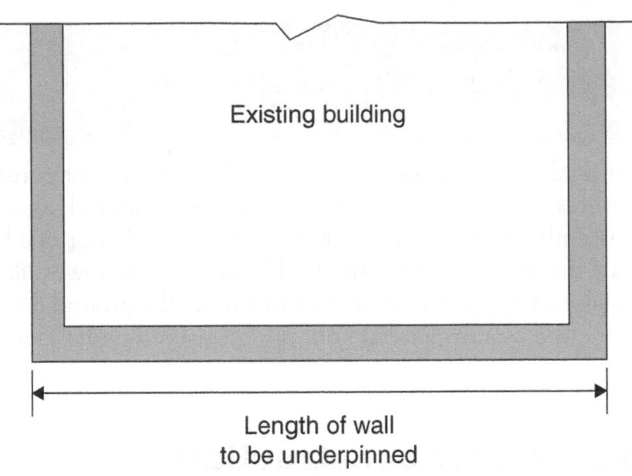

Figure 15.6 Measurement of underpinning

Example 15.4 Formulation of descriptions for underpinning

Code	Element/sub-elements/components	Qty	Unit	Rate	Total
SUBSTRUCTURE					
				£p	£
1	**SUBSTRUCTURE**				
1.1	**Substructure**				
1.1.2	*Specialist foundations*				
	Notes:				
	(1) Underpinning to adjoining property: an allowance for the extra cost of disposing of contaminated material arising from excavations has been included in the construction risk allowance				
	(2) Underpinning to adjoining property: method of underpinning based on structural engineer's initial advice				
1.1.2.1	Underpinning; in stages as specified by structural engineer (assumed 1.50m long); from one side of existing wall and foundation, including excavating preliminary trench, partial backfill, partial disposal, earthwork support, cutting away projecting foundation, prepare underside of existing foundation, compact base of pits, plain in situ concrete 20N/mm² – 20mm aggregate, formwork, brickwork in cement mortar (1:3), damp-proof course, wedge and pin to underside of existing with slates; 1500mm high, one brick wall; commencing 2.00m below ground level		m		
1.1.2	**TOTAL – Specialist foundations: to element 1.1 summary**				

15.6 Lowest floor construction (sub-element 1.1.3)

This sub-element deals with the measurement of the lowest floor construction, which is defined by the rules as being the 'entire lowest floor [structural] assembly'. However, the lowest floor assembly will not necessarily be the ground floor of a building. If, for example, there is a basement beneath the building, then the lowest floor assembly will be the base slab to the basement. In this instance, the ground floor to the building is to be treated as a suspended floor and measured under element 2.2 (Upper floors).

15.6.1 Lowest floor construction

The lowest floor construction consists of the whole floor assembly, including all drainage beneath and assembly (i.e. within the footprint of the building). In the case of a reinforced concrete bed the assembly includes all excavation and earthworks, hardcore, blinding material, concrete beds, thickenings to concrete beds, reinforcement, protection boards, formwork, worked finishes to unset concrete, and damp-proof membranes (DPMs) and insulation (alternatively insulation can be measured with internal floor finishes). Similarly, a timber suspended floor assembly will include all floor boarding, wall plates, damp-proof courses, sleeper walls, concrete bed and hardcore, as well as all excavation and earthworks. Screeds, post-installation treatments to floor boarding (e.g. sanding and polishing) and site-applied finishings to floors are measured under element 3.2 (Finishes to floors).

The lowest floor construction is measured superficially, in square metres, with the composition of the floor construction, including the thickness of each sub-component, stated in the description. Descriptions for reinforced concrete slabs, or beds, are to include reinforcement content (in kg/m^3), together with details of concrete additives and surface treatments. The area measured is the surface area of the floor construction, measured to the internal face of the external perimeter walls. The methodology for measuring the gross internal floor area (GIFA) of a building is used to interpret the area to be measured.

Care must be taken when calculating the area for ground floor construction when pile caps and ground beams are involved. This is because the depth of pile caps and ground beams are measured from the top of the concrete base slab (i.e. the top of the pile cap or ground beam) to the bottom of the pile cap or ground beam. The depth includes the thickness of the concrete base slab. Therefore, the area measured for ground floor construction must be adjusted to compensate for the area of the lowest floor construction displaced by pile caps and ground beams. Failure to adjust the area might result in significant over-measurement and over-pricing of the sub-element.

Ramps and the like which are an integral part of the lowest floor construction are also measured superficially, in square metres, as 'extra over' items. The area is measured in the same way as the lowest floor construction.

Lift pits and the like are also described as 'extra over' items. They are measured as enumerated items (nr), with the overall size (internal dimensions, measured on plan) stated in the description.

Retaining walls at changes in levels within the floor construction are measured linear, in metres, with their extreme length over all obstructions measured.

Descriptions for retaining walls are to include details of their composition, height, overall thickness and reinforcement content (in kg/m³), together with the thickness of each sub-component. The height of the retaining wall is measured from the top of the bed to the underside of the attached slab.

Forming swimming pool tanks, within the lowest floor construction, is measured superficially, in square metres, and described as an 'extra over' item. The area is measured on plan, measured to the internal face of the walls to the swimming pool. Details of its composition, together with the overall dimensions, are to be given in the description. Other similar facilities are measured in the same way.

Designed joints at junctions between beds and external walls or retaining walls, and where used to divide the bed into bays, are measured linear, in metres. The extreme length, over all obstructions, is measured. The height and composition of design joints are to be stated in the description.

Sample descriptions of typical lowest floor construction components are shown in Example 15.5.

Example 15.5 Formulation of descriptions for lowest floor construction

SUBSTRUCTURE					
Code	**Element/sub-elements/components**	**Qty**	**Unit**	**Rate**	**Total**
				£p	**£**
1	**SUBSTRUCTURE**				
1.1	**Substructure**				
1.1.3	*Lowest floor construction*				
1.1.3.1	Ground slab construction; excavation to reduce levels, disposal of surplus excavated material off site, level and compact, 250mm thick hardcore bed blinded with sand, 1200 gauge polythene DPM; 300mm thick reinforced concrete slab, (30N/mm³ – reinforcement rate 65kg/m³), floated finish		m²		
1.1.3.2	Extra over for forming ramp		m²		
1.1.3.3	Extra over for forming steps		m²		
1.1.3.4	Extra over for forming lift pits; internal size, 3.40m × 2.60m × 1.40m deep		nr		
1.1.3.5	Retaining walls at changes in levels; reinforced concrete, 200mm thick (reinforcement rate 120kg/m³), including formwork to both sides; 1.30m high		m		
1.1.3.6	Designed joints; in 350mm thick concrete bed		m		
	Note:				
	Where no information is available about internal drainage, quantification can be based on area of the ground floor construction				
1.1.3.7	Drainage below ground; excavation, disposal of surplus excavated material off site, vitrified clay pipes and fittings, backfill and compact with granular material, not exceeding 1.50m deep, 100 to 150mm nominal size		m		

Example 15.5 *(Continued)*

SUBSTRUCTURE					
Code	**Element/sub-elements/components**	**Qty**	**Unit**	**Rate**	**Total**
				£p	**£**
1.1.3.8	Internal manholes; excavation, disposal of surplus excavated material off site, earthwork support, compact base, in situ concrete 20 N/mm² – 20mm aggregate base, formwork, reinforced concrete shaft rings, cover slabs, clay channels, benching, step irons, medium duty cover and frame; internal diameter 1350mm, depth from cover to invert not exceeding 3.00m		nr		
1.1.3	**TOTAL – Lowest floor construction: to element 1.1 summary**				

15.6.2 Drainage

Drainage situated beneath the lowest floor assembly is also measured as part of the ground floor construction. Drainage pipelines are measured linear, in metres, with their extreme length measured over all fittings, branches and the like. The average depth of trench, the type (e.g. clay, plastics or concrete) and the nominal size of pipe, together with the material used for beds, haunchings and surrounds (e.g. granular material or concrete) are to be stated in the description. Fittings, branches and the like are not measured separately, so the cost manager will need to make allowance in the unit rates applied to drainage pipelines for these ancillary items. However, the rules recognise that gullies, floor outlets and the like can be cost significant items. Therefore, they are to be separately measured as enumerated items (nr), with the nature of the sub-component described.

Internal manholes, catch-pits, petrol interceptors and the like are measured as enumerated items (nr). Descriptions are to include the type and composition of the sub-component (e.g. brick-built manhole, concrete manhole, proprietary shaft rings or proprietary item), together with their approximate depth.

The measurement of drainage systems is fragmented, with the different parts of the system being measured under various elements and sub-elements. These different facets, together with the applicable elements and sub-elements under which they are measured, are summarised in Table 19.1 in Chapter 19 ('Services (group element 5)').

15.7 Basement excavation (sub-element 1.1.4)

Basement excavation consists of bulk excavation required for the construction of floors below ground level. Basement excavation and the disposal of excavated material are both measured in cubic metres. The volume of both items is calculated by multiplying the area of the basement by the average depth of excavation. The area of basement excavation is measured to the external face of the external perimeter walls (basement retaining walls), and the depth of basement excavation is measured from the average existing ground level, or adjusted ground level (i.e. where a new ground level has been established following the removal of topsoil or the like), whichever is applicable, to the formation level (underside) of the basement base slab construction. The description for the disposal of excavated material should state that the material is 'inert'. This is because there are different classifications of waste material (i.e. inert, non-hazardous or hazardous), and the cost of disposing of contaminated material is significantly higher than that for disposing of inert material.

Preparatory groundworks in adjusting site levels are measured under sub-element 8.1.2 (Preparatory groundworks).

Where measured, the disposal of contaminated excavated material classified as either non-hazardous or hazardous is dealt with as an 'extra over' item, with the volume measured in cubic metres. However, where the extent or classification of contaminated material is unknown, an allowance should be included in the allowance for construction risks by the cost manager. Similarly, if there is a risk of obstructions, bands of hard material or the like being encountered within the ground, then the cost manager should make adequate allowance within the construction risk allowance for the extra cost of breaking through obstructions.

In the case of an embedded retaining wall, the piled wall will normally be put in place before the basement is excavated. Therefore, the area to be excavated should be measured to the internal face of the piled wall. This is to avoid double counting, as the disposal of the excavated material arising from piling operations is measured under sub-element 1.1.5 (Basement retaining walls). Again, the area is multiplied by the depth to determine the volume of basement excavation.

Quantities given for excavation and the disposal of excavated material is the bulk before excavating (i.e. the same volume as for basement excavation), with no allowance made for subsequent variations to bulk or for extra space to accommodate earthwork support. Allowance must be made by the cost manager for variations to bulk, and for any extra space required to accommodate earthwork support, when computing the unit rates used for excavation and disposal of excavated material.

Earthwork support in connection with basement excavations is measured as a discrete item. This is because it can be a cost-significant item, particularly if steel sheet piling or the like is used. Earthwork support is measured superficially, in square metres – length multiplied by height. It is measured to the face of all basement excavations (i.e. the external face of the basement retaining wall) to the full depth. The depth, or height, of earthwork support is measured in the same way as the depth of the basement excavation.

Figure 15.7 comprises a typical plan and section of a basement produced by a structural engineer.

Using the information provided in Figure 15.7, the cost manager can calculate initial measurement, and supporting calculations are as own in Example 15.6.

From the initial measurements shown in Example 15.6, the cost manager can determine a cost target for basement excavation as shown in Example 15.7.

Additional excavation required to facilitate construction of basement retaining walls (e.g. where open excavation method is to be employed) is measured superficially, in square metres. The area is measured in the same way as earthwork support. Descriptions for additional excavation are to include the method of excavation and the nature of the material to be used to backfill the over-excavation (e.g. selected excavated material or granular material), together with details of what is to be done with any surplus excavated material. Additional excavation is only measured when an open, or similar, method of excavation has been expressly specified by the structural engineer. Figure 15.8 illustrates the method of measuring additional excavation.

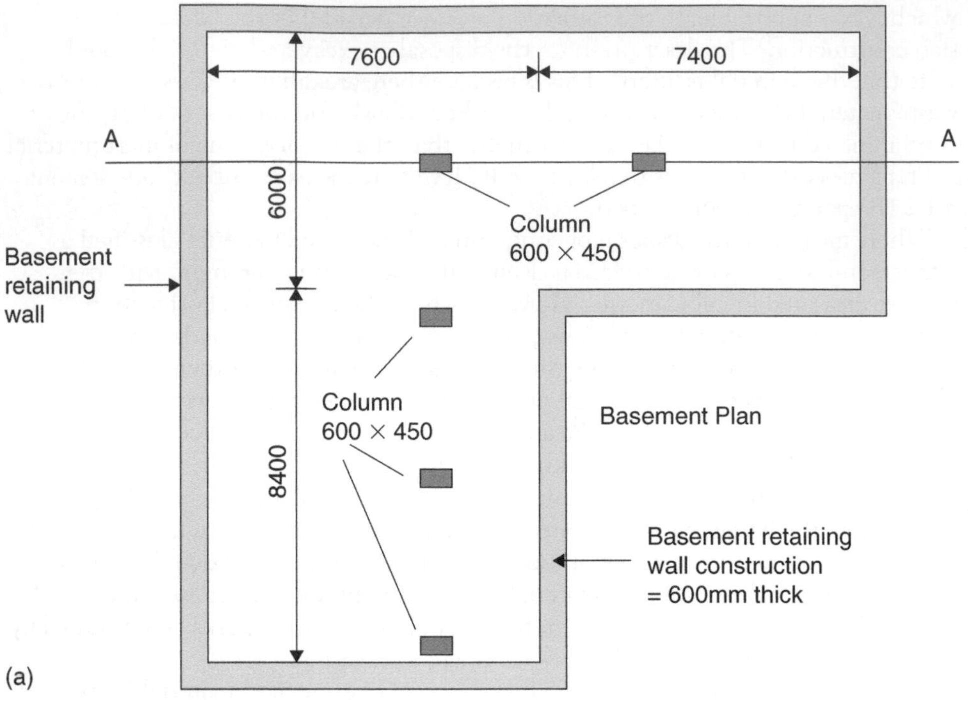

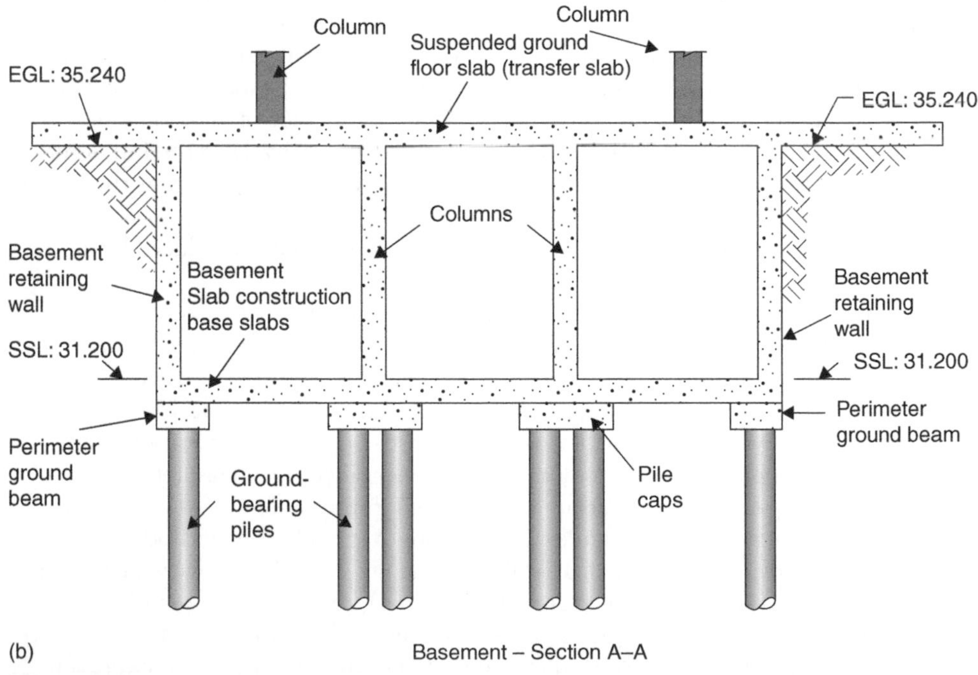

Figure 15.7 Basement (a) plan, (b) section (A–A)

Example 15.6 Record of initial measurement and quantification of basement excavation

			Calculation of plan dimensions of basement excavation:	
			Length / Width	
			7.600 / 8.400	
			7.400 / 6.000	
			15.000 / 15.400	
			ADD	
			Basement retaining wall	
			2/600 — 1.200 / 1.200	
			15.200 / 15.600	
			Calculation of depth of basement excavation:	
			Adjusted ground level — 35.240	Measurement rules for components C2 states:
			LESS	
			Top of base slab (SSL) — 31.200	The depth of basement excavation shall be measured from the formation level to the average of the existing ground level or adjusted ground level (i.e. where a new ground level has been established following the removal of topsoil or the like).
			4.040	
			ADD	
			Concrete bed — 300	
			Concrete blinding — 100	
			Heave protection board — 150	
			Sand blinding — 50	
			600	
			Total depth (m) — 4.640	
	15.20 15.60 4.64	1,172.62	Basement excavation; average depth 4.60m ξ	
	Ddt			
	7.40 8.40 4.64	(288.42)	Disposal of excavated material; inert; off site	Measurement rules for components C4 states:
		884.20	m^3	The quantity given for disposal of excavated material is the bulk before excavating and no allowance is made for subsequent variations to bulk or for extra space to accommodate earthwork support.

Example 15.6 *(Continued)*

	884.20	m^3	Extra for disposal of contaminated excavated material; non-hazardous; off site
		884.20	$m^3 \times 20\% = 176.84\ m^3$
	884.20	m^3	Extra for disposal of contaminated excavated material; hazardous; off site
		884.20	$m^3 \times 10\% = 88.42\ m^3$
2/	15.20 4.64	141.05	Earthwork support, steel sheet piling; extracted
2/	15.60 4.64	144.76	
		285.82	m^2

Example 15.7 Formulation of descriptions and computation of cost target for basement excavation

SUBSTRUCTURE					
Code	**Element/sub-elements/components**	**Qty**	**Unit**	**Rate**	**Total**
				£p	**£**
1	**SUBSTRUCTURE**				
1.1	**Substructure**				
	Notes:				
	(1) Breaking through obstructions: an allowance for the extra cost of breaking through or excavating has been included in the construction risk allowance				
	(2) Disposal of contaminated material: allowance based on Land Remediation Plan produced by environmental/geotechnical consultant				
	(3) Dewatering: measured under element 0.4 (Specialist groundworks)				
	(4) Piling mat/platforms: included with piling mat/ platform measured under sub-element 1.1.2 (Specialist foundations)				
1.1.4	*Basement excavation*				
1.1.4.1	Basement excavation; average depth 4.50m	884	m^3	6.00	5,304
1.1.4.2	Disposal of excavated material; inert material; off site	884	m^3	25.00	22,100

Example 15.7 *(Continued)*

SUBSTRUCTURE					
Code	**Element/sub-elements/components**	**Qty**	**Unit**	**Rate**	**Total**
				£p	**£**
1.1.4.3	Extra over disposal of excavated material for disposal of contaminated excavated material; non-hazardous material; off site; (based on 20% of volume of basement excavation)	177	m³	48.00	8,496
1.1.4.4	Extra over disposal of excavated material for disposal of contaminated excavated material; hazardous material; off site; (based on 10% of volume of basement excavation)	88	m³	157.00	13,816
1.1.4.5	Earthwork support, steel sheet piling, using silent vibratory hammer, extraction of sheet piles on completion; (including provision of all plant for installation and extraction)	295	m²	35.00	10,325
1.1.4	**TOTAL – Basement excavation: to element 1.1 summary**				**60,041**

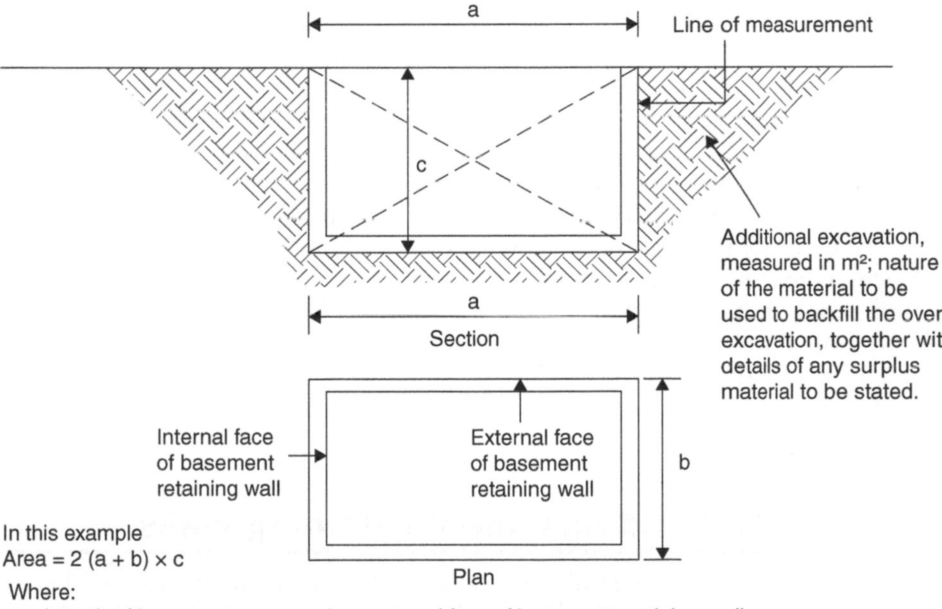
a
Line of measurement
c
Additional excavation, measured in m²; nature of the material to be used to backfill the over-excavation, together with details of any surplus material to be stated.
a
Section
Internal face of basement retaining wall
External face of basement retaining wall
b
Plan
In this example
Area = 2 (a + b) × c
Where:
a = Length of basement measured on external face of basement retaining wall.
b = Width of basement measured on external face of basement retaining wall.
c = Total depth of basement measured on external face of basement retaining wall.

Figure 15.8 Measurement of additional excavation

15.8 Basement retaining walls (sub-element 1.1.5)

NRM 1 divides basement retaining walls into two categories, namely:

- basement retaining walls; and
- embedded retaining walls.

15.8.1 Basement retaining walls

The rules classify basement retaining walls as 'external basement retaining walls in contact with earthwork up to and including the damp-proof course'.

Basement retaining walls can be measured and described as linear items, in metres, or measured superficially, in square metres. Where measured as a linear item, the wall is to be measured on its centre line. When measured superficially, the area measured is the surface area of the exposed face of the retaining walls (i.e. the length of the wall multiplied by the height of the wall). The height of the wall is measured from the top of the base slab/bed, or the top of the basement retaining wall base/toe, to the level at which the basement retaining wall connects with the external wall above ground (i.e. at the level at which the external wall changes from being a retaining wall to a non-retaining wall).

Descriptions for basement retaining walls are to include details of all key sub-components (e.g. excavation and earthworks, concrete foundations and walls, waterproofing, protection boards, masonry and integral drainage systems). The reinforcement content, in kg/m^3, is to be stated for reinforced concrete components, together with details of the formwork finish (e.g. basic finish, fine finish or patterned finish). Likewise, details of any masonry reinforcement and fixing systems are also to be described.

Ground-bearing piles integral to the basement retaining wall design are to be measured in accordance with the rules for piled foundations (sub-element 1.1.2 Specialist foundations), and included in the cost target for piled foundations.

Where more than one type of retaining wall construction is specified, each type of retaining wall construction is to be measured and described separately.

Typical sub-components and principal dimensions to be stated for a basement retaining wall are shown in Figure 15.9.

Illustrative descriptions for typical basement retaining wall components are shown in Example 15.8.

15.8.2 Embedded basement retaining walls

Embedded basement retaining walls consist of shoulder-to-shoulder piles, or other vertical construction, which are subsequently partially excavated on one side to form basement retaining walls. These retaining walls obtain their stability from the embedded lower portion.

When measuring embedded retaining walls, there are a number of cost-significant activities and components that need to be considered by the cost manager. They include the following:

- piling mats and piling platforms;
- piling plant, including moving plant to new piling positions;

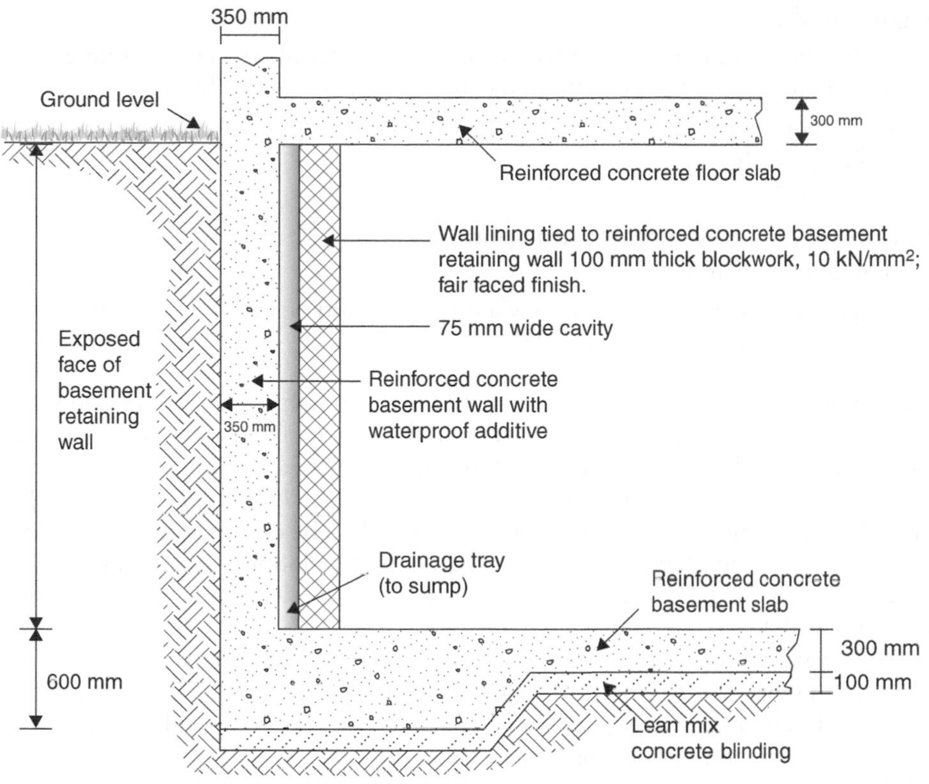

Figure 15.9 Sub-components and the principal dimensions for a typical basement retaining wall

Example 15.8 Formulation of descriptions basement retaining walls

SUBSTRUCTURE					
Code	**Element/sub-elements/components**	**Qty**	**Unit**	**Rate**	**Total**
				£p	**£**
1	**SUBSTRUCTURE**				
1.1	**Substructure**				
1.1.5	*Basement retaining walls*				
1.1.5.1	Retaining wall; comprising: 300mm thick reinforced concrete outer wall, including reinforcement (reinforcement content 200kg/m³) and formwork to both sides; 50mm drainage cavity; with 160mm thick blockwork inner wall tied to concrete outer wall		m²		
1.1.5.2	Retaining wall; comprising: 300mm thick reinforced concrete outer wall, including reinforcement (reinforcement content 180kg/m³) and formwork to both sides; 20mm thick mastic asphalt tanking, 30mm thick cement: sand grout; with 102.5mm thick facing brick inner wall (PC Sum £330/1,000 supply only)		m²		
1.1.5	**TOTAL – Basement retaining walls: to element 1.1 summary**				

- guide walls;
- piled walls, including the disposal of excavated material arising from piling, cutting off tops of piles and trimming and cleaning faces of piles;
- permanent steel sheet piling, including cutting off surplus lengths;
- diaphragm walls;
- ground anchors;
- temporary works to support piled walls;
- capping beams; and
- components used in the drainage, waterproofing and facing of the wall (e.g. masonry or spray-applied concrete).

Piling mats, piling platforms, piling rigs and moving piling rigs in connection with the installation of piled walls are measured and described in the same way as those associated with piled foundations (refer to paragraph 15.3.2). Care must be taken to avoid duplicating the provision for piling mats when a building comprises both ground-bearing piles and, say, contiguous piled walls. The cost manager will need to decide which sub-element, the piling mat, or piling platform, is to be measured – either sub-element 1.1.2 (Specialist foundations) or sub-element 1.3.2 (Basement retaining walls).

Guide walls are measured linear, in metres. Where known, the guide wall installation method is to be stated in the description. Guide walls are often required for the installation of secant piled and other embedded retaining wall systems to ensure correct alignment at capping beam level.

Piles are measured as enumerated items (nr), with the type, diameter, total length (exposed length plus embedded length), exposed length and embedded length stated in the description. The disposal of excavated material arising from piling is measured in cubic metres. Again, quantities given for disposal of excavated material are measured as the bulk before excavating, with no allowance made for subsequent variations to bulk. The cost manager will need to make allowance for variations to bulk in the unit rate used for the disposal of excavated material arising from piling. Cutting off the tops of concrete piles is measured as enumerated items (nr), with the diameter of the pile stated in the description. This work involves cutting off excess lengths of piles, cutting out concrete tops to piles and preparing the pile heads (including reinforcement) in readiness to receive capping beams, suspended floor slabs and the like.

Contiguous bored pile walls and the like are measured linear, in metres, with the type, diameter, total length, exposed length and embedded length stated in the description. The disposal of excavated material arising from contiguous bored piling is measured in the same way as ground-bearing piles. However, cutting off the tops of piles in connection with contiguous bored pile walls is measured linear, in metres, with the diameter of the pile stated in the description. Alternatively, cutting off the tops of piles can be measured as enumerated items (nr) in the same way as piles.

Pile tests are to be measured as items, with the nature of the testing given in the description (e.g. load tests and integrity tests).

Diaphragm walls are measured superficially, in square metres, measured on the centre line of the wall. Descriptions for diaphragm walls are to include the depth of excavation (m), the thickness of the wall (mm or m) and the reinforcement content in kg/m^2. The disposal of excavated material arising from the excavation of diaphragm walls is measured in cubic metres, in the same way as for piles. Diaphragm walls are used to construct retaining walls for deep basements, underpasses and the like, with the walls being constructed before excavation. The basic process consists of cutting a deep, narrow trench in the ground. To support the sides during excavation it is filled with a Bentonite suspension (which is a thixotropic clay, mixed with water to form a slurry). This permits continued excavation working through the suspension. Once any loose material has been removed from the bottom of the trench, the reinforcement cage for the wall is lowered into position through the Bentonite and fixed. Concrete is then placed using a tremie pipe, displacing the Bentonite as the trench is filled. Once the concrete has gained sufficient strength, the material, now contained by the diaphragm walls, can be excavated. Some form of temporary propping system will generally be required, to resist the lateral pressure of the soil behind the wall, until the permanent floor or roof structure has been completed. The exposed face of the diaphragm wall will be irregular, mirroring the side of the trench, and may require some secondary lining.

Ground anchors, whether permanent or temporary, are measured as enumerated items (nr). Descriptions for ground anchors should include the type and size of the anchor along with its construction method. Ground anchors are used to overcome any tendency for the structure to float during construction.

Works in trimming and cleaning the faces of piled walls and diaphragm walls, in preparation for receiving other work, is measured superficially, in square metres.

Permanent steel sheet piling is measured superficially, in square metres. Both the total area (in m^2) and total driven area (in m^2) of steel sheet piling are to be stated in the description. Steel sheet piling providing temporary support to basement excavations is dealt with under sub-element 1.1.4 (Basement excavation). Cutting off surplus lengths of steel sheet piling is measured linear, in metres.

Temporary works, other than temporary ground anchors, required to facilitate the construction of embedded retaining walls are to be identified and measured as discrete items. The nature of the temporary works is to be given in the description.

Capping beams are measured separately as linear items, in metres, measured on the centre line of the component. The description for capping beams is to include their composition, including the reinforcement content in kg/m^2 and overall size (i.e. width and depth – cross-sectional dimensions). Each different size of capping beam is to be measured separately.

Where there is a risk of obstructions, bands of hard material or the like being encountered within the ground, then the cost manager should make sufficient allowance within the construction risk allowance to cover the extra cost of breaking through such obstructions. Likewise, the cost manager should also consider including an allowance to cover the extra cost of disposing of contaminated material where there is a possibility that it be encountered.

Masonry walls and concrete walls used to face embedded piled walls and diaphragm walls, including integral drainage systems, tanking and waterproofing, protection boards, linings and other retaining wall components are measured superficially, in square metres. The length of the component is measured on the centre line. The height of the wall is measured from the top of the base slab/bed or the top of the basement retaining wall base/toe to the level at which the basement retaining wall connects with the external wall above ground (i.e. at the level at which the external wall changes from being a retaining wall to a non-retaining wall). To obtain the area of the component, the length of the wall is multiplied by the height of the wall. Descriptions for basement retaining wall components are to include the kind of material. The reinforcement content, in kg/m^3, is to be stated for reinforced concrete components, together with details of the formwork finish (e.g. basic finish, fine finish or patterned finish). Details of any masonry reinforcement and fixing systems are also to be described.

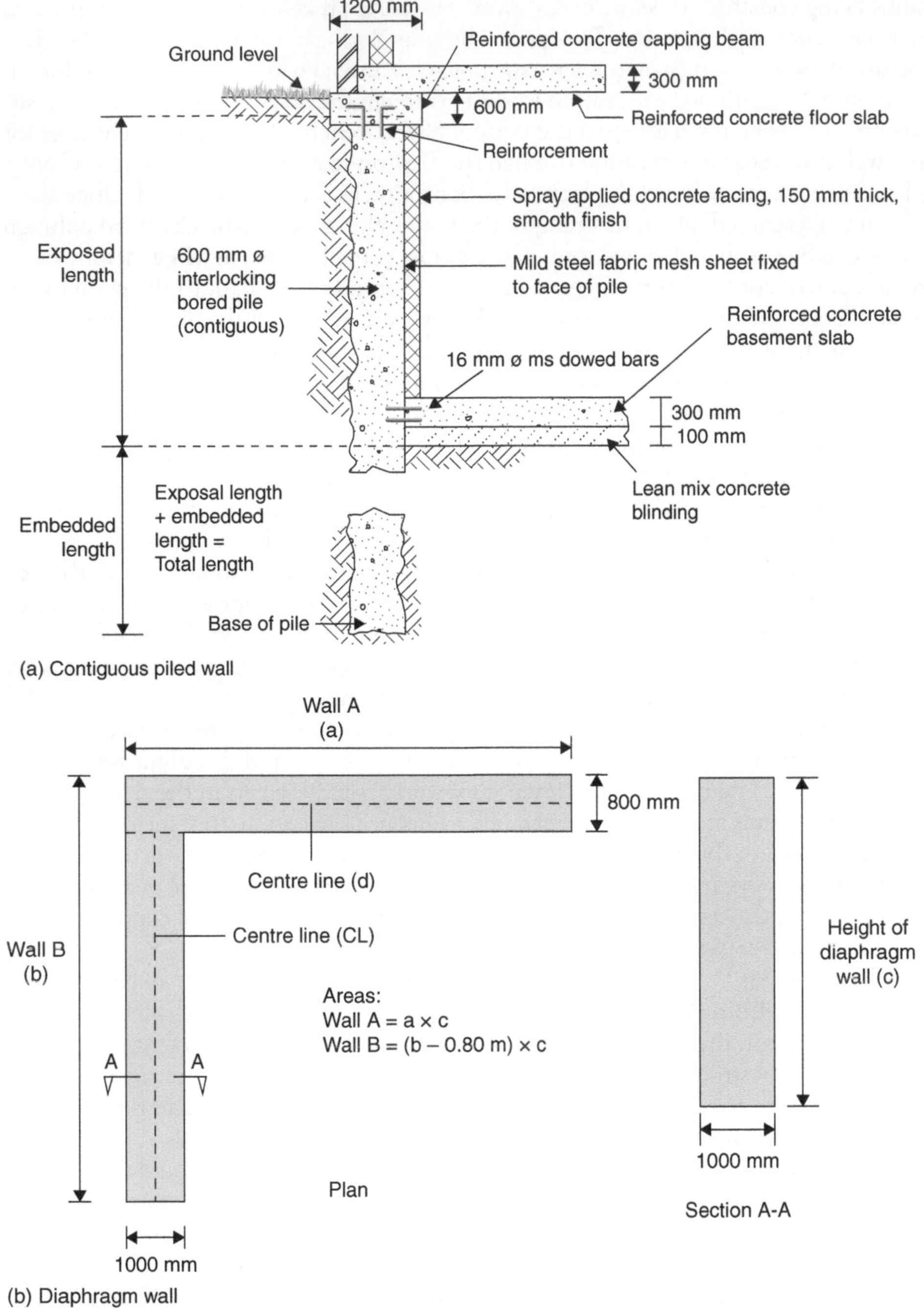

Figure 15.10 Sub-components and principal dimensions for typical types of embedded basement retaining walls

Concrete applied by spray or gun is also measured superficially, in square metres, with the method of application, type of material, thickness and details of any reinforcement stated in the description.

Figure 15.10 shows the sub-components and principal dimensions to be stated for embedded basement retaining walls – contiguous piled and diaphragm walls.

Examples of descriptions for contiguous piled and diaphragm wall components are shown in Example 15.9.

Example 15.9 Formulation of descriptions for embedded basement retaining walls

SUBSTRUCTURE					
Code	**Element/sub-elements/components**	**Qty**	**Unit**	**Rate**	**Total**
				£p	**£**
1	**SUBSTRUCTURE**				
1.1	**Substructure**				
1.1.5	*Basement retaining walls*				
	Example 1 – Contiguous piled wall				
1.1.5.1	Piling mat/platforms; imported granular material; 450mm thick; taking up and disposing on completion of piling operations		m²		
1.1.5.2	Piling plant; rig		item		
1.1.5.3	Moving piling rig to pile position		nr		
1.1.5.4	Guide walls; to align 600mm diameter piled walls		m		
1.1.5.5	Contiguous bored pile walls; 600mm diameter; total length 20m		m		
1.1.5.6	Disposal of excavated material arising from piling		m³		
1.1.5.7	Cutting off tops of concrete piles: 600mm diameter		nr		
1.1.5.8	Pile tests; integrity tests		item		
1.1.5.9	Trimming and cleaning faces of piles		m²		
1.1.5.10	Capping beams; reinforced concrete 35N/mm² – 20mm aggregate reinforced concrete (reinforcement rate 120kg/m³), formwork; capping beam size 2.00m wide × 1.25m deep		nr		
1.1.5.11	Concrete applied by spray or gun; 150mm thick		m²		
1.1.5.12	Steel fabric mesh, lightweight shot fixed to piles		m²		
	Example 2 – Diaphragm wall				
1.1.5.13	Diaphragm walls; reinforced concrete (reinforcement rate 90kg/m³); 750mm thick; including guide walls		m²		
1.1.5.14	Walls; 160mm thick blockwork, pointed one side; (internal lining to diaphragm wall)		m²		
1.1.5.15	Diaphragm walls; precast concrete; 350mm thick; including guide walls		m²		
1.1.5.16	Ground anchors; cable type, including grouting; to 350mm thick diaphragm walls; overall length 1.80m		nr		
1.1.5	**TOTAL – Basement retaining walls: to element 1.1 summary**				

CHAPTER 16

Superstructure (group element 2)

Introduction

This chapter:

- provides step-by-step guidance on how to measure and describe components forming the superstructure;
- contains worked examples showing how to measure key dimensions; and
- gives examples illustrating how to formulate descriptions for components.

16.1 Method of measurement

The rules for measuring the superstructure are found in group element 2 under NRM 1, which is divided into the following eight elements:

2.1 Frame;
2.2 Upper floors;
2.3 Roof;
2.4 Stairs and ramps;
2.5 External walls;
2.6 Windows and external doors;
2.7 Internal walls and partitions; and
2.8 Internal doors.

16.2 Frame (element 2.1)

Element 2.1 sets out the rules of measurement for frames, and is subdivided into six sub-elements as follows:

2.1.1 Steel frames;
2.1.2 Space frames/decks;
2.1.3 Concrete casings to steel frames;

2.1.4 Concrete frames;
2.1.5 Timber frames; and
2.1.6 Specialist frames.

16.2.1 Steel frames (sub-element 2.1.1)

Structural steel frames comprise all components that form an integral part of the frame. These include joists, girders, purlins, columns, beams, composite columns and beams, braces, struts, trusses, splices, base plates and all fittings and fixings – including holding-down bolts and holding-down bolt assemblies. Floor members, roof members, trusses and steel decking which cannot be separated from the structural steel framing of a building are also included. Grouting of holding-down bolts and grouting under base plates are treated as part of the structural steel frame.

Concrete casings to structural steel members are measured under sub-element 2.1.3 (Casings to steel frames).

Floor members, roof members, trusses and steel decking forming the roof construction which can be separated from the structural steel frame are measured under sub-element 2.3.1 (Roof structure).

Where a structural steel frame is being considered, it is normal for the structural engineer to use computer programs to generate the design solution. Therefore, the cost manager should be able to obtain the total mass of the structural steel frame, including fittings and components, from the structural engineer. Thus, structural steel frames are measured in tonnes (t).

It is important that the cost manager ascertains if solid or perforated members, or castellated members, are being proposed by the structural engineer, as this can have an impact on the cost (e.g. the extra cost of cutting the perforations, and larger beam sections). What is more, when building up a unit rate, in addition to the costs of fabrication and permanent erection of the structural steel frame, the cost manager will need to consider costs associated with any trial erection of the structural steel frame.

Fire protection and factory-applied paint systems to structural steel frames are also measured in tonnes, in the same way as the structural steel frame.

Component descriptions for steel frames are shown in Example 16.1.

16.2.2 Space frames and decks (sub-element 2.1.2)

Space frames and decks are truss-like, lightweight rigid structures constructed from interlocking struts in a geometric pattern. Space frames and decks usually utilise a multidirectional span and are often used to accomplish long spans with few supports. They derive their strength from the inherent rigidity of the triangular frame; flexing loads (bending moments) are transmitted as tension and compression loads along the length of each strut. Most often their geometry is based on platonic solids. The simplest form is a horizontal slab of interlocking square pyramids built from aluminium or steel tubular struts. In many ways this looks like the horizontal jib of a tower crane repeated many times to make it wider. A stronger, purer form is composed of interlocking tetrahedral pyramids in which all the struts have unit length. More technically this is referred to as an isotropic vector matrix or, in a single

Example 16.1 Formulation of descriptions for steel frames

SUPERSTRUCTURE					
Code	**Element/sub-elements/components**	**Qty**	**Unit**	**Rate**	**Total**
				£p	**£**
2	**SUPERSTRUCTURE**				
2.1	**Frame**				
2.1.1	*Steel frames*				
2.1.1.1	Fabricated steelwork; erected on site, comprising universal columns and beams with bolted connections; solid members; including fittings		t		
2.1.1.2	Fabricated steelwork; erected on site, comprising universal columns and beams with bolted connections; perforated members; including fittings		t		
2.1.1.3	Fire protection; intumescent paint, 60 minutes rating; applied at works		t		
2.1.1	**TOTAL – Steel frames: To element 2.1 summary**				

unit width, an octet truss. More complex variations change the lengths of the struts to curve the overall structure or may incorporate other geometrical shapes.

Space frames and decks comprise all components that form an integral part of the frame or deck, including support framework (e.g. beams and columns), fittings and fixings. They are measured superficially, in square metres, based on the area of the upper floors. The rules require the area of upper floors to be measured using the principals of measurement for ascertaining the gross internal floor area. Descriptions for space frames should include the size of the unit trays, and the distance between the top chord and the bottom chord (m). Similarly, descriptions for space decks should include the size of the unit/fascia trays, and the distance between the top of the standard unit (the unit/fascia tray) and the tie bars (m). Figure 16.1 shows the sub-components and principal dimensions to be stated for space frames and decks.

Space frames and decks forming the roof construction are measured under sub-element 2.3.1 (Roof structure).

Fire protection and factory-applied paint systems to structural steel frames are also measured in square metres, in the same way as the space frame or deck, with the type of protection or coating described.

Typical component descriptions for space frames and decks are shown in Example 16.2.

16.2.3 Concrete casings to steel frames (sub-element 2.1.3)

Concrete casings to columns and beams for structural or protective purposes, including fire protection, are measured in linear metres. The length of a casing to a column is the distance between the top of the slab, bed, pile cap or ground beam from which the column starts to the soffit of the beam attached to the next suspended floor slab. Where a flat slab is employed (i.e. no beams), the distance to

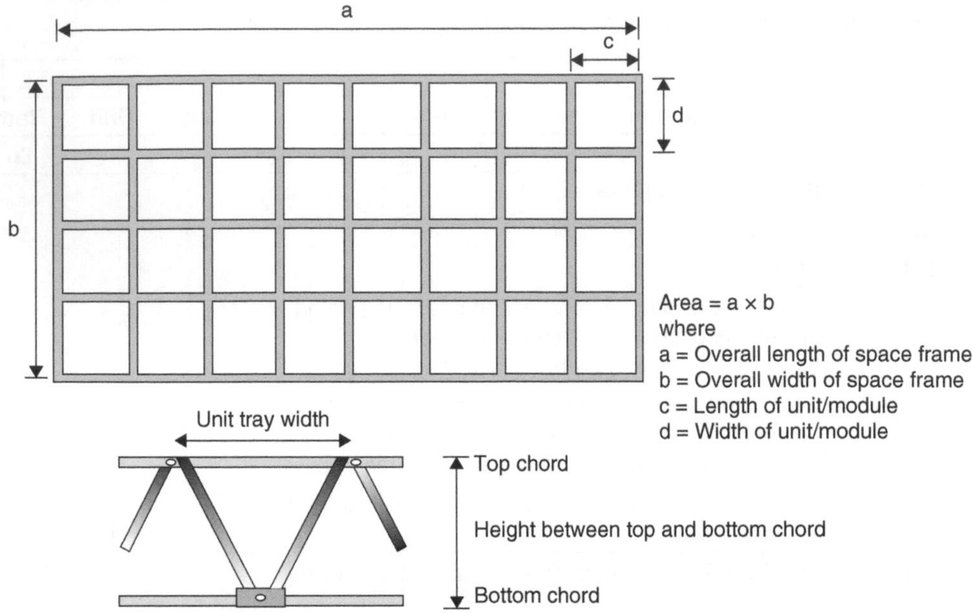

Figure 16.1 Sub-components and the principal dimensions for space frames and decks

Example 16.2 Formulation of descriptions for space frames and decks

SUPERSTRUCTURE					
Code	**Element/sub-elements/components**	**Qty**	**Unit**	**Rate**	**Total**
				£p	**£**
2	**SUPERSTRUCTURE**				
2.1	**Frame**				
2.1.2	*Space frames/decks*				
	Example 1 – Space frame				
2.1.2.1	Space frame; steel construction, distance between top chord and bottom chord 900mm		m²		
2.1.2.1	Space frame; steel construction, using 1,200mm × 1,200mm pyramidal units; distance between tie bars and unit/fascia trays 730mm		m²		
	Example 2 – Space deck				
2.1.2.3	Space deck; steel construction, using 1,200mm × 1,200mm pyramidal units; distance between tie bars and unit/fascia trays 730mm		m²		
2.1.2.4	Fire protection; intumescent paint, 60 minutes rating; applied on site		m²		
2.1.2	**TOTAL – Space frames/decks: to element 2.1 summary**				

Example 16.3 Formulation of descriptions for concrete casings to steel frames

SUPERSTRUCTURE					
Code	**Element/sub-elements/components**	**Qty**	**Unit**	**Rate**	**Total**
				£p	**£**
2	**SUPERSTRUCTURE**				
2.1	**Frame**				
2.1.3	*Concrete casings to steel frames*				
2.1.3.1	In situ concrete casings to steel column; formwork, fine finish; column size 300mm × 600mm		m		
2.1.3.2	In situ concrete casings to steel beam; formwork, fine finish; beam size 225mm × 450mm		m		
2.1.3	**TOTAL – Concrete casings to steel frames: to element 2.1 summary**				

the soffit of the suspended slab is measured. Where only part of a column is encased, the length of column encased is measured. The length of a casing to a beam is the actual length of beam encased in concrete, measured on the centre line of the beam.

The description for concrete casings is to comprise the size (i.e. cross-sectional dimensions) of the column or beam, the number (nr) of columns or beams, together with the grade of concrete, particulars of any specialist grades and additives, type of formwork finish and details of any formed finishes. Details of any reinforcement should also be stated.

Sample component descriptions for concrete casings to columns and beams are shown in Example 16.3.

16.2.4 Concrete frames (sub-element 2.1.4)

Concrete frames comprise beams, columns, blade columns and the like. They also include concrete walls and core walls that form an integral part of the structural frame.

Columns are measured in linear metres. The length of a column is the distance between the top of the slab, bed, pile cap or ground beam from which the column starts to the soffit of the beam attached to the next suspended floor slab. Where columns larger than the attached beams or a flat slab are employed (i.e. no beams), the distance to the soffit of the suspended slab is measured. The method of measuring these different scenarios of column height is illustrated in Figures 16.2 and 16.3. The description for columns is to comprise the number (nr) of columns, the size (i.e. cross-sectional dimensions) of the column, together with the grade of concrete, particulars of any specialist grades and additives, the reinforcement rate (kg/m^3), the type of formwork finish and details of any formed finishes. The size of the column grid should also be stated.

Beams are also measured in linear metres. The length is the actual length of beam, measured on the centre line of the beam. The depth of the beam is measured from the underside of the suspended floor slab to the soffit, or bottom, of the beam – irrespective of whether the beam is integral to the suspended floor slab or supporting a suspended floor slab of different construction (e.g. precast concrete

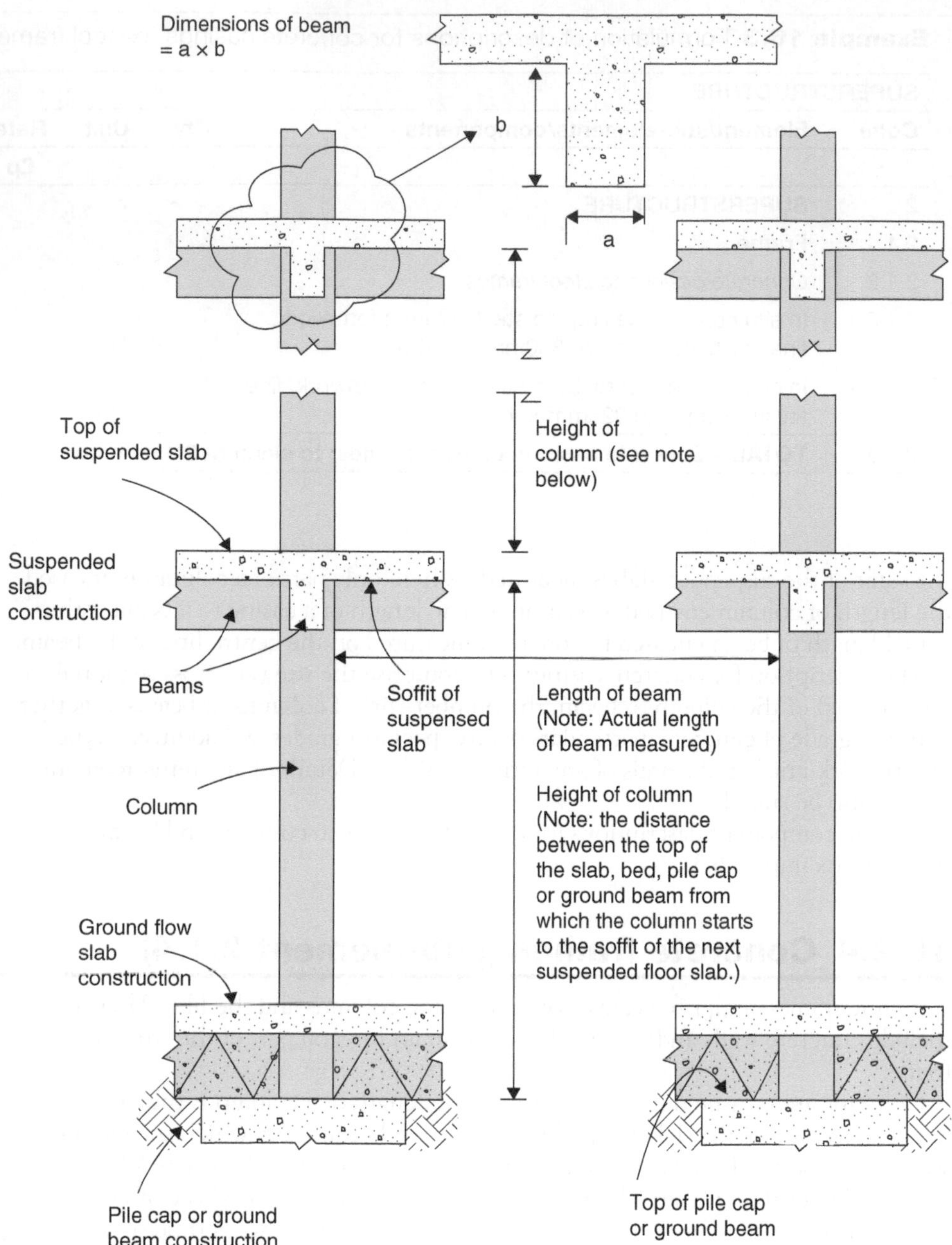

Figure 16.2 Measurement of concrete frames – where attached beams narrower than columns

decking system) – see Figures 16.2 and 16.3. The number (nr) of beams, the size (i.e. cross-sectional dimensions) of the beam, together with the grade of concrete, particulars of any specialist grades and additives, the reinforcement rate (kg/m³), the type of formwork finish and details of any formed finishes are to be stated in the description.

Care must be taken when ascertaining the cross-sectional area of attached beams (i.e. those attached to the slab). This is because the way in which structural

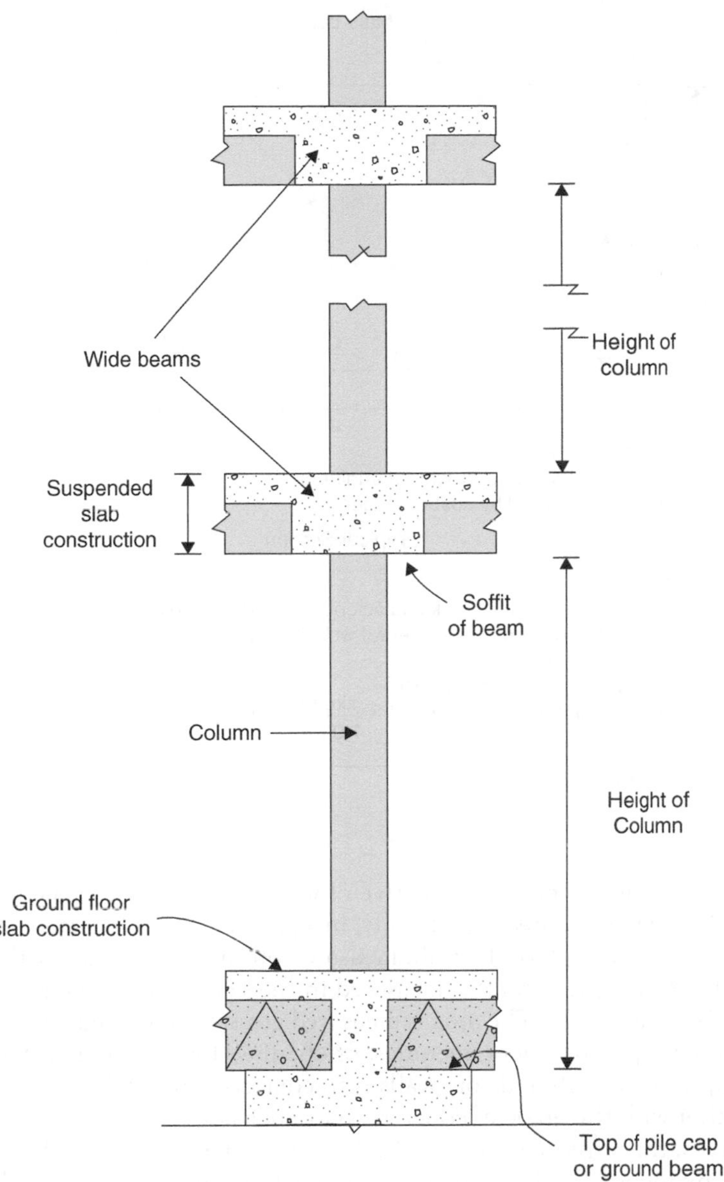

Figure 16.3 Measurement of concrete frames – where attached beams wider than columns

engineers specify and cost managers measure the depth of beams are different. Structural engineers will normally specify the depth of a beam as being from the top of the structural slab to the soffit of the beam, while the cost manager will measure the depth of a beam from the soffit of the suspended slab to the soffit of the beam. The rules recognise this problem and state that 'an appropriate allowance made in the unit rate [by the cost manager] for beam reinforcement that is integrated within the concert floor' (see Figure 16.4). Where this problem is not recognised by the cost manager, there is a risk that the cost of beams will be underestimated.

Walls and core walls forming an integral part of the structural assembly are

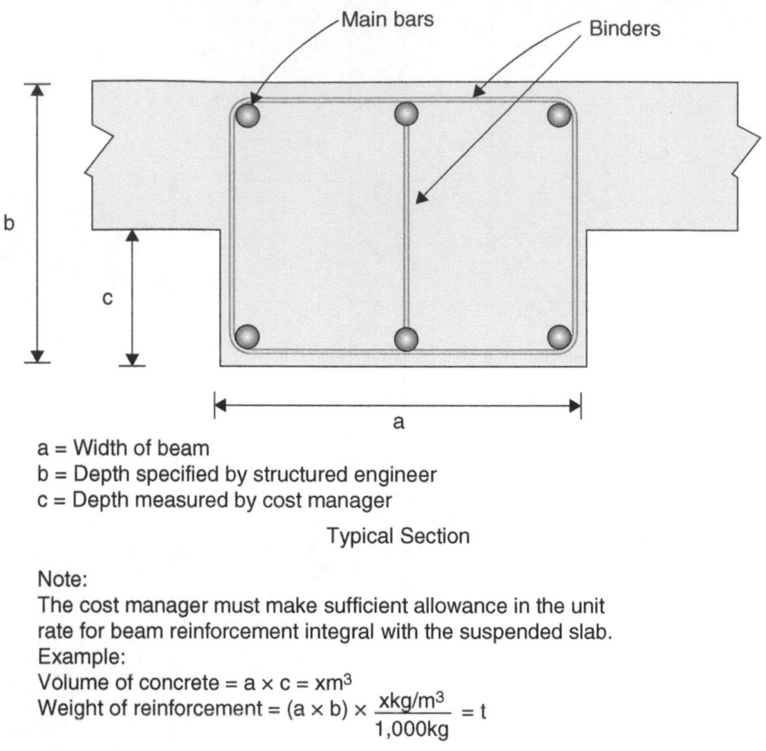

Main bars
Binders
b
c
a
a = Width of beam
b = Depth specified by structured engineer
c = Depth measured by cost manager
Typical Section
Note:
The cost manager must make sufficient allowance in the unit rate for beam reinforcement integral with the suspended slab.
Example:
Volume of concrete = a × c = xm3
Weight of reinforcement = (a × b) × xkg/m3 / 1,000kg = t

Figure 16.4 Depth of attached beams

included in this sub-element, irrespective of whether or not the frame itself is concrete. They are measured superficially, in square metres. The area is derived by multiplying the length of the wall, measured on the centre line, by the height of the wall. The height of walls is measured in the same way as columns (refer to Figures 16.1 and 16.2). The area is measured gross, with no deductions made for the window openings, door openings, or openings for screens and the like. The description for walls and core walls is to comprise the thickness of the wall, together with the grade of concrete, particulars of any specialist grades and additives, the reinforcement rate (kg/m^3), the type of formwork finish and details of any formed finishes. The number of sides requiring formwork should also be stated.

Formed openings in walls and core walls are enumerated (nr), measured as 'extra over' items. Although measured 'extra over', the cost of forming an opening might result in a cost saving (i.e. because the cost of walling filling for the opening is greater than the cost of forming the opening itself).

Designed joints to walls and core walls are measured in linear metres. The composition of the joint and the thickness are to be stated in the description.

No adjustment is made to any concrete component for the volume of concrete displaced by steel reinforcement.

Descriptions for typical concrete frame components are shown in Example 16.4.

Example 16.4 Formulation of descriptions for concrete frames

SUPERSTRUCTURE					
Code	**Element/sub-elements/components**	**Qty**	**Unit**	**Rate**	**Total**
				£p	**£**
2	**SUPERSTRUCTURE**				
2.1	**Frame**				
2.1.4	*Concrete frames*				
	Column				
2.1.4.1	Columns; reinforced concrete, grade 35N/mm², 350mm × 500mm (reinforcement rate 185kg/m³); formwork, fine finish		m		
2.1.4.2	Columns; reinforced concrete, grade 35N/mm², 350mm × 750mm (reinforcement rate 185kg/m³); formwork, fine finish		m		
	Beams				
2.1.4.3	Beams; reinforced concrete, grade 35N/mm², 1,200mm × 450mm (reinforcement rate 200kg/m³); formwork, fine finish		m		
	Walls				
2.1.4.4	Walls; reinforced concrete, 250mm thick (reinforcement rate 90kg/m³); formwork to both sides, fine finish		m²		
2.1.4.5	Extra over walls 250mm thick for forming opening in walls for doors; 1.20m × 2.10m		nr		
2.1.4.6	Walls; reinforced concrete, 350mm thick, concrete grade 35N/mm² (reinforcement rate 90kg/m³); formwork to both sides, fine finish		m²		
2.1.4.7	Designed joints; in 250mm thick concrete wall		m		
2.1.4.8	Designed joints; in 350mm thick concrete wall		m		
2.1.4	**TOTAL – Concrete frames: to element 2.1 summary**				

16.2.5 Timber frames (sub-element 2.1.5)

Timber frames include panel systems and laminated timber structures. They comprise all structural members, timber frame panels and fittings and fixings.

Timber frames are measured superficially, in square metres, based on the area of the upper floors – measured using the principles of measurement for ascertaining the gross internal floor area. The composition of the timber frame, including any treatments to timbers and applied fire-retarding paint systems, is to be stated in the description.

Floor members, roof members, trusses and structural wall members, including wall linings and floor boarding, which cannot be separated from the structural timber frame are also included. Where the roof structure is to be treated as part of the structural timber frame, the area of the roof is to be included in the measurement for the timber frame. It is recommended that the method of measuring roof structures be applied in this situation (see sub-element 2.3.1: Roof structure –

Example 16.5 Formulation of descriptions for timber frames

SUPERSTRUCTURE					
Code	**Element/sub-elements/components**	**Qty**	**Unit**	**Rate**	**Total**
				£p	**£**
2	**SUPERSTRUCTURE**				
2.1	**Frame**				
2.1.5	*Timber frames*				
2.1.5.1	Timber frames; structural timber frame system incorporating external and internal load-bearing wall panels, roof trusses and associated materials		m^2		
2.1.5	**TOTAL – Timber frames: to element 2.1 summary**				

measurement rules for components C1, C2 and C3). That is, the area of the roof is the area of the roof on plan, measured to the extremities of the eaves; and for flat roofs with parapet walls, the area is measured to the internal face of the parapet walls to the roof.

A description for a timber frame structure is shown in Example 16.5.

When building up a unit rate, the cost manager will need to consider costs associated with any trial erection of the timber frame as well as the cost of permanent erection.

16.2.6 Specialist frames (sub-element 2.1.6)

Portal frames, specialist proprietary and modular lightweight steel framing systems and any other specialist framing systems are treated as specialist frames. They are measured superficially, in square metres – measured in the same way as timber frames. The type of specialist framing system is to be stated in the description, together with details of any applied fire-protective coatings and paint systems. A description for a typical specialist frame is illustrated in Example 16.6.

Example 16.6 Formulation of descriptions for specialist frames

SUPERSTRUCTURE					
Code	**Element/sub-elements/components**	**Qty**	**Unit**	**Rate**	**Total**
				£p	**£**
2	**SUPERSTRUCTURE**				
2.1	**Frame**				
2.1.6	*Specialist frames*				
2.1.6.1	Steel propped portal frame; cold rolled purlin sections, surface treatments, including factory-applied fire-protective coatings and paint system		m^2		
2.1.6	**TOTAL – Specialist frames: to element 2.1 summary**				

16.3 Upper floors (element 2.2)

Element 2.2 subdivides the measurement of upper floors into the following three sub-elements:

2.2.1 Floors;

2.2.2 Balconies; and

2.2.3 Drainage to balconies.

This element covers suspended floors, podium slabs forming roofs to basements and balcony floors that form an integral part of the suspended floors.

16.3.1 Measurement generally

Upper floors are measured superficially, in square metres, with the area measured being the area of the upper floor, or total area of all upper floors where the building is of more than two storeys. The area is measured in accordance with the principles for measuring the gross internal floor area. Sloping surfaces, such as tiered terraces, are measured flat on plan.

Upper floors include galleries, tiered terraces, walkways and internal bridges. Balcony floors and roofs to internal buildings that are an integral part of the upper floor construction are also included. They are to be separately identified and included with the type of upper floors to which they relate. Purpose-made balconies that are not an integral part of the upper floor construction are included in sub-element 2.2.2 (Balconies).

Where the upper floor consists of more than one type of suspended floor construction (e.g. a combination of reinforced concrete, post-tensioned concrete and a composite decking system), each type of suspended floor construction is measured separately in accordance with the applicable rules. Likewise, if a suspended floor slab is of a single type of construction (e.g. reinforced concrete) but includes varying thicknesses (e.g. part 200mm thick and part 300mm thick), each thickness of slab is treated as though it were of a different construction and the areas measured accordingly (see Figure 16.5).

Suspended slabs forming the roof structure are to be measured under sub-element 2.4.1 (Roof structures).

16.3.2 Concrete floors (sub-element 2.2.1)

Concrete floors encompass reinforced concrete and post-tensioned concrete suspended floor slabs. The sub-element is subdivided into four cost-significant components:

- suspended floor slabs;
- edge formwork;
- designed joints; and
- surface treatments.

Descriptions of suspended concrete floor slabs are to include the type of floor construction (solid, waffle or trough slabs), the thickness of the slab (mm), the

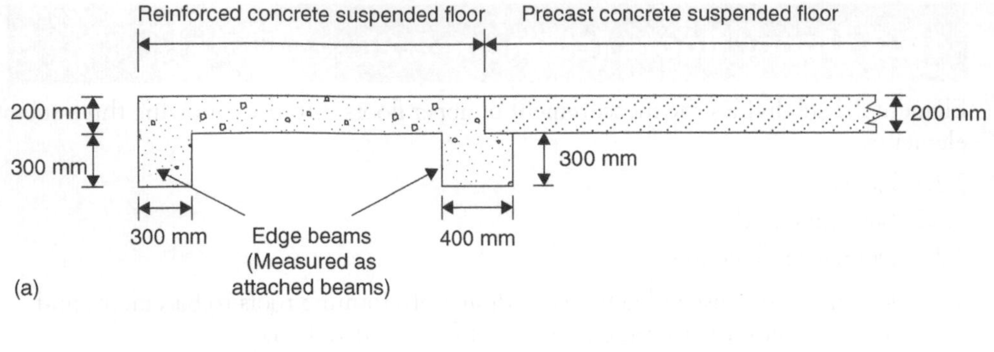

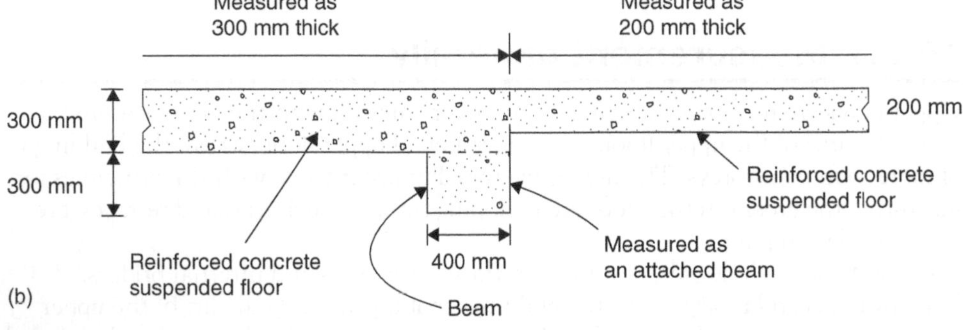

Figure 16.5 Measurement of upper floor comprising different forms of construction (a) Upper floor comprising more than one type of construction (b) Upper floor comprising varying thicknesses of same construction

concrete strength, the reinforcement rate (in kg/m³) and the type (quality) of formwork finish. Rates for both reinforcement and post-tensioning strands will be required for post-tensioned concrete slabs. Details of permanent formwork, including profiled sheet metal decking, should also be stated, as well as the floor loadings (e.g. institutional floor loading of 3.5 + 1kN/m²).

Other cost-significant components include edge formwork, designed joints and surface treatments:

- Edge formwork, including upstands to profiled sheet metal decking, is measured linear, in metres, stating the type (quality) of formwork finish. The height of the formwork should also be stated.
- Designed joints are measured linear, in metres. Where known, the type, thickness and depth of the joint should also be given in the description.
- In situ surface treatments (e.g. surface hardeners applied to unset concrete) and worked finishes (e.g. tamped, brush or power float finishes) are measured superficially, in square metres. The type of surface treatment is to be described.

Flat slabs have a major weakness, in that they are vulnerable to punching shear failure at the junctions of slabs and columns. To overcome this problem, punching shear reinforcement systems are often introduced by the structural engineer within a concrete slab to provide additional reinforcement around column heads, protecting the slab from the effects of punching shear. These can be proprietary systems or purpose designed. Although not specifically required by the rules, where viewed as a cost-significant item, the cost manager should measure punching shear reinforcement as a separate item. When quantified separately, it is suggested that punching shear reinforcement be measured as an enumerated (nr) item, with the composition stated in the description.

Descriptions for typical components for a selection of different concrete floor systems are shown in Example 16.7.

Example 16.7 Formulation of descriptions for concrete floors

SUPERSTRUCTURE					
Code	**Element/sub-elements/components**	**Qty**	**Unit**	**Rate**	**Total**
				£p	**£**
2	**SUPERSTRUCTURE**				
2.2	**Upper floors**				
2.2.1	*Floors*				
	Concrete floors				
	Example 1 – Composite construction				
2.2.1.1	Composite suspended floor slab,170mm overall thickness; comprising reinforced concrete on permanent formwork system ('Holorib' decking, 'Rideck decking' or the like), 1.20mm thick; reinforcement content 130kg/m³; 5.00kN/mm² loading		m²		
2.2.1.2	Edge formwork; upstand to permanent formwork system ('Holorib' decking, 'Rideck decking' or the like) – to perimeter of slab, 170mm high		m		
2.2.1.3	Finishes to unset concrete; brush finish		m²		
	Example 2 – Flat slab construction				
2.2.1.4	Suspended floor slab; reinforced concrete, grade C35, 230mm thick (reinforcement rate 95kg/m³); formwork to soffit, fine finish		m²		
2.2.1.5	Edge formwork; 230mm high; fine finish		m		
2.2.1.6	Punching shear reinforcement; to column heads		nr		
2.2.1.7	Finishes to unset concrete; brush finish		m²		
	Example 3 – Post-tensioned concrete construction				
2.2.1.8	Suspended floor slab; post-tensioned concrete, 40N/mm², 210mm thick (reinforcement rate 65kg/m³, and 25kg/m for PT strands); formwork to soffit, fine finish		m²		
2.2.1.9	Edge formwork; 210mm high; fine finish		m		
2.2.1.10	Finishes to unset concrete; brush finish		m²		
2.2.1	**TOTAL – Floors: to element 2.2 summary**				

16.3.3 Precast and composite decking systems (sub-element 2.2.1)

This sub-element deals with suspended floors constructed of precast concrete and prestressed concrete plank and decking systems, as well as composite decking systems. Typical systems are:

- solid, hollow, tee or other section precast, and prestressed, concrete plank and slab decks;
- composite decks of precast or prestressed concrete beams with filler blocks of precast concrete, in situ concrete or other material;
- composite decks comprising in situ concrete on precast or prestressed concrete planks;
- hollow tile decks consisting of in situ concrete on clay, concrete or precast concrete filler blocks.

Descriptions of precast and composite decking systems are to include the type of floor construction, the overall thickness of the deck, the span and the floor loadings. Example 16.8 shows a typical description for a common precast decking system.

Structural screeds applied to precast or composite decking systems are included under sub-element 2.2.1 – see paragraph 16.3.5 (Structural screeds (sub-element 2.2.1)).

Example 16.8 Formulation of descriptions for precast and composite decking systems

SUPERSTRUCTURE					
Code	**Element/sub-elements/components**	**Qty**	**Unit**	**Rate**	**Total**
				£p	**£**
2	**SUPERSTRUCTURE**				
2.2	**Upper floors**				
2.2.1	*Floors*				
	Precast/composite decking systems				
2.2.1.1	Suspended floor slab; prestressed precast flooring planks, 170mm thick, 1200mm wide planks		m²		
2.2.1	**TOTAL – Floors: to element 2.2 summary**				

16.3.4 Timber floors (sub-element 2.2.1)

Descriptions of timber suspended floor construction are to include the type (e.g. rigid sheet flooring or timber board flooring, and softwood or hardwood) and thickness of flooring and thermal insulation. A description of a typical timber suspended floor construction is shown in Example 16.9.

Preparatory works for applied finishes, such as power sanding in readiness for staining, is covered by sub-element 3.2.1 (finishes to floors). Floor coverings, such as carpets vinyl tiles, ceramic tiles, as well as applied finishes are also measured under sub-element 3.2.1 (Floor finishes).

Example 16.9 Formulation of descriptions for timber floors

SUPERSTRUCTURE					
Code	**Element/sub-elements/components**	**Qty**	**Unit**	**Rate**	**Total**
				£p	**£**
2	**SUPERSTRUCTURE**				
2.2	**Upper floors**				
2.2.1	*Floors*				
	Timber floors				
2.2.1.1	Softwood joisted floors; with t&g chipboard floor boarding		m^2		
2.2.1	**TOTAL – Timber floors: to element 2.2 summary**				

16.3.5 Structural screeds (sub-element 2.2.1)

A structural screed is a layer of concrete that is placed as the top or outer layer of preformed concrete flooring. To be considered to be a structural screed instead of a non-structural screed this layer must included reinforcing fibres or sub-components like a steel fabric mesh.

Structural screeds are measured superficially, in square metres. The area measured is the area to which the structural screed is to be applied. Descriptions of structural screeds are to include the thickness (mm) of screed, the reinforcement type, and details of any worked finishes and surface treatments (e.g. surface hardeners or non-slip inserts). Example 16.10 is typical of a description for a structural screed.

Example 16.10 Formulation of descriptions for structural screeds

SUPERSTRUCTURE					
Code	**Element/sub-elements/components**	**Qty**	**Unit**	**Rate**	**Total**
				£p	**£**
2	**SUPERSTRUCTURE**				
2.2	**Upper floors**				
2.2.1	*Floors*				
	Structural screeds				
2.2.1.1	Structural screed, 100mm thick, cement: sand with two layers of lightweight mesh reinforcement		m^2		
2.2.1	**TOTAL – Floors: to element 2.2 summary**				

16.3.6 Balconies (sub-element 2.2.2)

Purpose-made balconies (internal and external), which are not an integral part of the upper floor construction (i.e. they are constructed from a different material to the upper floor), are covered under this sub-element. Balconies include bolt-on frames,

Example 16.11 Formulation of descriptions for balconies

SUPERSTRUCTURE					
Code	**Element/sub-elements/components**	**Qty**	**Unit**	**Rate**	**Total**
				£p	**£**
2	**SUPERSTRUCTURE**				
2.2	**Upper floors**				
2.2.2	*Balconies*				
2.2.2.1	Balconies, purpose-made, galvanised steel framework, hardwood decking, stainless steel handrails and balusters, with reinforced glass balustrades, overall floor area 3.60m × 1.60m (5.76m²)		nr		
2.2.2	**TOTAL – Balconies: to element 2.2 summary**				

floor decking, soffit panels, integral drainage and drainage trays, and balustrades and handrails. They are enumerated (nr), with the composition and floor area (m²) of the balcony stated in the description. Example 16.11 shows a description for purpose-made balconies.

16.3.7 Drainage to balconies (sub-element 2.2.3)

Piped disposal systems for taking rainwater from balconies to the first underground drain connection are measured under this sub-element, irrespective of whether the system is internally or externally located. Drainage to balconies is divided into two cost-significant components, namely rainwater pipes and floor outlets. Rainwater pipes are measured linear, in metres, with the type of rainwater pipe stated in the description. The length of rainwater pipe measured is the extreme length, measured over all fittings, branches and the like. Floor outlets are simply enumerated (nr), with the type stated. A typical description for balcony drainage is shown in Example 16.12.

Example 16.12 Formulation of descriptions for drainage to balconies

SUPERSTRUCTURE					
Code	**Element/sub-elements/somponents**	**Qty**	**Unit**	**Rate**	**Total**
				£p	**£**
2	**SUPERSTRUCTURE**				
2.2	**Upper floors**				
2.2.3	*Drainage to balconies*				
2.3.3.1	Rainwater pipes; uPVC, 68mm diameter; including fittings		m		
2.2.3.2	Floor outlets; PC Sum £55.00 each		nr		
2.2.3	**TOTAL – Drainage to balconies: to element 2.2 summary**				

16.4 Roof (element 2.3)

The measurement of roofs is divided into the following six sub-elements:

2.3.1 Roof structure;
2.3.2 Roof coverings;
2.3.3 Specialist roof systems;
2.3.4 Roof drainage;
2.3.5 Rooflights, skylights and openings; and
2.3.6 Roof features.

16.4.1 Roof structure (sub-element 2.3.1)

The rules subdivide the roof structure into three cost-significant components, which are as follows:

- pitched roofs;
- dormers; and
- flat roof.

Roofs to basements that also act as a podium slab are treated as upper floors and measured under element 2.2 (Upper floors).

Roof structures to housings (e.g. to roof-top lift motor rooms and plant rooms) are measured separately

Pitched roof structures

Roofs can be classified as either:

- pitched – pitch over 10°; or
- flat – pitch from 0° to 10°.

For design purposes, roof pitches over 70° are normally classified as walls.

Pitched roof structures can take many different forms. Basic forms include lean-to roofs, mono-pitch roofs, gable end roofs, hipped end roofs, mansard roofs, northlight roofs, flat top girders or any combination of more than one of the different forms. Common materials for pitched roof structures are timber and steel. Components comprise wall plates, hip and valley rafters, ridge boards, ceiling joists, purlins, collars, struts, binders, hangers, boarding, trusses and all connectors, bolts and other fixings. They are measured superficially, in square metres, with the area of the roof on plan, to the extremities of the eaves, being measured. The composition of the pitched roof structure, the design load (kN/m^2), span (m) and pitch (°) are to be stated in the description.

Dormers to pitched roofs are enumerated (nr) and measured as 'extra over' the roof structure. Although not specifically requested by the rules, it is recommended that the area of the dormer (i.e. the area of the dormer on plan) is given in the description.

Flat roof structures

Flat roof structures can be constructed from a variety of materials, including timber, in situ concrete, precast concrete, prestressed concrete, composite decking systems, space decks and metal decking systems. Components of a timber flat roof structure comprise wall plates, ceiling joists, firrings, strutting, decking and all connectors, bolts and other fixings. Concrete slab and decking systems include formwork, reinforcement, filler units, fixing slips, clips, fixings and grouting of joints, as applicable. Flat roof structures are measured superficially, in square metres. However, the rules distinguish between the measurement of flat roofs with and without parapet walls as follows:

- for flat roof structures with parapet walls – the area measured is the area of the roof on plan, measured to the internal face of the parapet walls enclosing the roof.
- for flat roof structures without parapet walls – the area measured is the area of the roof on plan, measured to the extremities of the eaves.

Descriptions of flat roof structures are to include the composition of the roof structure, the design load (kN/m^2) and span (m).

Specialist roof structures

The rules do not specifically address the measurement of shell roof structures such as space decks, space frames (e.g. single and double layer grids and geodesic domes), space decks, domes (e.g. rotational, pendentive and translational domes), barrel vaults (e.g. single, intersecting, stepped and northlight barrel vaults) and shells formed with hyperbolic paraboloids. It is recommended that such specialist roof structures are measured in the same way as pitched roof structures. That is, measured superficially, in square metres, with the area of the roof on plan, to the extremities of the eaves, being measured. Descriptions for specialist roof structures should include the type and construction of the roof structure, together with the principal dimensions. For example, descriptions for space decks should include the size of the unit trays, and the distance between the top of the standard unit (the unit tray) and the tie bars (m). The distance between the top chord and the bottom chord (m) should be included in descriptions for space frames. The sub-components and principal dimensions to be stated for space decks and space frames are shown in Figure 16.1 (refer to paragraph 16.2.2 – Space frames and decks (sub-element 2.2.1)).

In the case of barrel vault roof structures, the width or chord (m), the overall rise (m) and outer radius (m) should be stated in the description. The area of the barrel vault roof should include edge beams and combined edge beams and gutters (see Figure 16.6).

Sample descriptions for conventional roof structure components are shown in Example 16.13.

16.4.2 Roof coverings (sub-element 2.3.2)

This sub-element comprises roof coverings, and is subdivided into five cost-significant components:

- roof coverings, non-structural screeds, thermal insulation and surface treatments;
- dormer roof coverings;

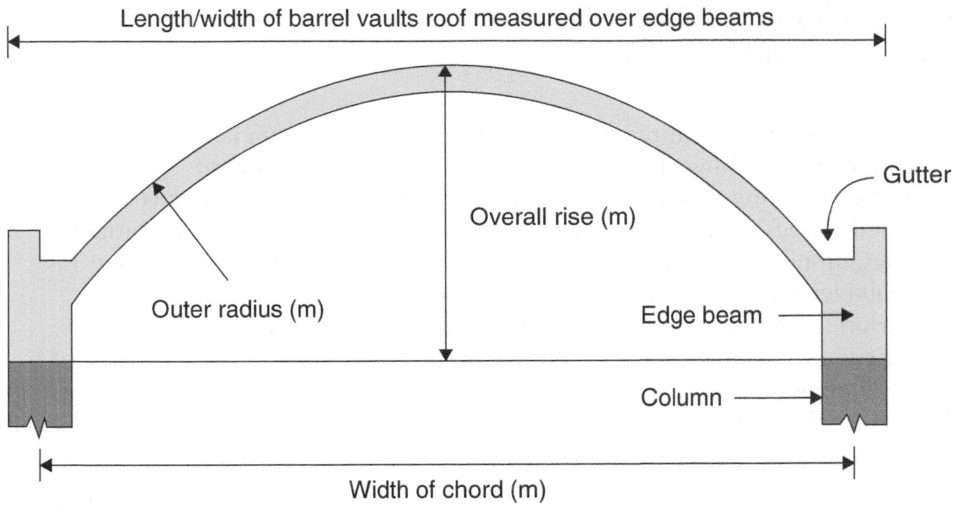

Figure 16.6 Sub-components and the principal dimensions for a barrel vault roof

Example 16.13 Formulation of descriptions for roof structures

Code	Element/sub-elements/components	Qty	Unit	Rate	Total
SUPERSTRUCTURE					
				£p	£
2	**SUPERSTRUCTURE**				
2.3	**Roof**				
2.3.1	*Roof structure*				
	Example 1 – Pitched roofs				
2.3.1.1	Trussed pitched roof structure; softwood root trusses (allowed 600mm centres) and wall plates; design load (TBC), pitch 45°, 9.00m span		m²		
2.3.1.2	Extra over pitched roof structure for forming dormers		m²		
	Example 2 – Flat roofs				
2.3.1.3	Flat roof structure; softwood, comprising roof joists (allowed for 200mm deep joists), wall plates, herringbone strutting and firrings; design load (TBC), 6.90m span		m²		
2.3.1.4	Flat roof structure; composite slab, 170mm overall thickness; comprising reinforced concrete on permanent formwork system ('Holorib' decking, 'Rideck decking' or the like), 1.20mm thick; reinforcement content 130kg/m³; 5.00kN/mm² loading		m²		
2.3.1.5	Extra for edge formwork; upstand to permanent formwork system ('Holorib' decking, 'Rideck decking' or the like) – to perimeter of slab, 170mm high		m		
2.3.1	**TOTAL – Roof structures: to element 2.3 summary**				

- eaves, verge treatment to pitched roofs;
- eaves treatment to flat roofs; and
- flashings.

Roof coverings include tiling, slating, mastic asphalt, liquid applied coatings, sheet coverings and thatching. They also include all ancillary components applicable to the roof-covering system (e.g. battening, underlay, vapour control layers, non-structural screeds, thermal insulation, ventilation tiles, hip and valley treatments, skirtings, upstands, verge treatments, surface treatments, flashings and edge trims). This sub-element also includes green roofs and roof garden systems, including protection layers, drainage layers, filter membranes and growing medium and planting.

The composition of the roof-covering system, including the type and thickness of thermal insulation is to be stated in the description. Coverings to roofs are measured superficially, in square metres, as follows:

- for pitched roofs – the surface area of the roof, to the extremities of the eaves;
- for flat roofs with parapet walls – the area of the roof on plan, measured to the internal face of the parapet walls enclosing the roof;
- for flat roofs without parapet walls – the area of the roof on plan, measured to the extremities of the eaves;
- for specialist roofs – the surface area of the roof, to the extremities of the eaves or to the internal face of the parapet walls enclosing the roof, whichever is applicable.

Coverings to dormers, including cladding to dormer cheeks, are measured superficially, in square metres, as 'extra over' the roof-covering system in which they occur.

In each case, the area of roof coverings displaced by rooflights, lens lights (pavement lights), skylights and openings that exceed 1.00m^2 are deducted.

While not called for by the rules, it is recommended that photovoltaic tiles, slates or profiled sheets which are an integral part of the roof-covering system be measured as 'extra over' the roof-covering system in which they occur. The actual surface area (m^2) of photovoltaic tiles, slates or profiled sheets is measured.

Roof paviours are another cost-significant item, which can be measured as 'extra over' the roof-covering system in which they occur. Again, the measurement is the actual surface area (m^2) of the roof paviours, with the pavement specification stated in the description.

Other cost-significant components include edge treatments, verge treatments and flashings:

- Edge and verge treatments (e.g. metal trims) are measured linear, in metres). The extreme length is measured, and the type (quality) of treatment is to be stated in the description. Treatments in connection with pitched and flat roofs are to be delineated.
- Flashings are measured linear, in metres, based on their extreme length. The type of material is to be given in the description. It is recommended that horizontal flashings and stepped flashings be measured and described separately.

Typical descriptions for roof structure components are shown in Example 16.14.

Example 16.14 Formulation of descriptions for roof coverings

Code	Element/sub-elements/components	Qty	Unit	Rate	Total
				£p	£
SUPERSTRUCTURE					
2	**SUPERSTRUCTURE**				
2.3	**Roof**				
2.3.2	Roof coverings				
	Example 1 – Pitched roofs				
2.3.2.1	Pitched roof coverings; comprising clay interlocking tiles, on batten and underlay; including thermal insulation to roof space, laid over ceiling joists		m²		
2.3.2.2	Extra over pitched roof coverings for coverings to dormers, including cladding to dormer cheeks		m²		
2.3.2.3	Eaves; comprising eaves framing, 25mm × 225mm softwood fascia, 50mm wide soffit and ventilator; painted		m		
2.3.2.4	Flashings; code 4 lead, horizontal		m		
2.3.2.5	Flashings; code 4 lead, stepped, including soakers code 3		m		
	Example 2 – Flat roofs				
2.3.2.6	Flat roof coverings; high performance bitumen felt roofing system; including thermal insulation and vapour control barrier, with chipping finish		m²		
2.3.2.7	Edge treatment; softwood splayed fillet, 25mm × 275mm softwood fascia, painted, with extruded aluminium alloy edge trim		m		
2.3.2.8	Flat roof coverings; single layer polymer roofing membrane with tapered insulation, on 75mm thick cement: sand screed, with solar reflective paint finish		m²		
2.3.2.9	Edge treatment; code 4 lead drip dressed into gutter, 25mm × 275mm softwood fascia, painted		m		
2.3.2.10	Landscaped roof; waterproofing and vapour equalisation layer, copper-lined bitumen membrane root barrier and waterproofing layer, separation and slip layers, protection layer, 50mm thick drainage board, filter fleece, insulation, Sedum vegetation blanket – low maintenance		m²		
2.3.2	**TOTAL – Roof coverings: to element 2.3 summary**				

16.4.3 Specialist roof systems (sub-element 2.3.3)

Specialist roof systems comprise patent glazing, glazed roofing systems and Perspex roofing systems, as well as all ancillary components, such as flashings, cover strips, integral drainage channels, perimeter treatments and the like. They are measured superficially, in square metres, with the area of the roof on plan measured. The

Example 16.15 Formulation of descriptions for glazed roofs

SUPERSTRUCTURE					
Code	**Element/sub-elements/components**	**Qty**	**Unit**	**Rate**	**Total**
				£p	**£**
2	**SUPERSTRUCTURE**				
2.3	**Roof**				
2.3.3	*Specialist roof systems*				
2.3.3.1	Roof cladding; patent glazing, with aluminium alloy bars; Georgian wired polished glass, 6mm thick; single tier; including opening lights and all flashings		m²		
2.3.3	**TOTAL – Specialist roof systems: to element 2.3 summary**				

composition of the glazed roof system is to be stated in the description. Example 16.15 illustrates a description for a specialist roof system.

16.4.4 Roof drainage (sub-element 2.3.4)

Roof drainage comprises two key components:

- gutters; and
- rainwater pipes.

Both gutters and rainwater pipes are measured linear, in metres, with the extreme length measured over all fittings, branches and the like. The length of rainwater pipes is taken from the point of connection to the gutter (i.e. the rainwater outlet) to the gulley or the rainwater shoe (measured over the shoe), whichever is applicable.

Allowance for costs in connection with the testing and commissioning of roof drainage systems are to be made in the unit rates applied to gutters and rainwater pipes by the cost manager.

Example 16.16 shows descriptions for roof drainage components.

Example 16.16 Formulation of descriptions for roof drainage

SUPERSTRUCTURE					
Code	**Element/sub-elements/components**	**Qty**	**Unit**	**Rate**	**Total**
				£p	**£**
2	**SUPERSTRUCTURE**				
2.3	**Roof**				
2.3.4	*Roof drainage*				
2.3.4.1	Gutters; uPVC, 112mm half round; including fittings		m		
2.3.4.2	Rainwater pipes; uPVC, 68mm diameter; including fittings		m		
2.3.4	**TOTAL – Roof drainage: to element 2.3 summary**				

16.4.5 Rooflights, skylights and openings (sub-element 2.3.5)

Roof lights, lens lights (pavement lights), skylights, roof hatches, smoke vents, roof vents, roof cowls and the like are measured as enumerated (nr) items. In addition to the component, the cost manager will need to consider the costs associated with forming the opening in the roof structure (e.g. kerbs around openings). However, such items can be measured and described separately with the roof structure (sub-element 2.3.1); as 'extra over' the roof structure in which they occur. Costs in connection with working roof coverings around the component are measured as part of roof coverings (sub-element 2.3.2).

The type and composition of the unit, together with any key dimensions, are to be given in the description. Example 16.17 illustrates typical component descriptions.

Example 16.17 Formulation of descriptions for rooflights, skylights and openings

SUPERSTRUCTURE					
Code	**Element/sub-elements/components**	**Qty**	**Unit**	**Rate**	**Total**
				£p	**£**
2	**SUPERSTRUCTURE**				
2.3	**Roof**				
2.3.5	*Rooflights, skylights and openings*				
	Example 1 – Component in connection with pitched roof				
2.3.5.1	Roof windows; composite timber and aluminium double glazed roof windows (Velux or similar); including trimming opening; 940mm × 1600mm		nr		
	Example 2 – Component in connection with flat roof				
2.3.5.2	Rooflight; double skin polycarbonate dome on upstand; 900mm × 900mm		nr		
2.3.5	**TOTAL – Rooflights, skylights and openings: to element 2.3 summary**				

16.4.6 Roof features (sub-element 2.3.6)

Roof features comprise turrets, wind vanes, spires, false chimneys, roof-top plant room screens and enclosures (i.e. complete structures, including wall louvres), fall arrest systems, access systems for cleaning roof, service walkways within roof voids, permanent roof edge protection (including balustrades, handrails and the like to roof edges and walkways).

The rules stipulate that such items are enumerated (nr), with the type of unit stated in the description. Nonetheless, in some instances it might be better to measure items such as roof-top plant room screens and enclosures, walkways within roof voids and permanent roof edge protection as linear items, in metres.

Descriptions for some common roof features are illustrated in Example 16.18.

Example 16.18 Formulation of descriptions for roof features

SUPERSTRUCTURE					
Code	**Element/sub-elements/components**	**Qty**	**Unit**	**Rate £p**	**Total £**
2	**SUPERSTRUCTURE**				
2.3	**Roof**				
2.3.5	*Roof features*				
2.3.5.1	Wind vane; PC sum £250		nr		
2.3.5.2	False chimneys; GRP chimney stacks, mid-ridge/ apex type, one pot, 650mm × 650mm × 900mm (height from top of ridge to top of stack)		nr		
2.3.5.3	Fall arrest systems; roof anchor system, stainless steel eyebolt and steel clamping system (number of anchor points measured)		nr		
2.3.5	**TOTAL – Roof features: to element 2.3 summary**				

16.5 Stairs and ramps (element 2.4)

The measurement of stairs and ramps is divided into the following sub-elements:

2.4.1 Stair/ramp structures;
2.4.2 Stair/ramp finishes;
2.4.3 Stair/ramp balustrades and handrails; and
2.4.4 Ladders/chutes/slides.

16.5.1 Stair/ramp structures (sub-element 2.4.1)

This sub-element comprises the structure associated with staircases (including spiral staircases and fire escape staircases), access ramps and the landings between floor levels. In situ concrete and precast concrete units include all concrete, reinforcement, formwork, worked finishes and, in the case of precast units, grouting between units. Stairs and ramps fabricated from steel, aluminium or timber include all off-site-applied coatings and paint systems. Stair and ramp structures are measured and described separately.

Stair structures are enumerated (nr), based on the number of storey flights (i.e. the number of staircases multiplied by the number of floors served. The type and composition of the stair structure is to be stated in the description, together with the vertical rise (m). The vertical rise is the distance from the top of the structural floor level to the top of the structural floor level at the next floor or roof level, whichever is applicable. Refer to Figure 8.10 in Chapter 8 ('Preparation of an initial elemental cost model – using the elemental method of estimating'). Ramp structures are measured and described in the same way as stair structures.

Example 16.19 is typical of component descriptions.

Example 16.19 Formulation of descriptions for stair/ramp structures

SUPERSTRUCTURE					
Code	**Element/Sub-elements/Components**	**Qty**	**Unit**	**Rate**	**Total**
				£p	**£**
2	**SUPERSTRUCTURE**				
2.4	**Stairs and ramps**				
2.4.1	*Stair/ramp structures*				
	Example 1 – Stairs				
2.4.1.1	Stair structure; reinforced concrete structure, with half landing; including reinforcement and formwork, fine finish; 3.00m rise		nr		
2.4.1.2	Stair structure; precast concrete structure, with half landing; 3.00m rise		nr		
2.4.1.3	Stair structure; softwood structure, straight flight; 2.60m rise		nr		
2.4.1.4	Stair structure; spiral staircase with solid wooden treads, metal balusters and tubular steel handrail, 2.00mm diameter, 3.00m rise		nr		
	Example 2 – Ramps				
2.4.1.5	Ramp structure; reinforced concrete structure, with half landing; including reinforcement and formwork, fine finish; 1.80m rise		nr		
2.4.1	**TOTAL – Stair/ramp structures: to element 2.4 summary**				

16.5.2 Stair/ramp finishes (sub-element 2.4.2)

Finishes to stairs and ramps encompass finishes to treads and risers, to landings between floor levels, to ramp surfaces and finishes to strings and soffits of staircases and ramps. They are measured as enumerated (nr) items in the same way as stair and ramp structures. The nature or quality of finishes is to be given in the description. In the case of ramps, the approximate area of floor finish should also be stated.

Example 16.20 shows typical descriptions of components.

16.5.3 Stair and ramp balustrades and handrails (sub-element 2.4.3)

Balustrades and handrails to stairs and ramps are measured as enumerated (nr) items in the same way as stair and ramp structures. They include balustrades and handrails to stairs, ramps and landings, as well as all off-site-applied coatings and paint systems. The type of balustrades and handrails is to be given in the description.

Sample component descriptions for balustrades and handrails to stairs and ramps are shown in Example 16.21.

Example 16.20 Formulation of descriptions for stair/ramp finishes

SUPERSTRUCTURE					
Code	**Element/sub-elements/components**	**Qty**	**Unit**	**Rate**	**Total**
				£p	**£**
2	**SUPERSTRUCTURE**				
2.4	**Stairs and ramps**				
2.4.2	*Stair/ramp finishes*				
	Example 1 – Stairs				
2.4.2.1	Stair finishes; terrazzo finishes to treads, risers and half landings; 3.00m rise		nr		
2.4.2.2	Stair finishes; rubber sheet finishes to treads, risers and half landings, including heavy duty aluminium alloy stair tread nosings; 3.00m rise		nr		
2.4.2.3	Stair finishes; carpets (PC £50/m^2), to treads and risers; straight flight; 2.60m rise		nr		
	Example 2 – Ramps				
2.4.2.4	Ramp finishes; edge fixed carpeting (PC £50/m^2); 1.80m rise; surface area approximately 9.00m^2		nr		
2.4.2	**TOTAL – Stair/ramp finishes: to element 2.4 summary**				

Example 16.21 Formulation of descriptions for stair/ramp balustrades and handrails

SUPERSTRUCTURE					
Code	**Element/sub-elements/components**	**Qty**	**Unit**	**Rate**	**Total**
				£p	**£**
2	**SUPERSTRUCTURE**				
2.4	**Stairs and ramps**				
2.4.3	*Stair/ramp balustrades and handrails*				
	Example 1 – Stairs				
2.4.3.1	Wall handrails; timber handrails on brackets, raking; to stairs and half landing; 3.00m rise		nr		
2.4.3.2	Combined balustrades and handrails; stainless steel balusters and handrails, with reinforced glass infill panels; to stairs and half landing; 3.60m rise		nr		
2.4.3.3	Combined balustrades and handrails; painted softwood balusters and varnished hardwood handrails; to straight flight; 2.60m rise		nr		
	Example 2 – Ramps				
2.4.3.4	Combined balustrades and handrails; stainless steel balusters and handrails, with reinforced glass infill panels; raking; 3.60m rise		nr		
2.4.3	**TOTAL – Stair/ramp balustrades and handrails: to element 2.4 summary**				

16.5.4 Ladders/chutes/slides (sub-element 2.4.4)

Fire escape ladders, fire escape chutes and slides, access ladders, loft ladders (including hatch doors where an integral part of the loft ladder) and the like are measured as enumerated (nr) items. The type of component is to be stated in the description.

Typical component descriptions for ladders, chutes and slides are shown in Example 16.22.

Example 16.22 Formulation of descriptions for ladders/chutes/slides

SUPERSTRUCTURE					
Code	Element/sub-elements/components	Qty	Unit	Rate	Total
				£p	£
2	**SUPERSTRUCTURE**				
2.4	**Stairs and ramps**				
2.4.4	*Ladders/chutes/slides*				
2.4.4.1	Ladders; wooden loft ladders, floor to ceiling height 3.00m, including frame, spring balanced mechanism, insulated trap door and trap door release pole		nr		
2.4.4.2	Ladders; metal concertina-style loft ladder, complete with insulated trap door, ceiling hole lining and trap door release pole		nr		
2.4.4.3	Chutes; fire escape chute, approx. 20 metres long		nr		
2.4.4	**TOTAL – Ladders/chutes/slides: to element 2.4 summary**				

16.6 External walls (element 2.5)

External walls include structural walls (i.e. those that are not formed by retaining walls) and non-structural walls that enclose the building. The rules divide the element into six cost-significant sub-elements:

2.5.1 External enclosing walls above ground floor level;
2.5.2 External enclosing walls below ground floor level;
2.5.3 Solar/rain screening;
2.5.4 External soffits;
2.5.5 Subsidiary walls, balustrades, handrails and proprietary balconies; and
2.5.6 Facade access/cleaning systems.

External walls below ground formed of retaining walls are measured under element 1.3 (Basement retaining walls).

16.6.1 External enclosing walls above and below ground floor level (sub-elements 2.5.1 and 2.5.2)

There are numerous different types of external enclosing wall systems. Examples include masonry cavity walls, curtain walling, structural glazed assemblies, profiled sheet cladding systems, rigid sheet cladding systems, panel walling systems, as well as many walling systems comprising a combination of different materials.

The rules for measuring external walls above and below ground floor level are the same. Nevertheless, the rules recommend that external walls above and below ground floor level are measured and described separately. Figure 16.7 illustrates the delineation of external walls above and below ground floor level.

External enclosing walls are measured superficially, in square metres, with the area measured on the centre line of the wall construction. Walls are measured gross with no deductions for door openings, window openings, openings for screens and the like. Descriptions for external walls are to fully describe the composition of the wall construction. In addition, details of any PC sums included in a unit rate (e.g. for facing bricks) are to be stated in the description. Where external walls consist of more than one type of walling (e.g. a combination of curtain walling and masonry walls), each type of walling is to be measured and described separately.

Parapet walls to roofs are measured as part of the external wall. Copings and cappings to parapet walls that are of a different material to the general walling (e.g. facing brickwork walling with precast concrete copings), although not specifically identified by the rules, should be treated as a cost-significant item (see Measurement rules for components, C5) and measured and described separately. It is recommended that copings and cappings be measured as linear items, in metres. The type of component should be stated in the description.

Gable ends are treated as external walls, as are chimney breasts that form an integral part of the external wall construction. However, walls to external walkways are measured under sub-element 2.5.5 (Subsidiary walls, balustrades, handrails, railings and proprietary balconies) – see sub-paragraph 16.6.4.

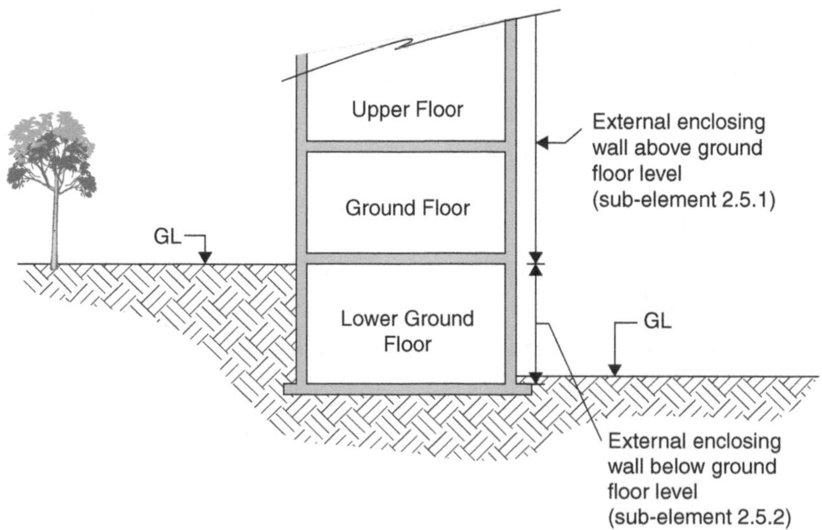

Figure 16.7 Delineation of external enclosing walls above and below ground floor level

Forming the openings in external walls for doors, windows, screens and the like is measured 'extra over' the type of wall construction to which they relate. They are measured as separate enumerated (nr) items, with the overall size of the opening given in the description. It is also necessary to consider the perimeter of the opening and describe the materials used at the head, jambs and sill. This is particularly important if they are of a different material to the external wall, as they are likely to be cost-significant. An example is the use of Portland stone in the head, jambs and sill to door and window openings where the external wall is of brick and block cavity construction.

Enhancing features incorporated in the external wall, such as plinths, cornices and ornamental bands, are measured 'extra over' the type of wall construction to which they relate. They are measured as separate linear items, in metres, describing the materials used. The width of the plinths, cornices and ornamental bands should also be stated. Decorative panels incorporated in external walls are also measured as an 'extra over' item, and can be either enumerated (nr), stating the overall size, or measured superficially, in square metres. Again, the materials used are to be described.

Similarly, quoins formed using a different material to the general walling, or formed of the same material but of a different size to the general walling, are also measured 'extra over' the type of wall construction to which they relate. Quoins are measured linear, in metres, on the vertical angle. The girth of the quoin should also be stated in the description.

Specialist components such as photovoltaic glazing or cladding panels, which are an integral part of the walling system, are to be measured superficially, in square metres, as 'extra over' the type of wall construction to which they relate. The composition of panel is to be specified in the description.

Finishes applied to external walls (e.g. coating systems, paint systems or tiles) are measured superficially, in square metres, describing the materials used. The surface area to which the finish is to be applied is measured.

Illustrative descriptions of components for external walls above ground floor level are shown in Example 16.23. The same approach is used to describe external walls defined as being below ground floor level.

Example 16.23 Formulation of descriptions for external enclosing walls above ground floor level

SUPERSTRUCTURE					
Code	**Element/sub-elements/components**	**Qty**	**Unit**	**Rate**	**Total**
				£p	**£**
2	**SUPERSTRUCTURE**				
2.5	**External walls**				
2.5.1	*External enclosing walls above ground floor level*				
2.5.1.1	External walls; cavity wall construction, 140mm thick block outer skin, 75mm thick insulation, 100mm thick lightweight block inner skin		m²		
2.5.1.2	Finishes applied to external walls; Sto external render system to block outer skin		m²		
2.5.1.3	External walls; cavity wall construction, facing brick (PC sum £330/1,000) outer skin, 75mm thick insulation, 100mm thick lightweight block inner skin		m²		

Example 16.23 *(Continued)*

SUPERSTRUCTURE					
Code	**Element/sub-elements/components**	**Qty**	**Unit**	**Rate**	**Total**
				£p	**£**
2.5.1.4	Extra over external walls for ornamental bands in facing bricks (PC sum £480/1,000), flush, horizontal		m		
2.5.1.5	Extra over external walls for quoins in facing bricks (PC sum £480/1,000), flush		m		
2.5.1.6	Extra over external walls for forming openings for windows; including lintels, head courses, damp-proof courses, cavity trays, closing cavities and all other work to soffits, sills and reveals of openings; opening 900mm × 1350mm		nr		
2.5.1.7	Extra over external walls for forming openings for external doors including lintels, head courses, damp-proof courses, cavity trays, closing cavities and all other work to soffits, sills and reveals of openings; opening 1,000mm × 2,100mm		nr		
	Notes: Based on the advice of the architect and building services engineer: (1) Approximately 27% of area of curtain walling system will comprise opening panes (included in unit rate) (2) Approximately 30% of area of curtain walling system will require acoustic treatment				
2.5.1.8	External walls; unitised curtain walling system, comprising high-performance double-glazed units with neutral solar control coating set in dark aluminium-framed window system. Dark spandrel zone to mask floor build up with aluminium edge detail (with provisions for mechanical inlet/outlet ventilation behind). Mixture of profiled metal panels formed out of extruded aluminium sections with insulated panels		m²		
2.5.1.9	Extra over external walls for acoustic treatment to facades		m²		
2.5.1.10	Extra over external walls for interstitial blinds		m²		
2.5.1	**TOTAL – External enclosing walls above ground floor level: to element 2.5 summary**				

16.6.2 Solar/rain screening (sub-element 2.5.3)

Exterior solar and rain screen systems are classified as:

- vertical (over cladding) systems; or
- horizontal systems (e.g. Brise Soleil).

Example 16.24 Formulation of descriptions for solar and rain screen cladding

SUPERSTRUCTURE					
Code	**Element/sub-elements/components**	**Qty**	**Unit**	**Rate**	**Total**
				£p	**£**
2	**SUPERSTRUCTURE**				
2.5	**External walls**				
2.5.3	*Solar/rain screening*				
2.5.3.1	Brise Soleil, including support framework; bright aluminium finish; horizontal		m		
2.5.3	**TOTAL – Solar/rain screening: to element 2.5 summary**				

Vertical solar and rain screen systems are measured superficially, in square metres, with the surface area of the over cladding system measured. Horizontal solar and rain screen systems are measured linear, in metres, along their extreme length. Each different type of system is to be measured and described separately.

The description for vertical and horizontal solar and rain screen systems is to fully depict the system to be used. A typical description of a horizontal solar system is shown in Example 16.24.

16.6.3 External soffits (sub-element 2.5.4)

External soffits to the underside of upper floors that project beyond the footprint of the floor below are measured superficially, in square metres, with the surface area of the soffit measured. When describing external soffits it is necessary to distinguish between false ceilings and demountable suspended ceilings. Details of any insulation, and whether it is fixed directly to the underside of the upper floor construction or laid on the ceiling, is also to be stated.

Finishes applied to the exposed surface of external soffits are also measured superficially, in square metres, with the surface area of the soffit to which the finish is applied measured. The description for applied finishes is to include the nature of the finish (e.g. plaster skim coat, render, roughcast, paint or specialist coating). Edge treatments such as cornices, covings and shadow gaps are to be measured and described separately. They are measured linear, in metres, along their extreme length over all obstructions, with the type of component stated in the description.

Access hatches and the like within external soffits are to be enumerated (nr), stating their size, make-up, and whether proprietary or purpose made in the description. A description for a typical external ceiling system is shown in Example 16.25.

16.6.4 Subsidiary walls, balustrades, handrails and proprietary balconies (sub-element 2.5.5)

Sub-element 2.5.5 addresses non-structural external walls which are an integral part of the building but do not enclose it. Examples include low-level walls (dwarf

Example 16.25 Formulation of descriptions for external soffits

SUPERSTRUCTURE					
Code	**Element/sub-elements/components**	**Qty**	**Unit**	**Rate**	**Total**
				£p	**£**
2	**SUPERSTRUCTURE**				
2.5	**External walls**				
2.5.4	*External soffits*				
2.5.4.1	External soffits; solid wood ceiling system comprising wooden panels fixed to an integrated metal suspension system		m^2		
2.5.4	**TOTAL – External soffits: to element 2.5 summary**				

walls) to external walkways and walls forming planters. It also deals with balustrades, handrails and railings fixed to low level walls, including those situated on parapet walls, and proprietary bolt-on balconies.

Low-level walls are measured linear, in metres, with their extreme length measured over all obstructions. The height, overall thickness and composition of the wall, including details of any copings or cappings, are to be included in the description. Walls forming planters are measured in the same way as low-level walls. The description for planters is to describe the composition of the complete planter, including protection layers, drainage layers, filter membranes and growing medium. Details of any planting should also be stated.

Combined balustrades and handrails, handrails and parapet railings are also measured and described as linear items, in metres. Descriptions for these items are to include the material, the overall size or height, and details of any coatings applied off site or on site.

Proprietary bolt-on balconies are enumerated (nr), with the type of balcony (e.g. 'Juliet' balconies) and details of any coatings applied off site or on site described.

Disposal installations taking surface water from external walkways, planters and proprietary bolt-on balconies are measured and described separately. Rainwater pipes are measured linear, in metres, and floor outlets are enumerated (nr). The diameter and type of rainwater pipe or floor outlet are described, together with details of any coatings applied off site or on site. Costs associated with the testing and commissioning of rainwater installations are to be incorporated within the unit rate applied to system.

Typical descriptions for parapet railings and proprietary bolt-on balconies are shown in Example 16.26.

16.6.5 Facade access/cleaning systems (sub-element 2.5.6)

Building maintenance units (BMUs) and other window, facade and roof access and cleaning equipment are treated as part of the external walls element. This is because the function of these items is to gain access to the external walls and windows for purposes of cleaning, maintenance and repair.

Example 16.26 Formulation of descriptions for subsidiary walls, balustrades, handrails and proprietary balconies

SUPERSTRUCTURE					
Code	**Element/sub-elements/components**	**Qty**	**Unit**	**Rate**	**Total**
				£p	**£**
2	**SUPERSTRUCTURE**				
2.5	**External walls**				
2.5.5	*Subsidiary walls, balustrades, handrails and proprietary balconies*				
2.5.5.1	Parapet railings; safety railing system, aluminium, untreated, curved posts, 1,100mm high		m		
2.5.5.2	Proprietary bolt-on balconies; Juliet balconies, mild steel, powder coated, to suit 1800mm wide opening		nr		
2.5.5	**TOTAL – Subsidiary walls, balustrades, handrails and proprietary balconies: to element 2.5 summary**				

Facade access and cleaning systems are measured and described as enumerated (nr) items, with the type of system is described.

Costs in connection with the testing and commissioning of mechanically and electrically operated facade access and cleaning systems are measured in conjunction with each component. Example 16.27 illustrates component descriptions for typical façade access and cleaning systems.

Example 16.27 Formulation of descriptions for facade access/cleaning systems

SUPERSTRUCTURE					
Code	**Element/sub-elements/components**	**Qty**	**Unit**	**Rate**	**Total**
				£p	**£**
2	**SUPERSTRUCTURE**				
2.5	**External walls**				
2.5.6	*Facade access/cleaning systems*				
2.5.6.1	Facade access systems; roof-mounted track-based standard building maintenance system (BMU)		nr		
2.5.6.2	Facade access systems; trackless standard building maintenance system (BMU)		nr		
2.5.6.3	Facade access systems; portable davit access systems, receiver sockets positioned at 2–3m intervals		nr		
2.5.6.4	Testing of facade access systems		%		
2.5.6.5	Commissioning of facade access systems		%		
2.5.6	**TOTAL – Facade access/cleaning systems: to element 2.5 summary**				

16.7 Windows and external doors (element 2.6)

The measurement of windows and external doors is divided into the following sub-elements:

2.6.1 external windows; and

2.6.2 external doors.

Forming openings for windows and external doors is measured under sub-elements 2.5.1 (External enclosing walls above ground floor level) and 2.5.2 (External enclosing walls below ground floor level), whichever is applicable. Similarly, forming openings for windows and external doors in existing external walls (i.e. cutting openings in existing work) is measured as spot items under sub-element 7.1.1 (Minor demolition works and alteration works).

16.7.1 External windows (sub-element 2.6.1)

External windows comprise the following key components:

- windows;
- louvres;
- shop fronts; and
- roller shutters, sliding shutters, grilles and the like to external windows and shop fronts.

Windows take account of openings in external walls for light and ventilation. Therefore, both windows and louvres are dealt with under this sub-element. However, glazed curtain walling systems and structural glazing assemblies are considered to be external wall components. Consequently, they are measured under sub-elements 2.5.1 (External enclosing walls above ground floor level) and 2.5.2 (External enclosing walls below ground floor level), as applicable.

Windows are measured superficially, in square metres, as composite units. The area occupied by the window unit is measured (i.e. the area of the window measured over all frames). The description for windows is to include the composition of the unit, the overall size of opening and the number of units (nr), together with details of any other sub-components which form an integral part of the component measured (e.g. ironmongery, linings, window boards, integral blinds, protective film, solar cladding, rainwater screens, canopies, external blinds and shutters). Curtains, curtain tracks, blinds, pelmets and the like which are not an integral part of the window unit are measured and described separately under sub-element 4.1.1 (General fittings, furnishings and equipment).

Today, it is common practice to apply sealants to the perimeter of window units. Therefore, the cost manager will need to consider this when building up a unit rate for the fabrication, supply and installation of window units. However, where it is envisaged that perimeter sealants will be applied by a works contractor different from those installing the window units it might be worth the cost manager treating the application of sealants as a separate measured item. This will aid post-cost control during procurement.

Louvred windows and shop fronts are measured in the same way as windows.

Roller shutters, sliding shutters, grilles and the like to external windows and shop fronts are enumerated (nr), with the type of component and overall size of the opening stated in the description.

The rules permit other cost-significant components (e.g. canopies above windows) to be measured and described separately. In such cases, the method of measurement is at the discretion of the cost manager.

Example component descriptions for windows and external doors are provided in Example 16.28.

Example 16.28 Formulation of descriptions for external windows

SUPERSTRUCTURE					
Code	**Element/sub-elements/components**	**Qty**	**Unit**	**Rate**	**Total**
				£p	**£**
2	**SUPERSTRUCTURE**				
2.6	**Windows and external doors**				
2.6.1	*External windows*				
2.6.1.1	Windows; uPVC windows, purpose made, double glazed, structural opening size 1,800mm × 1,350mm		m²		
2.6.1.2	Extra over windows for protective film; security film, clear, scratch resistant		m²		
2.6.1.3	Windows; uPVC windows, purpose made, double glazed, tinted glass; structural opening size 600mm × 1,170mm		m²		
2.6.1.4	Extra over windows for protective film; solar control combined security film, silver		m²		
2.6.1.5	Louvre; polyester powder coated aluminium or anodised finish; structural opening size 600mm × 1,900mm		m²		
2.6.1.6	Louvres; single bank blade system; polyester powder coated aluminium or anodised finish		m²		
2.6.1.7	Shop fronts; flat facade, glass in aluminium framing, double-leaf manual centre door unit		m²		
2.6.1	**TOTAL – External windows: to element 2.6 summary**				

16.7.2 External doors (sub-element 2.6.2)

External doors comprise the following key components:

- external doors;
- revolving doors;
- shop front doors (which are not an integral part of a shop front window unit);
- roller shutters, sliding shutters and the like to external door openings;
- garage doors;
- grilles; and
- architraves.

Example 16.29 Formulation of descriptions for external doors

SUPERSTRUCTURE					
Code	**Element/sub-elements/components**	**Qty**	**Unit**	**Rate £p**	**Total £**
2	**SUPERSTRUCTURE**				
2.6	**Windows and external doors**				
2.6.2	*External doors*				
2.6.2.1	External doors; single aluminium door and frame, double glazed, including ironmongery, structural opening size 1,000mm × 2,100mm		nr		
2.6.2.2	Revolving door; 2,000 mm diameter; clear laminated glazing; 4 nr wings; glazed walls; height 2,500mm		nr		
2.6.2.3	Revolving door; 2,000 mm diameter; clear laminated glazing; 4 nr wings; glazed walls; 2 nr passing doors; height 2,500mm		nr		
2.6.2.4	Shop front doors; single aluminium door and frame, double glazed, including ironmongery, structural opening size 1,800mm × 2,100mm		nr		
2.6.2.5	Roller shutters; continuous steel curtain roller shutter doors; powder coated, including thermal and acoustic insulation, vision panels, personnel entry doors, electronic operation, and secure locking mechanisms; structural opening size 5,000mm × 6,000mm		nr		
2.6.2.6	Architraves; 25mm × 70mm, hardwood, splayed; varnished		m		
2.6.2	**TOTAL – External doors: to element 2.6 summary**				

External doors comprise entrance doors, entrance doors with side screens, patio doors and garage doors. They are measured as composite items and as such include door frames, linings, ironmongery, glazed vision panels, as well as painting and decorating. External doors are measured and described as enumerated (nr) items. Descriptions for external doors are to include the type of door (e.g. entrance door, patio doors, external fire escape door, revolving door, shop front door, roller shutter, sliding shutter and garage door), the composition, the number of door leaves, size of each door leaf and overall size of the door opening (measured over frames), together with details of any other sub-components specific to the component measured (e.g. frame, lining, ironmongery, glazed vision panels, fly screens, integral blinds, protective film, solar cladding and rainwater screens).

Grilles are measured as enumerated (nr) items, with the type of grille and overall size of the door opening given in the description.

Canopies associated with external doors are also measured as enumerated (nr) items.

Architraves to external doors are to be measured linear, in metres, with the type of architrave and details of any applied finishes (e.g. paint, staining or varnish) stated in the description.

Example 16.29 shows typical component descriptions.

16.8 Internal walls and partitions (element 2.7)

The measurement of internal walls and partitions is divided into the following sub-elements:

2.7.1 Walls and partitions;
2.7.2 Balustrades and handrails;
2.7.3 Movable room dividers; and
2.7.4 Cubicles.

16.8.1 Walls and partitions (sub-element 2.7.1)

Internal walls and fixed partitions are measured superficially, in square metres. The length is measured on the centre line of the wall construction, and height is the actual height of the wall. The composition and thickness of walls and partitions are to be described. Each type of wall or partition is to be measured and described separately.

As with external walls, internal walls and partitions are measured gross with no deductions for door openings, screens and the like. Forming the openings in internal walls and partitions for doors, screens and the like is measured 'extra over' the type of wall or partition construction to which they relate. Openings are measured as enumerated (nr) items, with the overall size of the opening stated in the description.

Example 16.30 is illustrative of descriptions for wall and partition components.

Example 16.30 Formulation of descriptions for walls and partitions

SUPERSTRUCTURE					
Code	**Element/sub-elements/components**	**Qty**	**Unit**	**Rate £p**	**Total £**
2	**SUPERSTRUCTURE**				
2.7	**Internal walls and partitions**				
2.7.1	*Walls and partitions*				
2.7.1.1	Internal walls; blockwork, lightweight blocks; 100mm thick		m²		
2.7.1.2	Extra over for forming openings in 100mm thick wall for doors; including lintel, structural opening 1,000mm × 2,100mm		nr		
2.7.1.3	Internal walls; blockwork, lightweight blocks; 140mm thick		m²		
2.7.1.4	Internal walls; blockwork, lightweight blocks; 200mm thick		m²		
2.7.1.5	Internal walls; blockwork, dense concrete blocks; 100mm thick; fair faced both sides		m²		
2.7.1.6	Fixed partitions; softwood stud and plasterboard partitions; 100mm thick; comprising single layer 13mm thick board each side; tape and filled joints		m²		

Example 16.30 *(Continued)*

SUPERSTRUCTURE					
Code	**Element/sub-elements/components**	**Qty**	**Unit**	**Rate**	**Total**
				£p	**£**
2.7.1.7	Extra over for forming openings in wall for doors; opening 1,000mm × 2,100mm		nr		
2.7.1.8	Fixed partitions; metal stud and plasterboard partitions; 100mm thick; comprising single layer 13mm thick board each side; tape and filled joints		m²		
2.7.1.9	Extra over for forming openings in wall for doors; opening 1,000mm × 2,100mm		nr		
2.7.1.10	Fixed partitions; fire-resistant (120 minutes) metal stud and plasterboard partitions; 100mm thick; comprising two layers 13mm thick Fireline board each side; cavity insulation; tape and filled joints		m²		
2.7.1.11	Extra over for forming openings in wall for doors; opening 1000mm × 2100mm		nr		
2.7.1	**TOTAL – Walls and partitions: to element 2.7 summary**				

16.8.2 Balustrades and handrails (sub-element 2.7.2)

Combined balustrades and wall-mounted handrails are measured as linear items, in metres. The description for these items is to include the composition, the overall size or height, together with the details of any coatings applied on site or off site. A typical description of a combined balustrade is shown in Example 16.31.

Example 16.31 Formulation of descriptions for balustrades and handrails

SUPERSTRUCTURE					
Code	**Element/sub-elements/components**	**Qty**	**Unit**	**Rate**	**Total**
				£p	**£**
2	**SUPERSTRUCTURE**				
2.7	**Internal walls and partitions**				
2.7.2	*Balustrades and handrails*				
2.7.2.1	Combined balustrades and handrails; mild steel balusters and handrails, horizontal, painted, 1,100mm high		m		
2.7.2	**TOTAL – Balustrades and handrails: to element 2.7 summary**				

16.8.3 Movable room dividers (sub-element 2.7.3)

Movable room dividers and partitions (i.e. used to temporarily divide internal spaces into separate compartments) are measured superficially, in square metres. The area occupied by the divider or partition is measured (i.e. the area of the divider or partition measured over all frames). The composition of the movable room divider or partition is to be described.

Sliding or folding doors, which form an integral part of an internal wall or fixed partition, are measured under element 2.8 (Internal doors).

Example 16.32 illustrates component descriptions.

Example 16.32 Formulation of descriptions for movable room dividers

SUPERSTRUCTURE					
Code	**Element/sub-elements/components**	**Qty**	**Unit**	**Rate**	**Total**
				£p	**£**
2	**SUPERSTRUCTURE**				
2.7	**Internal walls and partitions**				
2.7.3	*Movable room dividers*				
2.7.3.1	Movable room dividers; movable wall system, wall panel to achieve an acoustic rating of Rw 57dB, 100mm thick		m^2		
2.7.3.2	Movable room dividers; movable partition wall, fire resisting (60 minutes), 117mm thick		m^2		
2.7.3	**TOTAL – Movable room dividers: to element 2.7 summary**				

16.8.4 Cubicles (sub-element 2.7.4)

This component deals with proprietary pre-finished cubicles, complete with doors, which are assembled on site. Cubicles can be either enumerated (nr), as individual or a range of cubicles, or measured superficially, in square metres, whichever is considered to be applicable. Where the area is measured, the area is the gross area of the cubicle partition, measured across the door. Descriptions are to include the type and composition of the cubicle, together with the number of cubicles and cubicle doors.

Cubicles formed of masonry or concrete walls and a door are measured under sub-elements 2.7.1 (Walls and partitions) and 2.8.1 (Internal doors) as appropriate. Finishes to cubicle walls (e.g. plasterwork and painting) are measured under sub-element 3.1.1 (Finishes to walls and columns).

Example 16.33 shows typical descriptions for proprietary pre-finished cubicles.

Example 16.33 Formulation of descriptions for cubicles

SUPERSTRUCTURE					
Code	**Element/sub-elements/components**	**Qty**	**Unit**	**Rate**	**Total**
				£p	**£**
2	**SUPERSTRUCTURE**				
2.7	**Internal walls and partitions**				
2.7.4	*Cubicles*				
2.7.4.1	Cubicles; partitions complete with doors; proprietary; pre-finished panels, standard colours, standard ironmongery, assembling and fixing; range of 1 nr open-ended cubicles; 1 nr panel; 1 nr door		nr		
2.7.4.2	Cubicles; partitions complete with doors; proprietary; pre-finished panels, standard colours, standard ironmongery, assembling and fixing; range of 3 nr open-ended cubicles; 3 nr panels; 3 nr doors		nr		
2.7.4	**TOTAL – Cubicles: to element 2.7 summary**				

16.9 Internal doors (element 2.8)

This element consists of a single sub-element:

2.8.1 Internal doors.

16.9.1 Internal doors (sub-element 2.8.1)

Internal doors comprise the following key components:

- Internal doors;
- Fire-resisting doors;
- Door sets;
- Composite door and sidelights/over panel units;
- Roller shutters, sliding shutters and grilles; and
- Architraves.

Internal doors are measured and described as enumerated (nr) items. They are measured as composite items and as such include door frames, linings, ironmongery, glazed vision panels, and painting and decorating. Descriptions for internal doors are to include the type (e.g. standard internal door, purpose-made internal door, door sets, internal fire escape door, cupboard doors, service and duct cupboard doors, composite door and sidelight and/or over panel units, hatch units, internal roller shutter, internal sliding shutter and grilles) and composition of the door. The number of door leaves, size of each door leaf and overall size of the door opening, details of any other sub-components specific to the component measured (e.g. frames, linings, ironmongery, vision panels, fly screens, integral blinds, protective film, solar cladding and rainwater screens) and details of any finishes applied after installation (e.g. painting or varnishing) are also to be given in the description.

Forming openings for internal doors is measured under sub-elements 2.8.1 (Walls and partitions). Similarly, forming openings for internal doors within existing internal walls and partitions (i.e. cutting openings in existing work) is measured as spot items under sub-element 7.1.1 (Minor demolition works and alteration works).

Architraves to internal doors are to be measured linear, in metres, with the type of architrave and details of any applied finishes (e.g. paint, staining or varnish) stated in the description. When dealing with internal doors, care must be taken by the cost manager to ensure that architraves are measured to both sides of the door opening. The need to measure architraves to both sides can easily be overlooked.

Example 16.34 shows typical component descriptions.

Example 16.34 Formulation of descriptions for internal doors

SUPERSTRUCTURE					
Code	**Element/sub-elements/components**	**Qty**	**Unit**	**Rate**	**Total**
				£p	**£**
2	**SUPERSTRUCTURE**				
2.8	**Internal doors**				
2.8.1	*Internal doors*				
2.8.1.1	Internal doors; flush doors; standard; non-fire rated; single leaf; solid core; painted; softwood frames; painting and ironmongery, opening size 1,000mm × 2,100mm	118	nr	235.00	27,730
2.8.1.2	Internal doors; flush doors; standard; fire rated; single leaf; solid core; with vision panel; softwood frames; painting and ironmongery, opening size 1,000mm × 2,100mm	46	nr	289.00	13,294
2.8.1.3	Internal doors; flush doors; standard; non-fire rated; double leaf; solid core; softwood frames; painting and ironmongery, opening size 1,600mm × 2,100mm	38	nr	475.00	18,050
2.8.1.4	Fire-resisting doors; flush doors; standard; fire rated (60 minutes); one and a half leaf; solid core; with vision panel; softwood frames; painting and ironmongery	96	nr	690.00	66,240
2.8.1.5	Architraves; MDF, painted	180	m	8.00	1,440
2.8.1	**TOTAL – Internal doors: to element 2.8 summary**				**126,754**

CHAPTER 17

Internal finishes (group element 3)

Introduction

This chapter:

- provides step-by-step guidance on how to measure and describe components forming:
  - internal wall finishes;
  - internal floor finishes; and
  - internal ceiling finishes; and
  - gives examples illustrating how to formulate descriptions for components.

17.1 Method of measurement

The rules for measuring internal finishes are found in group element 3 in NRM 1, which is divided into the following three elements:

- 3.1 Wall finishes
- 3.2 Floor finishes
- 3.3 Ceiling finishes.

Internal finishes applied to walls, floors and ceilings are generally measured superficially, in square metres. However, painting and decorating works to repetitive building types (e.g. residential dwellings), or buildings containing a range of similar room types (e.g. hotel rooms or student accommodation), can be enumerated (nr). Where buildings or rooms are to be enumerated, the type and size of building, dwelling or room is to be stated in the description. For example, residential dwellings would be described by the number of bedrooms, whilst hotel rooms can be described by the number of persons for which the room is designed (e.g. single, double or family room).

Each different type of wall, floor and ceiling finish is to be measured and described separately.

17.2 Wall finishes (element 3.1)

This element comprises the following single sub-element:

3.1.1 Wall finishes

17.2.1 Wall finishes (sub-element 3.1.1)

Wall finishes applied to internal wall surfaces, including surfaces of internal columns, internal walls and internal partitions are measured under this sub-element. They include in situ coatings (e.g. plaster, render and roughcast), sprayed coatings, plasterboard linings (including joint systems, joint reinforcing scrim and plaster scrim coat), or other sheet linings (e.g. timber panelling and pre-finished linings), wall tiling, decorative sheet coverings and painting. Secondary components such as picture rails, dado rails, impact and bumper guards, protection strips, and corners protectors to internal walls and columns are also dealt with under this sub-element. Wall finishes to staircase areas (stairwells) are measured under this sub-element.

Finishes to walls and columns are measured superficially, in square metres. The surface area to which the finish is to be applied is measured, with no deductions for door openings, window openings, openings for screens and the like from the gross area measured. Descriptions for finishes to walls are to include the type and composition of wall finish. In addition, details of any prime cost sums included in a unit rate (e.g. for supplying only ceramic wall tiles) are to be stated in the description.

Plasterboard and other sheet linings that form an integral part of an internal partition system are regarded as part of the internal partition system (e.g. metal stud and plasterboard partitions) and measured under sub-element 2.8.1 (Walls and partitions). Finishes to the sides (walls) of swimming pool tanks, including the linings to the tank, are measured under sub-element 3.2.1 (Finishes to floors).

Picture rails, dado rails and the like are measured linear, in metres, along their extreme length over all obstructions. The type of component is to be stated in the description.

Proprietary impact and bumper guards, protection strips, corner protectors to walls, partitions and columns can be enumerated (nr) or measured linear, in metres. Where measured linear, they are measured along their extreme length over all obstructions. Vertical protection is measured on the vertical angle. The type and nature of guards or protection, and whether they are horizontal or vertical, is to be given in the description.

Illustrative descriptions for typical components for wall finishes are shown in Example 17.1.

17.3 Floor finishes (element 3.2)

The measurement of floor finishes is divided into two sub-elements as follows:

3.2.1 Finishes to floors; and

3.2.2 Raised access floors.

Example 17.1 Formulation of descriptions for finishes to walls

INTERNAL FINISHES					
Code	**Element/sub-elements/components**	**Qty**	**Unit**	**Rate**	**Total**
				£p	**£**
3	**INTERNAL FINISHES**				
3.1	**Wall finishes**				
3.1.1	*Finishes to walls*				
	Example 1 – Typical components				
3.1.1.1	Dry plasterboard lining on metal battens; Gyproc wall board, 9.5mm thick, skim coat		m²		
3.1.1.2	Dry plasterboard lining on metal battens; Gyproc wall board, 9.5mm thick, for direct decoration		m²		
3.1.1.3	Plaster; two coats lightweight plaster		m²		
3.1.1.4	Painting; one mist coat and two coats emulsion paint		m²		
3.1.1.5	Fabric wall coverings (PC sum £35/m²)		m²		
3.1.1.6	Timber boarding/panelling; hardwood, t&g jointed, on and including battens		m²		
3.1.1.7	Ceramic wall tiles; medium quality (PC sum £65/m²)		m²		
3.1.1.8	Proprietary impact bumper guards; wall guards (PC sum £90/m) 150mm wide, horizontal		m		
3.1.1.9	Proprietary impact bumper guards; flush-mounted corner guards (PC sum £60/m), vertical		m		
	Example 2 – Complete rooms				
	Painting; one mist coat and two coats emulsion paint:				
3.1.1.10	Room size; floor area not exceeding 5.00m²		nr		
3.1.1.11	Room size; floor area 5.00m² to 10.00m²		nr		
3.1.1.12	Room size; floor area 10.00m² to 15.00m²		nr		
3.1.1.13	Room size; floor area 15.00m² to 20.00m²		nr		
	Example 3 – Complete residential properties				
	Painting; one mist coat and two coats emulsion paint:				
3.1.1.14	One-bedroom apartment		nr		
3.1.1.15	Two-bedroom apartment		nr		
3.1.1.18	Two-bedroom maisonette		nr		
3.1.1.17	Three-bedroom house		nr		
3.1.1	**TOTAL – Wall finishes: to element 3.1 summary**				

The definition of internal floor finishes also includes floor finishes to the following:

- external walkways and the like that form an integral part of the building superstructure;
- entrance ways to buildings which are not fully enclosed (e.g. entrance to a place of worship); and
- open-ended structures that are considered to be part of the building superstructure.

Floor finishes to such areas should be annotated accordingly in the cost plan.

17.3.1 Finishes to floors (sub-element 3.2.1)

Sub-element 3.2.1 deals with the complete make-up of floor finish above the structural floor. It includes substrata such as non-structural screeds (i.e. separate screeds, unbonded screeds with damp-proof membrane and thermal insulation, and floating screeds on resilient layers or quilts), chemical surface hardeners, latex self-levelling screeds and floating floors. Floor finishes include in situ floor finishes (e.g. granolithic or terrazzo flooring), tiled floor finishes (e.g. ceramic, quarry and mosaic tiling), wood and composition block flooring, flexible sheet flooring, carpeting, and floor painting and sealing. Specialist flooring systems (e.g. timber sprung floors to squash courts) and floor coverings are also covered by this sub-element. Skirtings and secondary components such as mat wells and mats, line markings and applied numerals and symbols are also dealt with under this sub-element. Finishes to stair treads and risers, ramps, and stair and ramp landings are measured under sub-element 2.4.2 (Stair/ramp finishes).

Finishes to floors, and specialist flooring systems, are measured superficially, in square metres. The surface area to which the finish is to be applied is measured. Descriptions for finishes to floors are to include the make-up of the complete floor finish from the structural floor to the applied finish. In addition, details of any PC sums included in a unit rate (e.g. for fixed edge carpeting) are to be stated in the description.

Skirtings and the like are measured linear, in metres along their extreme length over all obstructions. No allowance is made for corner joints, as a suitable allowance for such is to be included in the unit rate applied. The type of skirting, and the height and details of any applied finishes (e.g. paint, staining or varnish) are stated in the description. Skirtings should be linked to the floor finish to which they relate within the cost plan.

Mat wells and mats are enumerated (nr). The type of mat well and mat, together with the overall size, is to be given in the description.

Finishes to the sides (walls) and floor of swimming pool tanks, including the linings to the tank, are separately measured and described. They are measured superficially, in square metres, with the surface area to which the finish is to be applied measured. Although not explicitly required by the rules, it might be pertinent for the cost manager to separately measure and describe finishes to sides and floors, as the composition is likely to be different. Descriptions for swimming pool tanks are to specify the complete make-up of the finish. Kerb tiles or kerb stones to the perimeter of swimming pool tanks should be treated as a cost-significant item (refer to measurement rule for components, C5), and measured and described as linear items, in metres.

Line markings for sports pitches within sports halls and the like, or to delineate car parking spaces within a basement car park, are measured linear, in metres. The type of material used and width of the line marking is to be stated in the description. Numerals and symbols are enumerated (nr). The actual numbers of digits forming each number are counted; for example, number 13 equals two numbers, 118 equals three numbers, and so on. The type of symbol (e.g. disabled parking symbols or directional arrows) is to be stated in the description. In the same way as for line markings, the description for numerals and symbols is also to state the material from which they are to be formed.

Example 17.2 shows component descriptions for floor finishes.

Example 17.2 Formulation of descriptions for finishes to floors

INTERNAL FINISHES					
Code	**Element/sub-elements/components**	**Qty**	**Unit**	**Rate**	**Total**
				£p	**£**
3	**INTERNAL FINISHES**				
3.2	**Floor finishes**				
3.2.1	*Finishes to floors*				
	Example 1 – Typical components				
3.2.1.1	Latex cement screeds, 3mm thick, one coat		m²		
3.2.1.2	Screeds, cement: sand screed, 65mm thick, on 100mm thick insulation		m²		
3.2.1.3	Floating floor; 18mm thick cement impregnated t&g chipboard, on battens, including insulation		m²		
3.2.1.4	Rubber studded tiles; 2.50mm thick		m²		
3.2.1.5	Rubber studded tiles; 4.00mm thick		m²		
3.2.1.6	Floor paint; hard-wearing semi-gloss finish, resistant to mild chemicals		m²		
3.2.1.7	Quarry tile flooring, (PC sum £35/m²)		m²		
3.2.1.8	Skirtings; quarry tile; 150mm high		m		
3.2.1.9	Ceramic tile flooring, (PC sum £20/m²)		m²		
3.2.1.10	Skirtings; ceramic tile; 100mm high		m		
3.2.1.11	Vinyl sheet flooring; heavy duty		m²		
3.2.1.12	Entrance mat well and matting		nr		
3.2.1.13	Vinyl sheet flooring; heavy duty (PC Sum £25.00/m²)		m²		
3.2.1.14	Skirtings; MDF, painted		m		
3.2.1.15	Edge fixed carpet; carpet (PC Sum £45.00/m²), with underlay		m²		
3.2.1.16	Skirtings; hardwood, varnished		m		
3.2.1.17	Line markings; standard basket ball court; painted		m		
3.2.1.18	Line markings; standard badminton court; painted		m		
3.2.1.19	Line markings; thermoplastic; 100mm wide		m		
3.2.1.20	Numerals; thermoplastic; not exceeding 1.00m²		nr		
	Example 2 – Complete rooms				
	Edge fixed carpet; carpet (PC Sum £45.00/m²), with underlay:				
3.2.1.21	Room size; floor area not exceeding 5.00m²		nr		
3.2.1.22	Room size; floor area 5.00m² to 10.00m²		nr		
3.2.1.23	Room size; floor area 10.00m² to 15.00m²		nr		
3.2.1.24	Room size; floor area 15.00m² to 20.00m²		nr		
3.2.1	**TOTAL – Finishes to floors: to element 3.2 summary**				

17.3.2 Raised access floors (sub-element 3.2.2)

Sub-element 3.2.2 covers proprietary raised access floor systems that are positioned on structural floors, including all associated factory-applied floor coverings. Raised access floor systems provide space between the structural floor and floor finish for the distribution of mechanical and electrical services. Sub-element 3.2.2 excludes floating floors, which are measured under sub-element 3.2.1 (Finishes to floors).

Raised access floor systems are measured superficially, in square metres, with the area to which the floor system is to be applied measured. The type of raised access floor system, the height above the structural slab and details of factory-applied floor coverings is to be given in the description, together with details of any cavity fire or air plenum barriers. Details of any PC sums included in a unit rate (e.g. for carpet tiles) are to be stated in the description.

If considered to be cost significant in their own right, cavity fire barriers, air plenum barriers, access panels, electrical and data outlet boxes, or any other key component can be measured and described separately as other cost-significant items. The method of measurement for such items is at the discretion of the cost manager (i.e. enumerated (nr), measured linear (m) or measured superficially (m^2)).

Skirtings are to be measured and described in the same way as those associated with finishes to floors.

Typical descriptions for the components of a raised access floor system are shown in Example 17.3.

Example 17.3 Formulation of descriptions for raised access floors

INTERNAL FINISHES					
Code	**Element/sub-elements/components**	**Qty**	**Unit**	**Rate**	**Total**
				£p	**£**
3	**INTERNAL FINISHES**				
3.2	**Floor finishes**				
3.2.2	*Raised access floors*				
3.2.2.1	Raised access floor; full access system, 150mm high overall, with pedestal supports; factory applied anti-static grade fibre bonded carpet		m^2		
3.2.2.2	Extra over raised access flooring for ramps; overall size 3.60m × 1.40m		nr		
3.2.2.3	Skirtings; MDF, painted		m		
3.2.2	**TOTAL – Raised access floors: to element 3.2 summary**				

17.4 Ceiling finishes (element 3.3)

For the purpose of measurement, ceiling finishes are divided into three sub-elements as follows:

3.3.1 Finishes to ceilings;

3.3.2 False ceilings; and

3.3.3 Demountable suspended ceilings.

17.4.1 Finishes to ceilings (sub-element 3.3.1)

Sub-element 3.3.1 deals with ceiling finishes that are applied directly to soffits, sides and soffits of beams, bulkheads and the like. Ceiling finishes, in this context, include dry lined plasterboard, pre-finished sheet linings, board and sheet linings, in situ coatings (e.g. plaster skim coats), sprayed monolithic coatings, specialist ceiling finishes, painting and decorating. Secondary components such as cornices and covings are also dealt with under this sub-element. Finishes to soffits of staircases and ramps, including soffits of landings between floors, are excluded and measured under sub-element 2.4.2 (Stair/ramp finishes).

Finishes to ceilings are measured superficially, in square metres, with the surface area to which the ceiling finish is to be applied measured. Descriptions for ceiling finishes are to show the make-up of the ceiling finish. Notwithstanding this, it is common practice to measure painting and decorating works separately, as such work is usually executed by a different works contractor to the works contractor applying the ceiling finishes. Details of any PC sums included in a unit rate are also to be stated in the description.

Cornices, covings and the like are measured linear, in metres, along their extreme length over all obstructions, with the type and composition of the component stated in the description. As with skirtings, dados and the like, no allowance is made for corners or returned ends to cornices as a suitable allowance for such is to be included in the unit rate applied. Cornices should be linked to the ceiling finish to which they relate within the cost plan.

Descriptions for typical components associated with ceiling finishes are shown in Example 17.4.

Example 17.4 Formulation of descriptions for finishes to ceilings

INTERNAL FINISHES					
Code	**Element/sub-elements/components**	**Qty**	**Unit**	**Rate**	**Total**
				£p	**£**
3	**INTERNAL FINISHES**				
3.3	**Ceiling finishes**				
3.3.1	*Finishings to ceilings*				
	Example 1 – Typical components				
3.3.1.2	Painted concrete soffits; primer/sealer, one mist coat and two coats emulsion paint		m²		
3.3.1.3	Plaster, painted with primer/sealer, one mist coat and two coats emulsion paint		m²		
3.3.1.4	Plasterboard to soffits; comprising 12.50mm thick Gyproc insulating lath with skim coat; one mist coat and two coats emulsion paint		m²		
3.3.1.5	Plasterboard to soffits; comprising 12.50mm thick Gyproc insulating lath with skim coat; Artex finish		m²		
3.3.1.6	Gyproc coving; 125mm girth		m		
	Example 2 – Complete rooms				
	Painting; one mist coat and two coats emulsion paint:				
3.3.1.7	Room size; floor area not exceeding 5.00m²		nr		

Example 17.4 *(Continued)*

INTERNAL FINISHES					
Code	**Element/sub-elements/components**	**Qty**	**Unit**	**Rate**	**Total**
				£p	**£**
3.3.1.8	Room size; floor area 5.00m² to 10.00m²		nr		
3.3.1.9	Room size; floor area 10.00m² to 15.00m²		nr		
3.3.1.10	Room size; floor area 15.00m² to 20.00m²		nr		
	Example 3 – Complete residential properties				
	Plasterboard to soffits; comprising 12.50mm thick Gyproc insulating lath with skim coat; one mist coat and two coats emulsion paint:				
3.3.1.11	One-bedroom apartment		nr		
3.3.1.12	Two-bedroom apartment		nr		
3.3.1.13	Two-bedroom maisonette		nr		
3.3.1.14	Three-bedroom house		nr		
3.3.1	**TOTAL – Finishings to ceilings: to element 3.3 summary**				

17.4.2 False ceilings (sub-element 3.3.2)

Sub-element 3.3.2 deals with non-demountable false ceilings using both purpose-made or proprietary systems, which are fixed direct to the underside of structural suspended floor slabs (e.g. dry lined plasterboard with plaster skim on timber or metal battens, with paint finish). They are designed to provide a space between the underside of the structural suspended floor slab and the false ceiling to accommodate the distribution of mechanical and electrical services.

False ceilings are measured superficially, in square metres, with the surface area to which the ceiling is to be applied measured. The composition of the false ceiling, including the support framework or system, and details of site-applied paint or decorations are to be given in the description, together with details of any cavity fire barriers. Details of any PC sums included in a unit rate are also to be stated in the description.

If considered to be cost significant in their own right, cavity fire barriers or any other key component can be measured and described separately, as other cost-significant items. The method of measurement for such items is at the discretion of the cost manager (i.e. enumerated (nr), measured linear (m) or measured superficially (m^2)).

Cornices, covings and the like are to be measured and described in the same way as those associated with finishes to ceilings.

Access hatches within false ceilings are to be enumerated (nr), with their composition, details of site-applied paint or decorations, and overall size stated in the description.

Example 17.5 shows typical descriptions for false ceilings.

Example 17.5 Formulation of descriptions for false ceilings

INTERNAL FINISHES					
Code	**Element/sub-elements/components**	**Qty**	**Unit**	**Rate**	**Total**
				£p	**£**
3	**INTERNAL FINISHES**				
3.3	**Ceiling finishes**				
	Note:				
	Painting ceilings: Measured under sub-element 3.3.2 (Finishes to ceilings).				
3.3.2	*False ceilings*				
3.3.2.1	Gyproc GypLyner Universal ceiling lining system; 60-minute fire resistance; for direct decoration, including cavity fire barriers; hangers average 300mm long		m²		
3.3.2.2	Extra over Gyproc GypLyner Universal ceiling for change in level; 600mm		m		
3.3.2.3	Edge treatments; shadow gap		m		
3.3.2.4	Gyproc coving; 125mm girth		m		
3.3.2.5	Access panels; 60-minute fire resistance; with lock; 600mm × 600mm		nr		
3.3.2	**TOTAL – False ceilings: to element 3.3 summary**				

17.4.3 Demountable suspended ceilings (sub-element 3.3.3)

Sub-element 3.3.3 deals with proprietary demountable suspended ceilings, including suspension systems, which are fixed direct to the underside of structural suspended floor slabs (e.g.).

Demountable suspended ceilings are measured superficially, in square metres, with the surface area to which the ceiling is to be applied measured. The type and composition of the demountable suspended ceiling, including the suspension system, are to be stated in the description, together with details of any cavity fire barriers and any insulation, acoustic or fire, fixed directly to the underside of the floor construction. Details of any PC sums included in a unit rate are also to be stated in the description. Again, if considered to be cost significant in their own right, cavity fire barriers or any other key component can be measured and described separately as other cost-significant items.

Shadow gaps and the like are measured linear, in metres, along their extreme length over all obstructions. The type and composition of the component is to be described. They should also be linked to the demountable suspended ceiling to which they relate in the cost plan.

Access hatches and panels within demountable suspended ceiling systems are to be measured and described in the same way as those associated with false ceilings.

Example 17.6 illustrates component descriptions for demountable suspended ceilings.

Example 17.6 Formulation of descriptions for demountable suspended ceilings

INTERNAL FINISHES					
Code	**Element/sub-elements/components**	**Qty**	**Unit**	**Rate**	**Total**
				£p	**£**
3	**INTERNAL FINISHES**				
3.3	**Ceiling finishes**				
3.3.3	*Demountable suspended ceilings*				
3.3.3.1	Suspended ceiling; medium quality (PC £30/m²), concealed grid; hangers 700mm long		m²		
3.3.3.2	Extra over suspended ceiling for vertical bulkhead		m²		
3.3.3.3	Edge trim; shadow gap		m		
3.3.3.4	Suspended ceiling; medium quality (PC £35/m²), exposed grid; hangers 400mm long		m²		
3.3.3	**TOTAL – Demountable suspended ceilings: to element 3.3 summary**				

CHAPTER 18

Fittings, furniture and equipment (group element 4)

Introduction

This chapter:

- provides step-by-step guidance on how to measure and describe fittings, furniture and equipment; and
- gives examples illustrating how to formulate descriptions for components.

18.1 Method of measurement

The rules for measuring furniture, fittings and equipment are found in group element 4 of NRM 1, which comprises a single element:

4.1 Fittings, furnishings and equipment.

18.2 Fittings, furnishings and equipment (element 4.1)

Element 4.1 (Fittings, furnishings and equipment) is subdivided into five sub-elements as follows:

4.1.1 General fittings, furnishings and equipment;
4.1.2 Domestic kitchen fittings and equipment;
4.1.3 Special-purpose fittings, furnishings and equipment;
4.1.4 Signs/notices;
4.1.5 Works of art;
4.1.6 Non-mechanical and non-electrical equipment;
4.1.7 Internal planting; and
4.1.8 Bird and vermin control.

All components measured under these sub-elements are measured as enumerated (nr) items, with the type, the principal details and sizes and, if applicable, the quality of the item stated in the description. Details of any prime cost (PC) sums included in a unit rate are also to be stated in the description.

Before commencing the measurement of components, it is essential that the cost manager determines from the employer the extent of fittings, furnishings and equipment which are to be included within the cost plan. This is because some employers will exclude some or all loose fittings, furnishings and equipment from the cost plan.

18.2.1 General fittings, furnishings and equipment (sub-element 4.1.1)

General fittings, furnishings and equipment are either fixed to the building fabric or provided loose within the building. Examples include reception desks, counters, mirrors, curtains, curtain tracks, blinds (which are not an integral part of a window or external door), loose carpets, tables and chairs, bedroom furniture, lockers, hand-held fire extinguishers, televisions and vending machines.

Regarding hand-held fire extinguishers, there is currently no legal requirement to have formal installation or commissioning carried out; however, impending changes in the applicable British Standard will require commissioning of hand-held fire extinguishers having to be carried out by a competent person. It is recommended, therefore, that the cost manager measures and describes commissioning of hand-held fire extinguishers as a separate item – as an 'extra over' item.

Example 18.1 illustrates descriptions for general fittings, furnishings and equipment.

Example 18.1 Formulation of descriptions for general fittings, furnishings and equipment

FITTINGS, FURNITURE AND EQUIPMENT					
Code	**Element/sub-elements/components**	**Qty**	**Unit**	**Rate**	**Total**
				£p	**£**
4	**FITTINGS, FURNITURE AND EQUIPMENT**				
4.1	**Fittings, furnishings and equipment**				
4.1.1	*General fittings, furnishings and equipment*				
	Note: Services in connection with Reception desk installation measured under group element 5 (Services)				
4.1.1.1	Reception desk; straight counter; 4.50m long; 3 person; (PC Sum £6,000/desk supply only)		nr		
4.1.1.2	Shelving support system; 1,000mm × 600mm × 1,850mm; fixing to masonry; (PC Sum £200/unit supply only)		nr		
4.1.1.3	Cloakroom racks; (PC Sum £360/unit supply only)		nr		

Example 18.1 *(Continued)*

FITTINGS, FURNITURE AND EQUIPMENT					
Code	**Element/sub-elements/components**	**Qty**	**Unit**	**Rate**	**Total**
				£p	**£**
4.1.1.4	Mat well; polished aluminium mat well and entrance mat; 2400mm × 1200mm		nr		
4.1.1.5	Fast food table and chairs (× 4 nr); fixing through tiling to concrete (PC Sum £550/unit supply only)		nr		
4.1.1.6	Blinds; vertical louvre blinds; 2,000mm wide; (PC Sum £75/blind supply only)		nr		
4.1.1.7	Office fire extinguisher pack; comprising 1 nr × 6 litre AFFF foam fire extinguisher, 1 nr × 2kg CO_2 fire extinguisher, 1 nr × composite extinguisher stand, 2 nr × fire extinguisher ID signs		nr		
4.1.1.8	Fire extinguisher; 5kg CO_2 carbon dioxide		nr		
4.1.1.9	Fire extinguisher cabinets; rotationally moulded; single cabinet with key lock and alarm; capacity 1 nr × 6kg/6 litre fire extinguisher		nr		
4.1.1.10	Extra over fire extinguisher for commissioning (initial service) and installation		nr		
4.1.1.11	Flammable liquid storage cabinets; 30-minute fire rating; high visibility yellow powder coat finish; removable spillage sump in cabinet's base, with 1 nr adjustable perforated shelf; size 915mm × 459mm × 459mm		nr		
4.1.1.12	Rotary alarm bell; die-cast aluminium dome, coated red; bell sounds 60dB alarm at up to 35 metres		nr		
4.1.1.13	Cigarette bin; wall mounted; stainless steel; (PC Sum £100/unit supply only)		nr		
4.1.1	**TOTAL – General fittings, furnishings and equipment: to element 4.1 summary**				

18.2.2 Domestic kitchen fittings and equipment (sub-element 4.1.2)

Domestic kitchen units, kitchen equipment (e.g. ovens, hobs, fridges and sinks) and kitchen fittings and accessories are measured under this sub-element.

Commercial catering equipment is excluded, as it is measured under element 5.2 (Services equipment).

Example 18.2 illustrates descriptions for domestic kitchen fittings and equipment.

Example 18.2 Formulation of descriptions for domestic kitchen fittings and equipment

FITTINGS, FURNITURE AND EQUIPMENT					
Code	**Element/sub-elements/components**	**Qty**	**Unit**	**Rate**	**Total**
				£p	**£**
4	**FITTINGS, FURNITURE AND EQUIPMENT**				
4.1	**Fittings, furnishings and equipment**				
4.1.2	*Domestic kitchen fittings and equipment*				
	Kitchen units, including base units, drawer units, worktops, sinks, taps, wall units and all associated fittings, excluding kitchen appliances (white goods):				
4.1.2.1	One-bedroom apartment; (PC Sum £3,000/kitchen supplied only)		nr		
4.1.2.2	Two-bedroom apartment; (PC Sum £3,750/kitchen supplied only)		nr		
4.1.2.3	Two-bedroom maisonette; (PC Sum £4,250/kitchen supplied only)		nr		
4.1.2.4	Three-bedroom house; (PC Sum £6,500/kitchen supplied only)		nr		
	Kitchen appliances:				
4.1.2.5	Cooker; slide-in; (PC Sum £350/nr supply only)		nr		
4.1.2.6	Refrigerator/freezer; freestanding; (PC Sum £350/nr supply only)		nr		
4.1.2.7	Dishwasher; freestanding; A-rated wash programme and AAA energy ratings; (PC Sum £300/nr supply only)		nr		
4.1.2.8	Washer dryer; freestanding; A Energy, A Wash, B Spin Dry energy ratings; (PC Sum £400/nr supply only)		nr		
4.1.2	**TOTAL – Domestic kitchen fittings and equipment: to element 4.1 summary**				**8,740**

18.2.3 Special-purpose fittings, furnishings and equipment (sub-element 4.1.3)

Special-purpose fittings, furnishings and equipment can be either fixed to the building fabric or provided loose within the building. They are classified as 'special purpose' in the sense that they are designed for the particular purpose of the building and are likely to be obtained from a specialist supplier or specialist contractor. Examples of special fittings, furnishings and equipment include: equipment for an operating theatre within a hospital, a dentist surgery or a bar within a hotel.

18.2.4 Signs and notices (sub-element 4.1.4)

This sub-element covers directories, notice boards, letters, signs, plaques, symbols and emblems of all kinds for identification and directional purposes within or attached to the building.

18.2.5 Works of art (sub-element 4.1.5)

Works of art covers objects d'art, ornamental features, decorative features and decorative panels. Descriptions are to include the nature of the works of art. In the case of decorative panels, the size should also be given in the description. Other features such as fish tanks, including fish tanks set into internal walls and partitions, are also measured and described under this sub-element.

18.2.6 Non-mechanical and non-electrical equipment (sub-element 4.1.6)

This sub-element covers non-mechanical and non-electrical equipment for use within, or to enter, the building, such as removable disabled access equipment and removable ladders.

18.2.7 Internal planting (sub-element 4.1.7)

Internal planting comprises the provision of natural and artificial planting within internal environments. It includes plant and shrub beds, as well as trees. Containers in which they are to be planted are also included.

Plant and shrub beds can be enumerated (nr), measured as linear, in metres, or measured as superficial items, in square metres. Descriptions are to include the composition of the bed, together with details of the method of containment (i.e. the type of container). Where measured linear, the length measured is the extreme length over all obstructions. When measured superficially, the surface area of the bed is measured.

Trees are enumerated (nr), with the type, age (if applicable) and height stated in the description. The method of containment is to also be specified in the description.

Maintenance works to be executed during the defects liability period (or, depending on which standard conditions of contract are applicable, rectification period or maintenance period) such as watering, feeding and maintenance of internal planting should be treated as a discrete item. It is suggested that maintenance works be measured as enumerated (nr) items, based on the number of occasions the task is to be completed.

Example 18.3 illustrates descriptions for internal planting.

18.2.8 Bird and vermin control (sub-element 4.1.8)

Sub-element 4.1.8 (Bird and vermin control) deals with installations and equipment used to repel, trap or otherwise control birds or vermin that may be of a danger to health. It covers wires, nets and traps, as well as electronic and sonic detection systems, and bird-repellent coatings.

Example 18.3 Formulation of descriptions for internal planting

FITTINGS, FURNITURE AND EQUIPMENT					
Code	**Element/sub-elements/components**	**Qty**	**Unit**	**Rate**	**Total**
				£p	**£**
4	**FITTINGS, FURNITURE AND EQUIPMENT**				
4.1	**Fittings, furnishings and equipment planting**				
4.1.7	*Internal planting*				
	Example 1 – General allowance (where no detailed information is available)				
4.1.7.1	Plant and shrub beds; with real planting; (PC Sum £15/m² supply only plants/shrubs)		m²		
4.1.7.2	Trees; real; in containers; (PC Sum £450/nr supply only)		nr		
	Example 2 – Specific requirements measured and described in detail				
4.1.7.3	Containerised trees; in GRP round tree tubs; (PC Sum £700/nr supply only)		nr		
4.1.7.4	Containerised planting; with real planting; GRP square trough planter 1,000mm × 1,000mm × 450mm high; (PC Sum £250/nr supply only)		nr		
4.1.7.5	Containerised planting; with real planting; GRP rectangular trough planter 1,000mm × 300mm × 300mm high; (PC Sum £150/nr supply only)		nr		
4.1.7.6	Containerised planting; with real planting; Philodendron 'Red Emerald' 1.7m; in GRP circular planter 380mm × 400mm high; (PC Sum £125/nr supply only)		nr		
4.1.7.7	Containerised planting; with real planting; *Beaucarnea recurvata* (pony tail palm), 1.8m high, multi stem; in GRP Circular bell shaped planter 660mm × 440mm high; (PC Sum £200/nr supply only)		nr		
4.1.7.8	Containerised planting; with real planting; *Dracaena fragrans* 'Janet Craig', 1.8m high; in circular brushed aluminium planter 380mm × 400mm high; (PC Sum £200/nr supply only)		nr		
4.1.7.9	Containerised planting; with artificial planting; Globe Thistle Bush, 450mm; in bowl planter; (PC Sum £120/nr supply only)		nr		
4.1.7.10	Containerised planting; with artificial planting; Box Spiral Topiary, 1.8m high; in circular brushed aluminium planter; (PC Sum £150/nr supply only)		nr		
4.1.7	**TOTAL – Internal planting: to element 4.1 summary**				

Wires, nets, traps, electronic detection systems and sonic detection systems are measured as enumerated (nr) items, with the type of treatment or system stated in the description. Bird-repellent coatings are measured superficially, in square metres. The description is to specify the nature of the coating and method of application.

Example 18.4 illustrates descriptions for bird and vermin control measures.

Example 18.4 Formulation of descriptions for bird and vermin control measures

SERVICES					
Code	**Element/sub-elements/components**	**Qty**	**Unit**	**Rate**	**Total**
				£p	**£**
4	**FITTINGS, FURNITURE AND EQUIPMENT**				
4.1	**Fittings, furnishings and equipment**				
4.1.8	*Bird and vermin control*				
4.1.8.1	Bird netting		m^2		
4.1.8.2	Bird spikes; polycarbonate spike		m		
4.1.8.3	Bird slope system; 350mm wide; including end caps		m		
	Note: Power supply in connection with bird dispersal system measured under group element 5 (Services)				
4.1.8.4	Bird dispersal system, including speakers; (PC Sum £1,000/system supply only)		nr		
4.1.8	**TOTAL – Bird and vermin control: to element 4.1 summary**				

CHAPTER 19

Services (group element 5)

Introduction

This chapter:

- provides step-by-step guidance on how to measure and describe mechanical and electrical services installations and equipment;
- summarises the main elements and components which need to be considered for mains services installations; and
- contains examples illustrating how to formulate descriptions for components.

19.1 Method of measurement

The rules for measuring services are found in group element 5 of NRM 1, which is divided into 14 elements as follows:

5.1 Sanitary installations;
5.2 Services equipment;
5.3 Disposal installations;
5.4 Water installations;
5.5 Heat source;
5.6 Space heating and air conditioning;
5.7 Ventilation;
5.8 Electrical installations;
5.9 Fuel installations;
5.10 Lift and conveyor installations;
5.11 Fire and lightning protection;
5.12 Communication, security and control systems;
5.13 Specialist installations; and
5.14 Builder's work in connection with services.

19.1.1 Measurement of areas served by services installations

Under NRM 1, many service installations are measured superficially, in square metres, based on the area served by the installation (i.e. the treated area). The rules require the area of each room or common area (i.e. circulation space) served by the installation to be measured separately, with the area of the room measured using the principals of measurement for ascertaining the gross internal floor area – see Figure 19.1. Where an installation serves more than one room, the area of each room is added to give the total area served.

Where service installations are to be measured for repetitive building types (e.g. residential dwellings), or for buildings containing a range of room types (e.g. hotel rooms or student accommodation), rather than being measured superficially, the services can be measured by enumerated items (nr). When this approach is adopted, the type and size of building, dwelling or room is to be stated in the description. For example, residential dwellings would be described by the number of bedrooms, whilst hotel rooms can be described by the number of persons for which the room is designed (e.g. single, double or family room).

When measuring mechanical and electrical service installations, the cost manager will need to consider the employer's requirements in respect of servicing (i.e. planned maintenance).

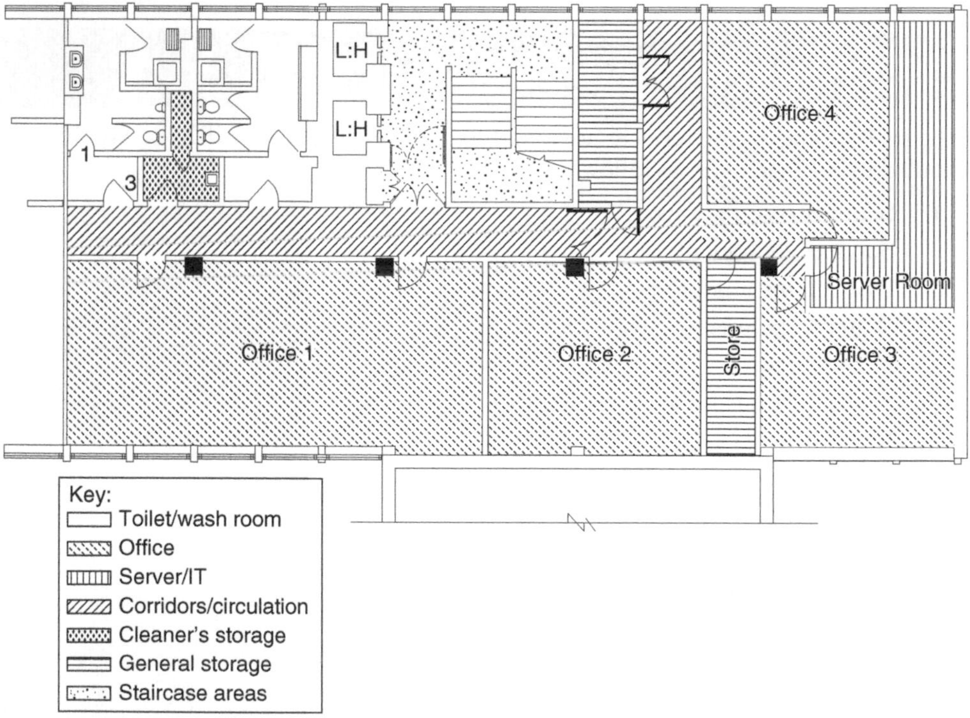

Figure 19.1 Measurement of the area served by the installations

19.1.2 Testing and commissioning of services

Testing and commissioning of mechanical and electrical services installations is measured in conjunction with each sub-element. This is because separate percentage rates might be applicable for different types of services installations. Apposite percentages for calculating cost targets for the testing and commissioning can be derived by the cost manager by comparing a number of published cost analyses or from 'in-house' sources of data.

The rules treat testing and commissioning as two separate cost centres.

19.2 Sanitary installations (element 5.1)

Included in this element are two sub-elements:

5.1.1 Sanitary appliances;

5.1.2 Sanitary ancillaries.

19.2.1 Sanitary appliances (sub-element 5.1.1)

Sanitary appliances are fixtures used in bathrooms, toilets, washrooms and kitchens for sanitary purposes. They are measured as enumerated (nr) items, with the type and quality of the appliance stated in the description.

Cold and hot water supplies to sanitary appliances are measured and described separately under element 5.4 (Water installations). Waste pipes and fittings, together with discharge and ventilation stacks to which the wastes are connected, are also dealt with separately under sub-element 5.3.1 (Foul drainage above ground). Typical component descriptions for sanitary installations are shown in Example 19.1.

Example 19.1 Formulation of descriptions for sanitary appliances

SERVICES					
Code	**Element/sub-elements/components**	**Qty**	**Unit**	**Rate**	**Total**
				£p	**£**
5	**SERVICES**				
5.1	**Sanitary installations**				
5.1.1	*Sanitary appliances*				
	Example 1 – Hotel				
5.1.1.1	Sanitary installations; range, including WC, wash-hand basin and bath (including bath panel and trim)		nr		
5.1.1.2	Sanitary installations; range, including WC, wash-hand basin, bidet and bath (including bath panel and trim)		nr		
5.1.1.3	Sanitary installations; shower over bath (shower screen measured separately)		nr		
5.1.1.4	Sanitary installations; range, including WC, wash-hand basin and shower (bath/shower screen measured separately)		nr		

Example 19.1 *(Continued)*

SERVICES					
Code	**Element/sub-elements/components**	**Qty**	**Unit**	**Rate**	**Total**
				£p	**£**
5.1.1.5	Testing of installations		%		
5.1.1.6	Commissioning of installations		%		
	Example 2 – Offices				
	Note:				
	No allowance for tea points as deemed to be part of occupier fit-out				
5.1.1.1	Sanitary appliances; WCs, low level, vitreous china pan; plastic seat		nr		
5.1.1.2	Sanitary appliances; WCs, disabled persons' use, low-level, vitreous china pan; plastic seat		nr		
5.1.1.3	Sanitary appliances; wall urinal, white vitreous china		nr		
5.1.1.4	Sanitary appliances; isolated division, white glazed vitreous china		nr		
5.1.1.5	Sanitary appliances; wash-hand basins, vanity type, including c.p. monobloc mixing tap		nr		
5.1.1.6	Sanitary appliances; 'Belfast' sink with pillar taps, (cleaner's sink)		nr		
5.1.1.7	Sanitary appliances; drinking fountain, vitreous china		nr		
5.1.1.8	Testing of installations		%		
5.1.1.9	Commissioning of installations		%		
5.1.1	**TOTAL – Sanitary appliances: to element 5.1 summary**				

19.2.2 Sanitary ancillaries (sub-element 5.1.2)

Sanitary ancillaries include a vast array of different components, from toilet paper holders to shower cubicles. These are measured and described as enumerated (nr) items, with type of fitting given in the description. Example 19.2 illustrates component descriptions for typical sanitary ancillaries.

Example 19.2 Formulation of descriptions for sanitary ancillaries

SERVICES					
Code	**Element/sub-elements/components**	**Qty**	**Unit**	**Rate**	**Total**
				£p	**£**
5	**SERVICES**				
5.1	**Sanitary installations**				
5.1.2	*Sanitary ancillaries*				

Example 19.2 *(Continued)*

SERVICES					
Code	**Element/sub-elements/components**	**Qty**	**Unit**	**Rate**	**Total**
				£p	**£**
	Example 1 – Hotel				
5.1.2.1	Shower screen, glass,		nr		
5.1.2.2	Bath/shower screen, glass		nr		
5.1.2.3	Toilet roll holder		nr		
5.1.2.4	Towel shelf		nr		
	Example 2 – Offices				
	Note:				
	Toilet roll holders are an integral part of WC cubicles; costs included therein				
5.1.2.1	Grab rail sets for disabled WC		nr		
5.1.2.2	Soap dispenser		nr		
5.1.2.3	Paper towel dispenser		nr		
5.1.2.4	Shower screen, glass		nr		
5.1.2.5	Sanitary macerators, electric		nr		
5.1.2	**TOTAL – Sanitary ancillaries: to element 5.1 summary**				

19.3 Services equipment (element 5.2)

Services equipment comprises communal and commercial catering equipment (i.e. catering equipment that has been designed to provide food and drink on a communal or commercial scale). It also covers other, free-standing and fixed, mechanical and electrical equipment, including specialist equipment (i.e. from vending machines, to dental surgery or hospital equipment). Domestic kitchen appliances, including sinks and free-standing appliances, are measured and described under sub-element 4.1.2 (Domestic kitchen fittings and equipment).

Components are measured as enumerated (nr) items, with the type of equipment described.

19.4 Disposal installations (element 5.3)

This element is divided into three sub-elements as follows:

5.3.1 Foul drainage above ground;

5.3.2 Chemical, toxic and industrial liquid waste drainage; and

5.3.3 Refuse disposal.

19.4.1 Measurement of disposal installations – summarised

The measurement of disposal installations is fragmented, with the different parts of the installation being measured under various elements and sub-elements. These different facets, together with the applicable elements and sub-elements under which they are measured, are summarised in Table 19.1.

Table 19.1 Measurement of disposal installations – summarised

Component	**Element**	*Sub-element*
Notes: Testing and commissioning of installations:		
1. Testing of installations	Testing is measured as a composite allowance for each individual sub-element. The costs of all components forming the cost target for the sub-element are totalled and a percentage allowance applied.	
2. Commissioning of installations	Commissioning is measured as a composite allowance for each individual sub-element. The costs of all components forming the cost target for the sub-element, excluding the cost of testing, are totalled and a percentage allowance applied.	
Sanitary installations and services equipment:		
Sanitary appliances	5.1: Sanitary installations	5.1.1: Sanitary appliances
Sanitary appliances within pods	6.1: Prefabricated buildings and building units	6.1.3: Pods
Services equipment	5.2: Services equipment	5.2.1: Services equipment
Surface water and foul water drainage above ground:		
Surface water drainage from balconies (where integral with upper floor construction) to first underground drainage connection or gully	2.2: Upper floors	2.2.5: Balconies
Surface water drainage from roof to first underground drainage connection or gully	2.3: Roof	2.3.4: Roof drainage
Surface water drainage, including floor outlets, from external walkways and the like to first underground drainage connection or gully	2.5: External walls	2.5.5: Subsidiary walls, balustrades, handrails, railings and proprietary balconies
Foul water drainage from sanitary appliance to connection within the lowest floor assembly	5.3: Disposal installations	5.3.1: Foul drainage above ground
Foul water drainage from services equipment to connection within the lowest floor assembly	5.3: Disposal installations	5.3.1: Foul drainage above ground

Table 19.1 *(Continued)*

Component	Element	Sub-element
Surface water and foul water drainage below ground:		
Connection to the statutory undertaker's foul and surface water sewer	8.6: External drainage	8.6.1: Surface water and foul water drainage
Surface water and foul water drainage (including manholes and the like) from first manhole beyond the external face of the building to the statutory undertaker's foul or surface water sewer	8.6: External drainage	8.6.1: Surface water and foul water drainage
Surface water and foul water drainage (including manholes and the like) below or within the lowest floor assembly to first manhole beyond the external face of the building	1.1: Substructure	1.1.3: Lowest floor construction
Surface water drainage integral to basement retaining walls to first underground drainage connection or gully	1.1: Substructure	1.1.4: Basement retaining walls
Surface water drainage integral to embedded basement retaining walls to first underground drainage connection or gully	1.1: Substructure	1.1.4: Basement retaining walls
Pumping stations, ejector stations, storage and retention tanks, and sewage treatment systems	8.6: External drainage	8.6.2: Ancillary drainage systems
Sustainable urban drainage schemes (SUDS):		
Sustainable urban drainage schemes (SUDS)	8.6: External drainage	8.6.2: Ancillary drainage systems
Chemical, toxic and industrial liquid waste drainage:		
Chemical, toxic and industrial liquid waste drainage from appliance or equipment to the external face of the building	5.3: Disposal installations	5.3.2: Chemical, toxic and industrial waste drainage
Chemical, toxic and industrial liquid waste drainage (including manholes, collection facilities, and the like) from first manhole beyond the external face of the building to point of disposal	8.6: External drainage	8.6.3: External chemical, toxic and industrial waste drainage
Land drainage:		
Land drainage	8.6: External drainage	8.6.4: Land drainage
Land drainage to parkland	8.6: External drainage	8.6.4: Land drainage

Table 19.1 *(Continued)*

Component	Element	*Sub-element*
Works to existing drainage installations:		
Works to existing drainage systems (including manholes and the like) within existing buildings	8.6: External drainage	8.6.1: Surface water and foul water drainage
Works to existing external drainage systems (including manholes and the like)	7.1: Minor demolition works and alteration works	7.1.1: Minor demolition works and alteration works
Temporary diversion of existing surface water and foul water drainage installations	9.4: Temporary diversion works	9.4.1: Temporary diversion works
Refuse disposal:		
Refuse chutes, incineration plant and the like	5.3: Disposal installations	5.3.1: Refuse disposal

19.4.2 Foul drainage above ground (sub-element 5.3.1)

Foul drainage above ground comprises waste pipes and fittings to sanitary appliances (e.g. WCs, wash-hand basins, kitchen sinks, baths and showers) and to services equipment (e.g. dishwashers and industrial washing machines), together with discharge and ventilation stacks to which the wastes are connected.

Systems include all foul drainage from sanitary installations and services equipment to the first underground connection. This connection is either within the ground floor assembly of the building (e.g. to a manhole within the ground floor construction) or outside the footprint of the building. Where the first underground connection is outside the footprint of the building, the drainage system is taken to the first manhole beyond the external face of the external wall to the building. Drainage within the ground floor assembly is measured under sub-element 1.1.3 (Lowest floor construction). Drainage installations beyond the external face of the external wall to the building to the point of disposal are measured under sub-element 8.6.1 (Surface water and foul water drainage).

The number of connections to sanitary appliances and items of services equipment are measured as enumerated (nr) items. The type of sanitary appliance or services equipment to which the drainage installation applies is stated in the description. Example 19.3 illustrates typical component descriptions for disposal installations located above ground.

Example 19.3 Formulation of descriptions for foul drainage above ground

SERVICES					
Code	Element/sub-elements/components	Qty	Unit	Rate	Total
				£p	£
5	**SERVICES**				
5.3	**Disposal installations**				
5.3.1	*Foul drainage above ground*				

Example 19.3 *(Continued)*

SERVICES					
Code	**Element/sub-elements/components**	**Qty**	**Unit**	**Rate**	**Total**
				£p	**£**
	Example 1 – Residential				
	Drainage to sanitary appliances:				
5.3.1.1	Range, comprising WC, wash-hand basin and bath, 3 points		nr		
5.3.1.2	Range, comprising WC, wash-hand basin, bidet and bath, 4 points		nr		
5.3.1.3	Range, comprising WC, wash-hand basin and shower, 3 points		nr		
5.3.1.4	Kitchen sink, dish washer and washing machine, 3 points		nr		
5.3.1.5	Testing of installations		%		
5.3.1.6	Commissioning of installations		%		
	Example 2 – Hotel				
	Drainage to sanitary appliances:				
	[Guest Rooms]				
5.3.1.1	Range, comprising WC, wash-hand basin and bath, 3 points		nr		
	[Front of House]				
5.3.1.2	WCs, 6 points		nr		
5.3.1.3	Range, comprising 12 urinals, 12 points		nr		
5.3.1.4	Range, comprising 8 wash-hand basins, 8 points		nr		
5.3.1.5	Testing of installations		%		
5.3.1.6	Commissioning of installations		%		
	Example 3 – Offices:				
	Drainage to sanitary appliances:				
5.3.1.1	Range, comprising 3 WCs, 3 points		nr		
5.3.1.2	Range, comprising 5 WCs, 5 points		nr		
5.3.1.3	Range, comprising 2 urinals, 2 points		nr		
5.3.1.4	Range, comprising 3 wash-hand basins, 3 points		nr		
5.3.1.5	Range, comprising 5 wash-hand basins, 5 points		nr		
5.3.1.6	Shower, 1 point		nr		
5.3.1.7	Provision for tea point, 1 point		nr		
5.3.1.8	Cleaner's sink, 1 point		nr		
5.3.1.9	Testing of installations		%		
5.3.1.10	Commissioning of installations		%		
5.3.1	**TOTAL – Foul drainage above ground: to element 5.3 summary**				

19.4.3 Chemical, toxic and industrial liquid waste drainage (sub-element 5.3.2)

Chemical, toxic and industrial liquid waste drainage installations, including glass drainage, comprise separate piped waste disposal systems for waste that needs special treatment, or separate storage, before disposal. Systems include all storage tanks and vessels, settlement tanks, effluent treatment plant, dosing equipment, sterilisation equipment and the like, together with all drainage pipes, waste pipes and pipe fittings. The number of connections is measured and described in the same way as for sanitary appliances and services equipment (see sub-paragraph 19.4.2).

Installations beyond the external face of the external wall to the building to the point of disposal are measured under sub-element 8.6.3 (External chemical, toxic and industrial liquid waste drainage).

19.4.4 Refuse disposal (sub-element 5.3.3)

Refuse disposal includes refuse input devices, refuse chutes, compacting and macerating plant, incineration plant and refuse collection equipment. They are enumerated (nr), with the type and composition of refuse disposal system described. Example 19.4. illustrates typical component descriptions for refuse disposal installations.

Example 19.4 Formulation of descriptions for refuse disposal

SERVICES					
Code	**Element/sub-elements/components**	**Qty**	**Unit**	**Rate**	**Total**
				£p	**£**
5	**SERVICES**				
5.3	**Disposal installations**				
5.3.3	*Refuse disposal*				
5.3.3.1	Static waste compaction system; with heavy duty open top bulk loading hopper (PC Sum £15,000/nr supply only)		nr		
5.3.3.2	Static waste compaction system; with fully enclosed integral hydraulic rear bin lift, including personnel safety enclosure with interlocking gates (PC Sum £28,000/nr supply only)		nr		
5.3.3.3	Balers; semi-automatic baler; with bin lift and full weather canopy (PC Sum £7,500/nr supply only)		nr		
5.3.3.4	Drum crushers; for reducing tins up to 60 litres (PC Sum £1,800/nr supply only)		nr		
5.3.3.5	Testing of installations		%		
5.3.3.6	Commissioning of installations		%		
5.3.3	**TOTAL – Refuse disposal: to element 5.3 summary**				

19.5 Water installations (element 5.4)

Element 5.4 deals with water installations and is divided into five sub-elements as follows:

5.4.1 Mains water supply;
5.4.2 Cold water distribution;
5.4.3 Hot water distribution;
5.4.4 Local hot water distribution; and
5.4.5 Steam and condensate distribution.

19.5.1 Measurement of water installations – summarised

As for drainage systems, the process of measuring mains water supplies and the distribution of water to user points is fragmented, with the parts of the installations dealt with by a number of different elements and sub-elements. Table 19.2

Table 19.2 Measurement of water installations – summarised

Works	*Element*	*Sub-element*
Notes: Testing and commissioning of installations:		
1. Testing of installations	Testing is measured as a composite allowance for each individual sub-element. The costs of all components forming the cost target for the sub-element are totalled and a percentage allowance applied.	
2. Commissioning of installations	Commissioning is measured as a composite allowance for each individual sub-element. The costs of all components forming the cost target for the sub-element, excluding the cost of testing, are totalled and a percentage allowance applied.	
Connection to the statutory undertaker's water main	8.7: External services	8.7.1: Water mains supply
Mains water supply from statutory undertaker's water main to point of entry into the building	8.7: External services	8.7.1: Mains water supply
Mains water supply from point of entry into the building to appliances, equipment and points of storage	5.4: Water installation	5.4.1: Mains water supply
Cold water distribution from storage to user points (i.e. sanitary appliances or services equipment)	5.4: Water installation	5.4.2: Cold water distribution
Hot water distribution from points of storage to user points (i.e. sanitary appliances or services equipment)	5.4: Water installation	5.4.3: Hot water distribution
Hot water generated at user point (e.g. instantaneous water heaters)	5.4: Water installation	5.4.4: Local hot water distribution
Steam distribution to and condensate return from services equipment	5.4: Water installation	5.4.5: Steam and condensate distribution

summarises these different facets, as well as the applicable elements and sub-elements under which they are measured.

19.5.2 Mains water supply (sub-element 5.4.1)

Provision of mains water supply (i.e. piped water supply systems that bring water from the statutory mains to the point of entry into the building) is measured and described under this sub-element. The mains water supply can be either enumerated (nr) or measured superficially, in square metres. Where measured superficially, the GIFA of the entire building is measured with the number of draw-off points given in the description.

Connections to the statutory undertaker's water main and the mains water supply from statutory undertaker's water main to point of entry into the building are both measured under sub-element 8.7.1 (Water main supply). Typical component descriptions for mains water supply installations are shown in Example 19.5.

Example 19.5 Formulation of descriptions for mains water supply

SERVICES					
Code	Element/sub-elements/components	Qty	Unit	Rate	Total
				£p	£
5	SERVICES				
5.4	Water installation				
5.4.1	*Mains water supply*				
	Example 1 – Enumerated				
5.4.1.1	Mains water supply	3	nr		
5.4.1.2	Testing of installations		%		
5.4.1.3	Commissioning of installations		%		
	Example 2 – Measured superficially				
5.4.1.1	Mains water supply; 3 nr draw-off points	11,906	m^2		
5.4.1.2	Testing of installations		%		
5.4.1.3	Commissioning of installations		%		
5.4.1	**TOTAL – Mains water supply: to element 5.4 summary**				

19.5.3 Cold water distribution (sub-element 5.4.2)

Cold water distribution systems deliver cold water to user points (i.e. to sanitary appliances or to services equipment) from the points of storage. Piped cold water systems can either be measured as enumerated (nr) items or measured superficially, in square metres.

When enumerated, the number of connections (i.e. the number of draw-off points) to sanitary appliances and items of services equipment are counted. The type of sanitary appliance or services equipment to which the cold water supply relates is to be stated in the description. Where measured superficially, the area serviced by the

Example 19.6 Formulation of descriptions for cold water distribution

SERVICES					
Code	**Element/sub-elements/components**	**Qty**	**Unit**	**Rate**	**Total**
				£p	**£**
5	**SERVICES**				
5.4	**Water installation**				
5.4.2	*Cold water distribution*				
5.4.2.1	Cold water distribution; 40 nr draw-off points		m^2		
5.4.2.2	Testing of installations		%		
5.4.2.3	Commissioning of installations		%		
5.4.2	**TOTAL – Cold water distribution: to element 5.4 summary**				

system is measured (refer to Figure 19.1), with the total number of draw-off points given in the description. Example 19.6 provides typical component descriptions for cold water distribution installations.

19.5.4 Hot water distribution (sub-element 5.4.3)

This covers the distribution of hot water from the points of storage to user points (i.e. to sanitary appliances or to equipment). Hot water distribution systems are measured and described in the same way as cold water distribution systems. Example 19.7 provides typical component descriptions for hot water distribution installations.

Example 19.7 Formulation of descriptions for hot water distribution

SERVICES					
Code	**Element/sub-elements/components**	**Qty**	**Unit**	**Rate**	**Total**
				£p	**£**
5	**SERVICES**				
5.4	**Water installation**				
5.4.3	*Hot water distribution*				
5.4.3.1	Hot water distribution; 30 nr draw-off points		m^2		
5.4.3.2	Testing		%		
5.4.3.3	Commissioning		%		
5.4.3	**TOTAL – Hot water distribution: to element 5.4 summary**				

19.5.5 Local hot water distribution (sub-element 5.5.4)

Instantaneous water heaters, shower heaters and storage water heaters (i.e. systems where hot water is generated in the vicinity of the sanitary appliance or services

Example 19.8 Formulation of descriptions for local hot water distribution

SERVICES					
Code	**Element/sub-elements/components**	**Qty**	**Unit**	**Rate**	**Total**
				£p	**£**
5	**SERVICES**				
5.4	**Water installations**				
5.4.4	*Local hot water distribution*				
5.4.4.1	Water heaters; instantaneous, 3.0kW		nr		
5.4.4.2	Testing of installations		%		
5.4.4.3	Commissioning of installations		%		
5.4.4	**TOTAL – Local hot water distribution: to element 5.4 summary**				

equipment being served) are measured under this sub-element. Works include the installation of any associated flue pipes and terminals. Water heaters and the like are measured and described as enumerated (nr) items, with the type and capacity of the unit stated. Example 19.8 provides typical component descriptions for local hot water distribution installations.

19.5.6 Steam and condensate distribution (sub-element 5.4.5)

Steam distribution pipelines to and condensate return pipelines from services equipment are measured in the same way as cold water and hot water distribution systems.

19.6 Heat source (element 5.5)

Plant that generates heat (i.e. heat sources), including gas-and oil-fired boilers, biomass fuel boilers, coal-fired boilers, electric boilers, wood pellet boiler plant, package steam generators, heat pumps and ground source heating are dealt with under this element. Central or combined heat and power (CHP) boiler plant is also covered under this element.

Heat source plant is measured and described as enumerated (nr) items, with the type of plant and the output in kilowatts (kW) given in the description. All ancillary plant, equipment and interconnecting pipelines (e.g. between header tank heat source) associated with the heat source plant are to be included in the item measured. Each type of heat source is measured and described separately. Typical component descriptions for various heat sources are illustrated in Example 19.9.

Example 19.9 Formulation of descriptions for heat sources

SERVICES					
Code	**Element/sub-elements/components**	**Qty**	**Unit**	**Rate**	**Total**
				£p	**£**
5	**SERVICES**				
5.5	**Heat source**				
5.5.1	*Heat source*				
5.5.1.1	Heat source; domestic water boiler, stove enamelled casing, gas fired, 15kW, wall hung		nr		
5.5.1.2	Heat source; commercial cast iron sectional floor standing boiler, 105kW		nr		
5.5.1.3	Heat source; ground source heat pump, including horizontal collectors, 48kW (includes all work in installing heat source)		nr		
5.5.1.4	Testing of installations		%		
5.5.1.5	Commissioning of installations		%		
5.5.1	**TOTAL – Heat source: to element 5.6 summary**				

19.7 Space heating and air conditioning (element 5.6)

Element 5.6 is divided into eight sub-elements as follows:

5.6.1 Central heating;
5.6.2 Local heating;
5.6.3 Central cooling;
5.6.4 Local cooling;
5.6.5 Central heating and cooling;
5.6.6 Local heating and cooling;
5.6.7 Central air conditioning; and
5.6.8 Local air conditioning.

19.7.1 Central heating (sub-element 5.6.1)

Central heating systems are those where the heat is generated at a central point and distributed via pipelines or ducts to the locations that are to be heated. The component measured includes the entire central heating system: from the point of connection to the heat source to the heat emitters being served by the heat source. The heat source is excluded from this sub-element, as it is measured and described separately under sub-element 5.5.1 (Heat source).

Central heating systems are measured superficially, in square metres, with the area serviced by the system measured (refer to Figure 19.1). However, central heating

Example 19.10 Formulation of descriptions for central heating

SERVICES					
Code	**Element/sub-elements/components**	**Qty**	**Unit**	**Rate**	**Total**
				£p	**£**
5	**SERVICES**				
5.6	**Space heating and air conditioning**				
5.6.1	*Central heating*				
5.6.1.1	Central heating; LTHW heating system		m²		
5.6.1.2	Testing of installations		%		
5.6.1.3	Commissioning of installations		%		
5.6.1	**TOTAL – Central heating: to element 5.6 summary**				

systems to repetitive building types or buildings containing a range of room types can be enumerated (nr) – refer to paragraph 19.1.1. Descriptions of central heating systems are to depict the type of heating installation, including the method of heat distribution and type of heat emission units (e.g. radiators, heated ceiling panels, radiant strips, fan convectors and skirting heaters). Where more than one system is employed, they are to be measured and described separately. (See Example 19.10.)

19.7.2 Local heating (sub-element 5.6.2)

Local heating is where heat is generated using a self-contained room heater or fire located in or adjacent to the space to be heated. Room heaters are measured as enumerated (nr) items, with the type of room heater stated in the description. Fires are also measured as enumerated (nr) items, with the type of fire and details of any proprietary chimneys or flues associated with fires, which are not an integral part of the building structure, described.

19.7.3 Central cooling (sub-element 5.6.3)

Central cooling systems are those where cooling is performed at a central point and distributed via pipelines or ducts to the locations that are to be cooled. The item measured is to include the entire central cooling system. Systems include fan coil systems, variable air volume (VAV) systems, variable refrigerant volume (VRV) systems, chillers, package chillers, central refrigeration plant and cooling towers.

Central cooling systems are measured superficially, in square metres, with the area serviced by the system measured (refer to Figure 19.1). However, central cooling systems to repetitive building types or buildings containing a range of room types can be enumerated (nr) – refer to paragraph 19.1.1. Descriptions of central cooling systems are to include the type of system, the main equipment employed, and the type of emission units (e.g. fan coil units or chilled beams). Where more than one cooling system is employed, they are to be measured and described separately.

19.7.4 Local cooling (sub-element 5.6.4)

Local cooling is where cooling is performed by a self-contained cooling unit located in or adjacent to the space to be cooled. Cooling units are measured as enumerated (nr) items, with the type of unit stated in the description.

19.7.5 Central heating and cooling (sub-element 5.6.5)

Combined central heating and cooling systems are where both heating and cooling is performed at a central point and distributed via pipelines or ducts to the locations that are to be treated. The heat source is measured under sub-element 5.5.1 (Heat source). Combined central heating and cooling systems include fan coil systems, air based systems – variable air volume (VAV) systems, reverse cycle heat pump systems or chillers.

Combined central heating and cooling systems are measured superficially, in square metres, with the area serviced by the system measured (refer to Figure 19.1). However, combined central heating and cooling systems to repetitive building types, or buildings containing a range of room types can be enumerated (nr) – refer to paragraph 19.1.1. Descriptions of central heating and cooling systems are to include the type of system, and the type of emission units (e.g. fan coil units). Where more than one system is employed, they are to be measured and described separately.

19.7.6 Local heating and cooling (sub-element 5.6.6)

Local heating and cooling is where both heating and cooling are performed using self-contained heating and cooling units located in or adjacent to the space to be treated. Combined heating and cooling units are measured as enumerated (nr) items, with the type of unit described.

19.7.7 Central air conditioning (sub-element 5.6.7)

Central air conditioning systems are where air treatment is performed at a central point and distributed via pipelines or ducts to the locations that are to be treated. The heat source is measured under sub-element 5.5.1 (Heat source). Central air conditioning systems include plenum air heating systems, variable air volume systems, dual-duct and induction systems, multi-zone systems and hybrid systems. The item measured is to include the entire central air conditioning system, including the air-handling units, chillers, all distribution pipelines and ducts, terminal units and emitters, and all ancillary components.

Central air conditioning systems are measured superficially, in square metres, with the area serviced by the system measured (refer to Figure 19.1). However, central air conditioning systems to repetitive building types, or buildings containing a range of room types can be enumerated (nr) – refer to paragraph 19.1.1. Descriptions of central air conditioning systems are to include the type of system, and the type of terminal or emission units. Where more than one system is employed, they are to be

Example 19.11 Formulation of descriptions for central air conditioning

SERVICES					
Code	**Element/sub-elements/components**	**Qty**	**Unit**	**Rate**	**Total**
				£p	**£**
5	**SERVICES**				
5.6	**Space heating and air conditioning**				
5.6.7	*Central air conditioning*				
5.6.7.1	Central air conditioning; variable air volume (VAV), ductwork, supply and extract, insulation and 4-pipe fan coil		nr		
5.6.7.2	Testing of installations		%		
5.6.7.3	Commissioning of installations		%		
5.6.7	**TOTAL – Central air conditioning: to element 5.6 summary**				

measured and described separately. Example 19.11 provides component descriptions for central air conditioning.

19.7.8 Local air conditioning (sub-element 5.6.8)

Local air conditioning is where air treatment is performed in or adjacent to the space to be treated using self-contained air conditioning units. Air conditioning units are measured as enumerated (nr) items, with the type of unit stated.

19.8 Ventilation (element 5.7)

This element is divided into three sub-elements as follows:

5.7.1 Central ventilation;
5.7.2 Local and special ventilation; and
5.7.3 Smoke extract/control.

19.8.1 Central ventilation (sub-element 5.7.1)

Ventilation systems are air movement systems, which either remove vitiated air from spaces or both remove vitiated air from spaces and supply fresh air from outside the building to the spaces. With the exception of filtration when required, no environmental control or air treatment is provided.

Central ventilation systems include air extract systems, and air supply and extract systems. The item measured is to include the entire ventilation system, including all distribution ductwork, fans, filters, grilles, pumps, controls and all other ancillary components. Systems are measured superficially, in square metres, with the area serviced by the system measured (refer to Figure 19.1). Descriptions of central ventilation systems are to include the type of system. Where more than one system is employed, they are to be measured and described separately.

19.8.2 Local and special ventilation (sub-element 5.7.2)

These are self-contained air movement units. Examples include extract fans to toilets and bathrooms, and kitchen ventilation, safety cabinet and fume cupboard extracts, dust collection units, anaesthetic gas extracts, cyclone systems, impulse fans and roof-mounted ventilation units. They are measured and described as enumerated (nr) items, with the type of unit or system stated in the description. Example 19.12 provides component descriptions for local and special ventilation installations.

Example 19.12 Formulation of descriptions for local and special ventilation

SERVICES					
Code	**Element/sub-elements/components**	**Qty**	**Unit**	**Rate**	**Total**
				£p	**£**
5	**SERVICES**				
5.7	**Ventilation**				
	[Residential]				
5.7.2	*Local ventilation*				
5.7.2.1	Toilet extractor fan ventilation; PIR activated with electronic timer, including ducting and wall grilles with back draft shutter; 100mm diameter; wall mounted		nr		
5.7.2.2	Wall extractor fan; humidity controlled, including ducting and wall grilles with back draft shutter; 100mm diameter; wall mounted		nr		
5.7.2.3	Bathroom extractor ventilation; pull-cord activated with timer, including ducting and wall grilles with back draft shutter; 125mm diameter; wall mounted		nr		
5.7.2.4	Testing of installations		%		
5.7.2.5	Commissioning of installations		%		
5.7.2	**TOTAL – Local ventilation: to element 5.7 summary**				

19.8.3 Smoke extract and control (sub-element 5.7.3)

Smoke extract and control systems comprise air movement and pressurisation systems for removing and controlling the build-up of smoke arising from a fire. The item measured is to include the entire ventilation system, including all distribution ductwork, fans, filters, grilles, controls and all other ancillary components. Examples include automatic smoke extract systems and automatic smoke compartmentalisation systems. Systems are measured in the same way as central ventilation systems (see sub-paragraph 19.8.1). Descriptions of smoke extract and control systems are to include the type of system. Where more than one system is employed, they are to be measured and described separately.

19.9 Electrical installations (element 5.8)

This element is divided into six sub-elements as follows:

5.8.1 Electrical mains and sub-mains distribution;
5.8.2 Power installations;
5.8.3 Lighting installations;
5.8.4 Specialist lighting installations;
5.8.5 Local electricity generation systems;
5.8.6 Earthing and bonding systems.

19.9.1 Measurement of electrical installations – summarised

Electrical installations are measured under a number of different elements and sub-elements. Table 19.3 identifies the various facets of electrical installations and shows the elements and sub-elements under which they are measured.

Table 19.3 Measurement of electrical installations – summarised

Works	*Element*	*Sub-element*
Notes: Testing and commissioning of installations:		
1. Testing of installations	Testing is measured as a composite allowance for each individual sub-element. The costs of all components forming the cost target for the sub-element are totalled and a percentage allowance applied.	
2. Commissioning of installations	Commissioning is measured as a composite allowance for each individual sub-element. The costs of all components forming the cost target for the sub-element, excluding the cost of testing, are totalled and a percentage allowance applied.	
Connection to the statutory undertaker's electrical main	8.7: External services	8.7.2: Electricity mains supply
Electrical mains HV distribution – from statutory undertaker's electrical main to on-site transformer sub-stations	8.7: External services	8.7.2: Electricity mains supply
Transformer sub-stations	8.7: External services	8.7.2: Electricity mains supply
Local electricity generation plant – internal	5.8: Electrical installations	5.8.5: Local electricity generation systems
Local electricity generation plant – external to building	8.7: External services	8.7.2: Electricity mains supply
Electrical mains LV distribution – to main switchgear panels within buildings	8.7: External services	8.7.2: Electricity mains supply
Electrical mains LV distribution – from main switchgear panels within buildings to, and including, area distribution boards	5.8: Electrical installations	5.8.1: Electricity mains and sub-mains distribution
Electrical mains distribution – from distribution boards, or area distribution boards, to sub-distribution boards – internal	5.8: Electrical installations	5.8.2: Power installations

Table 19.3 *(Continued)*

Works	*Element*	*Sub-element*
Electrical sub-mains distribution – from sub-distribution boards to mechanical and electrical plant or equipment, including external features – external to building	8.7: External services	8.7.4: Electricity distribution to external plant and equipment
Electrical sub-mains distribution – from sub-distribution boards to mechanical and electrical plant or equipment (including final connections), power outlets or other terminations – internal	5.8: Electrical installations	5.8.2: Power installations
Electrical sub-circuit installations – from sub-distribution boards to light fittings	5.8: Electrical installations	5.8.3: Lighting installations; or 5.8.4: Specialist lighting installations
Final connections to mechanical and electrical plant or equipment – internal to the building	5.8: Electrical installations	5.8.2: Power installations
Final connections to mechanical and electrical plant or equipment, including external features – external to the building	8.7: External services	8.7.4: Electricity distribution to external plant and equipment

19.9.2 Electrical mains and sub-mains distribution (sub-element 5.8.1)

This sub-element deals with the distribution of electricity from the main switchgear panels within the building to the area distribution boards. The item measured is to include the high-voltage (HV) switchgear panel, low-voltage (LV) switchgear panel and distribution boards. The mains distribution is measured superficially, in square metres, with the gross internal floor area of the entire building measured.

Connections to the statutory undertaker's electricity mains and the HV and LV electrical mains distribution from statutory undertaker's electrical main to the main switchgear panels within the building are measured under sub-element 8.7.1 (Electricity mains supply). The provision of transformer sub-stations, if required, is dealt with under the same sub-element. Example 19.3 provides component descriptions for electrical mains and sub-mains suppliers provided by statutory undertakers.

Example 19.13 Formulation of descriptions for electrical mains and sub-mains distribution

SERVICES					
Code	**Element/sub-elements/components**	**Qty**	**Unit**	**Rate**	**Total**
				£p	**£**
5	**SERVICES**				
5.8	**Electrical installations**				
5.8.1	*Electrical mains and sub-mains distribution*				
5.8.1.1	Mains switchgear; HV		m²		
5.8.1.2	Mains and sub-mains distribution; including LV switchgear and distribution boards		m²		

Example 19.13 *(Continued)*

SERVICES					
Code	**Element/sub-elements/components**	**Qty**	**Unit**	**Rate**	**Total**
				£p	**£**
5.8.1.3	Testing of installations		%		
5.8.1.3	Commissioning of installations		%		
5.8.1	**TOTAL – Electrical mains and sub-mains distribution: to element 5.8 summary**				

19.9.3 Power installations (sub-element 5.8.2)

Power installations encompass sub-circuit power installations, from distribution boards, or area distribution boards, to sub-distribution boards to the point of termination at power outlets, at fuse connections or at other terminations, including final connections to permanent mechanical and electrical plant or equipment located inside the building.

Power installations systems are measured superficially, in square metres, with the area serviced by the system measured (refer to Figure 19.1). However, power installations to repetitive building types or buildings containing a range of room types can be enumerated (nr) – refer to paragraph 19.1.1. Descriptions of power installations are to include the purpose of the system. Where more than one installation is employed, they are to be measured and described separately. Example 19.14 illustrates typical component descriptions for electrical power installations.

Example 19.14 Formulation of descriptions for power installations

SERVICES					
Code	**Element/sub-elements/components**	**Qty**	**Unit**	**Rate**	**Total**
				£p	**£**
5	**SERVICES**				
5.8	**Electrical installations**				
5.8.2	*Power installations*				
	Example 1 – Hotel				
	Power installations; LV:				
	[Guest rooms and corridors]				
5.8.2.1	Guest room, single bed		nr		
5.8.2.2	Guest room, double bed		nr		
5.8.2.3	Corridors		m²		
	[Front of house areas]				
5.8.2.4	Power installations; LV		m²		
	[Back of house areas]				
5.8.2.5	Power installations; LV		m²		
5.7.2.6	Testing of installations		%		

Example 19.14 *(Continued)*

SERVICES					
Code	**Element/sub-elements/components**	**Qty**	**Unit**	**Rate**	**Total**
				£p	**£**
	Example 2 – Offices				
5.8.2.1	Power installations; LV		m^2		
5.8.2.2	Testing of installations		%		
5.8.2.3	Commissioning of installations		%		
	Example 3 – Residential				
	Power installations; LV:				
5.8.2.2	One-bedroom apartment		nr		
5.8.2.3	Two-bedroom apartment		nr		
5.8.2.4	Two-bedroom maisonette		nr		
5.8.2.5	Three-bedroom house		nr		
5.8.2.6	Testing of installations		%		
5.8.2.7	Commissioning of installations		%		
5.8.2	**TOTAL – Power installations: to element 5.8 summary**				

19.9.4 Lighting installations (sub-element 5.8.3)

These include sub-circuit power installations from sub-distribution boards terminating at light switches and lighting points. Installations dealt with include general internal lighting and emergency lighting. Lighting fixed to the facade of the building, including roof-top lighting, is also measured under this sub-element. It is suggested that aircraft warning lights and the like are treated as specialist lighting installations and measured under sub-element 5.8.4 (Specialist lighting installations).

Lighting installations systems are measured in the same way as power installations (see sub-paragraph 19.9.3). Descriptions of lighting installations are to include the type of installation. Again, where more than one installation is employed, they are to be measured and described separately. Typical component descriptions for lighting installations are shown in Example 19.15.

Example 19.15 Formulation of descriptions for lighting installations

SERVICES					
Code	**Element/sub-elements/components**	**Qty**	**Unit**	**Rate**	**Total**
				£p	**£**
5	**SERVICES**				
5.8	**Electrical installations**				
5.8.3	*Lighting installations*				
	Example 1 – Office				
5.8.3.1	General lighting installation		m^2		

Example 19.15 *(Continued)*

SERVICES					
Code	**Element/sub-elements/components**	**Qty**	**Unit**	**Rate**	**Total**
				£p	**£**
5.8.3.2	Emergency lighting installation		m²		
5.8.3.3	Uninterrupted power supply (UPS)		m²		
5.8.3.4	External lighting installation; to facades of building		m²		
5.8.3.5	Testing of installations		%		
5.8.3.6	Commissioning of installations		%		
	Example 2 – Residential				
	General lighting installation:				
5.8.3.1	One-bedroom apartment		nr		
5.8.3.2	Two-bedroom apartment		nr		
5.8.3.3	Two-bedroom maisonette		nr		
5.8.3.4	Three-bedroom house		nr		
5.8.3.5	Testing of installations		%		
5.8.3.6	Commissioning of installations		%		
5.8.3	**TOTAL – Lighting installations: to element 5.8 summary**				

19.9.5 Specialist lighting installations (sub-element 5.8.4)

Specialist lighting installations include illuminated display signs, aircraft warning lights, studio lighting, auditorium lighting, arena lighting, operating theatres and other specialist lighting installations. They can be either measured as enumerated (nr) items or superficially, in square metres. When measured superficially, the area serviced by the system is measured (refer to Figure 19.1). Each installation is to be measured and described separately, with the type of specialist lighting installation stated in the description. Typical component descriptions for specialist lighting installations are illustrated in Example 19.16.

Example 19.16 Formulation of descriptions for specialist lighting installations

SERVICES					
Code	**Element/sub-elements/components**	**Qty**	**Unit**	**Rate**	**Total**
				£p	**£**
5	**SERVICES**				
5.8	**Electrical installations**				
5.8.4	*Specialist lighting installations*				
5.8.4.1	Illuminated display signs; 'Woollard House' (PC sum £8,500)		nr		

Example 19.16 *(Continued)*

Code	Element/sub-elements/components	Qty	Unit	Rate	Total
SERVICES					
				£p	£
5.8.4.2	Testing of installations		%		
5.8.4.3	Commissioning of installations		%		
5.8.4	**TOTAL – Specialist lighting installations: to element 5.8 summary**				

19.9.6 Local electricity generation systems (sub-element 5.8.5)

Sub-element 5.8.5 covers local electricity generation equipment (i.e. emergency and standby generator plant) and transformation devices (i.e. systems using the natural elements to generate electricity). Example 19.17 shows sample component descriptions for local electricity generation systems.

Example 19.17 Formulation of descriptions for local electricity generation systems

Code	Element/sub-elements/components	Qty	Unit	Rate	Total
SERVICES					
				£p	£
5	**SERVICES**				
5.8	**Electrical installations**				
5.8.5	*Local electricity generation systems*				
	[Local generation equipment]				
5.8.5.1	Standby generators; packaged standby diesel generating set; 750 kVA		nr		
5.8.5.2	Standby generators; prefabricated drop-over acoustic housing; for 750 kVA generating set		nr		
5.8.5.3	Testing of installations		%		
5.8.5.4	Commissioning of installations		%		
	[Transformation devices]				
5.8.5.5	Wind turbines; direct-drive wind generator, 0.50kW output, 1.10m diameter rotor, wall mounted	6	nr		
5.8.5.6	Photovoltaic devices; ultra-high efficiency solar cells; PV modules comprising electrically interconnected crystalline silicon solar cells permanently laminated within a pottant and encapsulated between a tempered glass cover plate and a back sheet; entire laminate secured within an anodised aluminium frame for structural strength; panel size 1,037mm × 527mm; including solar charge controllers and high-capacity 'wet' deep cycle batteries; peak power 75Wp; surface area of units approximately 60m²; roof mounted	110	nr		

Example 19.17 *(Continued)*

SERVICES					
Code	**Element/sub-elements/components**	**Qty**	**Unit**	**Rate**	**Total**
				£p	**£**
5.8.5.7	Testing of installations		%		
5.8.5.8	Commissioning of installations		%		
5.8.5	**TOTAL – Local electricity generation systems: to element 5.8 summary**				

Emergency and standby generator plant, irrespective of the type of fuel to operate the plant, is measured as enumerated (nr) items. The output in kilowatts (kW) or in kilovolt ampere (kVA) should be given in the description. Acoustic housings to generator plant can also be measured under this sub-element. They should be enumerated (nr), with the type stated (e.g. prefabricated drop-over acoustic housing).

Energy transformation devices include wind turbines, roof-top wind energy systems, photovoltaic devices (e.g. surface-mounted cells, panels and modules) and solar collectors; they are measured and described as enumerated (nr) items. Descriptions for transformation devices are to include the type of device and the output in kilowatts. For photovoltaic devices, the surface area, in square metres, is also to be stated.

19.9.7 Earthing and bonding systems (sub-element 5.8.6)

Earthing and bonding systems provide for the transfer of electrical current to the earth, to protect people, buildings, structures, plant and equipment in the case of fault within the electricity supply. They are also used as a means of protecting against interference from electromagnetic fields and electromagnetic forces.

Earthing and bonding systems are measured superficially, in square metres, with the area serviced by the system measured (refer to Figure 19.1). However, earthing and bonding systems to repetitive building types or buildings containing a range of room types can be enumerated (nr) – refer to paragraph 19.1.1. Descriptions of earthing and bonding systems are to include the purpose of the system. Where more than one system is employed, they are to be measured and described separately.

Lightning protection is measured under sub-element 5.11.3 (Lightning protection).

Example 19.18 shows sample component descriptions for earthing and bonding systems.

Example 19.18 Formulation of descriptions for earthing and bonding systems

SERVICES					
Code	**Element/sub-elements/components**	**Qty**	**Unit**	**Rate**	**Total**
				£p	**£**
5	**SERVICES**				
5.8	**Electrical installations**				
5.8.6	*Earthing and bonding systems*				

Example 19.18 *(Continued)*

SERVICES					
Code	**Element/sub-elements/components**	**Qty**	**Unit**	**Rate**	**Total**
				£p	**£**
	Example 1 – Office				
5.8.6.1	Earthing and bonding		m²		
5.8.6.2	Testing of installations		%		
5.8.6.3	Commissioning of installations		%		
	Example 2 – Residential				
	Earthing and bonding:				
5.8.6.1	One-bedroom apartment		nr		
5.8.6.2	Two-bedroom apartment		nr		
5.8.6.3	Two-bedroom maisonette		nr		
5.8.6.4	Three-bedroom house		nr		
5.8.6.5	Testing of installations		%		
5.8.6.6	Commissioning of installations		%		
5.8.6	**TOTAL – Earthing and bonding systems: to element 5.8 summary**				

19.10 Fuel installations (element 5.9)

This element is divided into two sub-elements as follows:

5.9.1 Fuel storage; and
5.9.2 Fuel distribution systems.

19.10.1 Fuel storage (sub-element 5.9.1)

Fuel storage tanks and piped supply systems distributing oil, petrol, diesel and liquefied petroleum gas (LPG) from storage tanks to user points are dealt with under this sub-element. Fuel storage tanks are enumerated (nr), with the type and purpose of the tank or vessel stated. Piped distribution systems are measured superficially, in square metres, with the total GIFA of the entire building measured. Distribution systems to repetitive building types or buildings containing a range of room types can be enumerated (nr) – refer to paragraph 19.1.1. Descriptions of piped supply systems are to include the purpose of the system. Where more than one installation is employed, they are to be measured and described separately.

19.10.2 Fuel distribution systems (sub-element 5.9.2)

This entails the distribution of fuel from the point of storage or from the point of the mains connection within the building to the user points, including pipelines, pipeline

Example 19.19 Formulation of descriptions for fuel distribution systems

SERVICES					
Code	**Element/sub-elements/components**	**Qty**	**Unit**	**Rate**	**Total**
				£p	**£**
5	**SERVICES**				
5.9	**Fuel installations**				
5.9.2	*Fuel distribution systems*				
	Example 1 – Office				
	[Gas installation]				
5.9.2.1	Piped distribution; gas		m²		
5.9.2.2	Testing of installations		%		
5.9.2.3	Commissioning of installations		%		
	Example 2 – Residential				
	[Gas installation]				
	Piped distribution; gas				
5.9.2.1	One-bedroom apartment		nr		
5.9.2.2	Two-bedroom apartment		nr		
5.9.2.3	Two-bedroom maisonette		nr		
5.9.2.4	Three-bedroom house		nr		
5.9.2.5	Testing of installations		%		
5.9.2.6	Commissioning of installations		%		
5.9.2	**TOTAL – Fuel distribution systems: to element 5.9 summary**				

ancillaries and fittings. Fuel distribution systems are measured superficially, in square metres, with the total gross internal floor area of the building measured. However, fuel distribution systems to repetitive building types or buildings containing a range of room types can be enumerated (nr) – refer to paragraph 19.1.1.

In addition to piped distribution systems for oil, petrol, diesel and liquefied petroleum gas, sub-element 5.9.2 also covers piped distribution systems for gas.

Connections to the statutory undertaker's gas main and the mains gas supply from statutory undertaker's main to point of mains connection within the building are measured under sub-element 8.7.5 (Gas main supply).

Typical component descriptions for fuel distributions systems are shown in Example 19.19.

19.11 Lift and conveyor installations (element 5.10)

This element is divided into ten sub-elements as follows:

5.10.1 Lifts and enclosed hoists;

5.10.2 Escalators;

5.10.3 Moving pavements;
5.10.4 Powered stair lifts;
5.10.5 Conveyors;
5.10.6 Dock levellers and scissor lifts;
5.10.7 Cranes and unenclosed hoists;
5.10.8 Car lifts, car stacking systems, turntables and the like;
5.10.9 Document handling systems; and
5.10.10 Other lift and conveyor installations.

19.11.1 Lifts and enclosed hoists (sub-element 5.10.1)

This sub-element covers all types and configurations of passenger and goods lifts, including fire-fighting lifts, platform lifts and wall climbing lifts. It addresses the complete lift installation, including lift cars, doors, motors, guides and counter balances, hydraulic and electric equipment, emergency lighting within the lift car, lift alarms and emergency call systems. (See Example 19.20.)

Sub-element 5.10.1 also covers enclosed hoists.

Lifts are enumerated (nr). Descriptions for lifts are to include:

- for passenger lifts – the number (nr) of people that the lift can accommodate (i.e. the capacity), the speed of the lift (m/sec – metres per second), the number and type of doors (e.g. two-panel centre opening doors), door heights (particularly if non-standard), the type of finishes to the lift car (e.g. manufacturer's standard car finish, brushed stainless steel) and the number of levels served.
- for goods lifts – the capacity (kg), the speed of the lift (m/sec), the number and type of doors (e.g. two-panel centre opening doors), door heights, the type of finishes to the lift car (e.g. manufacturer's standard car finish, painted).

Other items which the cost manager will need to consider and, where required, make allowance for in the cost plan (allocated to the appropriate sub-element) include:

- enhanced finishes to the lift car (e.g. centre mirror, flat ceiling and carpets);
- provision of fire-fighting controls (to provide a fire-fighting lift);
- glass lift cars and doors;
- painting of the entire lift pit;
- heating, cooling and ventilation to lift motor rooms;
- lighting to lift shafts and lift motor rooms; and
- intercom linked to reception desk and security.

When measuring lift and other types of conveyor installations, the cost manager will need to consider the employer's requirements in respect of independent insurance inspections, warranties and servicing (i.e. planned maintenance).

Enclosed hoists include service hoists, dumb waiters and the like. Service hoists and dumb waiters can be found in restaurants, hospitals, hotels and laundry businesses, to name a few. They are generally used to aid the movement of food, cutlery or any other similar items between floors.

Hoists are enumerated (nr), stating the capacity (kg) and number (nr) of levels served in the description. Any specific performance and quality requirements should also be stated.

Example 19.20 Formulation of descriptions for lifts and enclosed hoists

SERVICES					
Code	**Element/sub-elements/components**	**Qty**	**Unit**	**Rate**	**Total**
				£p	**£**
5	**SERVICES**				
5.10	**Lift and conveyor installations**				
5.10.1	*Lifts and enclosed hoists*				
	Example 1 – Lifts				
	Note:				
	Rates include for or independent insurance inspections and 12 month warranty and servicing				
5.10.1.1	Passenger lifts; electrically operated; to take 8 persons; speed of 1.6m/sec; manufacturer's enhanced car finish, comprising wood lining to walls, centre mirror, flat ceiling and carpet; two-panel centre opening doors, door height 2.10m; serving 6 levels		nr		
5.10.1.2	Passenger lifts – with fire-fighting control; electrically operated; to take 8 persons; speed of 1.60m/sec; manufacturer's enhanced car finish, comprising wood lining to walls, centre mirror, flat ceiling and carpet; two-panel centre opening doors, door height 2.10m; serving 6 levels		nr		
5.10.1.3	Goods lifts; electrically operated; to take 2,250kg; speed of 1.00m/sec; manufacturer's standard car finish, painted; two-panel centre opening doors, door height 2.10m; serving 6 levels		nr		
5.10.1.4	Testing of installations		%		
5.10.1.5	Commissioning of installations		%		
	Example 2 – Enclosed hoists				
	Note:				
	Rates include for or independent insurance inspections and 12 month warranty and servicing				
5.10.1.1	Food hoist; electrically operated; double decker; to take 100kg; including: rise and fall shutters, hinged doors, stainless steel finish; serving 2 levels		nr		
5.10.1.2	Testing of installations		%		
5.10.1.3	Commissioning of installations		%		
5.10.1	**TOTAL – Lifts and enclosed hoists: to element 5.10 summary**				

19.11.2 Escalators (sub-element 5.10.2)

Escalators are measured as enumerated (nr) items. Descriptions are to include the number of flights served, the angle of rise (in degrees (°)) and the width of the step (in metres). The rise (i.e. the distance between finished floor levels in metres) should also be stated. In addition, the type of finishes (i.e. to side panels and soffit, details of any lighting (e.g. under-step lighting, balustrade lighting and skirting lighting), emergency stop buttons, together with details of any guards or other protection measures are to be specified. Example 19.21 illustrates typical component descriptions for an escalator installation.

Example 19.21 Formulation of descriptions for escalators

SERVICES					
Code	**Element/sub-elements/components**	**Qty**	**Unit**	**Rate £p**	**Total £**
5	**SERVICES**				
5.10	**Lift and conveyor installations**				
5.10.2	*Escalators*				
	Note: Rates include for or independent insurance inspections and 12 month warranty and servicing				
5.10.2.1	Escalator; 1 nr floors served, angle of rise 30 degrees, rise 3.80m, step width 0.80m; standard material and finish, stainless steel panel balustrade, black, synthetic rubber, handrail; automatic start/stop; motor energy controller; under-handrail lighting; dress guards to skirt panels		nr		
5.10.2.2	Testing of installations		%		
5.10.2.3	Commissioning of installations		%		
5.10.2	**TOTAL – Escalators: to element 5.10 summary**				

19.11.3 Moving pavements (sub-element 5.10.3)

Moving pavements (sometimes called travelators, moving walkways, conveyors or auto walks) are measured as enumerated (nr) items. Descriptions are to include:

- for horizontal moving pavements – the length of pavement and the width of pallet (in metres);
- for inclined moving pavements – the length of pavement, the width of pallet and the rise (in metres). The angle (in degrees) is also to be stated.

In addition, the type of finishes to side panels, details of any lighting (e.g. skirting lighting), emergency stop buttons, together with details of any guards or other protection measures are to be stated. Example 19.22 illustrates typical component descriptions for moving pavements.

Example 19.22 Formulation of descriptions for moving pavements

SERVICES					
Code	**Element/sub-elements/components**	**Qty**	**Unit**	**Rate**	**Total**
				£p	**£**
5	**SERVICES**				
5.10	**Lift and conveyor installations**				
5.10.3	*Moving pavements*				
5.10.3.1	Horizontal moving walkway; overall length 3.60m, pallet width 1.00m; standard material and finish; balustrade panel clear, 10mm tempered safety glass with black, synthetic rubber, handrail		nr		
5.10.3.2	Inclined moving walkway; overall length 3.60m, pallet width 0.80m, angle of rise 12 degrees; standard material and finish; balustrade panel clear, 10mm tempered safety glass with black, synthetic rubber, handrail		nr		
5.10.3.3	Testing of installations		%		
5.10.3.4	Commissioning of installations		%		
5.10.3	**TOTAL – Moving pavements: to element 5.10 summary**				

19.11.4 Powered stair lifts (sub-element 5.10.4)

Stair lifts are devices that are used to carry a person up and down staircases. Powered stair lifts are enumerated (nr), with the type of stair lift stated (e.g. seated, standing or perched stair lifts) in the description.

Powered inclined platform lifts and the like, which accommodate wheelchairs can also be included within this sub-element.

19.11.5 Conveyors (sub-element 5.10.5)

These are systems for the mechanical conveyance of goods between two or more points within a building. Goods conveyors are enumerated (nr). Descriptions are to include the purpose of the conveyor (e.g. food, packaging and baggage-handling systems), the type of conveyor (e.g. modular belt conveyors, PVC and polyurethane (Pu) belt conveyors, accumulating line shaft roller conveyors, pallet roller conveyors or magnetic conveyors) and the length and the width of the conveyor in metres. Gravity roller conveyors, which are non-powered manual handling systems, can also be measured under this sub-element.

People conveyors, although referred to under sub-element 5.10.5 (Conveyors), are commonly referred to as moving walkways and are better dealt with under sub-element 5.10.3 (Moving pavements) – refer to sub-paragraph 19.11.3.

19.11.6 Dock levellers and scissor lifts (sub-element 5.10.6)

Dock levellers and scissor lifts are measured and described as enumerated (nr) items. Descriptions are to include:

- for dock levellers – the type of unit (e.g. hydraulic dock leveller or mechanically operated dock leveller), the total rise (or lift – in metres), the static and dynamic load capacity (in tonnes or kilonewtons), and the overall dimensions of the platform;
- for scissor lifts – the type of unit (e.g. scissor lift table, high-capacity scissor lift, dock scissors lift, tilter (used for tilting heavy loads) and high travel hydraulic lifts), the total rise (or lift – in metres), and the load capacity (in tonnes or kilonewtons).

19.11.7 Cranes and unenclosed hoists (sub-element 5.10.7)

Cranes and unenclosed hoists for the lifting and moving of heavy goods and equipment are enumerated (nr). Descriptions are to include:

- for cranes – the type of crane (e.g. jib crane), load capacity (in tonnes or kilonewtons), and the jib length (in metres);
- for travelling cranes – the type of crane (e.g. single girder, double girder and underslung overhead travelling crane, or single girder wall travelling crane), the load capacity in tonnes or kilonewtons, and the span and the length of crane track in metres;
- for unenclosed hoists – the type of the hoist (e.g. electric wire rope hoists and electric chain hoists) and the load capacity (in tonnes or kilonewtons).

Hoists for moving people with disabilities are measured under sub-element 5.10.10 (Other lift and conveyor installations).

19.11.8 Car lifts, car stacking systems, turntables and the like (sub-element 5.10.8)

Vehicle-lifting, storage and moving systems, including car lifts, car stacking systems, sliding platforms and vehicle turntables, are measured under this sub-element. Hydraulic vehicle lifts and commercial vehicle lifts used in garages and workshops for vehicle servicing and repair are also measured here. Systems are measured and described as enumerated (nr) items. Descriptions are to include:

- for car lifts – the number of floors served and the load capacity (in tonnes);
- for vehicle lifts – the type (e.g. two-post, for post or scissor lifts) and the load capacity (in tonnes);
- for car stacking systems – the capacity in terms of the number of cars (e.g. double-stacking systems or triple-stacking systems);
- for sliding platforms – the capacity in terms of the number of cars, and the type of platform (i.e. longitudinal or transverse platforms); and

- for vehicle turntables – the type of vehicle to be accommodated (e.g. car or lorry), the turning radius (in metres) and the load capacity (in tonnes).

19.11.9 Document-handling systems (sub-element 5.10.9)

Document-handling systems include rail-based conveyor systems for in-house material and documents transport, pneumatic tube systems and other specialist in-house material or document delivery systems. Warehouse picking systems (i.e. automated storage and retrieval systems and material-handling equipment) are also included under this sub-element.

Document-handling systems and warehouse picking systems are enumerated (nr), with the type and extent of the system given in the description.

19.11.10 Other lift and conveyor installations (sub-element 5.10.10)

The cost manager can allocate lift and conveyor installations that are not addressed by any other sub-element to this sub-element. Examples of other installations include:

- people hoists – hoists for transferring people with disabilities (e.g. ceiling track hoists);
- paternoster lifts (also called 'orbitors' or continuous elevators) – use the paternoster principle of picking up going 'up' and dropping off going 'down'. Paternoster lifts are designed to handle loads continuously with loading and unloading facility at a number of different levels;
- pallet lifts – used to transport heavy materials between different levels (e.g. materials with individual payloads of 3 tonnes);
- pendulum bucket conveyors – carry granular materials, powders and delicate materials;
- carton elevators (also called 'box lifters' or tote elevators) – carry unit loads such as cartons, cases, boxes, sacks, tote bins and pallets vertically between two different levels, for example, packaging up to a high-level conveying system or from first floor down to ground level;
- continuous interlinked bucket conveyor systems; and
- reciprocating vertical hoists.

19.12 Fire and lightning protection (element 5.11)

This element is divided into three sub-elements as follows:

5.11.1 Fire-fighting systems;
5.11.2 Fire suppression systems; and
5.11.3 Lightning protection.

19.12.1 Fire-fighting systems (element 5.11.1)

Fire-fighting systems include fire hose reels, dry risers, wet risers and fire and smoke protection curtains. Control components associated with fire-fighting systems are deemed to be included in the item measured. Example 19.23 sample component descriptions for a typical fire-fighting system.

Example 19.23 Formulation of descriptions for fire-fighting systems

SERVICES					
Code	**Element/sub-elements/components**	**Qty**	**Unit**	**Rate**	**Total**
				£p	**£**
5	**SERVICES**				
5.11	**Fire and lightning protection**				
5.11.1	*Fire-fighting systems*				
5.11.1.1	Fire hose reels; comprising hinged automatic fire hose reel, 30m hose, hose guide and fire hose reel cover		nr		
5.11.1.2	Testing of installations		%		
5.11.1.3	Commissioning of installations		%		
5.11.1	**TOTAL – Fire-fighting systems: to element 5.11 summary**				

Generally, fire-fighting systems are measured superficially, in square metres, with the area serviced by the system measured (refer to Figure 19.1). Dry risers, wet risers and fire hose reels and the like are enumerated (nr). Descriptions for fire-fighting systems are to explain the type of installation.

Hand-held fire-fighting equipment, including fire extinguishers and fire blankets are measured under sub-element 4.1.1 (General fittings, furnishings and equipment), and manual call points are measured under sub-element 5.12.1 (Communication systems).

19.12.2 Fire suppression systems (element 5.11.2)

Fire suppression systems include sprinkler installations, deluge systems, gas fire suppression systems and foam fighting systems. Control components associated with fire suppression systems are deemed to be included in the item measured. Example 19.24 shows sample component descriptions for various typical fire suppression systems.

Example 19.24 Formulation of descriptions for fire suppression systems

SERVICES					
Code	**Element/sub-elements/components**	**Qty**	**Unit**	**Rate**	**Total**
				£p	**£**
5	**SERVICES**				
5.11	**Fire and lightning protection**				
5.11.2	*Fire suppression systems*				

Example 19.24 *(Continued)*

SERVICES					
Code	**Element/sub-elements/components**	**Qty**	**Unit**	**Rate**	**Total**
				£p	**£**
5.11.2.1	Sprinkler installation		m^2		
5.11.2.2	Inert gas fire suppression system		m^2		
5.11.2.3	Testing of installations		%		
5.11.2.4	Commissioning of installations		%		
5.11.2	**TOTAL – Fire suppression systems: to element 5.11 summary**				

19.12.3 Lightning protection (sub-element 5.11.3)

Lightning protection systems are also measured superficially, in square metres, with the area serviced by the system measured (refer to Figure 19.1). The type of installation is to be stated in the description.

19.13 Communication, security and control systems (element 5.12)

This element is divided into three sub-elements as follows:

5.12.1 Communication systems;
5.12.2 Security systems; and
5.12.3 Central control/building management systems.

19.13.1 Communication systems (sub-element 5.12.1)

These include, amongst other things, telecommunication systems, data transmission, paging and emergency call systems, public address systems, conference audio facilities, radio systems, fire detection and alarm systems, systems giving early warning of water or liquid or leakage, clocks and flexitime installations. (See Example 19.25.)

In the main, communication systems are measured superficially, in square metres, with the area serviced by the system measured (refer to Figure 19.1). However, communication systems to repetitive building types, or buildings containing a range of room types, can be enumerated (nr) – refer to paragraph 19.1.1. Irrespective of how measured, the type of communication system is to be given in the description. Where more than one system is employed, they are to be measured and described separately.

Equipment bias installations such as projection systems, clocks, card systems, flexitime installations, televisions, and the like are measured as enumerated (nr) items. Again, the type of system is to be described.

Example 19.25 Formulation of descriptions for communication systems

SERVICES					
Code	**Element/sub-elements/components**	**Qty**	**Unit**	**Rate**	**Total**
				£p	**£**
5	**SERVICES**				
5.12	**Communication, security and control systems**				
5.12.1	*Communication systems*				
	Example 1 – Office				
5.12.1.1	Telecommunication systems; telephone containment only		m²		
5.12.1.2	Data transmission systems; containment only		m²		
5.12.1.3	Digital clocks; 240V, 50Hz supply; PC Sum £600.00/nr; flush mounted		nr		
5.12.1.4	Public address system		m²		
5.12.1.5	Fire and smoke detection and alarm system		m²		
5.12.1.6	Testing of installations		%		
5.12.1.7	Commissioning of installations		%		
	Example 2 – Residential				
	Telecommunication systems; telephone containment, fascia plates and cabling:				
5.12.1.1	One-bedroom apartment		nr		
5.12.1.2	Two-bedroom apartment		nr		
5.12.1.3	Two-bedroom maisonette		nr		
5.12.1.4	Three-bedroom house		nr		
	Data transmission systems; containment, fascia plates and cabling:				
5.12.1.5	One-bedroom apartment		nr		
5.12.1.6	Two-bedroom apartment		nr		
5.12.1.7	Two-bedroom maisonette		nr		
5.12.1.8	Three-bedroom house		nr		
	Entertainment systems; containment for entertainment system, (excluding entertainment system and speakers):				
5.12.1.9	One-bedroom apartment		nr		
5.12.1.10	Two-bedroom apartment		nr		
5.12.1.11	Two-bedroom maisonette		nr		
5.12.1.12	Three-bedroom house		nr		
5.12.1.13	Satellite; communal satellite dish		nr		
	Television systems; containment, fascia plates and cabling, (excluding television):				
5.12.1.14	One-bedroom apartment		nr		
5.12.1.15	Two-bedroom apartment		nr		

Example 19.25 *(Continued)*

SERVICES					
Code	**Element/sub-elements/components**	**Qty**	**Unit**	**Rate**	**Total**
				£p	**£**
5.12.1.16	Two-bedroom maisonette		nr		
5.12.1.17	Three-bedroom house		nr		
5.12.1.18	Testing of installations		%		
5.12.1.19	Commissioning of installations		%		
5.12.1	**TOTAL – Communication systems: to element 5.12 summary**				

19.13.2 Security systems (sub-element 5.12.2)

Observation and access control installations are measured under this sub-element. Such installations include surveillance equipment (e.g. CCTV), security detection and alarm equipment, access control systems, burglar and security alarms, audio and visual door entry systems and other security systems. Example 19.26 provides sample component descriptions for common security system installations.

Example 19.26 Formulation of descriptions for security systems

SERVICES					
Code	**Element/sub-elements/components**	**Qty**	**Unit**	**Rate**	**Total**
				£p	**£**
5	**SERVICES**				
5.12	**Communication, security and control systems**				
5.12.2	*Security systems*				
	Example 1 – Office				
5.12.2.1	Access control system		m^2		
5.12.2.2	Detection and alarm systems		m^2		
5.12.2.3	Surveillance systems		m^2		
5.12.2.4	Testing of installations		%		
5.12.2.5	Commissioning of installations		%		
	Example 2 – Residential				
	Door entry systems; video and audio entry (connect to TV/home phone)		nr		
5.12.2.1	One-bedroom apartment		nr		
5.12.2.2	Two-bedroom apartment		nr		
5.12.2.3	Two-bedroom maisonette		nr		
5.12.2.4	Three-bedroom house		nr		
5.11.2.5	Testing of installations		%		
5.11.2.6	Commissioning of installations		%		
5.12.2	**TOTAL – Security systems: to element 5.12 summary**				

Security systems can be measured superficially, in square metres, or as enumerated (nr) items. Where measured superficially, the area serviced by the system is measured (refer to Figure 19.1). The type of system is to be given in the description.

Where security systems are to serve repetitive building types, or buildings containing a range of room types, they can be enumerated (nr) in the same way as communication systems. The type of security system is to be given in the description. Where more than one system is employed, they are to be measured and described separately.

19.13.3 Central control/building management systems (sub-element 5.12.3)

This sub-element deals with central control panels and building management systems for mechanical and electrical systems and installations (i.e. control systems which, from a central remote location, provide means for controlling and reporting on the performance of the operational systems of a building). Example 19.27 illustrates component descriptions for a typical central control / building management system.

Example 19.27 Formulation of descriptions for central control/building management systems

SERVICES					
Code	**Element/sub-elements/components**	**Qty**	**Unit**	**Rate**	**Total**
				£p	**£**
5	**SERVICES**				
5.12	**Communication, security and control systems**				
5.12.3	*Central control/building management systems*				
5.12.3.1	Building management system (BMS)		m^2		
5.12.3.2	Testing of installations		%		
5.12.3.3	Commissioning of installations		%		
5.12.3	**TOTAL – Central control/building management systems: to element 5.12 summary**				

Central control and building management systems are measured superficially, in square metres, with the area serviced by the system measured (refer to Figure 19.1). The type of system is to be given in the description. Where more than one system is employed, they are to be measured and described separately.

19.14 Specialist installations (element 5.13)

This element is divided into five sub-elements as follows:

5.13.1 Specialist piped supply installations;

5.13.2 Specialist refrigeration systems;

5.13.3 Specialist mechanical installations;
5.13.4 Specialist electrical/electronic installations; and
5.13.5 Water features.

19.14.1 Specialist piped supply installations (sub-element 5.13.1)

Specialist piped supply systems include medical and laboratory gas supply systems, centralised vacuum cleaning systems, treated water systems, swimming pool water treatment, compressed air systems, vacuum systems. Systems can be either measured as enumerated (nr) items or superficially, in square metres. When measured superficially, the area serviced by the system is measured (refer to Figure 19.1). Each type of system is to be measured and described separately, with the type of system stated in the description. Testing and commissioning of the installations are measured separately, based on a percentage addition of the combined total cost of all specialist piped supply installations.

19.14.2 Specialist refrigeration systems (sub-element 5.13.2)

Specialist refrigeration systems include cold rooms, ice pads and any other specialist refrigeration system. Systems are measured and described in the same way as specialist piped supply systems. Testing and commissioning of the installations are measured separately, based on a percentage addition of the combined total cost of all specialist refrigeration systems.

19.14.3 Specialist mechanical installations (sub-element 5.13.3)

This sub-element deals with mechanical installations of a specialist nature, for example, wave machines, saunas and sauna equipment, Jacuzzis and swimming pools. Systems are measured and described in the same way as for sub-element 5.13.2 (specialist refrigeration systems). Testing and commissioning of the installations are measured separately, based on a percentage addition of the combined total cost of all specialist mechanical installations. Example 19.28 shows sample component descriptions for typical specialist mechanical installations.

Example 19.28 Formulation of descriptions for specialist mechanical installations

SERVICES					
Code	**Element/sub-elements/components**	**Qty**	**Unit**	**Rate**	**Total**
				£p	**£**
5	**SERVICES**				
5.13	**Specialist installations**				
5.13.3	*Specialist mechanical installations*				

Example 19.28 *(Continued)*

SERVICES					
Code	**Element/sub-elements/components**	**Qty**	**Unit**	**Rate**	**Total**
				£p	**£**
5.13.3.1	Sauna; internal; infrared, for 6 people, 'L-shaped' seating arrangement; curved spruce panelling and clear glass door; fitted with black light curved ceramics far infrared emitters; with full size colour light panel, electric aromatherapy dispenser with oils, full spectrum light panel, 5-layer air purifier, electric ioniser; external dimensions 2,200mm × 1,800mm × 1,975mm high		nr		
5.13.3.2	Jacuzzi; mosaic tiled overflow spa—Martinique 2.5m, with balance tank – 1,000 litre capacity, 1 nr air ring in seat		nr		
5.13.3.3	Testing of installations		%		
5.13.3.4	Commissioning of installations		%		
5.13.3	**TOTAL – Specialist mechanical installations: to element 5.13 summary**				

19.14.4 Specialist electrical/electronic installations (sub-element 5.13.4)

Sub-element 5.13.4 covers specialist electrical and electronic installations such as bespoke equipment for radio and television studios, recording studio equipment, communal television aerials and satellite systems, home cinema, multi-room audio and video systems, and automated curtains and blinds. Systems are measured as enumerated (nr) items, and the type of studio, and the composition of equipment within, are described. Testing and commissioning of the installations are measured separately, based on a percentage addition of the combined total cost of all specialist electrical and electronic installations.

19.14.5 Water features (sub-element 5.13.5)

Internal water features such as fountains and waterfalls are measured as enumerated (nr) items, with the type of feature stated in the description. Details of filtration equipment, nutrient treatment and equipment, and controls are to be described. Testing and commissioning of the installations are measured separately, based on a percentage addition of the combined total cost of all water features. Example 19.29 shows sample component descriptions for typical water features.

Example 19.29 Formulation of descriptions for water features

Code	Element/sub-elements/components	Qty	Unit	Rate	Total
SERVICES					
				£p	£
5	**SERVICES**				
5.13	**Specialist installations**				
5.13.5	*Water features*				
5.13.5.1	Floor fountain; Harmony floor fountain by Rebecca; centre mount with rustic copper frame		nr		
5.13.5.2	Wall fountain; Deep Creek Falls slate wall fountain by Rebecca; three panel with company logo; antique black copper frame		nr		
5.13.5.3	Extra for mock-up of company logo		nr		
5.13.5.4	Testing of installations		%		
5.13.5.5	Commissioning of installations		%		
5.13.5	**TOTAL – Water features: to element 5.13 summary**				

19.15 Builder's work in connection with services (element 5.14)

Element 5.14 deals with builder's work in connection with the installation of mechanical and electrical services internal to the building. Builder's work comprises:

- bases for mechanical and electrical plant and equipment;
- fuel bunds and the like to storage tanks and storage vessels;
- supports to storage tanks and storage vessels;
- holes, mortices, sinkings and chases;
- ducts, pipe sleeves and the like;
- stopping and sealing holes;
- fire-resistant stopping and fire sleeves;
- fire breaks;
- on-site painting of, and other anti-corrosive treatments to, mechanical services equipment, fuel storage tanks and vessels, supports and pipelines;
- identification labelling and colour coding of services installations and systems; and
- other builder's work in connection with mechanical and electrical services.

Builder's work in connection with the installation of external mechanical and electrical services is measured under sub-element 8.7.11 (Builder's work in connection with external services).

Builder's work in connection with mechanical and electrical services within the building is measured superficially, in square metres. The area measured is the gross internal floor area for the entire building. However, whenever applicable, the rules require the cost of builder's work to be apportioned under the following three headings:

- general areas (i.e. all areas other than landlord areas and plant room areas);
- landlord areas (i.e. the part of the building which will be managed by the landlord or managing agent appointed by the building owner); and
- plant room areas.

The cost target for builder's work in connection with services internal to the building is calculated as follows:

Step 1: Determine the total GIFA for the building.

Step 2: Apportion the total GIFA for the building into general areas, landlord areas and plant room areas, as applicable.

Note:
The sum of general areas, landlord areas and plant room areas is to equal the total GIFA for the building.

Step 3: Determine the unit rates to be applied to general areas, landlord areas and plant room areas.

Step 4: Ascertain the estimated cost of builder's work in connection with services to general areas by multiplying the proportion of GIFA applicable to general areas by the unit rate determined for builder's work in connection with services to general areas.

Step 5: Repeat step 4 for landlord areas.

Step 6: Repeat step 4 for plant room areas.

Step 7: Ascertain the total cost target for cost of builder's work in connection with services by adding together the estimated costs of builder's work in general areas, landlord areas and plant room areas.

Notwithstanding the above, cost-significant items such as fuel bunds to storage tanks are measured and described as enumerated (nr) items.

A typical build-up of a cost target for builder's work in connection with the installation of mechanical and electrical services internal to the building is shown in Example 19.30.

Example 19.30 Calculation of cost target for builder's work in connection with services

SERVICES					
Code	Element/sub-elements/components	Qty	Unit	Rate	Total
				£p	£
5	**SERVICES**				
5.14	**Builder's work in connection with services**				
5.14.1	*General builder's work*				
	Example 1 – Office				
5.14.1.1	Builder's work in general areas	5,478	m²	15.00	82,170
5.14.1.2	Builder's work in landlord areas	596	m²	15.00	8,940
5.14.1.3	Builder's work in plant rooms	157	m²	35.00	5,495

Example 19.30 *(Continued)*

SERVICES					
Code	**Element/sub-elements/components**	**Qty**	**Unit**	**Rate**	**Total**
				£p	**£**
	Example 2 – Residential				
5.14.1.4	Builder's work in general areas	855	m^2	19.00	15,390
5.14.1	**TOTAL – Builder's work in connection with services: to element 5.14 – summary**				**111,995**

Should the estimated cost of builder's work in connection with services need to be broken down further, then estimated costs can be determined separately for each type of service (e.g. sanitary installations, services equipment, disposal installations, water installations and so on). The cost of builder's work for each service is calculated in the same way as described above.

CHAPTER 20

Prefabricated buildings and building units (group element 6)

Introduction

This chapter:

- defines prefabricated buildings;
- distinguishes between prefabricated buildings and building units;
- provides step-by-step guidance on how to measure and describe prefabricated buildings and building units, including pods; and
- contains examples illustrating how to formulate descriptions for components.

20.1 Method of measurement

The rules for measuring prefabricated buildings and building units are found in group element 6 of NRM 1, which comprises a single element: 6.1 (Prefabricated buildings and building units). This element is subdivided into three sub-elements as follows:

6.1.1 Complete buildings;

6.1.2 Building units; and

6.1.3 Pods.

Prefabricated building products and systems are used in a variety of market sectors, including residential, educational, commercial, agricultural and industrial applications. Modular buildings, modular units, prefabricated buildings and manufactured buildings are all terms used to describe modular construction building products where a building is preassembled in a factory before being delivered and installed at the building site. Unlike conventional buildings, modular buildings are constructed in sections (generally in an assembly line fashion), transported to the site, and put together by a specialist provider on the building site. Modular construction uses many traditional construction materials, but because a modular building construction product is constructed in a factory, the costs and delivery dates are by and large more predictable.

Manufactured buildings are constructed entirely from proprietary components, which are designed to provide specific building configurations (i.e. building systems). The building is designed and erected on site by the specialist provider.

20.2 Complete buildings (sub-element 6.1.1)

These are buildings where all, or part, of the superstructure is constructed of modular buildings or manufactured buildings (e.g. agricultural buildings, warehouses, aircraft hangers, garages, farm buildings, golf driving ranges and modular houses). The substructure, below-ground drainage and provision of mains services are usually completed in readiness for the delivery and erection of the building system.

Complete buildings are measured superficially, in square metres. The area measured is the gross internal floor area of the entire building, or the part of the building which is constructed using a prefabricated building system. Descriptions for complete buildings should include the principal dimensions, the type of building, and composition of the prefabricated building system (see Example 20.1).

Example 20.1 Formulation of descriptions for complete buildings

COMPLETE BUILDINGS AND BUILDING UNITS					
Code	**Element/sub-elements/components**	**Qty**	**Unit**	**Rate**	**Total**
				£p	**£**
6	**PREFABRICATED BUILDINGS AND BUILDING UNITS**				
6.1	**Prefabricated buildings and building units**				
6.1.1	*Complete buildings*				
	Note:				
	Bases and provision of drainage and services measured and described elsewhere (included under applicable element)				
6.1.1.1	Complete building; 34-place nursery building; internal dimensions 14.28m × 9.95m; accommodation based on National Day Care Standards: entrance lobby, staff room, 3 nr nursery rooms, staff toilets (2 nr × WCs and wash-hand basins), children's toilets (3 nr × WCs and range of 3 nr wash-hand basins) and kitchenette (including single drainer stainless steel sink with mixer taps, 2 nr × 1.00m long base units and worktops, and single 1.00m × 0.90m high cupboard over); construction: steel-framed portakabin duplex modular building system; comprising plastic coated steel profile roof, plastic coated steel external finish, uPVC double glazed windows, external steel security doors, and vinyl faced plasterboard internal finish; corridors, offices and nursery rooms carpeted, vinyl floor coverings to toilets and kitchenette; including drainage and services installations within the building		m²		
6.1.1.2	On-site testing of installations		%		
6.1.1.3	On-site commissioning of installations		%		
6.1.1	**TOTAL – Complete buildings: to element 6.1 summary**				

Enabling works such as substructures, below-ground drainage and provision of mains services are measured and described in accordance with the rules of measurement of that particular aspect of the building (e.g. group elements 1 and 8).

The on-site testing and commissioning of mechanical and electrical services within the prefabricated building are measured as separate items, being measured as a percentage addition of the mechanical and electrical services costs associated with the complete building. The steps in calculating the cost target for on-site testing are as follows:

Step 1: Determine the percentage addition to be applied in respect of on-site testing.

Step 2: Determine the cost of mechanical, electrical and drainage installations within the complete building.

Step 3: Ascertain the cost target for on-site testing by multiplying the sum total of the mechanical, electrical and drainage installations within the complete building by the percentage applicable for on-site testing.

Note:
The equation for calculating the cost target for on-site testing is:

$$t = s \times p$$

Where:
s = sum total of the mechanical, electrical and drainage installations within the complete building
p = percentage addition for on-site testing
t = estimated cost of on-site testing.

The cost target for on-site commissioning is calculated in the same way as for on-site testing. However, it should be noted that the cost target ascertained for on-site testing is not included as part of the sum total of the mechanical, electrical and drainage installations within the complete building.

Note:
The equation for calculating the cost target for on-site commissioning is:

$$c = s \times p$$

Where:
s = sum total of the mechanical, electrical and drainage installations within the complete building
p = percentage addition for on-site commissioning
c = estimated cost of on-site commissioning.

20.3 Building units (sub-element 6.1.2)

Building units are factory-assembled individual modular units (e.g. apartments, hotel rooms, student accommodation, prison cells and offices). The modular units may be room-sized or parts of larger spaces which are combined together to form complete buildings, such as residential buildings, hotels and students' halls of residence. The units comprise the internal skin to external walls, party walls, windows, external doors, partitions and internal doors. The internal fit-out, sanitary appliances and fittings, finishes and building services would normally be installed and commissioned in the units in the factory. External facade claddings and roof treatments are often installed on site after the placement of the modular units. The substructure, below-ground drainage, provision of mains services, external facade cladding, roof treatments, rainwater installations, common areas, staircases and final services connections will normally be carried out on site by a number of different subcontractors other than the provider of the modular units.

Building units are measured superficially, in square metres, with the area measured being the gross internal floor area of the modular unit. Descriptions for building units should include the principal dimensions and the type and composition of the building unit. Each different type and configuration of building unit is to be measured and describe separately, with the number (nr) of identical units stated in the description. Example 20.2 illustrates component descriptions for a typical building unit.

On-site testing and commissioning of mechanical and electrical services within building units are measured in the same way as for complete buildings.

Example 20.2 Formulation of descriptions for building units

COMPLETE BUILDINGS AND BUILDING UNITS					
Code	**Element/sub-elements/components**	**Qty**	**Unit**	**Rate**	**Total**
				£p	**£**
6	**PREFABRICATED BUILDINGS AND BUILDING UNITS**				
6.1	**Prefabricated buildings and building units**				
6.1.2	*Building units*				
6.1.2.1	Hotel bedroom units, single bedroom; internal dimensions 6.00m × 2.46m; construction: light steel frame with walls, floors and ceilings fully boarded to provide required thermal and acoustic performance and fire rating; modules fully fitted out to include bathrooms, wall (including full height tiling to Bathroom), floor (including tiling to bathroom – painted softwood skirtings) and ceiling finishes; fixed furniture (bed head and side tables, wardrobe, desk, and vanity unit, mirror and fittings to bathroom); including drainage and services installations within the unit; (246 nr identical units)		m^2		
6.1.2.2	On-site testing of installations		%		
6.1.2.3	On-site commissioning of installations		%		
6.1.2	**TOTAL – Building units: to element 6.1 summary**				

20.4 Pods (sub-element 6.1.3)

These are modular, pre-finished bathroom, toilet and shower room units. Now very much part of the modern methods of construction embraced by the Northern European Construction Industry, the pod solution is now recognised as a viable alternative to using traditional construction in almost every type of development. With traditional construction, a multitude of trades need to be organised to realise the bathroom design. This requires a high degree of supervision and management on site to ensure correct sequencing and quality of work from plumbers, electricians, tilers, floor layers, sealant applicators, decorators, glaziers, carpenters and other specialists. Pods do away with most of these issues and although the capital cost might not be less than site-constructed bathrooms, toilets and shower rooms, it is argued that savings from waste and improved quality result in less defects and better performance in use. More significantly, reduced construction times can be achieved, which can result in earlier income streams from the building for the owner. Conversely, it could be argued that there is a risk of placing an order for a large volume of pods with a single manufacturer. This is because pods are important and high-cost components. Should the manufacturer not perform, or go out of business, there would be significant cost and time risks to the building project. For this reason, traditional construction using a multitude of trades might be considered the preferred option. Risks such as this must be considered and weighed up by the project team.

There is no such thing as a standard pod. However, there will be cost benefits in specifying to manufacturers' 'standards' or in large volumes. Depending on the complexity and manufacturer a minimum order of 10 to 100 pods may be required for a reasonable final cost. Larger quantities allow further spreading of set-up costs, for example, making a prototype and tooling up.

The range of options for pod specifications are limitless, but it is worth bearing in mind that the capital costs of the bathroom pods are influenced by the structure, size, shape and quality of finishes and sanitary appliances. Typically they include:

- Structure: this can be timber or steel frame, reinforced glass fibre, precast concrete panels or a composite panel system.
- Internal linings and finishes: wall finishes are typically ceramic tiles on a moisture-resistant or waterproof lining. Tanking might be an option. The floor to wall, bath edge and shower tray wall are the key junctions to resolve to prevent water penetrating the lining and structure.
- Sanitary appliances: basins and WCs are typically vitreous china, baths enamelled steel. Shower enclosures are usually made from safety glass with an aluminium frame.
- Services: this includes all mechanical and electrical services installations within the pod, off-site testing and commissioning of the pod, and final on-site connections. Cold and hot water supplies up to the point of connection within the pods are measured separately under element 5.4 (Water installations) as applicable. Power supplies to pods are measured under sub-element 5.8.2 (Power installations).
- Distribution and waste pipework: these are normally plastic waste and supply pipework, or copper supply pipes. Discharge and ventilation stacks to which the wastes are connected are measured separately under sub-element 5.3.1 (Foul drainage above ground).
- Fixtures and fittings: the range is vast and may include mirrors, soap holders and glass shelves.

Pods are simply enumerated (nr), with the type and quality of the pod interior stated in the description. The type and number of sanitary appliances, as well as details of any cost-significant sanitary fittings, should also be given. Example 20.3 illustrates component descriptions for a typical pod.

Example 20.3 Formulation of descriptions for pods

SERVICES					
Code	**Element/sub-elements/components**	**Qty**	**Unit**	**Rate**	**Total**
				£p	**£**
6	**PREFABRICATED BUILDINGS AND BUILDING UNITS**				
6.1	**Prefabricated buildings and building units**				
6.1.3	*Pods*				
6.1.3.1	Prefabricated bathroom pods; comprising WC, wash-hand basin, bath (including bath panel and trim), and shower over bath; fully fitted out, including coloured glazed ceramic tiles to all walls, and sanitary fittings; installed		nr		
6.1.3.2	On-site testing of installations		%		
6.1.3.3	On-site commissioning of installations		%		
6.1.3	**TOTAL – Pods: to element 6.1 summary**				

Where not an integral part of the pod design, fire-resistant stopping to pods, including fire sleeves, are measured under sub-element 5.14.1 (Builder's work in connection with services).

On-site testing and commissioning of mechanical and electrical services within pods are measured in the same way as for complete buildings.

CHAPTER 21

Work to existing buildings (group element 7)

Introduction

This chapter:

- provides step-by-step guidance on how to measure and describe:
  - the removal of existing components;
  - alteration works;
  - repairs to existing building engineering services installations;
  - the insertion of damp-proof courses within existing walls;
  - fungus and beetle eradication;
  - facade retention works;
  - the cleaning of existing surfaces; and
  - the renovation of existing components.
- explains how to measure new building works within an existing building, including internal fit-out works; and
- contains examples illustrating how to formulate descriptions for components.

21.1 Method of measurement

The rules for measuring work to existing buildings are found in group element 7 of NRM 1, which is divided into six elements as follows:

7.1 Minor demolition works and alteration works;
7.2 Repairs to existing services;
7.3 Damp-proof courses/fungus and beetle eradication;
7.4 Facade retention;
7.5 Cleaning existing surfaces; and
7.6 Renovation works.

21.1.1 New building works within an existing building

New building works within an existing building (e.g. internal fit-out works to an existing building) are measured in accordance with the rules for new works appropriate to the component being measured. For example:

- replacement roof coverings are measured under sub-element 2.4.2 (Roof coverings);
- replacement windows and external doors are measured under element 2.6 (Windows and external doors);
- new walls and partitions dividing existing spaces are measured under element 2.7 (Internal walls and partitions);
- new internal doors to existing or new door openings are measured under element 2.8 (Internal doors);
- redecoration and new finishes are measured in accordance with the appropriate internal finishes element under group element 3 (Internal finishes);
- fittings, furnishings and equipment installed with an existing building are measured under group element 4 (Fittings, furnishings and equipment); and
- new mechanical electrical services installations within an existing buildings are measured in accordance with the appropriate services element under group element 5 (Services).

NRM 1 simply requires new building works within an existing building to be measured under a heading of 'Works to existing buildings' (see measurement rules for components throughout rules). Additionally, the rules stipulate that works arising out of party wall awards or agreements must be separately identified. Again, this can be done by way of clearly annotating such works in the cost plan.

21.1.2 Major demolition works and removal of toxic or hazardous materials

Major demolition works and the removal of toxic or hazardous materials (e.g. asbestos) come under facilitating works. They are measured in accordance with the rules for sub-element 0.2.1 (Demolition works) and sub-element 0.1.1 (Toxic or hazardous material removal), respectively.

21.2 Minor demolition works and alteration works (element 7.1)

Element 7.1 comprises a single sub-element as follows:

7.1.1 Minor demolition works and alteration works.

This sub-element deals with the removal of existing components from within an existing building. Such works include the stripping out of existing services installations, sanitary appliances, fixtures and fittings, skirtings, dado rails, floor coverings, suspended ceilings, internal doors and partitions. Such works are commonly referred to as 'soft strip'. It also includes the removal of existing roof coverings, windows and doors, as well as minor demolition works such as taking down internal masonry walls or demolishing a minor part of an existing building.

In addition, sub-element 7.1.1 contends with minor alteration works and isolated repairs. Alteration and repair works encompass many different types of work. Examples include:

- alteration works:
  - inserting a new structural beam to support the structure following the demolition of a load-bearing wall;
  - forming an opening in existing cavity wall construction for a new window;
  - filling existing openings in wall construction where a door or window has been removed; and
  - cutting back a chimney breast.
- repair works:
  - patch repairs to roof coverings and internal wall, floor and ceiling finishes;
  - rebuilding existing piers or columns; and
  - re-glazing.

For the purpose of measurement, the rules categorise components as: spot items, minor demolition works, removal and alteration works. In general, components can be itemised or measured as enumerated (nr), linear (in metres), or superficial (in square metres) items. For components measured linear, the length is the extreme length measured over all obstructions. Where measured superficially, the area is the surface area of the component, with no deduction for voids. In effect, the rules leave the choice of category and method of measurement to the discretion of the cost manager.

Descriptions for spot items, minor demolition works, removal works and alteration works are to describe the nature of the works. Details of any cost-significant new components should also be given. Example 21.1 illustrates component descriptions for typical minor demolition works and alternation works items.

Example 21.1 Formulation of descriptions for minor demolition works and alteration works

WORK TO EXISTING BUILDINGS					
Code	**Element/sub-elements/components**	**Qty**	**Unit**	**Rate**	**Total**
				£p	**£**
7	**WORK TO EXISTING BUILDINGS**				
7.1	**Minor demolition works and alteration works**				
7.1.1	*Minor demolition works and alteration works*				
7.1.1.1	Stripping roof coverings; slates; set aside for reuse		m²		
7.1.1.2	Demolishing chimney; brick, to below roof level, including sealing existing flues and making good roof coverings to match existing; approximately 800mm × 1,100mm × 1,200mm above roof		nr		
7.1.1.3	Removing existing chimney pots for reuse; demolishing defective chimney stack to roof level; rebuilding chimney stack using new facing bricks to				

Example 21.1 *(Continued)*

WORK TO EXISTING BUILDINGS

Code	Element/sub-elements/components	Qty	Unit	Rate £p	Total £
	match existing; providing new lead flashings; parge and core flues, resetting chimney pots including flaunching; approximately 800mm × 1,100mm × 1,200mm above roof		nr		
7.1.1.4	Demolishing internal walls; half brick thick, plastered both sides		m²		
7.1.1.5	Demolishing softwood stud partition; including plasterboard linings to both sides; approximately 150mm thick overall		m²		
7.1.1.6	Removing softwood floor construction; approximately 200mm thick		m²		
7.1.1.7	Removing infected t&g floor boarding		m²		
7.1.1.8	Removing pipe casings		m		
7.1.1.9	Remove internal doors and frame or lining, including cutting out fixings; cut out and remove flooring, fill opening and bond to existing wall, plaster wall both sides, 19mm × 100mm softwood skirting both sides to match existing; single door opening; 100mm thick blockwork wall		nr		
7.1.1.10	Remove internal doors and frame or lining in timber stud partition, including cutting out fixings; cut out and remove flooring, fill opening with timber studwork and plasterboard and plaster skim to both sides; 19mm × 100mm softwood skirting both sides to match existing; double door opening		nr		
7.1.1.11	Taking out casement windows and frames; disposing of arisings off site; approximate size 1.80m × 1.30m		nr		
7.1.1.12	Taking out complete staircase; timber, including handrail and baluster		nr		
7.1.1.13	Taking up existing edged fixed carpet; including underlay		m²		
7.1.1.14	Hacking up cement screed; approximately 65mm thick		m²		
7.1.1.15	Hacking off existing plaster; to brickwork or blockwork walls; renewing in repairs, including dubbing out		m²		
7.1.1.16	Form new single door opening in existing internal wall; cut opening, provide and build-in concrete lintel over, quoin up jambs; opening size approximately 1.00m × 2.10m; (new door and frame measured separately – element 2.H)		nr		

Example 21.1 *(Continued)*

WORK TO EXISTING BUILDINGS					
Code	**Element/sub-elements/components**	**Qty**	**Unit**	**Rate**	**Total**
				£p	**£**
	Removing sanitary appliances; complete with all associated services, overflows and waste pipes; making good all holes and other work disturbed in readiness for redecoration:				
7.1.1.17	WC suites		nr		
7.1.1.18	Wash-hand basins		nr		
7.1.1.19	Range of six wash-hand basins		nr		
7.1.1.20	Three-stall urinal		nr		
7.1.1.21	Taking down gutters and supports; 112mm diameter uPVC		m		
7.1.1.22	Overhauling existing gutters; cutting out existing joints; adjusting brackets to correct falls; remaking joints; 112mm diameter uPVC		m		
7.1.1	**TOTAL – Minor demolition works and alteration works: to element 7.1 summary**				

21.3 Repairs to existing services (element 7.2)

Element 7.2 comprises a single sub-element as follows:

7.2.1 Existing services.

This sub-element deals with the refurbishment of existing services plant, equipment and installations (e.g. repairs to sanitary appliances, overhauling boilers, upgrading lifts and recommissioning entire installations). Repairs to plant and equipment are measured as enumerated (nr) items. Overhauling entire mechanical and electrical systems (e.g. heating installations, ventilation systems and electrical installations) is measured superficially, in square metres. The nature of the component or type of installation to be repaired or overhauled is to be given in the description. (See Example 21.2.)

Example 21.2 Formulation of descriptions for repairs to existing services

WORK TO EXISTING BUILDINGS					
Code	**Element/sub-elements/components**	**Qty**	**Unit**	**Rate**	**Total**
				£p	**£**
7	**WORK TO EXISTING BUILDINGS**				
7.2	**Repairs to existing services**				
7.2.2	*Existing services*				
7.2.2.1	WC pans; renewing seat and cover		nr		

Example 21.2 *(Continued)*

WORK TO EXISTING BUILDINGS					
Code	**Element/sub-elements/components**	**Qty**	**Unit**	**Rate**	**Total**
				£p	£
7.2.2.2	WC pans; renewing pan connector		nr		
7.2.2.3	WC cistern; renewing fittings		nr		
7.2.2.4	WC cistern; overhauling		nr		
7.2.2.5	Clearing blockages; to shower, including removing and refixing tray		nr		
7.2.2.6	Clearing blockages; to soil stack; 4 storeys		nr		
	Fault finding; including minor renewals or repairs and any necessary testing:				
7.2.2.7	Shower and shower circuit		nr		
7.2.2.8	Immersion and immersion circuit		nr		
7.2.2.9	Boiler circuit and boiler controls		nr		
7.2.2.10	Central heating controls		nr		
7.2.2.11	Shower unit electrics; renewing		nr		
7.2.2.12	Storage water heaters; electric; overhauling		nr		
7.2.2.13	Overhaul and check consumer control unit		nr		
	Draining down, refilling and venting domestic heating and hot water installations, adding corrosion inhibitor to heating installation; testing and setting the system back to work; one visit:		nr		
7.2.2.14	One-bedroom apartment		nr		
7.2.2.15	Two-bedroom apartment		nr		
7.2.2.16	Two-bedroom maisonette		nr		
7.2.2.17	Three-bedroom house		nr		
	Gas installation, complete check, test and report:		nr		
7.2.2.18	One-bedroom apartment		nr		
7.2.2.19	Two-bedroom apartment		nr		
7.2.2.20	Two-bedroom maisonette		nr		
7.2.2.21	Three-bedroom house		nr		
7.2.2.22	Radiator; removing existing and reconnecting, including all necessary adjustments to pipework; adding corrosion inhibitor; venting		nr		
7.2.2.23	Radiator and brackets; removing existing and renewing; double convector panel; 1,200mm × 600mm		nr		
7.2.2.24	Kitchen extractor fans; domestic; overhauling		nr		
7.2.2.25	Cooker hood; industrial; servicing; cleaning throughout, replacing disposable filters; renewing minor parts as necessary		nr		
7.2.2.26	Convector heaters; servicing; renewing minor parts as necessary		nr		

Example 21.2 *(Continued)*

WORK TO EXISTING BUILDINGS					
Code	**Element/sub-elements/components**	**Qty**	**Unit**	**Rate**	**Total**
				£p	**£**
7.2.2.27	Water tanks; cleaning and removing all scale from water tank, approximately 1,500 litre capacity; including draining down and refilling		nr		
7.2.2.28	Combination hot water storage units; indirect, with factory applied insulation; capacity: 150 litres; removing existing and renewing		nr		
7.2.2.29	Fire alarm installations; complete check, test and report; including carrying out minor repairs		nr		
7.2.2.30	Public address installations; complete check, test and report; including carrying out minor repairs		nr		
7.2.1	**TOTAL – Existing services: to element 7.2 summary**				

Decontaminating existing services prior to removal (e.g. boilers and fuel storage tanks) is measured under facilitating works – sub-element 0.1.1 (Toxic or hazardous material removal).

21.4 Damp-proof courses/fungus and beetle eradication (element 7.3)

Element 7.3 is divided into two sub-elements as follows:

7.3.1 Damp-proof courses; and

7.3.2 Fungus and beetle eradication.

21.4.1 Damp-proof courses (sub-element 7.3.1)

The prevention of rising damp in existing walls entails the insertion of a damp-proof course into the wall. Typical methods include injecting chemical or mortar damp-proof courses, or inserting mechanical damp-proof courses. Remedial works in inserting damp-proof courses into existing construction are measured linear, in metres. The extreme length of the wall to be treated is measured over all obstructions, with the type of damp-proof course stated in the description. (See Example 21.3.)

21.4.2 Fungus and beetle eradication (sub-element 7.3.2)

Sub-element 7.3.2 deals with opening up existing work to expose timbers affected by fungus or wood-boring infestation, cutting out defective

Example 21.3 Formulation of descriptions for damp-proof courses

WORK TO EXISTING BUILDINGS					
Code	**Element/sub-elements/components**	**Qty**	**Unit**	**Rate**	**Total**
				£p	**£**
7	**WORK TO EXISTING BUILDINGS**				
7.3	**Damp-proof courses/fungus and beetle eradication**				
7.3.1	*Damp-proof courses*				
7.3.1.1	Silicone injection damp-proofing; 450mm centres; making good brickwork; half brick thick; horizontal		m		
7.3.1.2	Specialist chemical transfusion damp-proof course system; one brick thick; horizontal		m		
7.3.1	**TOTAL – Damp-proof courses: to element 7.3 summary**				

Example 21.4 Formulation of descriptions for fungus and beetle eradication

WORK TO EXISTING BUILDINGS					
Code	**Element/sub-elements/components**	**Qty**	**Unit**	**Rate**	**Total**
				£p	**£**
7	**WORK TO EXISTING BUILDINGS**				
7.3	**Damp-proof courses/fungus and beetle eradication**				
7.3.2	*Fungus and beetle eradication*				
7.3.2.1	Removing cobwebs, dust and roof insulation; de-frass; treating exposed joists/rafters timber with two coats proprietary insecticide and fungicide; by spray application		m²		
7.3.2.2	Treating existing timber floor boarding; with two coats proprietary insecticide and fungicide; by spray application		m²		
7.3.2.3	Lifting existing timber floor boarding as necessary; treating floors with two coats proprietary insecticide and fungicide; by spray application; refixing floor boards		m²		
7.3.2	**TOTAL – Fungus and beetle eradication: to element 7.3 summary**				

timbers and applying preservative treatment (e.g. anti-fungi solutions and insecticide treatments). Reinstatement works should be measured and described under sub-element 7.1.1 (Minor demolition works and alteration works). (See Example 21.4.)

Eradication treatment is measured superficially, in square metres. The area measured is the surface area of the surface to be treated, with no deduction for voids. The nature of the works and treatment method are to be described.

21.5 Facade retention (element 7.4)

Element 7.4 comprises a single sub-element as follows:

7.4.1 Facade retention.

This sub-element deals with the provision of temporary or semi-permanent supports to unstable structures and facades that are to be retained and integrated into the new building (e.g. a series of raking shores upholding a single wall). Example 21.5 shows typical component descriptions for facade retention works.

Example 21.5 Formulation of descriptions for facade retention

WORK TO EXISTING BUILDINGS					
Code	**Element/sub-elements/components**	**Qty**	**Unit**	**Rate**	**Total**
				£p	**£**
7	**WORK TO EXISTING BUILDINGS**				
7.4	**Facade retention**				
7.4.1	*Facade retention*				
	Design and installation of facade retention measures:				
7.4.1.1	Raking shores; triple raker		nr		
7.4.1.2	Dead shore; approximately 3.00m high		nr		
7.4.1.3	Inspection and maintenance of facade retention measures		Item		
	Note:				
	Remedial works/making good to existing structures following removal of temporary facade retention measures measured and described elsewhere under sub-element 7.1 (Minor demolition and alteration works)				
	Removal of facade retention measures, including temporary bases:				
7.4.1.4	Raking shores; triple raker		nr		
7.4.1.5	Dead shore; approximately 3.00m high		nr		
7.4.1	**TOTAL – Facade retention: to element 7.1 summary**				

Temporary or semi-permanent supports to structures adjacent to the site on which the building is being built, including party walls, are measured under sub-element 0.3.1 (Temporary supports to adjacent structures).

Support structures are measured as enumerated (nr) items, with the type of support stated in the description. Components measured are deemed to include cutting holes in existing structures to take the support structure, as well as all foundations. Although not specifically stated by the rules, the removal of support structures is best measured under this sub-element.

Costs in connection with the inspection and maintenance of facade retention measures will also need to be considered and allowed for by the cost manager.

Although not specifically mentioned by the rules, such costs can be significant, especially where they are likely to be in place for a lengthy period. Furthermore, considerable additional costs can also arise where there is a risk of the temporary works impacting on adjoining properties; this is particularly where road, rail networks or other operational environments are involved. The cost manager will need to consider how to address such costs in the cost plan. It is suggested that the inspection and maintenance of facade retention measures are initially treated as a risk item and transferred to sub-element 7.4 (Facade retention) when the actual requirements have been determined.

21.6 Cleaning existing surfaces (element 7.5)

Element 7.5 is divided into two sub-elements as follows:

7.5.1 Cleaning; and

7.5.2 Protective coatings.

21.6.1 Cleaning (sub-element 7.5.1)

Cleaning existing surfaces encompasses removing efflorescence, stains, soot, graffiti, vegetation, algae, bird droppings and the like from existing internal and external surfaces (e.g. from walls, floors, ceilings, roofs, windows and doors).

Work in cleaning and removing stains and deposits from existing surfaces is measured superficially, in square metres. The area measured is the surface area of the surface to be cleaned, with no deduction for voids. The method of cleaning is to be stated in the description (e.g. cleaning by washing, abrasive blasting and chemical treatment).

21.6.2 Protective coatings (sub-element 7.5.2)

Protective coatings comprise specialist painting and coating systems applied to existing internal and external surfaces. They are measured superficially, in square metres. The area measured is the surface area of the surface to be coated, with no deduction for voids. The nature of coating is to be stated in the description (e.g. lime washing, colourless coatings and anti-graffiti coatings).

Bird-repellent coatings to new surfaces are measured under sub-element 4.1.8 (Bird and vermin control).

21.7 Renovation works (element 7.6)

Element 7.6 is divided into five sub-elements as follows:

7.6.1 Masonry repairs;

7.6.2 Concrete repairs;

7.6.5 Metal repairs;

7.6.4 Timber repairs; and

7.6.5 Plastics repairs.

Renovation works will have normally been identified as a result of a condition survey carried out by a building surveyor or structural engineer. In some instances, specialist contractors might have been involved with the survey. Specialist contractors, such as concrete repair specialists, will normally provide a comprehensive list of the repairs that need to be undertaken. Therefore, the cost manager should be able to measure and describe the repair work identified within the resulting condition survey report – by way of schedules of work and photographs. In some instances, the cost manager might need to carry out a visual site inspection to clarify the requirements. With condition surveys and surveys by specialist contractors, it is essential that the cost manager correctly interprets the information provided, taking particular care to identify any omitted information which will have a significant impact on the cost of the building project. Obviously, where the cost manager is concerned that the extent of works specified in reports is inadequate, it is essential that sufficient provision to deal with any unforeseen or unidentified works arising is included in the appropriate risk allowance.

Owing to the varied nature of renovation works that can be encountered, the rules leave the method of measurement to the discretion of the cost manager. Consequently, repairs to components can be measured as enumerated (nr), linear (in metres), or superficial (in square metres) items. For components measured linear, the length is the extreme length measured over all obstructions. Where measured superficially, the surface area of the repair is measured. Descriptions are to identify the component to be renovated and the nature of the repair works.

Where a component is to be replaced in its entirety (e.g. a window, a door, roof coverings or cladding), the removal of the component is measured under sub-element 7.1.1 (Minor demolition works and alteration works). The replacement of the component is to be measured under the appropriate element or sub-element, e.g. replacement windows will be measured under sub-element 2.6.1 (External windows).

21.6.1 Masonry repairs (sub-element 7.6.1)

This covers repairs to brickwork, blockwork and stonework. Typical items include cutting out and replacing isolated, or patches of, bricks, blocks and stones, stitching brickwork where cracks have appeared, plastic repairs, redressing stonework, inserting new wall ties into existing cavity walls and repointing. (See Example 21.6.)

Example 21.6 Formulation of descriptions for masonry repairs

WORK TO EXISTING BUILDINGS					
Code	**Element/sub-elements/components**	**Qty**	**Unit**	**Rate**	**Total**
				£p	**£**
7	**WORK TO EXISTING BUILDINGS**				
7.6	**Renovation works**				
7.6.1	*Masonry repairs*				
7.6.1.1	Cutting out defective facing bricks; PC £350.00/1,000 bricks; facing and pointing one side to match existing; isolated bricks; half brick thick		nr		

Example 21.6 *(Continued)*

WORK TO EXISTING BUILDINGS					
Code	**Element/sub-elements/components**	**Qty**	**Unit**	**Rate**	**Total**
				£p	**£**
7.6.1.2	Cutting out defective facing bricks; PC £350.00/1,000 bricks; facing and pointing one side to match existing; small patches; half brick thick		m²		
7.6.1.3	Cutting out staggered cracks in brickwork facing bricks and repointing to match existing along joints		m		
7.6.1.4	Cutting out raking cracks in brickwork; stitching in new facing bricks, PC £350.00/1,000; facing and pointing one side to match existing; half brick thick		m²		
7.6.1.5	Removing defective parapet wall, 600mm high; rebuilding in new facing bricks, PC £350.00/1,000, with precast concrete coping stone; one-brick-thick parapet wall		m		
7.6.1.6	Taking down brick external skin of cavity wall and setting aside for reuse; removing existing insulation, and cutting out defective wall ties; cleaning salvaged facing bricks and rebuilding external skin, pointed one side, including inserting replacement wall ties and new cavity insulation; half-brick thick		m²		
7.6.1	**TOTAL – Masonry repairs: to element 7.6 summary**				

21.6.2 Concrete repairs (sub-element 7.6.2)

This sub-element deals with cutting out and repairing defective concrete, which includes the treatment of exposed, or replacement of defective, reinforcement. With this type of work, it is more than likely that one or more concrete repair specialists will have surveyed the building or structure concerned. As a result, the cost manager should be able to measure and describe the concrete repair works from the specialists' survey reports.

21.6.3 Metal repairs (sub-element 7.6.3)

This comprises the renovation of existing metal components, including straightening and restoring components. Renovation includes repairs to architectural metalwork (e.g. welding and replacing missing components), structural steel members (e.g. roof members and structural beams), windows, doors, frames and linings, roof lights, staircases, handrails and balustrades.

21.6.4 Timber repairs (sub-element 7.6.4)

Repairs to timber components, including repairs to structural members (e.g. roof members and structural beams) are dealt with under this sub-element. Typical works include the cutting out of defective timber and piecing-in new, replacement of defective sub-components (e.g. a complete rafter) and resin repairs to timbers.

21.6.5 Plastics repairs (sub-element 7.6.5)

This encompasses repairs to plastics components, such as uPVC windows and doors, rooflights and cladding.

CHAPTER 22

External works (group element 8)

Introduction

This chapter:

- provides step-by-step guidance on how to measure and describe external works; and
- contains examples illustrating how to formulate descriptions for components.

22.1 Method of measurement

The rules for measuring external works are found in group element 8 of NRM 1, which is divided into eight elements as follows:

8.1 Site preparation works;
8.2 Roads, paths, pavings and surfacing;
8.3 Soft landscaping, planting and irrigation systems;
8.4 Fencing, railings and walls;
8.5 External fixtures;
8.6 External drainage;
8.7 External services; and
8.8 Minor building works and ancillary buildings.

22.1.1 Works undertaken outside the curtilage of the site

NRM 1 stipulates that works to be undertaken outside the curtilage of the site are to be measured and described separately. This can be done by way of clearly annotating such works in the cost plan.

22.1.2 Contaminated land and material

The treatment of contaminated land and the disposal of contaminated ground material are both dealt with under facilitation works – sub-element 0.1.2 (Contaminated land).

22.2 Site preparation works (element 8.1)

This element is divided into two sub-elements as follows:

8.1.1 Site clearance; and

8.1.2 Preparatory groundworks.

Major demolition works and site preparation works of a specialist nature are measured and described under group element 0 (Facilitating works).

22.2.1 Site clearance (sub-element 8.1.1)

This sub-element covers preparatory works required to clear existing site vegetation, trees and the like, as well as the application of herbicides over the site, before commencement of construction works. It also deals with minor demolition works (i.e. the demolition of outbuildings, sheds and the like). Specialist works involving the removal or chemical treatment of highly invasive plants or vegetation such as Japanese knotweed, giant hogweed, Japanese seaweed and giant kelp are measured under sub-element 0.1.3 (Eradication of plant growth).

The timing of site clearance work will need to be considered, as will its impact on the construction programme. This is because such works are often linked to planning conditions, in which the planning authority is likely to dictate the time of year when the vegetation clearance and tree felling can be carried out (e.g. 'after 1 November and before 31 March to avoid the bird-breeding season and compatible with the reptile mitigation works'). This often results in such works being procured ahead of the main construction works (i.e. as part of a separate enabling works contract).

Clearing existing site vegetation is measured superficially, in square metres, with the actual area of the site covered by vegetation measured. The type and nature of the vegetation to be removed (e.g. shrubs and undergrowth) should be stated in the description, together with the method to be used (i.e. mechanical or hand clearance). Cutting and stripping grassed areas for turves can be treated as a cost-significant item under measurement rule C4. It is recommended that such works be measured in the same way as clearing vegetation.

Taking down and disposing of trees and the grubbing up and disposal of tree stumps, including the roots, are separately measured and described as enumerated (nr) items. They are measured as separate items because they are two discrete activities, which are often carried out at different times and by different works contractors. The method of disposing of tree arisings should be considered: for example, can the trees simply be converted into chippings, or will the logs, or proportion of the logs, be required to be used within the building project (e.g. used as wildlife habitats)? Descriptions for tree and stump removal should include the indicative girth of the trees to be felled.

Temporary fencing or other measures required to safeguard 'protected trees', or other existing trees that are to be kept as part of the finished scheme, is measured as an enumerated (nr) item, with the type of protection measures described.

Minor demolition works under this sub-element comprise small outbuildings (e.g. sheds, workshops and bicycle racks). Works can be itemised or measured as enumerated (nr) items, with the type of outbuilding stated in the description. Other items, such as taking down free-standing walls and the like can be treated as other cost-significant items (i.e. measured in accordance with measurement rule C4).

The application of herbicides is measured superficially, in square metres, with the actual area to which the herbicide is applied measured. Whether herbicides are applied by hand or by mechanical means should be stated in the description, as this will undoubtedly impact on cost.

Descriptions for typical site clearance works are shown in Example 22.1.

Example 22.1 Formulation of descriptions for site clearance

EXTERNAL WORKS					
Code	**Element/sub-elements/components**	**Qty**	**Unit**	**Rate**	**Total**
				£p	**£**
8	**EXTERNAL WORKS**				
8.1	**Site preparation works**				
8.1.1	*Site clearance*				
8.1.1.1	Clearing generally; light fencing and gates, general rubbish and debris		m²		
8.1.1.2	Clearing generally; light fencing and gates		m²		
8.1.1.3	Clearing vegetation, including shrubs and hedges; heavy weed growth; dispose of arisings off site		m²		
8.1.1.4	Taking down trees; dispose of arisings off site; not exceeding 1.00m girth		nr		
8.1.1.5	Taking down trees; dispose of arisings off site; 1.00 to 2.00m girth		nr		
8.1.1.6	Removing tree stumps and roots; dispose of arisings off site; not exceeding 1.00m girth		nr		
8.1.1.7	Removing tree stumps and roots; dispose of arisings off site; 1.00m to 2.00m girth		nr		
8.8.1.8	Tree protection; group of 7 nr trees; using 'Herras' type fencing or chain link fencing; 1.80m high; approx. length: 18.00m		item		
8.1.1.9	Taking down single-storey timber workshop; dispose off site; plan area approximately 2.40m × 5.40m		nr		
8.8.1.10	Demolishing; two-storey brick outbuilding, with timber joisted suspended floor and timber flat roof; including grubbing up foundations; dispose of arisings off site; size 9.60m × 3.20m on plan		nr		
8.1.1.11	Taking down; chain link fencing and gates; dispose of arisings off site; 1.80m high		m		
8.1.1.12	Demolishing; free-standing brick screen wall, 225mm thick; including grubbing up foundations; dispose of arisings off site; approximately 1.50m high		m		
8.1.1.13	Applying herbicides		m²		
8.1.1	**TOTAL – Site clearance: to element 8.1 summary**				

22.2.2 Preparatory groundworks (sub-element 8.1.2)

Excavation and earthworks to form new site contours, and to adjust existing site levels, together with the grubbing up of existing building substructures, hardstandings (including roads, pavings, paths, car park surfaces and the like), underground drainage and storage tanks are measured under this sub-element (see Example 22.2).

Forming new site contours (grading) and adjusting existing site levels is measured superficially, in square metres, with the surface area that is to be reprofiled measured. Preparatory works involving the removal or treatment of contaminated material from the ground prior to the commencement of construction works are measured under sub-element 0.1.2 (Contaminated land).

Breaking out, or grubbing up, of existing substructures (including ground slabs, strip foundations, basement retaining walls and the like) is measured superficially, in square metres, with the surface area of the building footprint measured (i.e. measured to the external face of the external walls). The nature of the work being removed is to be stated in the description. Although not identified as a discrete component by the rules, the extraction of existing piles can be treated as a cost-significant component and separately measured and described. The extraction of existing piles can be measured as either enumerated items (e.g. for discrete ground-bearing piles), or as linear or superficial items (e.g. for a continuous piled wall).

Grubbing up foundations to old garden walls or the like, and breaking out existing retaining structures, can also be treated as other cost-significant components and separately measured and described. Such works are probably best measured as linear items, in metres, measured along their extreme length.

Breaking out existing hardstandings and hard pavings (including concrete, bituminous bound material, brick, block and other hard materials) is measured superficially, in square metres, with the surface area of the component to be removed measured. The nature of the work being removed is to be described.

The disposal of materials arising from breaking out existing substructures, wall foundations, retaining structures, hardstandings, hard pavings and drainage is deemed to be included in the items measured. It should also be noted that certain materials will be classified as contaminated (e.g. bituminous bound material), which will attract a higher disposal charge.

Grubbing up redundant foul and surface water drainage runs is measured linear, in metres, measured along their extreme length. The type and size of drain should be described.

Grubbing up manholes, soakaways, catch pits, interceptors and the like is enumerated (nr), with the type of component stated in the description. Similarly, filling disused manholes, shafts and the like is also measured as enumerated (nr) items, with the filling material to be used stated in the description (e.g. lean mix concrete or granular material).

Removing existing underground storage tanks, including disposal and decontamination, is measured as enumerated (nr) items. The description is to include the type of tank to be removed (e.g. septic tank).

Example 22.2 Formulation of descriptions for preparatory groundworks

EXTERNAL WORKS					
Code	**Element/sub-elements/components**	**Qty**	**Unit**	**Rate**	**Total**
				£p	**£**
8	**EXTERNAL WORKS**				
8.1	**Site preparation works**				
8.1.2	*Preparatory groundworks*				
8.1.2.1	Cutting and stripping turves; store on site for reuse		m²		
8.1.2.2	Forming new site contours; excavating to reduce levels; average 350mm deep; disposal of excavated material off site; including grading excavated surface to receive subsequent treatments; heavy soils		m²		
8.1.2.3	Forming new site contours; filling to raise levels; average 250mm deep; 'inert' vegetable soil, obtained off site; including compacting and grading filled surface to receive subsequent treatments; heavy soils		m²		
8.1.2.4	Forming new site contours; filling to raise levels; average 150mm deep; 'inert' topsoil, obtained off site; including compacting and grading filled surface to receive subsequent treatments; loamy topsoil		m²		
8.1.2.5	Forming landform; from 'inert' imported recycled topsoil; mounds to falls and grades to receive turf or seeding treatment (measured separately); volume 150m³ spread over 300m² area; varying thickness: 600mm to 150mm		m²		
8.1.2.6	Breaking out existing substructures; by machine		m²		
8.1.2.7	Breaking out existing pavings; reinforced concrete slabs, approximately 300mm thick; dispose off site		m²		
8.1.2.8	Grubbing up old drainage pipelines; concrete pipes, including concrete surround		m		
8.1.2.9	Grubbing up old manholes; brick construction		nr		
8.1.2.10	Filling disused manholes; with inert material		nr		
	Note: Removal of hazardous material from underground oil storage tanks measured under sub-element 0.1.1 (Toxic and hazardous material removal)				
8.1.2.11	Removing existing underground oil storage tanks; filling voids with inert material		nr		
8.1.2	**TOTAL – Site preparation works: to element 8.1 summary**				

22.3 Roads, paths, pavings and surfacings (element 8.2)

This element is divided into two sub-elements as follows:

8.2.1 Roads, paths and pavings; and
8.2.2 Special surfacings and pavings.

22.3.1 Roads, paths and pavings (sub-element 8.2.1)

This sub-element covers all types of roads, paths and pavements (i.e. large paved areas such as hardstandings, amenity areas and car parks). It also covers non-specialist surfacings and pavings used for sports and general amenities, and perforated pavings providing protection to grassed areas (e.g. to form roads, paths and car parking areas).

Roads and paths are both measured as linear items, in metres, measured along their extreme length over all obstructions. Descriptions for roads and paths are to include the width, together with details of their make-up (including details of kerbs, kerb channels and path edgings) and overall thickness of the road or path construction.

Pavements are measured superficially, in square metres, with the composition of the pavement construction given in the description. Where considered a cost-significant item, sub-components such as paving slab cycle stands should be described as an 'extra over' item to the component to which they relate.

Roundabouts, road crossings (including zebra crossings and pelican crossings), steps, ramps and tree grilles are measured and described as enumerated items, and the main sub-components of these components are to be stated in the description. For roundabouts, the type and approximate size should also be given.

Traffic-calming measures can be either measured as enumerated (nr – e.g. traffic island) or linear items (m – e.g. proprietary bolt-down speed ramps or bumps). The type and, if applicable, the main sub-components of the traffic-calming measure are to be included in the description. Speed ramps, or bumps, formed of the same material as the road should be measured as an 'extra over' item, and linked to the item to which they relate.

Vehicle protection barriers and bumpers are measured linear, in metres, with their composition stated in the description.

Pavement markings can be either be measured as enumerated (nr) or linear items, in metres. Line markings on roads (e.g. double yellow lines or central hazard warning lines) or to delineate car parking spaces within a car park, are measured linear along their extreme length over all obstructions. Numerals and symbols are enumerated. The actual numbers of digits forming each number are counted; for example, number 13 equals two numbers, 116 equals three numbers, and so on. The type of symbol (e.g. rumble strip markings, directional arrows and disabled parking symbols) is to be stated in the description. The description for pavement markings is to state the material from which they are to be formed (e.g. chlorinated rubber paint or thermoplastic).

Repairs to existing roads, paths and pavings can be either measured as enumerated (nr), linear, in metres, or superficial items, in square metres. The nature of the repair is to be stated in the description.

Example 22.3 provides descriptions for common roads, paths and paving works items.

Example 22.3 Formulation of descriptions for roads, paths and pavings

EXTERNAL WORKS					
Code	**Element/sub-elements/components**	**Qty**	**Unit**	**Rate**	**Total**
				£p	**£**
8	**EXTERNAL WORKS**				
8.2	**Roads, paths and pavings**				
8.2.1	*Roads, paths and pavings*				
8.2.1.1	Roads; 150mm thick reinforced concrete road bed, with fibre board expansion joints, with sealant, at 30m centres; on 150mm thick blinded hardcore; 155mm × 255mm kerbs to both sides, including foundations haunched one side; including all excavation and disposing of surplus excavated material off site; 4.90m wide		m		
8.2.1.2	Paths; 600mm × 600mm × 50mm thick concrete slab pavings, exposed aggregate; on blinded Type 1 granular material, 100mm thick; including all excavation and disposing of surplus excavated material off site; 1.60m wide		m		
8.2.1.3	Paths; 50mm thick cedec gravel, watered and rolled; on 25mm thick sand blinding and geofabric; on 150mm thick hardcore; fixed timber edging to both sides; including all excavation and disposing of surplus excavated material off site; 0.60m wide		m		
8.2.1.4	Steps; form steps to cedec gravel path; approximately		nr		
8.2.1.5	Paved areas; York stone pavings, 38mm thick, laid random rectangular patterns; on blinded Type 1 granular material, 150mm thick; including all excavation and disposing of surplus excavated material off site		m²		
8.2.1.6	Paved areas; pedestrian deterrent paving; chamfered studs, 600mm × 600mm × 60mm thick; on blinded Type 1 granular material, 150mm thick; including all excavation and disposing of surplus excavated material off site		nr		
8.2.1.7	Roundabouts; mini-roundabout; domed; standard size		nr		
8.2.1.8	Traffic calming; surface speed cushion; 1,600mm × 2,000mm		nr		
8.2.1.9	Traffic calming; raised table, rubber; flat top length 2.80m; overall length 4.00m		nr		
8.2.1.10	Traffic calming; surface-mounted traffic control plates; bolted to concrete		nr		
8.2.1.11	Tree grilles; decorative cast iron square tree grille; 1,000mm × 1,000mm		nr		

Example 22.3 *(Continued)*

EXTERNAL WORKS					
Code	**Element/sub-elements/components**	**Qty**	**Unit**	**Rate**	**Total**
				£p	**£**
8.2.1.12	Vehicle protection barriers; sectional barrier (designed for low-speed applications), mild steel, galvanised; including foundations for posts		m		
8.2.1.13	Pavement markings; thermoplastic; 100mm wide; line		m		
8.2.1.14	Pavement markings; thermoplastic; 100mm wide; intermittent line		m		
8.2.1.15	Repairs to existing roads; filling pot hole; not exceeding 1.00m²		nr		
8.2.1.16	Repairs to existing roads; cracks; filling with latex rubber bitumen emulsion		m		
8.2.1	**TOTAL – Roads, paths and pavings: to element 8.2 summary**				

22.3.2 Special surfacings and pavings (sub-element 8.2.2)

Surfacings and pavings specially and specifically designed for outdoor sporting activities and general amenities are covered by this sub-element. Special surfacings and pavings designed for outdoor use include:

- sheet and liquid applied surfacings (e.g. synthetic rubber, granulated rubber, plastics and fibre);
- synthetic tufted surfacings for ski slopes; and
- proprietary coloured no fines concrete and clay/shale surfacings and pavings.

Special surfacings and pavings are measured superficially, in square metres, with the composition of the pavement given in the description. Pavement markings on special surfacings and pavings (e.g. marking out sports pitches) are deemed to be included in the item measured. Typical component descriptions for specialist surfacing and pavings items are illustrated in Example 22.4.

Example 22.4 Formulation of descriptions for special surfacings and pavings

EXTERNAL WORKS					
Code	**Element/sub-elements/components**	**Qty**	**Unit**	**Rate**	**Total**
				£p	**£**
8	**EXTERNAL WORKS**				
8.2	**Roads, paths and pavings**				
8.2.2	*Specialist surfacings and pavings*				
8.2.2.1	Safety surfacing; 35mm thick 'Ruberflex', coloured; on 40mm thick macadam base; on blinded Type 1 granular material, 150mm thick; including all		m²		

Example 22.4 *(Continued)*

EXTERNAL WORKS					
Code	**Element/sub-elements/components**	**Qty**	**Unit**	**Rate**	**Total**
				£p	**£**
	excavation and disposing of surplus excavated material, and applying herbicide to substrate				
8.2.2.2	Safety surfacing around play equipment; 150mm thick bark particles, on 150mm thick hardcore bed; including applying herbicide to substrate	48	m²	16.00	
8.2.2	**TOTAL – Specialist surfacings and pavings: to element 8.2 summary**				

22.4 Soft landscaping, planting and irrigation systems (element 8.3)

This element is divided into three sub-elements as follows:

8.3.1 Seeding and turfing;

8.3.2 External planting; and

8.3.3 Irrigation systems.

22.4.1 Seeding and turfing (sub-element 8.3.1)

This sub-element deals with the preparation of soil and seeding or turfing to form lawns, parklands and other general grassed areas. It also covers seeding and turfing to retaining structures. Typical component descriptions for seeding and turfing works are shown in Example 22.5.

Example 22.5 Formulation of descriptions for seeding and turfing

EXTERNAL WORKS					
Code	**Element/sub-elements/components**	**Qty**	**Unit**	**Rate**	**Total**
				£p	**£**
8	**EXTERNAL WORKS**				
8.3	**Soft landscaping, planting and irrigation systems**				
8.3.1	*Seeding and turfing*				
	Note: Imported vegetable soil for turfing or seeding measured elsewhere under sub-element 8.1.2 (Preparatory groundworks)				
8.3.1.1	Grassed areas; imported turf		m²		
8.3.1.2	Grassed areas; preserved turf from stack on site		m²		

Example 22.5 *(Continued)*

EXTERNAL WORKS					
Code	**Element/sub-elements/components**	**Qty**	**Unit**	**Rate**	**Total**
				£p	**£**
8.3.1.3	Grassed areas; seeding		m²		
8.3.1.4	Reinforced grass; turf-reinforcing mesh		m²		
	Note:				
	All excavation, disposal of surplus excavated material and preparatory works measured elsewhere under sub-element 8.1.2 (Preparatory groundworks)				
8.3.1.5	Reinforced grass paving grid; plastic; grids filled with a sand–soil rootzone; seeding		m²		
	Marking-out of grass sports pitches:				
8.3.1.6	Football, 114.00m × 72.00m		nr		
8.3.1.7	Rugby union, 156.00m × 81.00m		nr		
8.3.1.8	Work to existing grassed areas; re-seeding		m²		
8.3.1.9	Maintenance of grassed areas; aerate ground with spiked aerator and apply fertiliser; (1 nr occasion)		m²		
8.3.1.10	Maintenance of grassed areas; initial cutting of grassed areas		m²		
8.3.1.11	Maintenance of grassed areas; subsequent cutting of grassed areas; period 52 weeks (say, 26 nr cuts)		m²		
8.3.1	**TOTAL – Seeding and turfing: to element 8.3 summary**				

Preparatory works for areas to be grassed include:

- applying of herbicides;
- providing of topsoil, including transporting from stockpiles or importing topsoil and spreading;
- cultivating topsoil, including removing stones and weeds;
- fine grading of topsoil;
- providing, spreading and working in manure, compost, mulch, fertilised, soil ameliorants and the like; and
- providing light mesh reinforcement.

Grassed areas are measured superficially, in square metres, with the surface area of the area to be grassed measured. The area is measured over all obstructions; however, areas of roads, paths and pavings bisecting grassed areas are to be deducted. The description for grassed areas is to include the nature of the preparatory work and the type of seed (e.g. general purpose lawn seed, fine leaf ornamental lawn seed or agricultural forage mixtures) or quality of turf to be used. Where seeding or turfing is to retaining structures, this is to be stated in the description.

Similarly, reinforced grass proprietary systems are also measured superficially, in square metres, in the same way as for grassed areas.

The initial marking-out of grass sports pitches (e.g. football, rugby, hockey and cricket pitches, and tennis courts) is measured as enumerated items.

Remedial works to existing grassed areas is measured superficially, in square metres, with the surface area to be treated measured. The description is to state the type of remedial work required (e.g. scarifying, forking, fertilising, applying weedkillers, local re-seeding or re-turfing).

Maintenance of grassed areas during the period for rectifying defects (or defects liability period or maintenance period, whichever term is applicable) includes:

- watering;
- replacement seeding and turfing; and
- initial grass-cutting work.

Maintenance of grassed areas is measured superficially, in square metres. The description for maintenance of grassed areas is to include the nature of the maintenance activities and the overall maintenance period, including the number of occasions on which the activity is required to be completed. Maintenance work might also include the re-marking-out of grass sports pitches during the defects liability period. Re-marking grass sports pitches should be measured as enumerated (nr) items in the same way as for the initial marking-out of grass sports pitches.

22.4.2 External planting (sub-element 8.3.2)

This sub-element deals with the preparation of soil and planting bulbs, corms, tubers, herbaceous plants, trees, hedges, shrubs and reed beds. It also covers planting to retaining structures.

Preparatory works for areas to be planted include:

- applying of herbicides;
- providing of topsoil, including transporting from stockpiles or importing topsoil and spreading;
- cultivating topsoil, including removing stones and weeds;
- fine grading of topsoil;
- forming raised and sunken beds, borders and the like;
- providing, spreading and working in manure, compost, mulch, fertiliser, soil ameliorants and the like; and
- overlays, including mulch matting, gravel, bark or other materials.

Planting of bulbs and shrubs, and planting of reed beds, are measured superficially, in square metres, with the surface area of the area to be planted measured. The area is measured over all obstructions; however, areas of roads, paths and pavings bisecting areas to be planted are to be deducted. The description for planted beds, and reed beds, are to include the nature of the preparatory work and the type of plants, together with details of any plant-retaining structures (e.g. support wires for climbing plants). Where planting is to retaining structures, or roof gardens, this should be stated in the description. Likewise, plants or reeds planted in prefabricated plant containers are to be so described.

Hedges are measured linear, with the species of hedge (e.g. bare root, hawthorn hedge, pyracantha, red berried, or privet hedge (*Ligustrum ovalifolium*)), the age of the hedge plants and details of any supporting fences stated in the description.

Trees are measured and described by enumerated (nr) items, stating the type, age (e.g. nursery stock or semi-mature trees), the and size (girth and height) of the tree, and details of any tree stakes, tree guards, wrapping and other protection measures. Trees planted in prefabricated tree containers are to be described accordingly.

Similarly, woodland planting is also measured superficially, in square metres, stating the type, age and size of the trees to be planted, and any tree protection measures.

Subsidiary items associated with trees, shrubs and plants such as support wires for climbers, tree stakes, tree guards, wrapping and labelling are to be referred to in the description of the measured item.

Protecting new planted areas with temporary fencing, boards, tarpaulins and the like should be treated as other cost-significant items and described and measured as either enumerated, linear, in metres, or superficial items, in square metres, as applicable.

Costs in connection with support wires for climbers, tree stakes, tree guards, wrapping, labelling and other protection of trees, shrubs and plants are to be allowed for in the items measured.

Works to existing trees that are normally carried out by a specialist (e.g. crown reduction, crown thinning, pollarding, and deadwood and branch removal) are measured by enumerated (nr) items, explaining what works are to be carried out in the description.

Maintenance of planting beds, reed beds, trees and woodland planting areas during the period for rectifying defects (or defects liability period or maintenance period, whichever term is applicable) includes:

- applying anti-desiccants;
- watering;
- replacement planting; and
- pruning.

Descriptions for maintenance works include the nature of the maintenance activities and the overall maintenance period, including the number of occasions on which the activity is required to be completed. Maintenance of planting beds, reed beds and woodland planting are measured superficially, with the surface area of the area to be maintained measured. The maintenance of trees is measured using enumerated (nr) items, whereas the maintenance of hedges is measured as linear items, in metres. Typical descriptions for external planting works are shown in Example 22.6.

Example 22.6 Formulation of descriptions for external planting

EXTERNAL WORKS					
Code	**Element/sub-elements/components**	**Qty**	**Unit**	**Rate**	**Total**
				£p	**£**
8	**EXTERNAL WORKS**				
8.3	**Soft landscaping, planting and irrigation systems**				
8.3.2	*External planting*				
8.3.2.1	Planting; general planting		m²		
8.3.2.2	Planting; dense planting plants		m²		
8.3.2.3	Planting; shrubbed planting		m²		

Example 22.6 *(Continued)*

EXTERNAL WORKS					
Code	**Element/sub-elements/components**	**Qty**	**Unit**	**Rate**	**Total**
				£p	**£**
8.3.2.4	Planting; shrubbed area, including allowance for small trees		m²		
8.3.2.5	Hedges; plants 0.75m to 1.00m high, average 3 plants per metre		m		
8.3.2.6	Trees; advanced nursery stock; (PC Sum £350/nr supply only), in pit		nr		
8.3.2.7	Trees; semi-mature trees; 5.00m to 8.00m high; (PC Sum £1,800/nr supply only), in pit		nr		
8.3.2.8	Tree surgery; crown reduction		nr		
8.3.2.9	Tree surgery; branch removal		nr		
8.3.2.10	Maintenance works to plants, shrubs and planting beds; remulch at start of planting season		m²		
8.3.2	**TOTAL – External planting: to element 8.3 summary**				

22.4.3 Irrigation systems (sub-element 8.3.3)

Irrigation systems to landscape-planted areas, or crop-planted areas, providing a water supply for growing purposes, are measured superficially, in square metres. The area measured is the surface area of the land serviced by the irrigation system. The composition of the irrigation system is to be given in the description. Sample descriptions of typical irrigation systems are shown in Example 22.7.

Example 22.7 Formulation of descriptions for irrigation systems

EXTERNAL WORKS					
Code	**Element/sub-elements/components**	**Qty**	**Unit**	**Rate**	**Total**
				£p	**£**
8	**EXTERNAL WORKS**				
8.3	**Soft landscaping, planting and irrigation systems**				
8.3.3	*Irrigation systems*				
8.3.3.1	Irrigation systems; drip irrigation with automatic controls; connected to grey water system; to raised planters		m²		
8.3.3.2	Irrigation systems; automatic irrigation system; to bowling green		m²		
8.3.3.3	Testing of installations		%		
8.3.3.4	Commissioning of installations		%		
8.3.3	**TOTAL – Irrigation systems: to element 8.3 summary**				

The testing and commissioning of irrigation systems are measured as separate items, being measured as a percentage addition to the irrigation system costs. The steps in calculating the cost target for testing are as follows:

Step 1 Determine the percentage addition to be applied in respect of testing.

Step 2 Ascertain the sum total of the cost targets for all components comprising the irrigation system.

Step 3 Ascertain the cost target for testing by multiplying the sum total of the components comprising the irrigation system by the percentage applicable for testing.

Note:
The equation for calculating the cost target for testing is:

$$t = a \times p$$

Where:
a = sum total of the components comprising the irrigation system
p = percentage addition for testing
t = estimated cost of testing.

The cost target for commissioning is calculated in the same way as for testing. However, it should be noted that the cost target ascertained for testing is not included as part of the sum total of the cost targets for all components comprising the irrigation system.

Note:
The equation for calculating the cost target for commissioning is:

$$c = a \times p$$

Where:
a = sum total of the components comprising the irrigation system (excluding the cost target for testing)
p = percentage addition for commissioning
c = estimated cost of commissioning.

22.5 Fencing, railings and walls (element 8.4)

This element is divided into four sub-elements as follows:

8.4.1 Fencing and railings;
8.4.2 Walls and screens;
8.4.3 Retaining walls; and
8.4.4 Barriers and guardrails.

22.5.1 Fencing and railings (sub-element 8.4.1)

Fencing and railings are measured linear, in metres, measured along their extreme length over all obstructions. Descriptions for fencing and railings are to include their composition and height. Gates are measured as enumerated (nr) items, stating their composition and height in the description. Typical descriptions for fencing and railings planting works are provided in Example 22.8.

Example 22.8 Formulation of descriptions for fencing and railings

Code	Element/sub-elements/components	Qty	Unit	Rate	Total
EXTERNAL WORKS					
				£p	£
8	**EXTERNAL WORKS**				
8.4	**Fencing, railings and walls**				
8.4.1	*Fencing and railings*				
8.4.1.1	Fencing; chain link, plastic coated; concrete posts, 1.80m high		m		
8.4.1.2	Fencing; timber close boarded; oak posts and gravel board; 1.80m high		m		
8.4.1.3	Railings; RoSPA-approved anti-trap bow top fencing, mild steel, high-visibility coating; 1.20m high		m		
8.4.1.4	Gates; RoSPA-approved self-closing gate; to match anti-trap bow top fencing; with a dog grid to prevent animal access; high visibility coating		nr		
8.4.1	**TOTAL – Fencing and railings: to element 8.4 summary**				

22.5.2 Walls and screens (sub-element 8.4.2)

Walls and screens are measured linear, in metres, measured along their extreme length over all obstructions. Descriptions for walls and screens are to include their composition and height. Again, gates are measured as enumerated (nr) items, stating their composition and height in the description. Component descriptions for walls and screens are shown in Example 22.9.

Example 22.9 Formulation of descriptions for walls and screens

Code	Element/sub-elements/components	Qty	Unit	Rate	Total
EXTERNAL WORKS					
				£p	£
8	**EXTERNAL WORKS**				
8.4	**Fencing, railings and walls**				
8.4.2	*Walls and screens*				
8.4.2.1	Walls; brickwork, 215mm thick, with brick on edge coping, including piers and foundations; (PC Sum £300/1,000 bricks supply only); above-ground height: 1.80m		m		
8.4.2.2	Screen walls; brickwork, 102.5mm thick, with precast concrete copings, including piers and foundations; (PC Sum £300/1,000 bricks supply only); above-ground height: 1.20m		m		
8.4.2.3	Gates; decorative steel, painted; single, 1.80m high; including posts; (PC Sum £250/nr supply only)		nr		

Example 22.9 *(Continued)*

EXTERNAL WORKS					
Code	**Element/sub-elements/components**	**Qty**	**Unit**	**Rate**	**Total**
				£p	**£**
8.4.2.4	Gates; timber courtyard gate, pair; 1.80m high; including posts; (PC Sum £1,200/nr supply only)		nr		
8.4.2	**TOTAL – Walls and screens: to element 8.4 summary**				

22.5.3 Retaining walls (sub-element 8.4.3)

This sub-element deals with retaining walls that are not an integral part of the building – external retaining walls. Types of retaining wall include in situ concrete retaining walls, concrete block retaining walls, grass concrete bank revetments, gabion walls, timber log retaining walls, crib walls and bank stabilisation erosion control mats.

Retaining walls are measured linear, in metres, along their extreme length over all obstructions. Descriptions for retaining walls are to include their composition, including copings and details of any drainage. Where applicable, the height of the retaining wall is also to be given. Piles, which are an integral part of the external retaining walls, are not measured separately, but are to be included in the description of the retaining wall construction. Items measured include all excavation and earthworks, including the disposal of all surplus excavated material.

Temporary works required to facilitate the construction of retaining walls are itemised, explaining the nature of temporary works (e.g. propping) in the description. Component descriptions for common retaining walls constructions are illustrated in Example 22.10.

Example 22.10 Formulation of descriptions for retaining walls

EXTERNAL WORKS					
Code	**Element/sub-elements/components**	**Qty**	**Unit**	**Rate**	**Total**
				£p	**£**
8	**EXTERNAL WORKS**				
8.4	**Fencing, railings and walls**				
8.4.3	*Retaining walls*				
8.4.3.1	Reinforced concrete retaining walls; including foundations, expansion joints, granular fill with 100mm land drain; profiled formwork finish to one side); above-ground height: 6.00m high		m		
8.4.3.2	Gabion walls; gabions packed with broken stone; 1.00m thick; including earth anchors); above-ground height: 1.50m high		m		

Example 22.10 *(Continued)*

EXTERNAL WORKS					
Code	**Element/sub-elements/components**	**Qty**	**Unit**	**Rate**	**Total**
				£p	**£**
8.4.3.3	Precast concrete crib walls; including stabilisation works and foundations); above-ground height: average 3.60m high		m		
8.4.3.4	Precast reinforced unit retaining walls; including foundations, proprietary precast concrete units, joints, reinforcement, granular fill with 100mm land drain, earth anchors; including foundations); above-ground height: 2.40m high		m		
8.4.3	**TOTAL – Retaining walls: to element 8.4 summary**				

22.5.4 Barriers and guardrails (sub-element 8.4.4)

Vehicle and pedestrian control barriers and guardrail systems, including associated gates, are measured as linear items, in metres, with their composition, including details of any parapets, stated in the description. Details of any site-applied paint or coatings should also be described.

Components such as hoop barriers, which are used for pedestrian and traffic route delineation, should be treated as other cost-significant items and measured using an appropriate unit.

(See Example 22.11.)

Example 22.11 Formulation of descriptions for barriers and guardrails

EXTERNAL WORKS					
Code	**Element/sub-elements/components**	**Qty**	**Unit**	**Rate**	**Total**
				£p	**£**
8	**EXTERNAL WORKS**				
8.4	**Fencing, railings and walls**				
8.4.4	*Barriers and guardrails*				
8.4.4.1	Vehicle control barrier; manual pole; boom length 5.50m		nr		
	Note:				
	Power supply in connection with automatic vehicle control barrier measured under group element 5 (Services)				
8.4.4.2	Vehicle control barrier; automatic raise arm; boom length 4.50m		nr		
8.4.4.3	Vehicle restraint systems; double-leaf-opening height restriction barrier, standard height, root fixed; (PC Sum £1,700/nr supply only); 7.00m wide		nr		

Example 22.11 *(Continued)*

EXTERNAL WORKS					
Code	**Element/sub-elements/components**	**Qty**	**Unit**	**Rate**	**Total**
				£p	**£**
8.4.4.4	Vehicle and pedestrian control systems; fixed hoop barriers, mild steel, galvanised; cast-in-to pavings; 1800mm long units		nr		
8.4.4	**TOTAL – Barriers and guardrails: to element 8.4 summary**				

22.6 External fixtures (element 8.5)

This element is divided into two sub-elements as follows:

8.5.1 Site/street furniture and equipment; and
8.5.2 Ornamental features.

22.6.1 Site/street furniture and equipment (sub-element 8.5.1)

Sub-element 8.5.1 covers furniture and equipment designed for use externally and which form an integral part of the external works. Items of site and street furniture and equipment are measured as enumerated (nr) items, with the type of component stated in the description. Examples of site and street furniture and equipment include turnstiles, bollards, benches, litter bins, directional signs, sculptures and playground equipment.

Site and street furniture and equipment that are to be used within the building are measured under sub-element 4.1.1 (Fittings, furnishings and equipment). Street furniture and equipment that is the responsibility of statutory undertakers (e.g. road signs and bus stops) are excluded.

Sample descriptions for site and street furniture and equipment are shown in Example 22.12.

Example 22.12 Formulation of descriptions for site/street furniture and equipment

EXTERNAL WORKS					
Code	**Element/sub-elements/components**	**Qty**	**Unit**	**Rate**	**Total**
				£p	**£**
8	**EXTERNAL WORKS**				
8.5	**Site and street furniture and equipment**				
8.5.1	*Site/street furniture and equipment*				
	[Site furniture and equipment]				
8.5.1.1	Reflected traffic signs; on steel posts; including foundation		nr		

Example 22.12 *(Continued)*

EXTERNAL WORKS					
Code	**Element/sub-elements/components**	**Qty**	**Unit**	**Rate**	**Total**
				£p	**£**
8.5.1.2	Benches; precast concrete seat with backrest, with hardwood slats; (PC Sum £1,200/nr supply only)		nr		
8.5.1.3	Litter bins; precast concrete; (PC Sum £650/nr supply only)		nr		
8.5.1.4	Cycle stands; single; galvanised steel		nr		
8.5.1.5	Bollards; ornamental steel; including foundation; (PC Sum £135/nr supply only)		nr		
8.5.1.6	Tree guards; mild steel, circular; (PC Sum £100/nr supply only); 1.80m high		nr		
	[Playground equipment]				
	Note:				
	Safety surfacing: Measured under sub-element 8.2.2 (Specialist surfacings and pavings)				
8.5.1.7	Climbing frame; igloo type, 3.60m × 3.80m on plan × 2.30m high		nr		
8.5.1.8	Slide; stainless steel, 3.80m long		nr		
8.5.1.9	Seesaw		nr		
8.5.1.10	Swings; three seats		nr		
8.5.1	**TOTAL – Site/street furniture and equipment: to element 8.5 summary**				

22.6.2 Ornamental features (sub-element 8.5.2)

Ornamental water features and other similar features are measured as enumerated (nr) items. Items measured include the provision of mains water supply and mains power. The type of feature is to be stated in the description.

22.7 External drainage (element 8.6)

This element is divided into four sub-elements as follows:

8.6.1 Surface water and foul water drainage;

8.6.2 Ancillary drainage systems;

8.6.3 External chemical, toxic and industrial liquid waste drainage; and

8.6.4 Land drainage.

Complete drainage installations comprise a number of components, which are measured under different elements and sub-elements. The various components, together with the element or sub-element under which they are measured, are summarised in Table 19.1 in Chapter 19 ('Services (group element 5)').

22.7.1 Testing and commissioning of external drainage installations

These components deal with the testing and commissioning of the works that have been measured under element 8.6 (External drainage):

8.6.1 surface water and foul water drainage;
8.6.2 ancillary drainage systems;
8.6.3 external laboratory and industrial liquid waste drainage; and
8.6.4 land drainage.

The rules treat testing and commissioning as two separate cost centres.

The testing and commissioning of each type of external drainage installation are measured separately as a percentage addition to the sum total of the applicable drainage installation. The steps in calculating the cost target for testing are as follows:

Step 1 Determine the percentage addition to be applied in respect of testing. An appropriate percentage for calculating cost targets for testing and commissioning can be derived by comparing a number of published cost analyses or from 'in-house' sources of data.

Step 2 Ascertain the sum total of the cost targets for all components comprising the applicable drainage installation.

Step 3 Ascertain the cost target for testing by multiplying the sum total of the components comprising the applicable drainage installation by the percentage applicable for testing.

Note:
The equation for calculating the cost target for testing is:

$$t = a \times p$$

Where:

a = sum total of the components comprising the applicable drainage installation
p = percentage addition for testing
t = estimated cost of testing.

The cost target for commissioning is calculated in the same way as for testing. However, it should be noted that the cost target ascertained for testing is not included as part of the sum total of the cost targets for all components comprising the applicable drainage installation.

Note:
The equation for calculating the cost target for commissioning is:

$$c = a \times p$$

Where:

a = sum total of the components comprising the applicable drainage installation (excluding the cost target for testing)
p = percentage addition for commissioning
c = estimated cost of commissioning.

22.7.2 Surface water and foul water drainage (sub-element 8.6.1)

This sub-element covers surface water and foul water drainage, both below and above ground, from the first manhole beyond the external face of the external wall to the building, to the statutory undertaker's sewer connection or other outfall (e.g. an on-site sewage treatment facility).

Connections to the statutory undertaker's surface water and foul water sewers are enumerated (nr).

Drainage runs below ground are measured linear, in metres, measured along their extreme length over all branches, fittings and the like. The description for drainage runs below ground is to include type of pipe (e.g. concrete, clay or plastic – uPVC), the approximate average depth of trench (e.g. 750mm, 1.00m, 1.25m or 2.50m) and the nominal size of pipe (e.g. 100mm, 150mm, 200mm, 450mm or 600mm). The composition of beds and surrounds, cradles and haunchings should also be given in the description, particularly if a cost-significant component, such as lean mix concrete, is to be used.

Drainage runs above ground are measured in the same way as drainage runs below ground. However, descriptions for drainage runs above ground are to include the type of pipe, the approximate average height of the drain above the ground level and the nominal size of pipe. Details of any site-applied coatings (e.g. painting or anti-corrosion treatments), and the nature of any supports for the above-ground drainage (e.g. earth embankments or steel supports), are also to be given in the description.

Prefabricated channels in roads, paths and pavements are measured as linear items, in metres, with the composition of the channel unit and nominal size stated in the description.

Manholes, inspection chambers, catch pits, soakaways, retention and storage tanks, cesspools and sceptic tanks, petrol interceptors and other similar components are to be measured as enumerated (nr) items. The composition and approximate depth of the component is to be stated in the description. The type of access cover and frame should also be stated (e.g. light duty, heavy duty, lockable or non-locking type). Packaged pumping stations are measured in the same way.

Alterations to existing drainage systems (e.g. breaking into an existing drain and inserting a branch connecting for a new drain, breaking out and renewing gullies, or raising or lowering existing gully gratings and frames) are measured as enumerated (nr) items, with the nature of the works described.

Works to existing manholes, or the like, are also measured as enumerated (nr) items, with the nature of the works stated in the description. Such works might include enlarging existing manholes, raising manholes, renewing manhole covers and frames, and connecting new drains to existing manholes.

Clearing existing drains and sealing redundant drains can be measured as either enumerated or linear items, in metres, with the method of cleaning (e.g. rodding), or sealing (e.g. filling the entire length of redundant drain with foam concrete or plugging end of drain with concrete), given in the description.

Filling disused manholes and the like, where not measured under sub-element 8.1.2 (Preparatory groundworks), can be measured as enumerated (nr) items under this sub-element, with the filling material to be used stated in the description (e.g. lean mix concrete or granular material). Typical descriptions for surface water and foul water drainage installations are provided in Example 22.13.

Example 22.13 Formulation of descriptions for surface water and foul water drainage

EXTERNAL WORKS					
Code	**Element/sub-elements/components**	**Qty**	**Unit**	**Rate**	**Total**
				£p	**£**
8	**EXTERNAL WORKS**				
8.6	**External drainage**				
8.6.1	*Surface water and foul water drainage*				
	[Work outside the curtilage of the site]				
	Connections to statutory undertaker's sewers:				
8.6.1.1	Foul water sewers		nr		
8.6.1.2	Surface water sewers		nr		
8.6.1.3	Drainage runs below ground; nominal size 300mm concrete pipes, on concrete bed and surround; including excavation and disposal of surplus excavated material off site; depth of trench not exceeding 4.50m		m		
8.6.1.4	Work to existing manholes; connecting new branch drain; not exceeding 300mm diameter; making good		nr		
	[Site works]				
8.6.1.5	Drainage runs below ground; nominal size 100/150mm clay/uPVC pipes, on granular bed and surround; including excavation and disposal of surplus excavated material off site; depth of trench not exceeding 3.00m		m		
8.6.1.6	Prefabricated channels; 'Aco' drain channel on concrete base and surround; stainless steel grating; 100mm wide		m		
8.6.1.7	Manholes; brick construction; concrete cover with heavy duty lockable access cover; not exceeding 1.50m deep				
	[Work to existing surface water and foul water drainage]				
8.6.1.8	Manholes; cutting out existing branch drain and channel; making good		nr		
8.6.1.9	Clearing existing drains; rodding; distance between access points not exceeding 30m length		nr		
8.6.1.10	Sealing redundant drains; ends with concrete plug; not exceeding 300mm diameter		nr		
	[Testing and commissioning]				
8.6.1.11	Testing of installations		%		
8.6.1.12	Commissioning of installations		%		
8.6.1	**TOTAL – Surface water and foul water drainage: to element 8.6 summary**				

22.7.3 Ancillary drainage systems (sub-element 8.6.2)

This sub-element deals with:

- Pumping stations, ejector stations, and storage and retention tanks (i.e. systems with a storage tank or vessel for the reception of foul water and sewage at one level, for transfer by pump to drains or sewers at a higher level);
- Sewage treatment systems (i.e. packaged sewage treatment plant or conventional sewage treatment plant); and
- Sustainable urban drainage schemes (SUDS) – SUDS techniques mimic natural systems for draining surface water and include porous surfaces, soakaways, infiltration trenches, filter drains, filter strips, swales, detention basins and purpose- built ponds and wetlands. As well as treating polluted surface water runoff, SUDS provide attenuation of surface water to reduce the impact of flow on watercourses. Well-designed SUDS help to protect and enhance ground water quality and can make a positive contribution to the amenity and wildlife value of a site. The inclusion of these systems within new development programmes is being encouraged in the United Kingdom by the government through legalisation and guidance policies.

Pumping stations, ejector stations, storage tanks and retention tanks, and sewage treatment systems are measured as enumerated (nr) items, with the type or, in the case of a conventional sewage treatment plant, the composition, stated in the description. Proprietary pumping stations can also be measured under sub-element 8.6.1 (Surface water and foul water drainage). Electricity supply to pumping stations, sewage treatment systems and the like is measured under sub-element 8.7.4 (Electricity distribution to external plant and equipment).

Sustainable urban drainage schemes (SUDS) are measured superficially, in square metres, with the surface of the land served by the SUDS measured. The composition of the system is to be stated in the description.

Component descriptions for ancillary drainage systems are shown in Example 22.14.

Example 22.14 Formulation of descriptions for ancillary drainage systems

EXTERNAL WORKS					
Code	**Element/sub-elements/components**	**Qty**	**Unit**	**Rate £p**	**Total £**
8	**EXTERNAL WORKS**				
8.6	**External drainage**				
8.6.2	*Ancillary drainage systems*				
8.6.2.1	Pumping stations; total storage capacity: approx. 1,335 litres		nr		
8.6.2.2	Sewage treatment; packaged sewage treatment plant; capacity 20 person, completely surrounded in concrete, including excavation and backfilling with excavated material, surplus excavated material disposed off site		nr		

Example 22.14 *(Continued)*

EXTERNAL WORKS					
Code	**Element/sub-elements/components**	**Qty**	**Unit**	**Rate**	**Total**
				£p	**£**
8.6.2.3	Sustainable urban drainage; GEOlight stormwater attenuation system; to hard paved areas		m²		
8.6.2.4	Testing of installations		%		
8.6.2.5	Commissioning of installations		%		
8.6.2	**TOTAL – Ancillary drainage systems: to element 8.6 summary**				

22.7.4 External chemical, toxic and industrial liquid waste drainage (sub-element 8.6.3)

Chemical, toxic and industrial waste drainage installations external to a building are measured in the same way as for surface water and foul water drainage. Equipment and plant associated with chemical, toxic or industrial waste drainage are measured as enumerated (nr) items, with the type of equipment or plant described.

22.7.5 Land drainage (sub-element 8.6.4)

Land drainage encompasses disposal systems for the drainage of waterlogged ground.

Land drains, including perforated or porous pipes and French drains, are measured as linear items, in metres. The description for land drains is to include the type of drain, the depth of trench and the nominal size of the pipe. In the case of 'French' drains, the composition of the drain is to be described.

Manholes and the like are measured and described in the same way as those measured under sub-element 8.6.1 (Surface water and foul water drainage).

Drainage blankets (e.g. a granular blanket to an embankment) are measured superficially, in square metres, with the surface of the land served by the blanket measured. The composition of the blanket is to be stated in the description.

Land drainage to parkland or the like is also measured superficially, in square metres, with the surface of the land served by the blanket measured. Because of the large areas involved, the unit of measurement for land drainage to parkland areas is hectares (ha). The description for land drainage to parkland areas is to include composition of the drainage system, together with the centres of main drain runs and the length of the lateral drains.

Typical descriptions for land drainage installations are provided in Example 22.15.

Example 22.15 Formulation of descriptions for land drainage

EXTERNAL WORKS					
Code	**Element/sub-elements/components**	**Qty**	**Unit**	**Rate**	**Total**
				£p	**£**
8	**EXTERNAL WORKS**				
8.6	**External drainage**				
8.6.4	*Land drainage*				
8.6.4.1	French drain; non-coilable perforated plastic pipes, 150mm diameter; wrapping pipes with filter fabric; shingle filling; 600mm wide × 900mm deep		m		
8.6.4.2	Manholes; precast concrete ring construction; concrete cover with heavy duty lockable access cover; not exceeding 2.00m deep		nr		
8.6.4.3	Drainage blankets; 'Aztex Draintube' system or similar		m²		
8.6.4.4	Land drainage to parkland; with laterals at 30m centres and main runs at 100m centres		ha		
8.6.4.5	Land drainage to sports fields; with laterals at 10m centres and main runs at 33m centres		ha		
8.6.4.6	Testing of installations		%		
8.6.4.7	Commissioning of installations		%		
8.6.4	**TOTAL – Land drainage: to element 8.6 summary**				

22.8 External services (element 8.7)

This element is divided into 11 sub-elements as follows:

- 8.7.1 Water mains supply;
- 8.7.2 Electricity mains supply;
- 8.7.3 External transformation devices;
- 8.7.4 Electricity distribution to external plant and equipment;
- 8.7.5 Gas mains supply;
- 8.7.6 Telecommunications and other communication connections;
- 8.7.7 External fuel storage and piped distribution systems;
- 8.7.8 External security systems;
- 8.7.9 External street lighting systems;
- 8.7.10 Local/district heating installations; and
- 8.7.11 Builder's work in connection with external services.

22.8.1 Testing and commissioning of external services

NRM 1 treats testing and commissioning as two separate cost centres.

Testing and commissioning in connection with each type of external service installation is measured in conjunction with the apposite services installation. They are measured separately as a percentage addition to the sum total of the applicable external services installation. The steps in calculating the cost target for testing are as follows:

Step 1 Determine the percentage addition to be applied in respect of testing. An appropriate percentage for calculating cost targets for the testing and commissioning can be derived by comparing a number of published cost analyses or from 'in-house' sources of data.

Step 2 Ascertain the sum total of the cost targets for all components comprising the applicable external services installation.

Step 3 Ascertain the cost target for testing by multiplying the sum total of the components comprising the applicable external services installation by the percentage applicable for testing.

Note:
The equation for calculating the cost target for testing is:

$$t = a \times p$$

Where:

a = sum total of the components comprising the applicable external services installation
p = percentage addition for testing
t = estimated cost of testing.

The cost target for commissioning is calculated in the same way as for testing. However, it should be noted that the cost target ascertained for testing is not included as part of the sum total of the cost targets for all components comprising the applicable external services installation.

Note:
The equation for calculating the cost target for commissioning is:

$$c = a \times p$$

Where:

a = sum total of the components comprising the applicable external services installation (excluding the cost target for testing)
p = percentage addition for commissioning
c = estimated cost of commissioning.

22.8.2 Water mains supply (sub-element 8.7.1)

Water installations consist of various components, which are measured under different elements and sub-elements. These components, together with the element or sub-element under which they are measured, are summarised in Table 19.2 in Chapter 19 ('Services (group element 5)').

Water mains supply comprises the piped water supply systems bringing water from the statutory undertaker's mains to point of entry into the building and the distribution to external user points (e.g. to external plant and equipment) and fire hydrants.

Connections to the statutory undertaker's water mains are enumerated (nr).

Water main pipelines from statutory undertaker's mains to water meters are measured as linear items. Pipelines are measure along their extreme length over all branches, fittings and the like. The description for water main pipelines is to include the nominal size of the pipe, together with details of any ground anchor blocks or the like. Excavating and backfilling of trenches for buried pipelines are to be included with service runs, together with any protective measures (e.g. ducts). Water main connections to external plant and equipment are measured as enumerated (nr) items, with the number of draw-off points stated in the description.

Rainwater-harvesting systems external to the building are measured as enumerated (nr) items. The description for rainwater-harvesting systems is to include the kind of system, together with details of any key components (e.g. collection pipelines, tanks, pumps, filters and controls). Grey water systems external to the building are measured in the same way as rainwater-harvesting systems. Example 22.16 provides typical component descriptions for water mains supply installations by statutory undertakers and other types of water supply installations.

Example 22.16 Formulation of descriptions for water mains supply

EXTERNAL WORKS					
Code	**Element/sub-elements/components**	**Qty**	**Unit**	**Rate**	**Total**
				£p	**£**
8	**EXTERNAL WORKS**				
8.7	**External services**				
8.7.1	*Water mains supply*				
	Example 1 – Connection from statutory undertaker's water main				
8.7.1.1	Connections to statutory undertaker's water main		nr		
8.7.1.2	Service runs; 75mm uPVC main, including excavation and backfilling with excavated material		m		
8.7.1.3	Testing of installations		%		
8.7.1.4	Commissioning of installations		%		
	Example 2 – Offices				
8.7.1.1	Rainwater harvesting; underground system, volume 3,300 litres, including excavation and backfilling with excavated material, surplus excavated material disposed off site		nr		
8.7.1.2	Grey water systems; 'Ecosure' or the like, 2,800 litre rainwater-harvesting tank; range of three interconnected, including excavation and backfilling with excavated material, surplus excavated material disposed off site		nr		
8.7.1.3	Testing of installations		%		
8.7.1.4	Commissioning of installations		%		

Example 22.16 *(Continued)*

EXTERNAL WORKS					
Code	**Element/sub-elements/components**	**Qty**	**Unit**	**Rate**	**Total**
				£p	**£**
	Example 3 – Residential				
8.7.1.1	Rainwater harvesting; above-ground system, 1,800 litre GRP rainwater-harvesting tank		nr		
8.7.1.2	Grey water systems; 'AquaCycle 900' domestic grey water recycling system or the like		nr		
8.7.1.3	Grey water systems; 'Ecoplay' system or the like		nr		
8.7.1.4	Testing of installations		%		
8.7.1.5	Commissioning of installations		%		
8.7.1	**TOTAL – Water mains supply: to element 8.7 summary**				

22.8.3 Electricity mains supply (sub-element 8.7.2)

Complete electrical installations comprise a large number of facets, which are measured under a variety of different elements and sub-elements. These facets, together with the element or sub-element under which they are measured, are summarised in Table 19.3 in Chapter 19 ('Services (group element 5)').

Electricity mains supply comprises the distribution of high-voltage electricity from statutory undertaker's supply to an on-site transformer station; the distribution of low-voltage electricity from the on-site transformer (or other supply intake) to the main switchgear panel within the building; and external installations for providing electricity, including emergency or standby generation plant.

Connections to the statutory undertaker's electricity main are enumerated (nr).

Service runs distributing HV electricity to on-site transformer sub-stations, including packaged sub-stations, are measured linear along their extreme length over all obstructions. Service runs distributing LV electricity to main switchgear panels within buildings are measured the same way as service runs distributing HV electricity. Excavating and backfilling of trenches for buried cables are to be included with service runs, together with any protective measures (e.g. ducts).

Transformer sub-stations are enumerated. Where not supplied by the statutory undertaker, sub-station buildings, and housings, are measured under sub-element 8.8.2 (Ancillary buildings and structures). Fenced or walled enclosures are measured under either sub-element 8.4.1 (Fencing and railings) or sub-element 8.4.2 (Walls and screens), as appropriate.

External electricity generation plant, including emergency or standby generation plant, is measured as enumerated (nr) items, with the type of plant described.

Component descriptions for mains electricity supply are shown in Example 22.17.

Example 22.17 Formulation of descriptions for electricity mains supply

EXTERNAL WORKS					
Code	**Element/sub-elements/components**	**Qty**	**Unit**	**Rate**	**Total**
				£p	**£**
8	**EXTERNAL WORKS**				
8.7	**External services**				
8.7.2	*Electricity mains supply*				
8.7.2.1	Connections to statutory undertaker's electricity main		nr		
8.7.2.2	Service runs; 600/1000 volt cables, including 100mm duct, excavation and backfilling with excavated material		m		
	Note:				
	Transformer base and enclosure measured separately under sub-element 8.8.2 (Ancillary buildings and structures)				
8.7.2.3	Transformer sub-stations; packaged, 2500kVA		nr		
8.7.2.4	Testing of installations		%		
8.7.2.5	Commissioning of installations		%		
8.7.2	**TOTAL – Electricity mains supply: to element 8.7 summary**				

22.8.4 External transformation devices (sub-element 8.7.3)

External transformation devices such as wind turbines and solar collectors, which are not an integral part of a building, are measured as enumerated items, with the type of transformation device stated in the description.

Photovoltaic devices are also enumerated (nr), with the surface area of tiles given in the description.

Playground equipment and sculptures, which act as a transformation device, are measured under sub-element 8.5.1 (Site/street furniture and equipment).

Component descriptions for external transformation devices are illustrated in Example 22.18.

Example 22.18 Formulation of descriptions for external transformation devices

EXTERNAL WORKS					
Code	**Element/sub-elements/components**	**Qty**	**Unit**	**Rate**	**Total**
				£p	**£**
8	**EXTERNAL WORKS**				
8.7	**External services**				
8.7.3	*External transformation devices*				

Example 22.18 *(Continued)*

EXTERNAL WORKS					
Code	**Element/sub-elements/components**	**Qty**	**Unit**	**Rate**	**Total**
				£p	**£**
	Note: Bases for transformation devices measured elsewhere under sub-element 8.7.11 (Builder's work in connection with external services)				
8.7.3.1	Wind turbines; 1.00kW output, 1.80m diameter rotor, on hinged free-standing tubular towers		nr		
8.7.3.2	Photovoltaic devices; panels; ground mounted, including support structure; total surface area of units 136m^2		nr		
8.7.3.3	Testing of installations		%		
8.7.3.4	Commissioning of installations		%		
8.7.3	**TOTAL – External transformation devices: to element 8.7 summary**				

22.8.5 Electricity distribution to external plant and equipment (sub-element 8.7.4)

This sub-element deals with sub-circuit power installations (i.e. low-voltage electricity or power) from sub-distribution boards to mechanical and electrical plant and equipment, or external features such as water features, located externally to the building, terminating at socket outlets, fuse connection units or other termination points. It also encompasses final electrical connections to mechanical and electrical plant and equipment.

Connections to the mechanical and electrical plant and equipment are enumerated (nr).

Service runs distributing LV electricity to mechanical and electrical plant or equipment are measured linear, in metres, along their extreme length over all obstructions. Excavating and backfilling of trenches for buried cables are to be included with service runs, together with any protective measures (e.g. ducts).

22.8.6 Gas mains supply (sub-element 8.7.5)

This sub-element covers piped gas supply systems taking gas from either the statutory undertaker's mains to the gas meter, or liquefied petroleum gas from external storage vessels to the distribution point, and the distribution of the gas supply to external user points, including connections to external mechanical plant and equipment.

Connections to the statutory undertaker's gas main are enumerated (nr).

Gas pipelines, or service runs, distributing natural gas, or LPG, to external mechanical plant or equipment are measured linear, in metres, along their extreme length over all obstructions. Excavating and backfilling of trenches for buried gas pipelines are to be included with service runs, together with any protective measures (e.g. ducts).

Example 22.19 Formulation of descriptions for gas mains supply

EXTERNAL WORKS					
Code	**Element/sub-elements/components**	**Qty**	**Unit**	**Rate**	**Total**
				£p	**£**
8	**EXTERNAL WORKS**				
8.7	**External services**				
8.7.5	*Gas mains supply*				
8.7.5.1	Connections to statutory undertaker's gas main		nr		
8.7.5.2	Service runs; 150mm gas pipe, including excavation and backfilling with excavated material		m		
8.7.5.3	Governing stations; gas (PC £21,500/nr supply only)		nr		
8.7.5.4	Testing of installations		%		
8.7.5.5	Commissioning of installations		%		
8.7.5	**TOTAL – Gas mains supply: to element 8.7 summary**				

Gas-governing stations (also referred to as gas-reducing stations, or gas pressure-reducing and metering stations) are enumerated (nr). Natural gas is normally transported via pipelines at high pressures. The purpose of a gas-governing station is to reduce the gas pressure so that the gas can be used by the end-user. Protective compounds to gas-governing stations are measured under sub-element 8.7.11 (Builder's work in connection with external services).

Example 22.19 provides typical component descriptions for the installation of gas mains supply by statutory undertakers.

22.8.7 Telecommunications and other communication connections (sub-element 8.7.6)

Connection of telecommunications systems, cable television and other communication systems from statutory undertaker's or other service provider's supply to the main distribution point within the building is dealt with under this sub-element. Component descriptions for telecommunication and other communication systems are shown in Example 22.20.

Example 22.20 Formulation of descriptions for telecommunications and other communication connections

EXTERNAL WORKS					
Code	**Element/sub-elements/components**	**Qty**	**Unit**	**Rate**	**Total**
				£p	**£**
8	**EXTERNAL WORKS**				
8.7	**External services**				
8.7.6	*Telecommunication and other communication systems*				

Example 22.20 *(Continued)*

EXTERNAL WORKS					
Code	**Element/sub-elements/components**	**Qty**	**Unit**	**Rate**	**Total**
				£p	**£**
	Example 1 – Offices				
8.7.6.1	Service runs; cables laid in 100mm duct, including excavation and backfilling with excavated material		m		
8.7.6.2	Testing of installations		%		
8.7.6.3	Commissioning of installations		%		
	Example 2 – Residential				
8.7.6.1	Telecommunication connections; (per dwelling)		nr		
8.7.6.2	Testing of installations		%		
8.7.6.3	Commissioning of installations		%		
8.7.6	**TOTAL – Telecommunication and other communication systems: to element 8.7 summary**				

Connections to telecommunications, cable television and other communication networks are separately measured as enumerated (nr) items, with the type of communication system stated in the description.

Service runs distributing telecommunications, cable television and other communication systems are measured linear, in metres, along their extreme length over all obstructions. They are measured from the point of connection with the service provider's supply to the main distribution point within the building. Excavating and backfilling of communication cable trenches for buried communication cables are to be included with service runs, together with any protective measures (e.g. ducts).

22.8.8 External fuel storage and piped distribution systems (sub-element 8.7.7)

This sub-element covers storage tanks and vessels external to the building, and piped supply systems distributing oil, petrol, diesel or liquefied petroleum gas from storage tanks or vessels to entry point within the building or to external plant and equipment. Tanks and vessels can either be above ground or underground.

Fuel storage tanks and vessels located external to the building are measured as enumerated (nr) items. Compounds, bases, bunds and supports that are not an integral part of the tank or vessel are measured under sub-element 8.7.11 (Builder's work in connection with external services). Descriptions for fuel storage tanks and vessels are to include the type, capacity and use of the tank or vessel, together with details of any paint systems or coating systems, including anti-corrosion treatments. Excavation and earthworks, bases and any other building works in connection with underground storage tanks and vessels is to be measured and described within the item.

Pipeline runs, taking fuel from storage tanks or vessels to the building, or to supply external plant and equipment, are measured linear, in metres. They are measured

Example 22.21 Formulation of descriptions for external fuel storage and piped distribution systems

EXTERNAL WORKS					
Code	**Element/sub-elements/components**	**Qty**	**Unit**	**Rate**	**Total**
				£p	**£**
8	**EXTERNAL WORKS**				
8.7	**External services**				
8.7.7	*External fuel storage and piped distribution systems*				
	Example 1 – Above-ground tank installation				
	Note:				
	LPG tank bases and storage compound measured separately under sub-element 8.7.11 (Builder's work in connection with external services)				
8.7.7.1	LPG storage tank, above ground; set in position; capacity: 4,000 litres		nr		
8.7.7.2	Service runs; galvanised steel pipelines, including supports		m		
8.7.7.3	Testing of installations		%		
8.7.7.4	Commissioning of installations		%		
	Example 2 – Underground tank installation				
8.7.7.1	LPG storage tank, underground; capacity: 4,000 litres, approximate size: 4.40m long × 1.70m high; including excavating pit and disposal of excavated material, 150mm thick reinforced concrete base on sand blinding, backfilling pit and around tank with 'imported' sand, warning mesh, marker pegs and surface cover		nr		
8.7.7.2	Service runs; steel pipelines, treated with 2 coats bituminous paint, wrapped in protective tape, surrounded in crack-free mortar; including excavation and backfilling with excavated material, and mechanical protection to pipelines; trench not exceeding 600mm depth		m		
8.7.7.3	Testing of installations		%		
8.7.7.4	Commissioning of installations		%		
8.7.7	**TOTAL – External fuel storage and piped distribution systems: to element 8.7 summary**				

along their extreme length over all obstructions, from the point of connection with the storage tanks or vessels to the main distribution point within the building, or to connections with external mechanical plant and equipment. Excavating and backfilling of trenches for buried fuel pipelines are to be included with service runs, together with any protective measures (e.g. ducts).

Example 22.21 provides typical component descriptions for external fuel storage and piped distribution systems.

22.8.9 External security systems (sub-element 8.7.8)

This sub-element covers external observation and access control installations, including:

- surveillance equipment (e.g. CCTV);
- security detection equipment;
- security alarm equipment;
- gate access control systems;
- gate entry systems (audio and visual);
- security lights and lighting systems; and
- other security systems.

Surveillance equipment, security detection equipment, security alarm equipment, security lights and lighting systems, and other security systems can be either treated as a discrete item, or measured as enumerated (nr) items. Descriptions are to include type and composition of system. Gate access control systems and gate entry systems are measured as enumerated items, with the type and composition stated in the description.

Serviced runs in connection with external security systems are measured linear, in metres, measured along their extreme length over all obstructions. Excavating and backfilling of trenches for buried cables are to be included with service runs, together with any protective measures (e.g. ducts).

(See Example 22.22.)

Example 22.22 Formulation of descriptions for external security systems

EXTERNAL WORKS					
Code	**Element/sub-elements/components**	**Qty**	**Unit**	**Rate**	**Total**
				£p	**£**
8	**EXTERNAL WORKS**				
8.7	**External services**				
8.7.8	*External security systems*				
8.7.8.1	CCTV cameras; complete system comprising 8 high-resolution night vision cameras; camera housings with heater/blower; including heavy duty brackets		nr		
8.7.8.2	CCTV cameras; outdoor dummy camera, weather resistant, blinking led light; camera housings with heater/blower; including heavy duty brackets		nr		
8.7.8.3	Gate access control; automatic rising bollards, natural finish stainless steel, with built-in flashing lights; underground hydraulic drive unit; controls, with digital keypad; 700mm raise height		nr		
8.7.8.4	Gate entry systems; gate to reception intercom		nr		
8.7.8.5	Testing of installations		%		
8.7.8.6	Commissioning of installations		%		
8.7.8	**TOTAL – External security systems: to element 8.7 summary**				

22.8.10 External street lighting systems (sub-element 8.7.9)

This sub-element covers external illumination systems. They include lighting to pedestrian areas, paths and roads, and illuminated traffic signs. External site and street lighting systems are measured as enumerated (nr) items. Descriptions for external street lighting systems are to include the type of system, together with the number of luminaries or lamps (e.g. light columns, bollards or illuminated traffic signs) stated. Typical component descriptions for external street lighting systems are provided in Example 22.23.

Example 22.23 Formulation of descriptions for external street lighting systems

EXTERNAL WORKS					
Code	**Element/sub-elements/components**	**Qty**	**Unit**	**Rate**	**Total**
				£p	**£**
8	**EXTERNAL WORKS**				
8.7	**External services**				
8.7.9	*External street lighting systems*				
8.7.9.1	External lighting to pedestrian areas; lamp-post lighting; (PC Sum £800/lamp-post supply only)		nr		
8.7.9.2	External lighting to pedestrian areas; LED light in motion floodlight, including control gear and control unit: (PC Sum £1,250/floodlight, control gear and control unit supply only)		nr		
8.7.9.3	External lighting to pedestrian areas; floodlights, ground mounted; (PC Sum £150/floodlight supply only)		nr		
8.7.9.4	External lighting to paths; drive-over lighting – decorative; 2 tonne maximum loading		nr		
8.7.9.5	External lighting to paths; buried exterior uplighter for MR16 lamps; 5 tonne maximum loading		nr		
8.7.9.6	External lighting to roads; heavy duty (hydraulically operated) base-hinged columns, 8.00m high, with high performance road lantern; column flange plate mounted, including bolting to ground		nr		
8.7.9.7	Illuminated traffic signs; interactive slow down signs; (PC Sum £500/nr supply only)		nr		
8.7.9.8	Illuminated traffic signs; interactive slow down signs, powered by wind turbine and solar panel; (PC Sum £900/nr supply only)		nr		
8.7.9.9	Service runs; cables; including excavation and backfilling with excavated material, and mechanical protection; trench not exceeding 750mm depth		m		
8.7.9.10	Testing of installations		%		
8.7.9.11	Commissioning of installations		%		
8.7.9	**TOTAL – External street lighting systems: to element 8.7 summary**				

Lighting systems fixed to the exterior of the building are measured separately under sub-element 5.8.3 (Lighting installations).

22.8.11 Local and district heating installations (sub-element 8.7.10)

This sub-element deals with local and district heating installations, including heat sources that are located external to the building. Local or district heating is the use of a centralised boiler installation to provide heat for a number of buildings. This can be a heat-only boiler, or the heat from a combined heat and power (CHP) plant.

Heat sources are measured as enumerated (nr) items, with the type and output (in kW) stated in the description.

Service runs in connection with local and district heating installations are measured linear, in metres. They are measured along their extreme length over all obstructions, from the point of connection with the external heat source to the main distribution point within the building. The type of pipe and type of thermal insulation, including method of protection (if applicable), are to be described. Ducts for external mains services (e.g. to take services under roads) are measured under sub-element 8.7.11 (Builder's work in connection with external services).

External heating ducts, and duct access covers, to local and district heating installations are measured linear, in metres, along their extreme length over all obstructions. The description is to state the composition and overall size of the duct.

22.8.12 Builder's work in connection with external services (sub-element 8.7.11)

Sundry builder's work in connection with the installation of external mechanical and electrical services comprises:

- ducts for services;
- supports to storage tanks and storage vessels;
- fuel bunds and the like to storage tanks and storage vessels;
- protective compounds, storage racks and the like associated with fuel storage areas;
- protective compounds in connection with transformer sub-stations and the like;
- bases for mechanical and electrical plant and equipment; and
- other builder's work in connection with external services.

Ducts for services are either measured as enumerated (nr) items or linear, in metres. Where measured linear, ducts are measured along their extreme length over all obstructions. The description is to state the purpose (e.g. water main, gas main, CCTV cables, or electrical cable) and the overall size of the duct. Groups of ducts can be measured as enumerated (nr) items, with the number of ducts that comprise the group stated in the description.

Supports to storage tanks and storage vessels can either be itemised or measured as enumerated (nr) items. Descriptions for supports to storage tanks and vessels are to include their composition, together with details of any paint systems or coating systems, including anti-corrosion treatments.

Fuel bunds and protective compounds, including fencing and storage racks, to storage tanks and storage vessels are also either itemised of measured as enumerated (nr) items. The composition of fuel bunds and protective compounds, including details of fencing and storage racks, is to be given in the description.

Protective compounds in connection with transformer sub-stations, gas-governing stations and the like are measured in the same way as fuel bunds and protective compounds.

Bases for larger items of mechanical and electrical plant and equipment are enumerated (nr), and their purpose (e.g. for pre-packaged chillers or boilers), composition and indicative size should be stated in the description.

Other builder's work in connection with external services consists of:

- holes, mortices, sinkings and chases;
- ducts, pipe sleeves and the like;
- stopping and sealing holes;
- fire-resistant stopping and fire sleeves;
- on-site painting of, and other anti-corrosive treatments to, externally located equipment, fuel storage tanks and vessels, supports and pipelines;
- identification labelling and colour coding of services installations and systems; and
- other sundry builder's work items in connection with external services.

The cost target for other builder's work is based on a percentage of the sum of the cost targets for external services and is calculated using the following steps:

Step 1 Determine the percentage to be applied in respect of builder's work in connection with external services.

Step 2 Ascertain the sum total of the cost targets for all external services sub-elements, excluding external drainage.

Note:
The equation for calculating the total cost target for other builder's work in connection with external services is:

$$b = \Sigma(c1 + c2 + c3 + c4 + c5 + c6 + c7 + c8 + c9 + c10 + c11)$$

Where:
Cost target for other builder's work in connection with:
c1 = water mains supply
c2 = electricity mains supply
c3 = external transformation devices
c4 = electricity distribution to external plant and equipment
c5 = gas mains supply
c6 = telecommunications and other communications system connections
c7 = external fuel storage and piped distribution systems
c8 = external security systems
c9 = site and street lighting systems
c10 = irrigation systems
c11 = local and district heating installations
b = cost target for other builder's work in connection with external services.

Step 3 Ascertain the cost target for builder's work in connection with external services by multiplying the sum total of the cost targets for all external services elements by the percentage applicable for builder's work.

Note:
The equation for calculating the cost target for builder's work in connection with external services is:

$$t = a \times p$$

Where:

a = total cost target for external services
p = percentage addition for builder's work in connection with external services
t = estimated cost of builder's work in connection with external services.

Percentage additions to be used for calculating cost targets for other builder's work in connection with external services can be ascertained from published cost analyses or from 'in-house' sources of data. A typical build-up of a cost target for builder's work in connection with external services is illustrated in Example 22.4.

Example 22.24 Formulation of descriptions and calculation of cost for builder's work in connection with external services (where same single percentage is to be applied to all types of services installations)

EXTERNAL WORKS						
Code	**Element/sub-elements/components**	**Qty**	**Unit**	**Rate**		**Totals**
				£p	**£**	**£**
8	**EXTERNAL WORKS**					
8.7	**External services**					
8.7.11	*Builder's work in connection with external services* Note: Bund gauges, bund alarms and pumps measured separately under sub-element 8.7.7 (External fuel storage and piped distribution systems)					
8.7.11.1	Fuel bund; for oil tanks; base construction: 150mm thick reinforced concrete on blinded hardcore, 100mm thick, including excavation and disposal of surplus material; wall construction: 150mm thick × 500mm high reinforced concrete; including GRP lining to base and walls; internal dimensions approximately 4.00m × 3.60m		item			9,000
8.7.11.2	Protective compound; for LPG tanks; base construction: 150mm thick reinforced concrete on blinded hardcore, 100mm thick, including excavation and disposal of surplus material; perimeter fencing: 1.80m high chain link fencing, with concrete posts, including 1 nr single lockable gate, with heavy-duty padlock; approximately 8.00m × 4.60m plan area		item			12,900
8.7.11.3	Bases; precast concrete; to wind turbines	6	nr	450.00		2,700

Example 22.24 *(Continued)*

Code	Element/sub-elements/components	Qty	Unit	Rate		Totals
EXTERNAL WORKS						
				£p	£	£
8.7.11.4	Bases; in situ reinforced concrete; for photovoltaic panels; approximately 5.40m × 2.40m	10	nr	1,300.00		13,000
	Water mains supply (sub-element 8.7.1):				30,900	
	Electricity mains supply (sub-element 8.7.2):				23,600	
	External transformation devices (sub-element 8.7.3):				13,000	
	Electricity distribution to external plant and equipment (sub-element 8.7.4):				27,900	
	Gas mains supply (sub-element 8.7.5):				10,800	
	Telecommunications and other communications system connections (sub-element 8.7.6):				5,960	
	External fuel storage and piped distribution systems (sub-element 8.7.7):				16,390	
	External security systems (sub-element 8.7.8):				21,960	
	Site and street lighting systems (sub-element 8.7.9):				32,000	
	Irrigation systems (sub-element 8.7.10):				0	
	Local and district heating installations (sub-element 8.7.11):				0	
	Total of external services sub-elements:				182,510	
8.7.11.5	Other builder's work in connection with external services:	3.50	%	of	182,510	6,388
8.7.11	**TOTAL – Builder's work in connection with external services: to element 8.7 summary**					**43,988**

Should the estimated cost of builder's work in connection with external services need to be broken down further, then estimated costs can be determined separately for each type of service (e.g. water mains supply, electricity mains supply, external transformation devices, and so on). The cost of builder's work in connection with each external service is calculated in the same way as described above. Different percentage allowances can be applied if required. Descriptions for components forming builder's work in connection with external services are shown in Example 22.25.

22.9 Minor building works and ancillary buildings (element 8.8)

This element is divided into three main sub-elements as follows:

8.8.1 Minor building works;

8.8.2 Ancillary buildings and structures; and

8.8.3 Underpinning to external site boundary walls.

Example 22.25 Formulation of descriptions and calculation of cost for builder's work in connection with external services (where either separate cost targets are required for each service installation or different percentages are applicable to each type of services installation)

EXTERNAL WORKS						
Code	**Element/sub-elements/ components**	**Qty**	**Unit**	**Rate**		**Totals**
				£p	**£**	**£**
8	**EXTERNAL WORKS**					
8.7	**External services**					
8.7.11	*Builder's work in connection with external services*					
	Note:					
	Bund gauges, bund alarms and pumps measured separately under sub-element 8.7.7 (External fuel storage and piped distribution systems)					
8.7.11.1	Fuel bund; for oil tanks; base construction: 150mm thick reinforced concrete on blinded hardcore, 100mm thick, including excavation and disposal of surplus material; wall construction: 150mm thick × 500mm high reinforced concrete; including GRP lining to base and walls; internal dimensions approximately 4.00m × 3.60m		item			8,960
8.7.11.2	Protective compound; for LPG tanks; base construction: 150mm thick reinforced concrete on blinded hardcore, 100mm thick, including excavation and disposal of surplus material; perimeter fencing: 1.80m high chain link fencing, with concrete posts, including 1 nr single lockable gate, with heavy-duty padlock; approximately 8.00m × 4.60m plan area		item			12,880
8.7.11.3	Bases; precast concrete; to wind turbines	6	nr	450.00		2,700
8.7.11.4	Bases; in situ reinforced concrete; for photovoltaic panels; approximately 5.40m × 2.40m	10	nr	1,300.00		13,000
	Other builder's work in connection with external services:					

Example 22.25 *(Continued)*

EXTERNAL WORKS						
Code	**Element/sub-elements/ components**	**Qty**	**Unit**	**Rate**		**Totals**
				£p	**£**	**£**
8.7.11.5	Water mains supply (sub-element 8.7.1):	4.00	%	of	30,900	1,236
8.7.11.6	Electricity mains supply (sub-element 8.7.2):	2.50	%	of	23,600	590
8.7.11.7	External transformation devices (sub-element 8.7.3):	1.50	%	of	13,000	195
8.7.11.8	Electricity distribution to external plant and equipment (sub-element 8.7.4):	1.00	%	of	27,900	278
8.7.11.9	Gas mains supply (sub-element 8.7.5):	2.50	%	of	10,800	270
8.7.11.10	Telecommunications and other communications system connections (sub-element 8.7.6):	2.50	%	of	5,960	149
8.7.11.11	External fuel storage and piped distribution systems (sub-element 8.7.7):	3.00	%	of	16,390	492
8.7.11.12	External security systems (sub-element 8.7.8):	5.00	%	of	21,960	1,098
8.7.11.13	External street lighting systems (sub-element 8.7.9):	6.50	%	of	32,000	2,080
8.7.11.14	Local and district heating installations (sub-element 8.7.10):	0.00	%	of	0	0
	Total of external services sub-elements:				182,510	
8.7.11	**TOTAL – Builder's work in connection with external services: to element 8.7 summary**					**43,988**

22.9.1 Minor building works (sub-element 8.8.1)

This sub-element covers the refurbishment of, and alterations to, existing separate external small ancillary buildings, such as boiler houses, including overhauling existing mechanical and electrical plant and equipment. It also covers repairs to existing fences, railings, walls, screen walls and retaining walls. Descriptions for minor building works items are shown in Example 22.26.

Refurbishment and alteration works to ancillary buildings can be either itemised or measured as enumerated (nr) items. Descriptions for refurbishment and alteration works are to include the nature of work to be executed.

Works in overhauling existing mechanical and electrical plant and equipment

Example 22.26 Formulation of descriptions for minor building works

EXTERNAL WORKS					
Code	**Element/sub-elements/components**	**Qty**	**Unit**	**Rate**	**Total**
				£p	**£**
8	**EXTERNAL WORKS**				
8.8	**Minor building works and ancillary buildings**				
8.8.1	*Minor building works*				
8.8.1.1	Existing boiler house; refurbishment, including minor repairs and redecoration internally and externally; (allowance)		item		
8.8.1.2	Existing boiler; overhauling		item		
8.8.1	**TOTAL – Minor building works: to element 8.8 summary**				

are also itemised or measured as enumerated (nr) items, with the nature of the work stated in the description.

22.9.2 Ancillary buildings and structures (sub-element 8.8.2)

Minor new small ancillary buildings and structures, both traditional-built and prefabricated buildings, are measured as enumerated (nr) items. Descriptions are to include the type of building and the overall size. Descriptions for ancillary buildings and structures are shown in Example 22.27.

Example 22.27 Formulation of descriptions for ancillary buildings and structures

EXTERNAL WORKS					
Code	**Element/sub-elements/components**	**Qty**	**Unit**	**Rate**	**Total**
				£p	**£**
8	**EXTERNAL WORKS**				
8.8	**Minor building works and ancillary buildings**				
8.8.2	*Ancillary buildings and structures*				
8.8.2.1	Cycle compound; with lockable gate and security canopy; capacity: 32 cycles; fixed to paved surface (paved surface measured elsewhere); (PC Sum £10,000/nr supply only)		nr		
8.8.2.2	Smoking shelter; capacity up to 8 nr people, 2 nr entrances; fixed to paved surface (paved surface measured elsewhere); dimensions 2.00m × 2.00m × 2.20m; (PC Sum £2,400/shelter supply only)		nr		
8.8.2	**TOTAL – Ancillary buildings and structures: to element 8.8 summary**				

22.9.3 Underpinning to external site boundary walls (sub-element 8.8.3)

Underpinning works to external site boundary walls is measured linear, in metres, with the nature and composition of the underpinning works stated in the description.

CHAPTER 23

Deriving unit rates for building components, sub-elements and elements

Introduction

This chapter:

- explains the hierarchical structure of cost data used by the construction industry;
- looks at the key factors affecting cost data;
- identifies the various sources of historical cost data;
- considers the pitfalls associated with both 'in-house' and published building cost data;
- provides worked examples of typical calculations to determine unit rates; and
- presents step-by-step examples of how historical cost data can be interpolated and adjusted for key factors.

23.1 Hierarchical structure of cost data

The hierarchical structure of cost data adopted by the construction industry is illustrated in Table 23.1.

In general, cost data ascertained at:

- Level 1 is used in the preparation of 'rough' order of cost estimates;
- Levels 1 to 4 are used in the preparation of order of cost estimates; and
- Levels 3 and 4 are used in the preparation of initial elemental cost models; and
- Levels 4 and 5 are used in the preparation of cost plans.

Cost data at levels 6 and 7 are used to build up unit rates for components and sub-components respectively.

Table 23.1 Hierarchical structure of cost data

Level		Composition	Cost data	Type of cost data
1	Tender price		Total cost	Cost/m² of GIFA; functional unit rates
2	Group element price	Substructure, superstructure, etc.	Simplified analysis	Cost/m² of GIFA
3	Element price	Foundations, basement excavation, basement retaining walls, ground floor construction, frame, upper floors, roof, etc.	Amplified analysis	Cost/m² of GIFA; element unit rates (EUQs)
4	Sub-element price	Standard foundations, piled foundations, underpinning, etc.	Detailed analysis	Cost/m² of GIFA; element unit rates (EUQs) for sub-elements
5	Component price	Strip foundations, isolated pad foundations, columns, beams, etc.	Sub-element analysis	Composite unit rates for components
6	Sub-component price	Concrete, reinforcement, formwork, subcontractor's on-costs, etc.	Component analysis	Composite unit rates for sub-components
7	Sub-sub-component price	Labour costs, material costs, plant costs and subcontractor's on-costs, including design costs	Sub-component analysis	'All-in' (composite) unit rates for labour, material, plant and subcontractor's on-costs
8	Constituent price	Build-up of labour rates for each grade of operative, material costs, plant costs and subcontractor's on-costs, including design costs	All-in rates	Rates for each grade of labour, materials costs, plant costs and subcontractor's on-costs, including design costs
9	Basic price	Basic costs that make up all-in labour rates, basic price of materials and plant costs	Basic prices	Components of labour rates, material costs and plant costs

23.2 Sources of building cost data

The most useful source of building cost data is considered to be that which has been generated 'in-house'. This is because the cost manager should be aware of the anomalies that might be inherent in its composition. Nonetheless, there are many other sources of cost data which the cost manager can call upon. Sources include:

- 'in-house' and published:
  - elemental cost analyses of complete building costs;
  - benchmark analysis of complete building costs and elemental costs; and
  - cost studies.

- published price books (e.g. for major works, minor works, mechanical and electrical engineering services, civil engineering, external works and landscape works, highway works and maintenance);
- estimates and quotations from:
  - contractors;
  - works contractors;
  - specialist contractors; and
  - suppliers.
- priced bill of quantities (i.e. both main contract and works contract packages);
- contract sum analyses; and
- indices (for differences in prices at different times and in different locations).

Bill of quantities were the major source of building cost information, but since the publication of the Latham Report of July 1994 and the report produced by Sir John Egan's Construction Task Force (*Rethinking Construction*) in 1998, there has been a significant move away from designer-led contract strategies in favour of contractor-led design strategies. As a result, bill of quantities as a means of obtaining a tender price have become outmoded. Consequently, meaningful building cost data has become harder for the cost manager to collect. This problem is recognised by NRM 1, which recommends that pricing documents should be structured to reflect the elements, sub-elements and components contained in the rules, so that the cost manager can gather cost data in a form that can be easily analysed and stored for use in the preparation of future cost estimates and cost plans (refer to Chapter 36: 'Analysing bids and collecting cost data').

Whichever source of building cost information is used, the building cost data obtained will need to be adjusted for, amongst other things, changes in general market price levels, location, variations in specification and differences in site conditions.

23.3 Factors affecting cost data

Besides the escalation or contraction of costs due to time, there are a number of other factors that can affect cost, including:

- **Size of the building** – obviously, the bigger the building the more expensive it will be. However, there should be benefits of scale and the unit rate (cost per m² of GIFA) should normally decrease with the increase in total GIFA.
- **Quality of the building** – the higher the specification of components the greater the cost, and vice versa.
- **Site conditions, including the vicinity of existing buildings and adjoining properties to construction operations** – these are specific to the actual site and might involve bad ground requiring enhanced foundations, obstructions in the ground requiring diversion or removal, problems with contaminated ground and problems with site access or space availability (especially in town and city centre sites).
- **Underlying market conditions** – in recession, the employer might be able to obtain a very good competitive price, but in boom periods prices are most likely to increase. The other key issue here is contractors' need for work. In times of

recession some contractors will submit low tenders simply in order to obtain turnover and a presence in the market. Alternatively, they might submit a 'non-compliant' bid (i.e. one that doesn't deliver the exact requirements of the employer) in order to put themselves in a position to negotiate.

- **Locality** – there are varying price levels throughout the United Kingdom. For example, it is more expensive to build in Central London than it is in the North East of England. Therefore, regional price levels need to be considered. The same is likely to apply in other countries.
- **Type of main contractor** – the nature of the main contractor and the extent to which he sub-lets work will impact on price levels. The greater the number of contractor tiers, the greater the cost. For example, it can be argued that a main contractor who directly employs the majority of operatives will build for less than a main contractor who sub-lets all works. This is because the number of contractor tiers is less and, as a consequence, there will be fewer tiers of overheads and profit.
- **Pre-construction and construction time** – the length of time taken to procure the building will impact on costs – principally those relating to time-related preliminaries and inflation.
- **Fixed price** – there is risk to the contractor in tendering a fixed price for the building project. Today, more and more employers require main contractors to offer a fixed lump sum price, which is not subject to fluctuations, for building projects of more than two years' duration.
- **Impact of changes in codes of practice –** the introduction of new codes of practice might impact on the design or site practices, thereby resulting in increased costs.
- **Impact of changes in statute** – changes in legislation since the cost data were captured might have an impact.
- **Unusual contract conditions** – extensive amendments to standard forms of contract, or an employer's bespoke terms and conditions of contract, might have a significant impact on cost due to the contractor's uncertainty and perception of risk; and
- **Liquidated damages** – if liquidated damages are set at a high level by the employer, this becomes a risk to the main contractor as well as potentially certain of his subcontractors. Consequently, the main contractor will most likely include a time and cost premium in his tender offer.

Adjusting unit rates to accommodate any of these factors is a matter for individual cost managers.

23.4 Pitfalls of using cost analysis and benchmark analysis

Care needs to be taken by the cost manager when using cost analyses and benchmark analyses of completed projects as:

- They relate to building projects that might not have been completed within budget or on time.
- They are associated with building projects built in different locations, with possible regional cost differences.

- They might not reflect the desired quality of a building project, as quality can mean different things to different people.
- No two similar building projects will cost exactly the same – there will always be a cost difference.
- The underlying market conditions will not necessarily be the same for any two building projects – the amount of work in the market will strongly influence the price level, including overheads and profit levels.
- The nature of risks transferred to the main contractor through the contract might not be known – whether or not the rates contain any premium for risk acceptance by the main contractor is not easily identifiable.
- The extent of costs in connection with specific limitations being imposed on the main contractor might not be readily identifiable, such as restrictions on construction methods, sequence of operations, phasing and restrictions to access and working (i.e. out of hours working).
- The effect on outturn costs that the type and nature of the employer will have (e.g. owner-developer, developer, public sector, experienced, inexperienced, etc.) might not be appreciated – they might be responsible for changing requirements during the construction phase, late decisions and interference.

23.5 Determining the cost per m² of gross internal floor area for building works

The cost target for building works for the building project is obtained by adjusting and updating the cost/m² of GIFA from cost analyses of previous building projects and multiplying the result by the total GIFA of the proposed new building.

Whichever source of cost analyses is used, the cost/m² of GIFA obtained will need to be adjusted for the difference in quantity (quantity adjustment); specification (quality adjustment); location (regional variation adjustment); and site conditions. It will also need to be updated to allow for changes in general market price level (time adjustment). Interpolation is the most common method used to ascertain an appropriate cost/m² of GIFA rate for building works.

23.5.1 Determination by interpolation

The interpolation method is where cost analyses of buildings of a similar type are studied. This method permits the differing sizes and standards between buildings to be examined and taken into account when considering the cost of the proposed building. Interpolation calls for the cost manager to use professional judgement aided by simple statistical measures. The complexity of the statistical measures used will depend on how many cost analyses are to be considered and the method used to build up the cost/m² of GIFA for building works. A simplistic approach is to adjust the total cost/m² of GIFA for building works from a range of cost analyses (holistic approach). A more complex, and possibly more accurate, approach is to consider each element in turn, adjust the cost/m² of GIFA for each element and build up the total cost/m² of GIFA for building works on an element-by-element basis. The simplistic approach is normally only suitable for 'rough' order of cost estimates.

Figure 23.1 describes the steps to determining a unit rate for building works using the interpolation method.

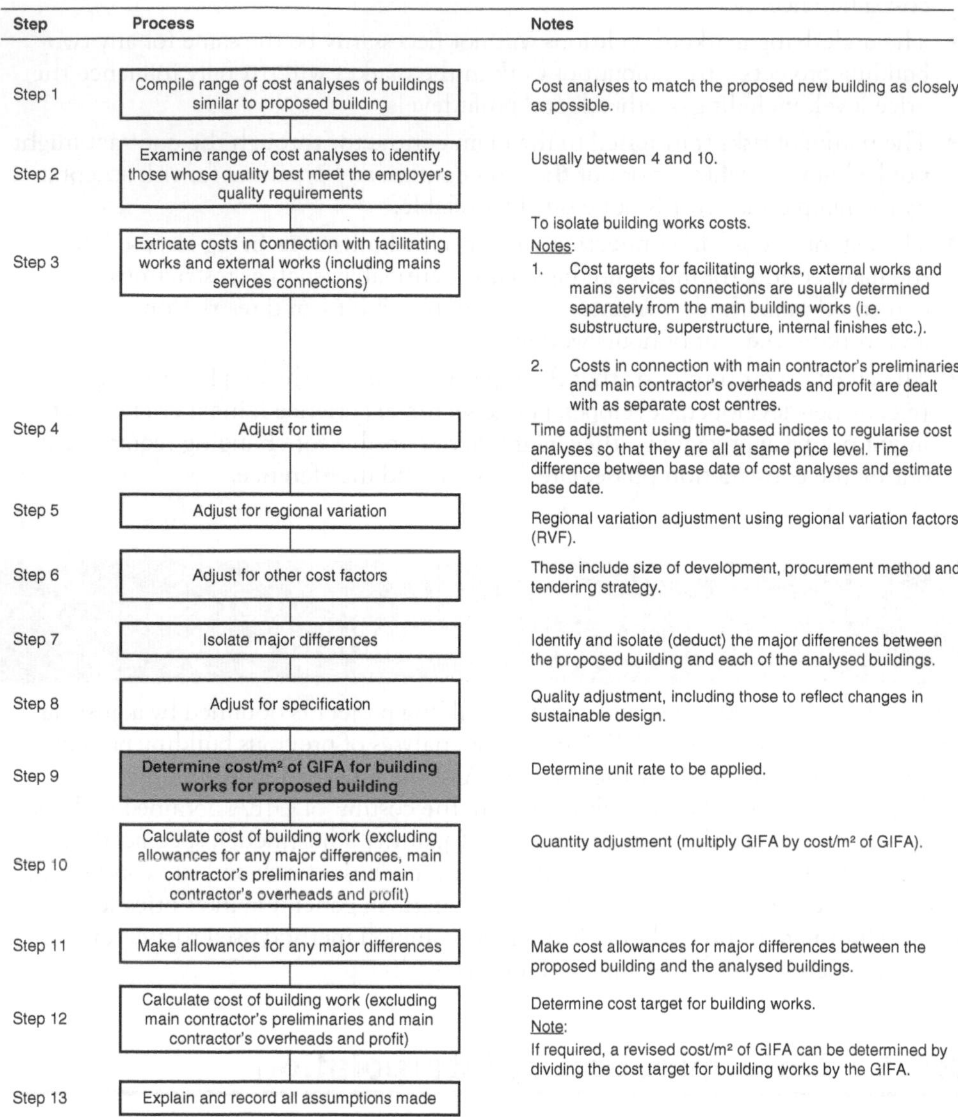

Step	Process	Notes
Step 1	Compile range of cost analyses of buildings similar to proposed building	Cost analyses to match the proposed new building as closely as possible.
Step 2	Examine range of cost analyses to identify those whose quality best meet the employer's quality requirements	Usually between 4 and 10.
Step 3	Extricate costs in connection with facilitating works and external works (including mains services connections)	To isolate building works costs. Notes: 1. Cost targets for facilitating works, external works and mains services connections are usually determined separately from the main building works (i.e. substructure, superstructure, internal finishes etc.). 2. Costs in connection with main contractor's preliminaries and main contractor's overheads and profit are dealt with as separate cost centres.
Step 4	Adjust for time	Time adjustment using time-based indices to regularise cost analyses so that they are all at same price level. Time difference between base date of cost analyses and estimate base date.
Step 5	Adjust for regional variation	Regional variation adjustment using regional variation factors (RVF).
Step 6	Adjust for other cost factors	These include size of development, procurement method and tendering strategy.
Step 7	Isolate major differences	Identify and isolate (deduct) the major differences between the proposed building and each of the analysed buildings.
Step 8	Adjust for specification	Quality adjustment, including those to reflect changes in sustainable design.
Step 9	**Determine cost/m² of GIFA for building works for proposed building**	Determine unit rate to be applied.
Step 10	Calculate cost of building work (excluding allowances for any major differences, main contractor's preliminaries and main contractor's overheads and profit)	Quantity adjustment (multiply GIFA by cost/m² of GIFA).
Step 11	Make allowances for any major differences	Make cost allowances for major differences between the proposed building and the analysed buildings.
Step 12	Calculate cost of building work (excluding main contractor's preliminaries and main contractor's overheads and profit)	Determine cost target for building works. Note: If required, a revised cost/m² of GIFA can be determined by dividing the cost target for building works by the GIFA.
Step 13	Explain and record all assumptions made	

Figure 23.1 Determination of unit rates for building works by interpolation

23.5.2 Time adjustment

The cost/m² of GIFA obtained from a cost analysis must be adjusted to take account of changes in prices levels between the base date of the cost analysis and the estimate base date (i.e. the date on which the order of cost estimate is prepared). Changes in price levels are measured from one time period to another using an appropriate price index. The most commonly used index is the 'all-in' TPI (tender price index). This covers new building work and reflects all sectors of the construction industry (e.g.

residential, hotels, commercial offices, retail, warehouses, hospitals and schools). It also covers both private and public sector construction projects. The index is based on a random sample of schemes and represents general trends of tender prices across all sectors. However, tender price indices are also produced in a number of other formats, including private sector TPI (covers non-housing schemes for private sector and public sector schemes where funded from the private sector), public sector TPI (covers non-housing schemes for the public sector), housing TPI and housing refurbishment TPI.

Changes in price levels are measured as the percentage change between two time periods. The equation is as follows:

$$\frac{\text{(The index at the later date − the index at the earlier date)}}{\text{The index at the earlier date}} \times 100 = \text{\% change}$$

To update the cost/m² of GIFA obtained from a cost analysis to the estimate base date, the percentage change is calculated in monetary terms and simply added to the cost/m² of GIFA obtained from a cost analysis. However, it should be noted that the percentage change could be positive or negative – resulting in the rate increasing or decreasing.

Use of an adjustment factor is another method of calculating the updated cost/m² of GIFA. In this case, the updated cost/m² of GIFA is determined by multiplying the cost/m² of GIFA obtained from a cost analysis by the factor resulting from:

$$\frac{\text{The index at the later date}}{\text{The index at the earlier date}} = \text{adjusting factor}$$

Examples 23.1 to 23.4 show how each method is applied.

Example 23.1 Time adjustment of cost/m² of GIFA using time-based indices

Assumptions:

- Cost/m² of GIFA at base date of cost analysis: £2,176.28/m² of GIFA
- Base date of cost analysis: February 2011
- Estimate base date: September 2013
- In-house TPI:
- February 2011: 223
- September 2013: 229 (forecast)

(a) Using percentage change:

$$\text{Percentage change} = \frac{(229–223)}{223} \times 100 = 2.69\%$$

Therefore:

Adjusted cost/m² of GIFA =
£2,176.28 + (2.69% of £2,176.28) =
£2,176.28 + £58.54 =
£2,234.82

Thus:

Adjusted cost/m² of GIFA = £2,234.82 (£2,234.82/m² of GIFA)

Example 23.1 *(Continued)*

(b) Using adjustment factor:

$$\text{Adjustment factor} = \frac{229}{223} = 1.0269$$

Therefore:

Adjusted cost/m² of GIFA =
£2,176.28 × 1.0269 =
£2,234.82

Thus:

Adjusted cost/m² of GIFA = £2,234.82 (£2,234.82/m² of GIFA)

Under NRM 1, possible changes in price levels from the estimate base date to the projected tender date, and from the projected tender date to the projected completion date of the building project, are dealt with as a discrete cost centre under group element 14 (Inflation). The approach to estimating the potential impact of inflation on a building project is explained in Chapter 31 ('Estimating the possible effects of inflation' (group element 14)).

23.5.3 Regional variation adjustment

Construction prices can change depending on regional workloads and local 'hot spots'. As a consequence, similar buildings constructed in different regions will produce different tender prices. In order to quantify the effects of regional variations on construction prices, the Building Cost Information Service and other organisations publish adjustment factors for different regions – regional variation factors (RVF). The regions selected are administrative areas and are not significant boundaries as far as the construction industry is concerned. In the United Kingdom they relate to standard statistical regions, counties and local authority districts.

The effect of regional variations on construction prices is measured as the percentage change between the RFV of the proposed new building and the RFV of the building analysed. The equation is as follows:

$$\frac{(\text{RVF factor for proposed new building} - \text{RVF for building analysed})}{\text{RVF for building analysed}} \times 100 = \%\ \text{change}$$

As for time adjustments, the percentage change is quantified in monetary terms and simply added to the cost/m² of GIFA. Equally, an adjustment factor can be used to update cost/m² of GIFA for regional variations. The updated cost/m² of GIFA is simply calculated by multiplying the cost/m² of GIFA obtained from a cost analysis by the factor resulting from:

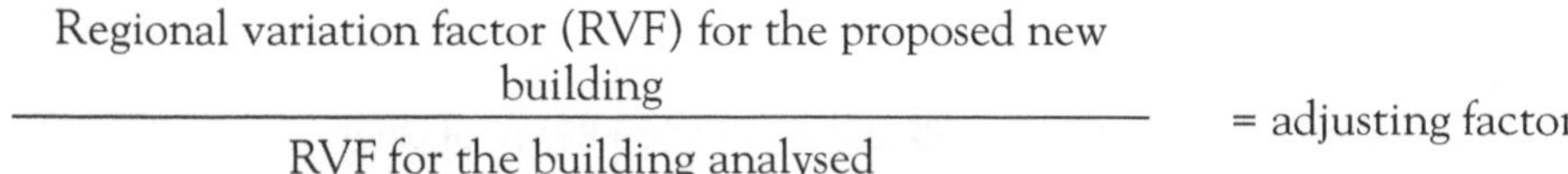

$$\frac{\text{Regional variation factor (RVF) for the proposed new building}}{\text{RVF for the building analysed}} = \text{adjusting factor}$$

Example 23.2 Regional adjustment using regional variation factors

Assumptions:

- Building works costs at estimate base date, unadjusted for regional variation: £2,234.82/m² of GIFA
- Base date of cost analysis: February 2011
- Estimate base date: September 2013
- Regions in which buildings are located:
- Building analysed: North east
- Proposed new building: South east
- Regional variation factors:
- North east: 0.97
- South east: 1.08

(a) Using percentage change:

$$\text{Percentage change} = \frac{(1.08-0.97)}{0.97} \times 100 = 11.34\%$$

Therefore:
Adjusted cost/m² of GIFA =
£2,234.82 + (11.34% of £2,234.82) =
£2,234.82 + £253.43 =
£2,488.25

Thus:
Adjusted cost/m² of GIFA = £2,488.25 (£2,488.25/m² of GIFA)

(b) Using adjustment factor:

$$\text{Adjustment factor} = \frac{1.08}{0.97} = 1.1134$$

Therefore:
Adjusted cost/m² of GIFA =
£2,234.82 × 1.1134 =
£2,488.25

Thus:
Adjusted cost/m² of GIFA = £2,488.25 (£2,488.25/m² of GIFA)

23.5.4 Other cost factor adjustments

When interpolating a unit rate to be applied to a proposed new building from an analysed building, there are a number of other factors that will need to be considered by the cost manager. For example, some variation in cost can be expected from differing procurement strategies (e.g. traditional, design and build, or management), type of contractor and tendering strategies (e.g. competitive or negotiated, single stage or two stage). For example, the building analysed might have been based on a single-stage negotiated tender, which might have resulted in higher rates than if the building had been procured through competition. Another factor might be the difference in size between the building analysed and the new proposed building. Better economies of scale should be expected from a larger development, even more so if extensive repetition was involved –

Example 23.3 Adjustment of cost/m² of GIFA for other cost factors

Assumptions:

- Updated cost/m² of GIFA: £2,488.25/m² of GIFA
- Percentage adjustment for tender strategy: 1.50%

(a) Using percentage adjustment:
= £2,488.25 × 1.50%

Therefore:
Adjusted cost/m² of GIFA =
£2,488.25 + (1.50% of £2,488.25) =
£2,488.25 + £37.32 =
£2,525.57

Thus:
Adjusted cost/m² of GIFA = £2,525.57 (£2,525.57/m² of GIFA)

(b) Using adjustment factor:
1 + 0.015 =
1.015

Therefore:
Adjusted cost/m² of GIFA =
£2,488.25 × 1.015 =
£2,525.57

Thus:
Adjusted cost/m² of GIFA = £2,525.57 (£2,525.57/m² of GIFA)

Example 23.4 Adjustment of cost/m² of GIFA for other cost factors – multiple

Assumptions:
Updated cost/m² of GIFA: £2,488.25/m² of GIFA

- Percentage adjustment for tender strategy: 1.50%
- Percentage adjustment for building size: -0.50%

(a) Using percentage adjustment:
Tender strategy = £2,488.25 × 1.50% = £37.33
= £2,488.25 + £37.33 = £2,525.58
Adjustment for building size = £2,525.58 × (–0.50%) = –£12.63
= £2,488.25 + £37.33 – £12.63 = £2,512.95

Therefore, adjusted cost/m² of GIFA:
= £2,512.95

(b) Using adjustment factor:
Adjustment for tender strategy = £2,488.25 × (1 + 0.015) × (1 – 0.005)
= £2,488.25 × 1.015 × 0.995
= £2,512.95

Therefore:
Adjusted cost/m² of GIFA = £2,512.95

thus reducing the unit cost to reflect the efficiency and repetition. Other price adjustment factors can be determined from benchmark data, indices or professional judgement.

Adjustments to unit rates for other factors are usually based on a percentage adjustment. To determine the cost adjustment, the updated cost/m^2 of GIFA is simply multiplied by the percentage adjustment. The equation is as follows:

Allowance for cost factor (£) = updated cost/m^2 of GIFA × percentage adjustment (%)

To update the cost/m^2 of GIFA, the cost factor allowance is simply added, or subtracted from, the cost/m^2 of GIFA.

Alternatively, the cost factor can be considered as an adjustment factor. The adjustment factor is ascertained as follows:

1 + percentage adjustment as a decimal (e.g. 1.50% equals 0.015 as a decimal)

The updated cost/m^2 of GIFA is determined by multiplying the updated cost/m^2 of GIFA by the adjusting factor.

Where adjustment for more than one cost factor is required, the percentage adjustments, or adjustment factors, for each cost factor are simply added together.

23.5.5 Isolate major differences

Major differences are items that are either missing from the analysed building, but will be required in the new building project, or included in the analysed building, but will not be required in the new building project. In order to isolate any major differences between the new proposed building and the analysed existing buildings, the cost manager will need to study the employer's brief for the proposed building and all the information given in the cost analyses. This does not include the design development risks as they are considered in a separate adjustment for the whole design.

For example, an analysed office building might have included for a 'Cat. B' fit-out (occupier fit-out), whereas a proposed new office building is to be fitted out to 'Cat. A' (developer fit-out), which will impact on the overall cost/m^2 of GIFA. In this case, the cost manager will need to isolate the costs associated with the 'Cat. B' fit-out and deduct them from the cost/m^2 of GIFA derived from the cost analysis. Conversely, if the building analysed included a 'Cat. A' fit-out and the proposed building was to 'Cat. B' fit-out, this will need to be recorded by the cost manager. Allowances for major factors which increase the cost of a building are usually made after the cost of building works based on the updated cost/m^2 of GIFA has been calculated (see paragraph 23.5.6). It is essential that any adjustments to the cost/m^2 of GIFA for major difference be made at the same price levels (i.e. at the estimate base date).

Other major differences might revolve around, for example, the nature of the ground conditions (i.e. the type of foundation construction), the proximity of adjoining properties, basements, or working hours.

23.5.6 Specification adjustment

The statement of quality in an employer's initial brief is often vague. Outline specification information prepared for cost analyses is equally vague. The differences in specification, or quality, will also vary from element to element. Consequently, it is difficult for the cost manager to identify with the specification information available in the employer's brief and cost analyses. It is often so vague that the relatively small differences in quality which remain after choosing the cost analyses, or benchmark data, cannot be allowed for until the employer's brief and specification information has been developed. Nevertheless, where it is considered that the cost analyses do not reflect the employer's desired quality requirements, the cost manager will need to make a quality adjustment.

Specification adjustments include those in connection with achieving different levels of sustainable design (e.g. changes to reflect the cost of attaining higher or lower environmental assessment ratings through BREEAM (Building Research Establishment Environmental Assessment Method) or the Code for Sustainable Homes or other applicable environmental assessment method).

Changes in specification, or quality, are usually based on a percentage adjustment. The percentage adjustment selected by the cost manager will reflect the increase or decrease in the cost of delivering the quality desired by the employer (e.g. +5.00% or −2.50%). To calculate the cost adjustment, the updated cost/m² of GIFA is simply multiplied by the percentage adjustment. The equation is as follows:

> Updated cost/m² of GIFA × specification adjustment (%) = allowance for specification adjustment (£)

To update the cost/m² of GIFA for specification, or quality, the allowance for specification adjustment is simply added to, or subtracted from, the cost/m² of GIFA.

Specification, or quality, adjustments can also be calculated using an adjustment factor. The updated cost/m² of GIFA is determined by multiplying the cost/m² of GIFA obtained from a cost analysis by the factor resulting from:

> 1 + specification adjustment as a decimal (e.g. 2.50% equals 0.025 as a decimal) = adjusting factor

Examples 23.5 and 23.6 show how each method is applied. For simplicity, it has been assumed that the same specification adjustment factor is applicable to all building work elements. The same method is used to adjust the cost/m² of GIFA for individual elements.

Example 23.5 Specification adjustment of cost/m² of GIFA

Assumptions:

- Updated cost/m² of GIFA: £2,512.95/m² of GIFA
- Allowance for higher specification: 3.50%

(a) Using percentage adjustment:

Adjusted cost/m² of GIFA =
£2,512.95 + (3.50% of £2,512.95) =
£2,512.95 + £87.96 =
£2,600.91

Thus:

Adjusted cost/m² of GIFA = £2,600.91 (£2,600.91/m² of GIFA)

(b) Using adjustment factor:

1 + 0.035 =
1.035

Therefore:

Adjusted cost/m² of GIFA =
£2,512.95 × 1.035 =
£2,600.91

Thus:

Adjusted cost/m² of GIFA = £2,600.91 (£2,600.91/m² of GIFA)

Example 23.6 Calculating the cost/m² of GIFA for proposed new building

Assumptions:

- Cost/m² of GIFA for building analysed: £2,176.28/m² of GIFA
- Time adjustment factor: 1.0269
- Regional variation adjustment factor: 1.1134
- Other price adjustment factors: 1.0099
- Specification or quality adjustment factor: 1.035
- Major differences: add £480.00/m² for 'Cat. B' fit-out (at current price levels)

$$\text{Cost/m}^2\text{ of GIFA}_{PB} = (£2{,}176.28 \times 1.0269 \times 1.1134 \times 1.0099 \times 1.035) + £480.00$$
$$= £2{,}600.84 + £480.00$$
$$= £3{,}080.84$$

Thus:

Cost/m² of GIFA for proposed new building (including 'Cat. B' fit-out) = £3,080.84 (£3,080.84/m² of GIFA)

Note:

The minor difference in the unit rate derived in the above example, when compared to the one previously calculated is due to rounding

23.5.7 Calculating the cost/m² of GIFA for proposed new building – holistic approach

The cost/m² of GIFA for the new proposed building is determined as follows:

Cost/m² of GIFAPB = (Cost/m² of GIFACA × T × RV × OPF × Q) + MDQ

Where:

Cost/m² of GIFAPB	=	Cost/m² of GIFA for proposed building
Cost/m² of GIFACA	=	Cost/m² of GIFA for building analysed
T	=	Time adjustment
RV	=	Regional variation adjustment
OPF	=	Other price adjustment (as required)
Q	=	Specification or quality adjustment
MD	=	Major differences.

The same method is used to adjust and update the cost/m² of GIFA for individual elements. To calculate the updated total cost/m² of GIFA for building works, the updated cost/m² of GIFA for each element is simply added together.

23.5.8 Quantity adjustment

The quantity adjustment is dealt with when the updated cost/m² of GIFA for building works is multiplied by the GIFA of the proposed new building:

Total cost of building works (excluding external works, facilitating works and items missing from the analysed building) =

GIFA of proposed new building × updated cost/m² of GIFA for building works

23.5.9 External works

Allowances for external works, facilitating works and items missing from the analysed building are added to give the total cost of building works.

23.6 Storey height

Storey heights of buildings are determined largely by the use of the building. Greater storey heights are than normal might be necessary to accommodate large mechanical and/or electrical plant, vehicular movements (e.g. loading bays), or for prestige reasons (e.g. office and hotel foyers). Higher storey heights do have the effect of increasing the costs of the vertical circulation elements initially, and also the future life costs in operating the building (i.e. running and maintenance costs). Buildings with high storey heights will cost more per square meter of floor area than comparable accommodation with lower storey heights. Therefore, when using rates based on the cost/m² of GIFA derived from cost analyses of previous building projects to estimate the cost of a new building, the cost manager must consider and make adjustment to the historical rates for the difference in storey heights (i.e. the different floor to floor heights). Failure by the cost manager to consider the cost effect of building storey heights could result in either the over or under cost estimate of the new proposed building.

23.7 Determining functional unit rates for building works

The interpolation method can also be used to determine functional unit rates for proposed buildings.

23.8 Estimating element unit rates (EURs)

The total building costs for the building project are obtained by adjusting and updating the cost/m² of GIFA from cost analyses of previously tendered building projects and multiplying the result by the total GIFA of the proposed new building. This gives the cost target for building works.

By taking existing cost analyses of representative and similar buildings, the cost manager can develop an initial elemental cost model and realistically set initial cost targets for group elements and elements and be reasonably confident that they represent practical cost targets for the cost planning stages. Element unit rates can then be derived from the cost targets established. Figure 23.2 shows typical elemental breakdowns of analysed buildings, illustrating the cost of group elements and elements as a cost/m² of GIFA and as a percentage of the total cost/m² of GIFA of building works.

Code	Group element/Element	Office Building 1		Office Building 2		Office Building 3		Office Building 4		Median
		£	%	£	%	£	%	£	%	%
0	**Facilitating works**	**0.00**	**0.00**	**0.00**	**0.00**	**0.00**	**0.00**	**0.00**		**0.00**
1	**Substructure**	**213.13**	**9.86**	**236.98**	**11.88**	**202.02**	**10.20**	**185.63**	**8.26**	**11.04**
2	**Superstructure**	**1,010.49**	**46.73**	**846.60**	**42.46**	**871.90**	**44.03**	**1,080.48**	**48.09**	**43.25**
2.1	Frame	209.54	9.69	205.52	10.31	211.66	10.69	221.80	9.87	10.50
2.2	Upper floors	137.35	6.35	120.15	6.03	123.74	6.25	129.67	5.77	6.14
2.3	Roof	42.59	1.97	34.46	1.73	35.49	1.79	28.92	1.29	1.76
2.4	Stairs and ramps	41.11	1.90	36.35	1.82	37.44	1.89	39.23	1.75	1.86
2.5	External walls	465.05	21.51	342.09	17.16	352.31	17.79	439.21	19.55	17.48
2.6	Windows and external doors	10.61	0.49	10.05	0.50	10.35	0.52	23.54	1.05	0.51
2.7	Internal walls and partitions	48.83	2.26	45.90	2.30	47.27	2.39	66.99	2.98	2.35
2.8	Internal doors	55.41	2.56	52.08	2.61	53.64	2.71	131.12	5.84	2.66
3	**Internal finishes**	**197.06**	**9.11**	**185.23**	**9.29**	**190.76**	**9.63**	**199.90**	**8.90**	**9.46**
3.1	Wall finishes	35.97	1.66	33.81	1.70	34.82	1.76	36.49	1.62	1.73
3.2	Floor finishes	95.39	4.41	89.66	4.50	92.34	4.66	96.76	4.31	4.58
3.3	Ceiling finishes	65.70	3.04	61.76	3.10	63.60	3.21	66.65	2.97	3.16
4	**Fittings, furnishings and equipment**	**43.62**	**2.02**	**41.01**	**2.06**	**42.23**	**2.13**	**52.76**	**2.35**	**2.10**

Table title rows: Cost analyses (at May 2013 price levels): Cost/m² of GIFA (£) of group element and element and percentage (%) of total cost/m² of GIFA of building works

5	**Services**	**697.97**	**32.28**	**684.23**	**34.31**	**673.42**	**34.01**	**727.80**	**32.40**	**34.16**
5.1	Sanitary installations	12.58	0.58	11.83	0.59	12.18	0.62	12.76	0.57	0.61
5.2	Services equipment	0.00	0.00	0.00	0.00	0.00	0.00	0.00	0.00	0.00
5.3	Disposal installations	7.85	0.36	7.38	0.37	7.60	0.38	7.96	0.35	0.38
5.4	Water installations	16.07	0.74	15.11	0.76	15.56	0.79	16.31	0.73	0.78
5.5	Heat source	21.75	1.01	29.60	1.48	21.06	1.06	22.07	0.98	1.27
5.6	Space heating and air conditioning	48.30	2.23	46.53	2.33	37.14	1.88	38.92	1.73	2.11
5.7	Ventilation	39.45	1.82	53.14	2.66	38.19	1.93	40.02	1.78	2.30
5.8	Electrical installations	352.65	16.31	331.48	16.62	341.38	17.24	357.73	15.92	16.93
5.9	Fuel installations	5.06	0.23	4.89	0.25	3.01	0.15	3.15	0.14	0.20
5.10	Lift and conveyor installations	93.67	4.33	103.76	5.20	90.68	4.58	95.02	4.23	4.89
5.11	Fire and lightning protection	30.63	1.42	17.60	0.88	29.65	1.50	31.07	1.38	1.19
5.12	Communication, security and control systems	42.50	1.97	39.95	2.00	41.14	2.08	43.11	1.92	2.04
5.13	Special installations	27.04	1.25	17.18	0.86	17.69	0.89	45.38	2.02	0.88
5.14	Builder's work in connection with services	13.00	0.60	17.61	0.88	18.14	0.92	27.06	1.20	0.90
6	**Prefabricated buildings and building units**	**0.00**	**0.00**	**0.00**	**0.00**	**0.00**	**0.00**	**0.00**	**0.00**	**0.00**
7	**Works to existing buildings**	**0.00**	**0.00**	**0.00**	**0.00**	**0.00**	**0.00**	**0.00**	**0.00**	**0.00**
8	**External works**	**0.00**	**0.00**	**0.00**	**0.00**	**0.00**	**0.00**	**0.00**	**0.00**	**0.00**
	Building works cost	**2,162.27**	**100.00**	**1,994.05**	**100.00**	**1,980.33**	**100.00**	**2,246.57**	**100.00**	**100.00**

Figure 23.2 Range of analysed buildings for estimating cost targets for group elements and elements

Box 23.1 Key definitions

- **Element** – a subdivision of a group element.
- **Element unit quantity (EUQ)** – a unit of measurement which relates solely to the quantity of the element.
- **Element unit rate (EUR)** – the unit rate applied to an element unit quantity. EURs are derived from cost information relating to one or more historical building projects. They are ascertained from a historical building project by calculating the total cost of an element and dividing it by the EUQ for the applicable element.
- **Group element** – used to describe the main facets of a building project.

23.8.1 Determining initial cost targets for group elements and elements

The cost manager can derive cost targets for group elements and elements by applying the appropriate percentage ascertained from the cost analysis, or range of cost analyses, to the total cost target for building works. This process is illustrated in Example 23.7.

Example 23.7 Calculating initial cost targets for group elements and elements

Assumptions:

- Building works costs, excluding external works and facilitating works: £20,356,250
- Percentage apportionment based on office building 4 (refer to Figure 23.1)
- Cost targets for external works and facilitating works are to be determined separately.

Initial cost targets for group elements and elements are calculated by multiplying the building works estimate (£) by the percentage apportionment (%). For example:

Substructure = £20,356,250 × 11.01% = £2,241,223

Frame = £20,356,250 × 10.47% = £2,131,299

Upper floors = £20,356,250 × 6.12 % = £1,245,803.

Code	Group element/Element	Percentage apportionment		Initial cost targets	
		Group element	Element	Group element	Element
		%	%	£	£
0	**Facilitating works**	**0.00**	0.00	**0**	0
1	**Substructure**	**11.01**	**11.01**	**2,241,223**	2,241,223
2	**Superstructure**	**43.11**		**8,775,580**	
2.1	Frame		10.47		2,131,299
2.2	Upper floors		6.12		1,245,803
2.3	Roof		1.75		356,234
2.4	Stairs and ramps		1.85		376,591

Example 23.7 *(Continued)*

Code	Group element/Element	Percentage apportionment		Initial cost targets	
		Group element	Element	Group element	Element
		%	%	£	£
2.5	External walls		17.42		3,546,059
2.6	Windows and external doors		0.51		103,817
2.7	Internal walls and partitions		2.34		476,336
2.8	Internal doors		2.65		539,441
3	**Internal finishes**	**9.46**		**1,919,595**	
3.1	Wall finishes		1.72		350,128
3.2	Floor finishes		4.57		930,281
3.3	Ceiling finishes		3.14		639,186
4	**Fittings, furnishings and equipment**	**2.09**	**2.09**	**425,446**	425,446
5	**Services**	**34.36**		**6,994,406**	
5.1	Sanitary installations		0.60		122,138
5.2	Services equipment		0.00		0
5.3	Disposal installations		0.38		77,354
5.4	Water installations		0.77		156,743
5.5	Heat source		1.27		258,524
5.6	Space heating and air conditioning		2.10		427,481
5.7	Ventilation		2.29		466,158
5.8	Electrical installations		16.88		3,436,135
5.9	Fuel installations		0.20		40,713
5.10	Lift and conveyor installations		4.88		993,385
5.11	Fire and lightning protection		1.19		242,239
5.12	Communication, security and control systems		2.03		413,231
5.13	Special installations		0.87		177,099
5.14	Builder's work in connection with services		0.90		183,206
6	**Prefabricated buildings and building units**	**0.00**		**0**	0
7	**Works to existing buildings**	**0.00**		**0**	0
8	**External works**	**0.00**		**0**	0
	Building works cost	**100.00**	100.00	**20,356,250**	20,356,250

23.8.2 Determining initial element unit rates

The cost of an element can be expressed in suitable units that relate solely to the quantity of an element itself. These units are referred to as element unit quantities. The cost of an EUQ is known as an element unit rate. The cost target for an element (i.e. the total cost of an element) is calculated by multiplying the EUQ by the EUR:

Cost target for an element = element unit quantity (EUQ) × Element unit rate (EUR)

Therefore, initial EURs can be derived for elements by simply dividing the cost target for each element by the applicable element unit quantity (EUQ):

$$\text{Element unit rate (EUR)} = \frac{\text{Cost target for element}}{\text{Element unit quantity (EUQ)}}$$

This methodology is demonstrated in Example 23.8.

Example 23.8 Calculating element unit rates

(a) Upper floors:

Assumptions:

- Cost target: £2,131,299
- Element unit quantity: 9,374m²

$$\text{EUR} = \frac{£2{,}131{,}299}{9{,}374\text{m}^2} = £227.37/\text{m}^2$$

(b) Roof:

Assumptions:

- Cost target: £356,234
- Element unit quantity: 845m²

$$\text{EUR} = \frac{£356{,}234}{845\text{m}^2} = £421.58/\text{m}^2$$

(c) External walls:

Assumptions:

- Cost target: £3,546,059
- Element unit quantity: 4,826m²

$$\text{EUR} = \frac{£3{,}546{,}059}{4{,}826\text{m}^2} = £734.79/\text{m}^2$$

(d) Space heating and air conditioning:

Assumptions:

- Cost target: £427,481
- Element unit quantity: 10,443m² (based on GIFA)

$$\text{EUR} = \frac{£427{,}481}{10{,}443\text{m}^2} = £40.94/\text{m}^2$$

23.9 Estimating unit rates for components

In cost planning, a 'unit rate' is established for each component measured. Components are 'composite items' (i.e. they include all materials, labour and plant necessary to construct and install them). The unit rate is then multiplied by the measured quantity to find the total cost of the component. The costs of all components that are part of a sub-element are summed to obtain the total estimated cost of the sub-component. Likewise, the costs of the sub-elements that form an element are summed to give the cost target for the said element, and so on.

Unit rates for components are determined by ascertaining the quantities of the main sub-components that constitute the component. For example, the cost of constructing a pile cap or ground beam will include: excavation, earthwork support, disposal of excavated material, blinding, concrete, reinforcement, formwork, extra excavation to and backfilling to provide working space and (in certain situations) heave protection board. All-in unit rates are used to estimate the cost of each individual sub-component. The cost of each sub-component is summed and adjusted to allow for minor miscellaneous items not considered to have been covered by the all-in rates and to include allowances for the subcontractor's preliminaries costs, the subcontractor's design fees and charges, the subcontractor's risk allowance and the subcontractor's overheads and profit.

An example of building up a unit rate for a pile cap is given (Figure 23.4) using the details shown in Figure 23.3.

The cost manager can easily set up unit rate build-up or calculation sheets such as the one illustrated in Figure 23.4 on a spreadsheet or database. The variables can be changed so that the unit rate of different sized components can be calculated by simply changing the variables (e.g. changes to pile cap dimensions and rate of reinforcement).

(a) Data collection for purpose of building up unit rate
(b) Calculation of unit rate

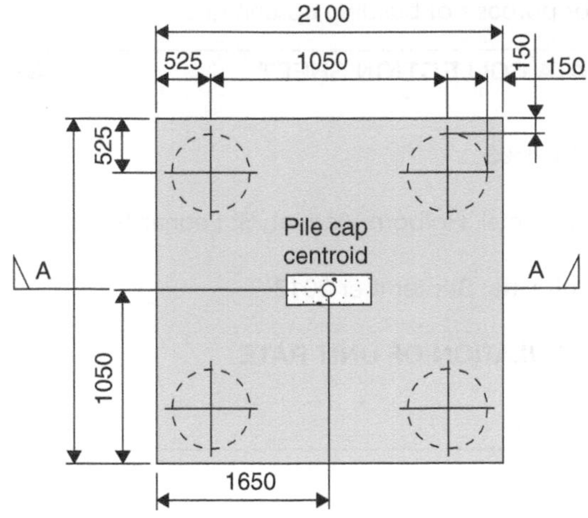

Pile Cap Type PC1. 750 mm diameter piles

Note:
Cellcore type leave protection board grade 10/15 to underside of pile caps

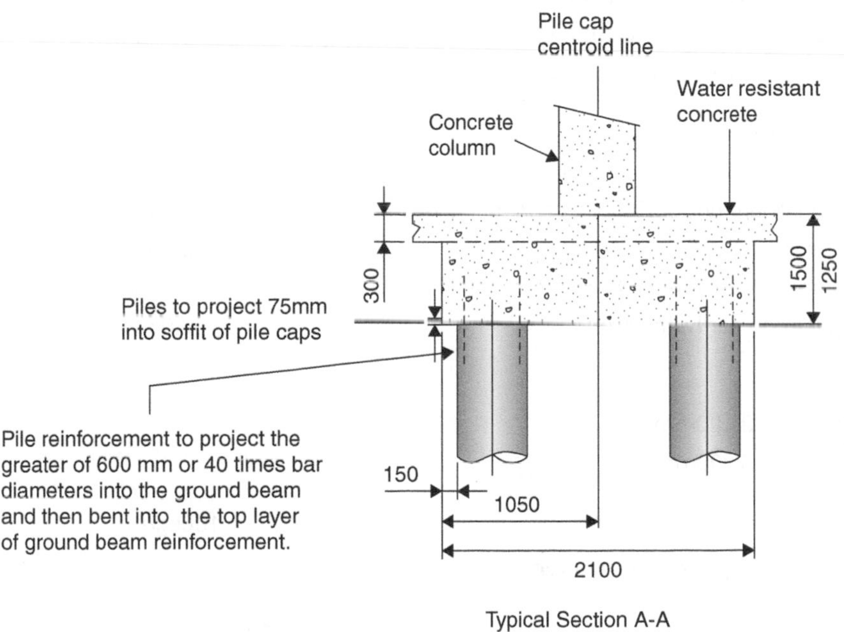

Typical Section A-A

Figure 23.3 Plan and section – pile cap

(a) Data collection for purpose of building up unit rate

UNIT RATE: DATA COLLECTION SHEET		**BRIGHT KEWESS PARTNERSHIP**
Project no.: DPB-0693		
Project: Woollard Hotel, Holborn Viaduct, St Leonards-on-Sea, East Sussex		
Base date of unit rate: September 2013		
DATA FOR CALCULATION OF UNIT RATE		
Component: Pile cap		
Assumptions		
Information taken from structural engineer's pile cap details		
Ground slab construction:		
RC base slab, Grade C32/40; thickness =		0.300m
Concrete; blinding, Grade C10 (lean mix)		0.100m
Heave protection board under RC ground slab; thickness =		0.150m
Sand blinding; laid on earth; thickness =		0.050m
RC pile cap thickness =		1.250m
RC Pile caps, Grade C32/40; thickness =		
Heave protection board under RC pile caps; thickness =		0.150m
Concrete; blinding, Grade C10 (lean mix)		0.100m
Heave protection board under RC base slab; thickness =		0.150m
Sand blinding; laid on earth; thickness =		0.050m
Pile cap sizes:		
PC1 = 2,100mm × 2,100mm		
PC2 = 2,100mm × 3,300mm		
Reinforcement content of RC pile cap =		100kg/m³
Calculation of total depth of excavation:		
Ground slab =	0.300m	
Pile cap	1.250m	
Add		1.550m
Sand, blinding	0.050m	
Heave protection board under pile cap	0.150m	
Add		1.550m

Figure 23.4 Calculation of unit rate – pile cap

Sand, blinding	0.050m	
Heave protection board under pile cap	0.150m	
Concrete; blinding	0.100m	
		0.300m
		1.850m
Less (see note)		
Ground slab	– 0.300m	
Sand, blinding	– 0.050m	
Heave protection board under ground slab	– 0.150m	
Concrete; blinding	– 0.100m	
		– (0.600)m
Total depth of excavation:		**1.250m**

Notes:

(1) Excavation associated with ground slab construction is to be measured with basement excavation or with ground slab construction where no basement

(2) Sand blinding, heave protection board and lean mix concrete blinding measured with RC base slab

(b) Calculation of unit rate

UNIT RATE: CALCULATION SHEET **BRIGHT KEWESS PARTNERSHIP**

Project no.: DPB-0693

Project: Woollard Hotel, Holborn Viaduct, St Leonards-on-Sea, East Sussex

Base date of unit rate: September 2013

CALCULATION OF UNIT RATE

Component: Pile cap type PC1

Pile cap data:

Overall dimensions:	(a)	Width	2.10m
	(b)	Length	2.10m
		Depths/thicknesses	
	(c)	Base slab	0.30m
	(d)	Pile cap	1.25m
	(e)	Excavation	1.25m
Reinforcement rate:	(f)		100.00 kg/m^3

Figure 23.4 *(Continued)*

Item	Sub-component	Formula (see note)	Qty	Unit	Rate £		Cost/nr £
1	Excavation (including earthwork support)	(a) × (b) × (e)	5.51	m^3	6.00		33.06
2	Disposal	(a) × (b) × (e)	5.51	m^3	38.00		209.38
3	Blinding; sand; 50mm thick	(a) × (b)	4.41	m^2	0.00		0.00
4	Heave protection board; 150mm thick	(a) × (b)	4.41	m^2	0.00		0.00
5	Concrete; blinding, Grade C10 (lean mix); 100mm thick	(a) × (b)	4.41	m^2	0.00		0.00
6	Reinforced concrete; Grade C32/40; in pile caps	(a) × (b) × (d)	5.51	m^3	140.00		771.40
7	Extra over for water resistant concrete	(a) × (b) × (d)	5.51	m^3	0.00		0.00
8	Reinforcement	[(a) × (b) × ((c) + (d))] × [(f)/1000]	0.68	t	960.00		652.80
9	Formwork; to sides of pile cap; basic finish	[2 × ((a) + (b))] × (d)	10.50	m^2	40.00		420.00
10	Excavating and filling of working space	[2 × ((a) + (b))] × (d)	10.50	m^2	7.00		73.50
	Sub-total						2,160.14
	Minor miscellaneous items and ancillary labours				1.50%		24.40
	Sub-total						2,192.54
	Subcontractor's preliminaries				15.00%		328.88
	Sub-total						2,521.42
	Subcontractor's design fees/charges				0.00%		0.00
	Sub-total						2,521.42
	Subcontractor's risk allowance				2.50%		63.04
	Sub-total						2,584.46
	Subcontractor's overheads and profit				10.00%		258.45
	Unit Rate:					£	**2,842.90**
	Unit Rate (rounded to nearest £):					£	**2,843**

Figure 23.4 *(Continued)*

CHAPTER 24

Calculating the building works estimate

Introduction

This chapter:

- explains how to calculate the building works estimate.

24.1 Composition of the building works estimate

Figure 24.1 shows the composition of the building works estimate.

Estimate	Group element	Facet	Chapter/reference
Estimate 1A	0: Facilitating works	*Cost target. Includes major demolition works.*	Chapter 14
	+		
Estimate 1B	1: Substructure	*Cost target.*	Chapter 15
	+		
Estimate 1C	2: Superstructure	*Cost target.*	Chapter 16
	+		
Estimate 1D	3: Internal finishes	*Cost target.*	Chapter 17
	+		
Estimate 1E	4: Fittings, furnishings and equipment	*Cost target.*	Chapter 18
	+		
Estimate 1F	5: Services	*Cost target.*	Chapter 19
	+		
Estimate 1G	6: Prefabricated buildings and building units	*Cost target.*	Chapter 20
	+		
Estimate 1H	7: Work to existing buildings	*Cost target.*	Chapter 21
	+		
Estimate 1I	8: External works	*Cost target. Includes cost of connections to statutory undertakers' mains services.*	Chapter 22

Figure 24.1 Composition of building works estimate

=

Estimate 1	**Building works estimate**	*Total cost of building works (excluding main contractor's preliminaries and main contractor's overheads and profit).*

Figure 24.1 *(Continued)*

24.2 Determining the building works estimate

The building works estimate is the summation of the cost targets for each group element. Figure 24.2 sets out how the building works estimate cost plan is calculated for a cost plan.

COST PLAN SUMMARY

Project: Woollard Hotel, Holborn Viaduct, St Leonards-on-Sea, East Sussex

Estimate base date: September 2013

Gross internal floor area (GIFA): 8,399m²

Number of Guest rooms (keys): 246

Code	Group element/Element	Element totals	Group element	Cost/m² of GIFA	Cost/Key
		£	£	£	£
0	Facilitating works		0	0.00	0.00
1	Substructure		2,480,300		
1.1	Substructure	2,480,300		295.31	10,082.52
2	Superstructure		5,841,312		
2.1	Frame	1,149,949		136.91	4,674.59
2.2	Upper floors	776,526		92.45	3,156.61
2.3	Roof	413,158		49.19	1,679.50
2.4	Stairs and ramps	281,000		33.46	1,142.28
2.5	External walls	2,629,284		313.05	10,688.15
2.6	Windows and external doors	85,523		10.18	347.65
2.7	Internal walls and partitions	176,211		20.98	716.30
2.8	Internal doors	329,661		39.25	1,340.09
3	Internal finishes		2,447,882		
3.1	Wall finishes	1,666,129		198.37	6,772.88
3.2	Floor finishes	415,163		49.43	1,687.65
3.3	Ceiling finishes	366,590		43.65	1,490.20
4	Fittings, furnishings and equipment		591,458		
4.1	Fittings, furnishings and equipment	591,458		70.42	2,404.30

Figure 24.2 Computing the building works estimate

Code	Group element/Element	Element totals	Group element	Cost/m² of GIFA	Cost/Key
		£	£	£	£
5	Services		4,033,645		
5.1	Sanitary installations	25,701		3.06	104.48
5.2	Services equipment	235,000		27.98	955.28
5.3	Disposal installations	68,575		8.16	278.76
5.4	Water installations	375,939		44.76	1,528.21
5.5	Heat source	230,049		27.39	935.16
5.6	Space heating and air conditioning	288,506		34.35	1,172.79
5.7	Ventilation	466,984		55.60	1,898.31
5.8	Electrical installations	498,806		59.39	2,027.67
5.9	Fuel installations	447,583		53.29	1,819.44
5.10	Lift and conveyor installations	342,595		40.79	1,392.66
5.11	Fire and lightning protection	532,413		63.39	2,164.28
5.12	Communication, security and control systems	261,797		31.17	1,064.22
5.13	Special installations	150,006		17.86	609.78
5.14	Builder's work in connection with services	109,691		13.06	445.90
6	Prefabricated buildings and building units		7,548,517		
6.1	Prefabricated buildings and building units	7,548,517		898.74	30,685.03
7	Works to existing buildings	0	0	0.00	0.00
8	External works		333,000		
8.1	Site preparation works	0		0.00	0.00
8.2	Roads, paths, pavings and surfacings	123,500		14.70	502.03
8.3	Soft landscaping, planting and irrigation systems	0		0.00	0.00
8.4	Fencing, railings and walls	0		0.00	0.00
8.5	External fixtures	0		0.00	0.00
8.6	External drainage	20,000		2.38	81.30
8.7	External services	189,500		22.56	770.33
8.8	Minor building works and ancillary buildings	0		0.00	0.00
	Building works estimate	**23,276,114**	**23,276,114**	**2,771.28**	**94,618.35**

Figure 24.2 *(Continued)*

CHAPTER 25

Main contractor's preliminaries (group element 9)

Introduction

This chapter:

- defines main contractor's preliminaries;
- distinguishes between main contractor's and subcontractor's preliminaries;
- explains how to deal with subcontractor's preliminaries;
- considers the principal items and factors that can significantly influence the cost of main contractor's preliminaries;
- describes the items which constitute main contractor's preliminaries items;
- demonstrates how to calculate an initial cost target for main contractor's preliminaries;
- outlines the dangers of using percentages derived from past projects to calculate cost targets for main contractor's preliminaries; and
- contains examples of typical calculations to determine the adequacy of cost targets for cost-significant main contractor's preliminaries items.

25.1 What are preliminaries?

Preliminaries are a term used to describe all the items of construction-related expenditure that cannot be attributed to individual items of building work. The term is used to describe these items because, as a rule, they appear at the beginning of a bill of quantities or employer's requirements. Preliminaries are a cost-significant element in most construction projects, and they are directly influenced by the choice of construction method more than any other element.

25.2 Estimation of preliminaries costs

Preliminaries costs form an ever increasing proportion of the total cost of building works, often representing 8% to 30% of the works cost estimate. This is due to such factors as:

- changes in approach of main contracting (i.e. from the main contractor executing the construction works using predominantly directly employed labour, to the main contractor adopting a management contractor approach using subcontractors to execute the works);
- the increasing complexity of building which leads to the need for more highly trained supervision;
- greater use of machines requires careful co-ordination (i.e. common-user plant); and
- increased standards of safety, health and welfare necessitating more and better site facilities.

It is important, therefore, that the cost manager fully understands the nature of the element and the factors that affect its level of pricing.

The cost of preliminaries will be influenced by choice of construction method more than any other element. However, at the time the cost plan is prepared, the main contractor's site organisation and method of construction is usually not known, since these are decisions made by the main contractor who has yet to be appointed. Consequently, the accurate estimation of preliminaries costs based on past building projects is made difficult. Nonetheless, the use of cost analyses and benchmark data provide the best sources of information from which to begin to estimate the cost of this element.

The estimated cost of preliminaries is often based on a percentage of the cost of building works. But cost managers often consider percentages in a cursory fashion, usually based on past projects and an interpretation of information available on the building project for which the cost plan is being produced. However, it is essential that cost managers consider the principal items and factors that can significantly influence the percentage used to estimate preliminaries costs. These principal items and factors are:

- **Location** – assessment of ease or difficulty of entry to and egress from the site, distance from major highways, requirements for temporary roads.
- **Space available on site** – limited on-site storage and space for facilities off site is likely to require special measures leading to increased costs.
- **Security** – temporary fencing, hoardings and gantries required for public safety and to protect the works from pilfering and vandalism are significant costs affected by the location of the boundary of the site.
- **Contract period** – many items in preliminaries are related directly to the length of the construction period (i.e. commonly referred to as 'time-related charges', e.g. management and staff, temporary accommodation, security staff, hire charges for mechanical plant and access scaffolding, and periodic fees and charges). Additionally, when an employer requires an early completion time, preliminaries will need to be increased to take account of overtime payments, weekend working and other premiums normally sought by the main contractor for such requirements.

- **Phasing** – building projects tend to have complex interactions. Therefore, for many employers, phasing the commencement and completion of the construction works is essential. Assume that a single building project comprises both the refurbishment of an existing and construction of a new manufacturing unit. However, the refurbishment works to the existing unit cannot be carried out at the same time as the new building because the employer's day-to-day operations must continue. This scenario is likely to result in the building project being phased. In this case, it is likely that the new unit will be completed first to allow the transfer of the employer's day-to-day operations to the new manufacturing unit before the refurbishment of the existing unit commences. Equally, a new permanent access road might be required to provide an essential early link between two locations before further construction work can continue on other parts of the building project. The upshot is that phasing construction works can add significant costs to a building project that the cost manager will need to consider when compiling the cost target for main contractor's preliminaries. Such costs can include the repositioning of site accommodation, dismantling and re-erection of mechanical plant, relocating hoardings and other temporary works. It is essential, therefore, that the cost manager scrutinises the proposed construction programme and ascertains the implications of any proposed phasing.
- **Plant** – many items of plant, particularly those provided by subcontractors, are directly related to the works in an element and are priced in that element. However, the costs of providing general items of mechanical and non-mechanical plant for the use of all operatives, including subcontractors' operatives, are to be included under main contractor's preliminaries. Such items are referred to as 'common-user plant'. Examples include tower cranes, mobile cranes, hoists, access plant and access scaffolding.
- **Fees and charges** – fees and charges where not paid by the employer (e.g. building control fees, oversailing fees, NHBC [National House-Building Council] Buildmark registration fees (providing post-completion warranty and insurance for residential schemes), and licences in connection with hoardings, scaffolding, crossovers, parking permits and parking bay suspensions).
- **Insurances** – the requirements for works insurance, public liability insurance, insurance in respect of adjoining owners and other insurances need to be judged on the basis of risks influencing the site.

NRM 1 distinguishes between main contractor's preliminaries and subcontractors' preliminaries. Main contractor's preliminaries are defined in the rules as 'items [that] cannot be allocated to a specific element, sub-element or component'. Subcontractors' preliminaries also deal with construction-related expenditure that cannot be attributed to individual items of building work, but only items that relate directly to the building work that is to be carried out by a subcontractor (e.g. the subcontractor's project-specific management staff, accommodation and storage, and mechanical plant for the sole use of the subcontractor).

25.3 Main contractor's preliminaries costs

Main contractor's preliminaries costs that need to be considered by the cost manager are listed in group element 9 of NRM 1 and come under two principal headings:

- employer's requirements; and
- main contractor's cost items.

25.3.1 Employer's requirements

These comprise:

- **Site accommodation** – accommodation, furniture and equipment, and telecommunications and IT systems that are separate from main contractor's site accommodation.
- **Site records** – documents specifically required by the employer (e.g. operation and maintenance manuals, health and safety file, and web-based document management systems).
- **Completion and post-completion requirements** – employer's requirements at handover (e.g. training of building user's staff in, and the provision of spare parts and tools for, the operation and maintenance of electrical services installations) and after practical completion (e.g. ongoing operation and maintenance of electrical services installations by specialist subcontractor).

25.3.2 Main contractor's cost items

Main contractor's costs items include:

- **Management and staff** – main contractor's project-specific employees (e.g. project manager or director, construction manager, element or trade package managers and quantity surveyors), employees in head office carrying out work directly associated with the project (where no allowance has been made within the main contractor's overheads), recruiting staff, staff training and development, temporary living accommodation, staff travel costs (e.g. to consultant's offices and overseas visits).
- **Site establishment** – materials and activities associated with setting up the construction site presence, including main contractor's common-user temporary site accommodation, storage, shelters and ablution facilities, furniture and equipment, temporary roads and hardstandings, brought-in services (e.g. catering) and other sundry items. This includes bringing to site, erecting, capital costs (i.e. if purchased), hire or rental charges, alterations and adaptations to site accommodation during construction works, and dismantling and removing from site on completion of the building project.
- **Temporary services** – temporary water, gas, electricity, telecommunications and drainage services required for carrying out the building works, including installation, distribution, operation, maintenance, charges and removal on completion of the building project.
- **Security** – security staff, security equipment, and temporary hoardings, fences, gates and the like to provide safe access to and around the site.
- **Safety and environmental protection** – works required to comply with the requirements of the CDM [Construction Design and Management] regulations, main contractor's safety programme, barriers and safety scaffolding (e.g. guard rails and edge protection), environmental protection (e.g. control of pollution).
- **Control and protection** – surveys, setting out the building, protection of the

finished works, provision of samples and mock ups, and the environmental control of the building (e.g. drying out and temporary heating or cooling).

- **Mechanical plant** – common-user mechanical plant and equipment used in construction operations (e.g. tower cranes, mobile cranes, hoists, mechanical access plant such as fork lifts and scissor lifts), concrete mixing plant, and other small plant and tools. This includes bringing to site; erecting; operating; maintaining; alterations and adaptations to facilitate construction works; and dismantling and removing the plant or equipment from site on completion of the building works. Mechanical plant also includes the construction, and removal, on completion, of temporary bases and temporary voids (e.g. in situations where a tower crane has to be situated within the building due to site constraints).
- **Temporary works** – common-user access scaffolding and common-user temporary structures and services installed to protect workers and/or the public, to provide safe access to and around the site and to provide temporary support to existing work being altered or extended (e.g. support scaffolding and propping, and floodlighting). Temporary works also include protection to trees and/or vegetation during building works. This includes bringing to site, erecting, altering and adapting during construction, and dismantling and removing the temporary works from site on completion of the building works.
- **Site records** – condition surveys, photographs, operation and maintenance manuals, as-built/installed drawings and compilation of health and safety file.
- **Completion and post-completion requirements** – works carried out to achieve completion of the building project and in the period following completion. Completion requirements include preparation of testing and commissioning plan(s) and handover plan(s); training of building user's staff in the operation and maintenance of the building engineering services installations; and pre-completion and final inspections. Post-completion services include the provision of supervisory staff, handymen and materials to deal with defects that arise during the period for rectifying defects (or defects liability period or maintenance period, whichever term is applicable).
- **Cleaning** – keeping the site clean during the construction, including cleaning the site accommodation; maintenance of site roads, paths and pavements (including adjoining public and private roads, paths and pavings); removing rubbish and waste materials; and final building clean prior to handover to the employer (or building user).
- **Fees and charges** – fees, including those in connection with building control, oversailing rights and building scheme registration fees that will not be paid directly by the employer. Charges include those relating to health and safety schemes, rates for temporary accommodation, licences in connection with temporary works (e.g. hoardings, scaffolding and gantries) and licences for crossovers, parking permits and parking bay suspensions.
- **Site services** – provision of multi-service gang(s) and the like.
- **Insurance, bonds, guarantees and warranties** – premiums paid for insurance policies taken out in relation to construction site activity (e.g. works-related insurances, public liability insurances, employer's liability insurances, employer's loss of liquidated damages and professional indemnity insurance; bonds; guarantees and warranties (e.g. collateral warranties, funder's warranties, and purchaser's and tenant's warranties).

Box 25.1 Key definitions

- **Main contractor's preliminaries** – items that cannot be allocated to a specific element, sub-element or component. Main contractor's preliminaries include the main contractor's costs associated with management and staff, site establishment, temporary services, security, safety and environmental protection, control and protection, common-user mechanical plant, common-user temporary works, the maintenance of site records, completion and post-completion requirements, cleaning, fees and charges, sites services and insurances, bonds, guarantees and warranties. Main contractor's preliminaries exclude costs associated with subcontractor's preliminaries, which are to be included in the unit rates applied to building works.
- **Subcontractor** – a contractor who undertakes specific work within the building project; known as specialist, works, trade, work package, and labour-only subcontractors.
- **Subcontractor's preliminaries** – preliminaries that relate specifically to building work that is to be carried out by a subcontractor. Costs associated with subcontractor's preliminaries are to be included in the unit rates applied to sub-elements and individual components.

25.4 Subcontractors' preliminaries costs

NRM 1 stipulates that subcontractors' costs in connection with preliminaries are to be included in the unit rates for sub-elements and components calculated by the cost manager. This is because they exclusively relate to construction work that is to be executed by a particular subcontractor and do not affect any other aspect of construction work. Suggested methods of dealing with subcontractors' preliminaries are shown in Chapter 23 ('Deriving unit rates for building components, sub-elements and elements').

25.5 Estimation of main contractor's preliminaries costs

The rules recognise that the cost estimating and cost checking of main contractor's preliminaries is an iterative process which is repeated at each cost-estimating and cost-planning stage. When preparing order of cost estimates and initial cost models, a percentage addition is to be added to the total cost of building works (i.e. the building works estimate) to cover the likely cost of main contractor's preliminaries. Equally, a percentage addition might be used to set the initial cost target at the commencement of cost planning. However, the rules advocate the cost manager undertaking cost checks on cost significant preliminary items.

As a rule, it is rare that main contractors provide a breakdown of the prices of items making up the main contractor's preliminaries. Most commonly, the total main contractor's preliminaries costs is given without any further information. This lack of information makes the estimation of main contractor's preliminaries costs in future building projects even more difficult. The exception to this rule is when a two-stage tendering process is used, as the provision of a detailed build-up of the main contractor's preliminaries costs is usually a prerequisite of progressing to the second stage. Where such information is available, the cost manager can analyse the make-up and distribution of the costs to obtain a better understanding of main contractor's preliminaries costs that can be used for future building projects.

The estimation of main contractor's preliminaries costs is considerably improved when a main contractor is appointed early in the development of the building project.

25.5.1 Estimation of the initial cost target

Main contractor's preliminaries costs estimated at the concept design of a building project are most commonly calculated by applying a percentage to the total cost of building works (i.e. to the building works estimate). This simple method is illustrated in Example 25.1. This percentage can be derived from cost analyses or benchmark data from previously completed building projects. However, there are dangers of using percentages for calculating main contractor's preliminaries costs that have been derived from past projects of which the cost manager must be aware. More often than not, a main contractor will consider the preliminaries and price only those items that have cost significance. Problems arise, however, as there are a number of ways in which main contractors price preliminary items within tender documents. Such methods include:

Example 25.1 Calculation of cost target for main contractor's preliminaries

Assume that the percentage to be applied in respect of main contractor's preliminaries is 13.00%. Thus, the cost target for main contractor's preliminaries is:

MAIN CONTRACTOR'S PRELIMINARIES					
Code	**Element/sub-elements/components**				**Totals**
				£	**£**
9	**MAIN CONTRACTOR'S PRELIMINARIES**				
	Group element 0: Facilitating works:			655,350	
	Group element 1: Substructure:			1,682,700	
	Group element 2: Superstructure:			5,990,070	
	Group element 3: Internal finishes:			2,300,688	
	Group element 4: Fittings, furnishings and equipment:			533,660	
	Group element 5: Services:			3,948,771	
	Group element 6: Prefabricated buildings and building units:			7,793,500	
	Group element 7: Work to existing buildings:			0	
	Group element 8: External works:			370,655	
	Building works estimate:			*23,275,394*	
	Main contractor's preliminaries:	13.00%	of	23,275,394	3,025,801
9	**TOTAL – Main contractor's preliminaries: to cost plan summary**				3,025,801

- showing no individually priced preliminary items, but giving a total only for the entire preliminaries section of the tender document;
- spreading prices attributable to preliminary items evenly or unevenly across the unit rates within the pricing document (i.e. included within the unit rates);
- pricing relevant preliminary items independently, showing a breakdown of fixed and time-related charges; and
- using various combinations of the above approaches.

Furthermore, it is often observed that there is a wide variation between costs for main contractor's preliminaries. Lower percentages might be because the costs have been allocated elsewhere in pricing documents by the contractor, or might reflect cost-cutting to obtain work or an inadequate assessment of the complexity of the work by the contractor. For these reasons, the reality of the sum for main contractor's preliminary items included in priced tenders might be questionable, as will be the consequent percentage derived – resulting in the estimated cost of main contractor's preliminaries inserted into a cost plan being somewhat hit and miss. Care, therefore, needs to be exercised by the cost manager when examining preliminary costs in cost analysis and benchmark data. It is essential for the cost manager to understand what the percentages for main contractor's preliminaries that are derived from past building projects actually include before using them. This also emphasises the importance of the cost manager completing cost checks on cost-significant preliminary items. Thus, the cost estimate for main contractor's preliminaries will eventually be based on a combination of both estimated fixed and time-related charges for cost-significant items and estimates based on percentages for lesser cost items.

Using a percentage rate to estimate main contractor's preliminaries costs for a proposed building project, the cost target is calculated as follows:

Step 1: Establish the building works estimate.

Step 2: Determine the percentage to be applied in respect of main contractor's preliminaries.

Step 3: Ascertain the cost target for main contractor's preliminaries by multiplying the building works estimate by the percentage applicable for main contractor's preliminaries.

Note:

The equation for calculating the cost target for main contractor's preliminaries is:

$$c = a \times p$$

Where:

a = building works estimate
p = percentage addition for main contractor's preliminaries
c = cost target for main contractor's preliminaries.

Where known, costs relating to site constraints, special construction methods, sequencing of works or other non-standard requirements should to be assessed and identified separately. If unknown, the cost manager will need to ascertain whether or not allowances ought to be made within the construction risk allowance for such eventualities. The cost manager can reappraise these risk allowances as more information about the site and the employer's phasing requirements become available.

This initial cost target for main contractor's preliminaries can be further broken down to aid future cost checking. Cost targets are determined for each of the principal items by percentage apportionment (e.g. the cost target for management and staff equates to 44.15% of the total cost target for the main contractor's preliminaries). Percentage splits are derived from an analysis of the main contractor's preliminaries of similar past building projects. An illustration of breaking down the cost target for main contractor's preliminaries by apportionment is shown in Example 25.2. By breaking down costs in this way, the cost-significant items can

Example 25.2 Initial apportionment of cost target for main contractor's preliminaries

MAIN CONTRACTOR'S PRELIMINARIES					
Code	**Element/sub-elements/components**				**Total**
				£	**£**
9	**MAIN CONTRACTOR'S PRELIMINARIES**				
	Total cost target for main contractor's preliminaries = £3,025,801 (Equates to 13.00% of building works estimate)				
9.1	**Employer's requirements**				
9.1.1	Site accommodation:	0%	of	3,025,801	0
9.1.2	Site records:	0%	of	3,025,801	0
9.1.3	Completion and post-completion requirements:	0%	of	3,025,801	0
9.2	**Main contractor's cost items**				
9.2.1	Management and staff:	44.15%	of	3,025,801	1,335,891
9.2.2	Site establishment:	10.50%	of	3,025,801	317,709
9.2.3	Temporary services:	2.00%	of	3,025,801	60,516
9.2.4	Security:	5.20%	of	3,025,801	157,342
9.2.5	Safety and environmental protection:	0.60%	of	3,025,801	18,155
9.2.6	Control and protection:	0.65%	of	3,025,801	19,667
9.2.7	Mechanical plant:	12.30%	of	3,025,801	372,174
9.2.8	Temporary work:	5.10%	of	3,025,801	154,316
9.2.9	Site works:	1.05%	of	3,025,801	31,771
9.2.10	Completion and post-completion requirements:	5.00%	of	3,025,801	151,290
9.2.11	Cleaning:	0.55%	of	3,025,801	16,642
9.2.12	Fees and charges:	2.50%	of	3,025,801	75,645
9.2.13	Site services:	0.80%	of	3,025,801	24,206
9.2.14	Insurance, bonds, guarantees and warranties	9.60%	of	3,025,801	290,477
	Percentage total check:	*100.00%*			
9	**TOTAL: Main contractor's preliminaries: to cost plan summary**				**3,025,801**

be quickly identified. In effect, the cost targets can be broken down further as is practically necessary to facilitate cost checking.

As the building project progresses and becomes more clearly defined, the cost target needs to be revisited and more finely tuned to reflect the better, more detailed information as it becomes available.

25.5.2 Checking of cost-significant items

Given that main contractor's preliminaries costs are a substantial cost of most building projects, NRM 1 recommends that cost targets are scrutinised by the cost manager at each cost-planning stage. This involves thorough cost checks by the cost manager of cost-significant cost items. Examples 25.3 to 25.5 show detailed build-ups for typical cost-significant main contractor's preliminaries items. Further guidance on estimating the cost of main contractor's preliminaries items can be found in most of the published cost guides.

The original cost target for the main contractor's management and staff was £1,335,891 (see initial apportionment of cost target for main contractor's preliminaries in Example 25.2). However, the cost check indicates that the initial cost target was a little too optimistic, with a more realistic cost target being £1,475,322 – i.e. there being a shortfall of about £139,431 in the original cost target (i.e. £1,475,322 less £1,335,891). In order to maintain the original cost limit (i.e. not to exceed the employer's authorised budget), the cost manager

Example 25.3 Checking adequacy of cost target for project-specific management and staff

PROJECT-SPECIFIC MANAGEMENT AND STAFF						
Serial No.	**Resources**	**Number of staff**	**Man hours per week**	**Number of weeks**	**Rate per week**	**Totals**
					£/p	**£**
	Assumptions:					
1	Project manager	1	24	78	982.20	76,612
2	Construction manager	1	40	78	1,425.00	111,150
3	Works package manager	8	40	78	1,311.00	818,064
4	Services co-ordination manager	1	40	60	1,397.00	83,820
5	Health and safety manager	1	8	78	250.00	19,500
6	Commercial manager	1	16	78	654.80	51,074
7	Quantity surveyor (senior)	1	40	78	1,462.00	114,036
6	Quantity surveyor (assistant)	1	40	72	918.00	66,096
9	Storeman	1	40	75	686.00	51,450
10	Secretary/administrator	1	40	72	634.00	45,648
11	Administrator (assistant)	1	40	72	526.00	37,872
TOTAL: Estimated cost of project-specific management and staff						**1,475,322**

Example 25.4 Checking adequacy of cost target for tower crane

TOWER CRANE			
Serial No.	**Item**		**Totals**
			£
	Assumptions:		
	1. Type: static, saddle jib tower crane, 40m radius, 7 tonnes maximum load. 2. Quantity: 1 nr. 3. Hire period: 46 calendar weeks. 4. Operator and banksman: 40-hour working week, with 0.50 hours overtime per day.		
	Notes: Costs relating to tower crane and operator obtained from Tall Cranes (UK) Ltd		
1.	**Fixed charges:**		
1.1	Base (1): construction of temporary base for tower crane, including piles and/or anchors – say:		16,000
1.2	Installation: bring tower crane to site, erection, testing and commissioning: static, saddle jib tower crane, 40m radius, 7 tonnes maximum load – say:		8,600
1.3	Power: connection to temporary electricity supply – say:		2,600
1.4	Dismantling: dismantling and removing from site – say:		3,100
1.5	Base (2): removal, including taking up, disposing and filling voids with granular material on completion – say:		6,000
	Total: fixed charge:	(f)	**36,300**
2.	**Time-related charges:**		
2.1	Hire charges: hire of crane, say 46 weeks @ £1,680 per week		77,280
2.2	Fuel: electricity, fuel and oil – say:		8,600
2.3	Crane operator: 46 weeks × 40 hours per week @ £19.00 per hour:		34,960
2.4	Crane operator overtime: 46 weeks × 0.50 hour per day × 5 days per week @ £28.50 per hour:		3,278
2.5	Banksman: 46 weeks × 40 hours per week @ £12.50 per hour:		23,000
2.6	Banksman overtime: 46 weeks × 0.50 hour per day × 5 days per week @ £18.25 per hour:		2,099
2.7	Periodic safety checks: included in hire charge for tower crane:		Included
	Total: time-related charges:	(t)	**149,216**
TOTAL:	**Estimated cost of tower crane (for 46 week period):**	(f + t)	**185,516**

Note:

$$\frac{£149{,}216}{46 \text{ weeks}} = £2{,}824 \text{ per week}$$

Example 25.4 *(Continued)*

TOWER CRANE		
Serial No.	**Item**	**Totals**
		£
	Amount by which estimated cost will increase or decrease per week if change in crane hire period:	**3,244**

Notes:

(1) Assume the crane hire period increased from 46 weeks to, say, 52 weeks, resulting in an additional 6 weeks' hire period. The increase in cost will be:

£3,244 × 6 weeks =
£19,464

Thus, the revised total cost of the tower crane will be:

£149,216 + £19,464 =
£168,680 (say £168,700)

(2) Congested sites or air trespass may dictate that the tower crane has to be a luffing type jib (i.e. the hook is suspended from the outer end of the jib and the jib angle varies to adjust the hook radius)

(3) To gauge the adequacy of the cost target for mechanical plant, the estimated cost of the tower is to be added to the estimated cost of other mechanical plant (e.g. goods and passenger hoists, folk lifts and concrete plan)

Example 25.5 Checking adequacy of cost target for hoardings

HOARDINGS			
Serial No.	**Item**		**Totals**
			£
	Assumptions:		
	1. Materials: 18mm thick chipboard, painted.		
	2. Height: 2.40m.		
	3. Quantity: 236m.		
	4. Gates: pair of gates × 2 nr.		
	5. Duration: 72 weeks.		
1.	**Fixed charges:**		
1.1	Installation: 236m × (say) £72.50/m:		17,110
1.2	Gates: say, 2 gates @ £600/pair, painting included:		1,200
1.3	Painting: One side; 236m × 2.40m high × (say) £8.50/m²:		9,629
1.4	Dismantling: dismantling and removing from site, 236m × (say) £10.00/m:		2,360
	Total: fixed charge:	(f)	**30,299**

Example 25.5 *(Continued)*

HOARDINGS			
Serial No.	**Item**		**Totals**
			£
2.	**Time-related charges:**		
2.1	Maintenance: say, 72 weeks × £75.00/week		5,400
	Total: time-related charges:	(t)	**5,400**
TOTAL:	**Estimated cost of hoardings (for 72 week period):**	(f + t)	**35,699**
	Note: $\frac{£5,400}{72 \text{ weeks}}$ = £75 per week		
	Amount by which estimated cost will increase or decrease if change in construction period:		**£75**

Notes:

(1) Assume the period that the hoarding is required increased from 72 weeks to, say, 78 weeks; resulting in an additional 6 weeks. The increase in cost will be:
£75 × 6 weeks =
£900
Thus, the revised total cost of the tower crane will be:
£5,400 + £900 =
£6,300

would need to transfer monies from another cost target. This might be able to be achieved through other main contractor's preliminaries items but, if not, the cost manager will need to ascertain another cost target from which to transfer the shortfall.

CHAPTER 26

Main contractor's overheads and profit (group element 10)

Introduction

This chapter:

- defines main contractor's overheads and main contractor's profit;
- describes how to deal with subcontractors' overheads and profit;
- explains how to calculate a cost target for main contractor's overheads and profit;
- contains worked examples showing how to calculate a cost target for main contractor's overheads and profit; and
- shows how to calculate the works cost estimate.

26.1 Main contractor's overheads and profit

This element comprises two elements: main contractor's overheads and main contractor's profit.

- Main contractor's overheads – the costs of head office set-up and administration proportioned to each contract by the main contractor. Costs associated with site-based management and staff, site accommodation, telephone and administration, IT and other site-based resources and facilities specific to the building project are included with main contractor's preliminaries costs.
- Main contractor's profit – the amount of net profit that the main contractor needs to achieve on the contract.

26.2 Subcontractors' overheads and profit

NRM 1 stipulates that subcontractor's overheads and profit are to be included in the unit rates for building works items ascertained by the cost manager. This is because rates for overheads and profit vary between different types of subcontractors. They are also only applicable to the construction work executed by a particular subcontractor; unlike main contractor's overheads and profit, they do not affect any other aspect of construction works. Suggested methods of dealing with subcontractors' overheads and profit are shown in Chapter 23 ('Deriving unit rates for building components, sub-elements and elements').

26.3 Estimation of main contractor's overheads and profit

The rules provide that main contractor's overheads and profit can be combined and treated as a single cost centre or treated as two separate cost centres (i.e. main contractor's overheads and main contractor's profit). However, most commonly the total estimated cost of main contractor's overheads and profit is treated as a single combined cost centre. The estimated cost of main contractor's overheads and profit is usually based on a percentage of the combined cost of building works and main contractor's preliminaries.

Where main contractor's overheads and profit are treated as a single cost centre (see Example 26.1), the estimated cost is derived as follows:

Step 1: Determine the percentage to be applied in respect of main contractor's overheads and profit.

Step 2: Determine the combined cost of the building works estimate and the cost target for main contractor's preliminaries.

Step 3: Ascertain the cost target for main contractor's overheads and profit by multiplying the combined cost of the building works estimate and the main contractor's preliminaries costs by the selected percentage addition for main contractor's overheads and profit.

Note:
The equation for calculating the estimated cost of main contractor's overheads and profit is therefore:

$$c = (a + b) \times p$$

Where:

- a = building works estimate (i.e. total estimated cost of the building works)
- b = cost target for main contractor's preliminaries
- p = percentage for main contractor's overheads and profit
- c = cost target for main contractor's overheads and profit.

Where main contractor's overheads and main contractor's profit are to be treated as two discrete cost targets, both items are calculated in the same way as described above. The cost targets for both items are simply added to give the total cost target for main contractor's overheads and profit (see Example 26.2).

Example 26.1 Calculation of cost target for main contractor's overheads and profit

Assume that the percentage to be applied in respect of main contractor's overheads and profit is 4.00%. Thus, the cost target for main contractor's overheads and profit is:

MAIN CONTRACTOR'S OVERHEADS AND PROFIT					
Code	**Element/sub-elements/components**				**Totals**
				£	**£**
10	**MAIN CONTRACTOR'S OVERHEADS AND PROFIT**				
	Building works estimate:			23,275,394	
	Group element 9: Main contractor's preliminaries:			3,025,801	
	Total: Building works estimate and cost target for main contractor's preliminaries:			*26,301,195*	
	Main contractor's overheads and profit:	4.00%	of	26,301,195	1,052,048
10	**TOTAL – Main contractor's overheads and profit: to cost plan summary**				**1,052,048**

Example 26.2 Calculation of separate cost targets for main contractor's overheads and main contractor's profit

Assume that the percentages to be applied in respect of main contractor's overheads and main contractor's profit are 2.60% and 3.00%, respectively. Thus, the cost target for main contractor's overheads and profit is:

MAIN CONTRACTOR'S OVERHEADS AND PROFIT					
Code	**Element/sub-elements/components**				**Totals**
				£	**£**
10	**MAIN CONTRACTOR'S OVERHEADS AND PROFIT**				
	Building works estimate:			23,275,394	
	Group element 9: Main contractor's preliminaries:			3,025,801	
	Total: Building works estimate and cost target for main contractor's preliminaries:			*26,301,195*	
	Main contractor's overheads:	1.20%	of	26,301,195	315,615
	Main contractor's profit:	2.80%	of	26,301,195	736,433
10	**TOTAL – Main contractor's overheads and profit: to cost plan summary**				**1,052,048**

CHAPTER 27

Calculating the works cost estimate

Introduction

This chapter:

- explains how to calculate the works cost estimate.

27.1 Composition of the works cost estimate

Figure 27.1 shows the composition of the works cost estimate.

Estimate	Process	Facet	Chapter reference
Estimate 1A	Building works estimate	Total cost of building works (the sum of the cost targets for all group elements or elements)	Chapter 24
	+		
Estimate 1B	9. Main contractor's preliminaries		Chapter 25
	+		
Estimate 1C	10. Main contractor's overheads and profit		Chapter 26
	=		
Estimate 1	**Works cost estimate**	Total cost of building works (including main contractor's preliminaries and main contractor's overheads and profit	

Figure 27.1 Composition of works cost estimate

27.2 Determining the works cost estimate

The works cost estimate is the summation of the building works estimate and the cost targets for main contractor's preliminaries and main contractor's overheads and profit. Figure 27.2 sets out how a works cost estimate within a cost plan is calculated.

COST PLAN SUMMARY

Project: Woollard Hotel, Holborn Viaduct, St Leonards-on-Sea, East Sussex

Estimate base date: September 2013

Gross internal floor area (GIFA): 8,399m²

Number of Guest rooms (keys): 246

Code	Group element/element	Element totals	Group element	Cost/m² of GIFA	Cost/Key
		£	£	£	£
0	Facilitating works		0	0.00	0.00
1	Substructure		2,480,300		
1.1	Substructure	2,480,300		295.31	10,082.52
2	Superstructure		5,841,312		
2.1	Frame	1,149,949		136.91	4,674.59
2.2	Upper floors	776,526		92.45	3,156.61
2.3	Roof	413,158		49.19	1,679.50
2.4	Stairs and ramps	281,000		33.46	1,142.28
2.5	External walls	2,629,284		313.05	10,688.15
2.6	Windows and external doors	85,523		10.18	347.65
2.7	Internal walls and partitions	176,211		20.98	716.30
2.8	Internal doors	329,661		39.25	1,340.09
3	Internal finishes		2,447,882		
3.1	Wall finishes	1,666,129		198.37	6,772.88
3.2	Floor finishes	415,163		49.43	1,687.65
3.3	Ceiling finishes	366,590		43.65	1,490.20
4	Fittings, furnishings and equipment		591,458		
4.1	Fittings, furnishings and equipment	591,458		70.42	2,404.30
5	Services		4,033,645		
5.1	Sanitary installations	25,701		3.06	104.48
5.2	Services equipment	235,000		27.98	955.28
5.3	Disposal installations	68,575		8.16	278.76
5.4	Water installations	375,939		44.76	1,528.21
5.5	Heat source	230,049		27.39	935.16
5.6	Space heating and air conditioning	288,506		34.35	1,172.79
5.7	Ventilation	466,984		55.60	1,898.31
5.8	Electrical installations	498,806		59.39	2,027.67
5.9	Fuel installations	447,583		53.29	1,819.44
5.10	Lift and conveyor installations	342,595		40.79	1,392.66
5.11	Fire and lightning protection	532,413		63.39	2,164.28

Figure 27.2 Computing the works cost estimate

Code	Group element/element	Element totals	Group element	Cost/m² of GIFA	Cost/Key
		£	£	£	£
5.12	Communication, security and control systems	261,797		31.17	1,064.22
5.13	Special installations	150,006		17.86	609.78
5.14	Builder's work in connection with services	109,691		13.06	445.90
6	Prefabricated buildings and building units		7,548,517		
6.1	Prefabricated buildings and building units	7,548,517		898.74	30,685.03
7	Works to existing buildings	0	0	0.00	0.00
8	External works		333,000		
8.1	Site preparation works	0		0.00	0.00
8.2	Roads, paths, pavings and surfacings	123,500		14.70	502.03
8.3	Soft landscaping, planting and irrigation systems	0		0.00	0.00
8.4	Fencing, railings and walls	0		0.00	0.00
8.5	External fixtures	0		0.00	0.00
8.6	External drainage	20,000		2.38	81.30
8.7	External services	189,500		22.56	770.33
8.8	Minor building works and ancillary buildings	0		0.00	0.00
	Building works estimate	23,276,114	23,276,114	2,771.28	94,618.35
9	Main contractor's preliminaries	3,025,081	3,025,081	360.17	12,297.08
	Subtotal	26,301,195	26,301,195	3,131.45	106,915.43
10	Main contractor's overheads and profit	1,052,522	1,052,522	125.32	4,278.54
	Works cost estimate	**27,353,717**	**27,353,717**	**3,256.77**	**111,193.97**

Figure 27.2 *(Continued)*

Estimating cost targets for non-building works items and risk allowances

CHAPTER 28

Project and design team fees (group element 11)

Introduction

This chapter:

- describes the categories of project and design team fees defined by NRM 1;
- highlights key issues that need to be considered by the cost manager when estimating;
- provides step-by-step guidance on estimating fees;
- contains worked examples illustrating how cost targets for consultants' fees, main contractor's pre-construction fees and main contractor's design fees are built up; and
- explains how to deal with subcontractors' design fees.

Figure 28.1 highlights where this part of the order of cost estimate fits within the order of cost estimate framework and provides a quick cross-reference to the other relevant chapters.

Estimate	Process	Notes	Chapter
Estimate 1	Works cost estimate	Total cost of building works (including main contractor's preliminaries and main contractor's overheads and profit)	Chapter 27
	+		
Estimate 2	**Project and design team fees estimate**		
	+		
Estimate 3	Other development and project costs estimate		Chapter 29
	+		
Estimate 4	Risk allowances estimate		Chapter 30
	+		
Estimate 5	Inflation estimate		Chapter 31
	=		
	COST LIMIT		Chapter 32
	VAT assessment		Chapter 33

Figure 28.1 Project and design team fees estimate in the context of the cost plan framework

28.1 Project and design team fees

This element comprises three items: consultants' fees, main contractor's pre-construction fees and main contractor's design fees.

- Consultants' fees – the cost of fees associated with project team, design team and other specialist consultants directly employed by the employer, including site investigation fees.
- Main contractor's pre-construction fees – the costs of fees associated with Pre-Construction Services Agreements (PCSA), sometimes called Pre-Contract Services Agreements.
- Main contractor's design fees – the cost of fees where design liability is to be transferred to the main contractor (i.e. where a design and build or other contractor-led design strategy is to be used) and all, or some, of the consultants within the design team are to be novated.

An extensive list of fees relevant to a building project is given in group element 11 of NRM 1.

The process for estimating the cost target for project and design team fees is illustrated in Figure 28.2.

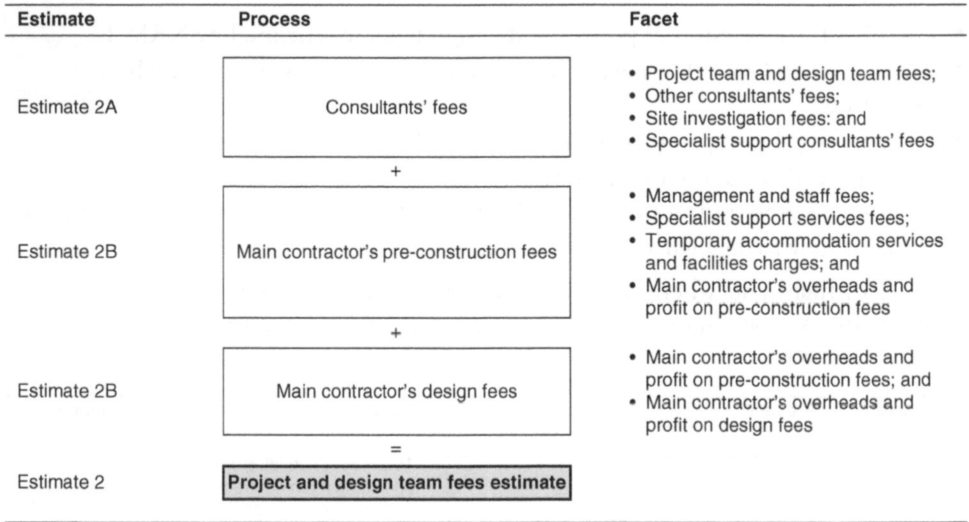

Figure 28.2 Process for estimating the cost target for project and design team fees

28.2 Estimation of consultants' fees (element 11.1)

To begin with, it will be necessary for the cost manager to determine how the employer wishes project and design team fees to be dealt with in the cost plan. This is because some employers might require the cost manager to only deal with construction costs. In contrast, others might require an allowance for all, or some aspects of, project and design team fees to be included in the cost plan.

Example 28.1 Calculation of initial cost target for consultants' fees based on a percentage

Assume that the percentage to be applied in respect of consultants' fees is 11.50%. Thus, the cost target for main consultants' fees is:

PROJECT AND DESIGN TEAM FEES					
Code	**Element/sub-elements/components**				**Totals**
				£	£
11	**PROJECT AND DESIGN TEAM FEES**				
11.1	**Consultants' fees**				
	Works cost estimate:			27,353,243	
	Consultants' fees:	11.50%	of	27,353,243	3,145,623
11.1	**TOTAL – Consultants' fees: to project and design team fees summary**				**3,145,623**

Consultants' fees at the concept stage of a building project are usually calculated as a percentage addition to the works cost estimate (i.e. the combined total of the building works estimate, main contractor's preliminaries estimate and main contractor's overheads and profit estimate). Initially, a single percentage can be used to cover all consultants' fees, but the cost of each consultant should be calculated separately as the building project develops. Percentages applied for consultants' fees are determined from analysis of past building projects. Where actual fees are known, the actual fee is to be included in the cost plan, based on either a percentage or lump sum, using whichever method on which the fee has been based. (See Example 28.1.)

When estimating consultants' fees, it is essential that the cost manager checks that the scope of services agreed between the employer and consultants is sufficient to complete the building project. Any requirement for additional services needs to be identified and allowed for in the cost plan (e.g. the cost of carrying out detailed reinforcement design would not normally be included in scope of services (or fee agreement) for the structural engineer, unless specifically requested by the employer). Thus, cost checks on consultants' fees by the cost manager must include checks on the sufficiency of the scope of services.

Where a contractor-led design contract strategy has been selected, at some time the responsibility of design will be transferred to the main contractor. This is likely to involve the novation of some, or all, of the designers. As a result, the residual fees will become the responsibility of the main contractor. The cost manager will, therefore, need to transfer the fees for residual design services from the cost centre for consultants' fees to the cost centre for main contractor's design fees (i.e. element 11.3). This course of action is further explained in paragraph 28.4. This emphasises the need of the cost manager to understand the scope of services included in each consultants' fee agreement.

Box 28.1 Key definitions

- **Novation** – cancelling an existing obligation and then creating a new obligation in its place. Therefore, in the case of designers being novated to the main contractor, novation will entail the substitution of an existing contract (with the employer) for a new contract (with the main contractor).

- **Pre-Construction Services Agreement (PCSA)** – sometimes referred to as a Pre-Contract Services Agreement enables employers to employ contractors before the main building contract commences.
- **Residual design services** – design services included in the consultant's scope of services that are still to be completed.

Irrespective of the contract strategy, the employer will require assurance that the works completed by the main contractor conform to the specification. To satisfy this requirement, it is usual for the employer to put in place a quality inspection team, which will normally include principal designers. However, this can prove difficult in the case of contractor-led design contract strategies, as it is likely that the principal designers have been novated to the main contractor. This is because there is, often, no contract in place between the employer and the design consultant. Nonetheless, it is not unusual today for the employer to retain the services of principal designers to carry out a monitoring and inspection role, even though they have been novated to the main contractor. Some main contractors are content for the design consultants to undertake quality inspections on behalf of the employer. Conversely, some consultants are reluctant, or refuse, to take on the quality inspection role for the employer whilst employed by the main contractor. Reasons for refusal often cited by consultants include: they are prohibited by their professional indemnity insurance and that there would be conflict of interest. In this situation, an alternative approach is for the employer to commission other apposite design consultants as part of the quality inspection team. It is necessary, therefore, for the cost manager to understand how the employer is going carry out the monitoring and inspection role (i.e. work in progress inspections, pre-completion inspections and completion inspections – to achieve practical completion and handover), so that sufficient allowance can be included in the cost plan.

The cost target for consultants' fees can initially be based on a percentage of the works cost estimate; with a single percentage being used to cover the cost of employing all consultants needed for the building project. Alternatively, the different consultants required for the project can be identified and individual cost targets can be determined for each in the same way. When identifying the different consultants needed, reference to the list in element 11.1 (Consultants' fees) of NRM 1 will provide a good starting point.

Using a percentage rate to estimate consultants' fees for a proposed building project, the cost target is calculated as follows:

Step 1: Establish the works cost estimate.

Step 2: Determine the percentage to be applied in respect of consultants' fees.

Step 3: Ascertain the cost target for consultants' fees by multiplying the works cost estimate by the percentage applicable for consultants' fees.

Note:

The equation for calculating the cost target for consultants' fees is:

$$f = a \times p$$

Where:

a = works cost estimate
p = percentage for consultants' fees
f = cost target for consultants' fees.

The employer will enter into formal appointments with the consultants as the project progresses. Therefore, as the details of the scope of services and fee agreements within the appointment documents become known to the cost manager, the adequacy of the initial cost target can be checked. Monitoring and reporting of expenditure can also begin. The cost manager will calculate the projected fee for consultants ascertained from their fee agreements. Fee agreements will be either based on a percentage fee, lump sum fee, time charge (e.g. an hourly or daily rate) or on any combination of the three. The revised fee estimates are then transferred to the cost plan and any compensating adjustments to other cost targets made and recorded. The cost manager must take care when calculating projected fees from consultants' appointment documents. This is because the scope of service for which a consultant is appointed might not cover all services necessary to complete the building project. Therefore, it is important that the cost manager has a sound understanding of the scope of services required of each consultant. Where there is a gap in service provision, the cost manager will need to make adequate allowance in the cost plan; either as an allowance within the overall cost target for consultants' fees, or as a risk allowance.

Example 28.2 illustrates a detailed build-up of cost target for consultants' fees.

Example 28.2 Detailed build-up of cost target for consultants' fees

Assumptions:

- Actual fees (A): ascertained from consultants' fee agreements.
- Estimated fees (Est): derived from interpolation of analyses of past building projects.

(a) Project team and design team fees:

PROJECT AND DESIGN TEAM FEES						
Code	**Element/sub-elements/components**					**Totals**
					£	**£**
11	**PROJECT AND DESIGN TEAM FEES**					
11.1	**Consultants' fees**					
	Works cost estimate:				27,353,243	
11.1.1	*Project team and design team fees*					
	A = Actual agreed fee. Est = Estimated fee.					
11.1.1.1	Project manager/contract administrator/employer's agent	A	1.95%	of	27,353,243	533,388
11.1.1.2	Architect (including disbursements)	A	3.60%	of	27,353,243	948,717
11.1.1.3	Quantity surveyor/cost manager	A	1.35%	of	27,353,243	369,269
11.1.1.4	Building services engineer	A	2.00%	of	27,353,243	547,065
11.1.1.5	Structural engineer	A	1.55%	of	27,353,243	423,975
11.1.1.6	Landscape architect	A	Lump sum			35,000
11.1.1.7	CDM co-ordinator	A	Lump sum			40,000
11.1.1	**TOTAL – Project team and design team fees: to Consultants' fees summary**					**2,933,414**

Example 28.2 *(Continued)*

(b) Other consultants' fees:

PROJECT AND DESIGN TEAM FEES					
Code	**Element/sub-elements/components**				**Totals**
				£	**£**
11	**PROJECT AND DESIGN TEAM FEES**				
11.1	**Consultants' fees**				
	Works cost estimate:			27,353,243	
	A = Actual agreed fee. Est = Estimated fee.				
11.1.2	*Other consultants' fees*				
11.1.2.1	Measuring surveyor	A	Lump sum		5,711
11.1.2.2	Geotechnical engineer	Est	Lump sum		15,000
11.1.2.3	Environmental consultant	Est	Lump sum		7,500
11.1.2.4	Ecologist	Est	Lump sum		7,500
11.1.2.5	Arboriculturist	Est	Lump sum		7,500
11.1.2.6	Party wall surveyor	Est	Lump sum		20,000
11.1.2.7	Acoustics consultant	Est	Lump sum		15,000
11.1.2.8	Facade consultant	Est	Lump sum		12,500
11.1.2.9	Fire consultant	Est	Lump sum		10,000
11.1.2.10	Building control consultant	Est	Lump sum		16,500
11.1.2.11	BREEAM assessor	A	Lump sum		10,000
11.1.2	**TOTAL – Other consultants' fees: to project and design team fees summary**				**127,211**

(c) Site investigation fees:

PROJECT AND DESIGN TEAM FEES					
Code	**Element/sub-elements/components**				**Totals**
				£	**£**
11	**PROJECT AND DESIGN TEAM FEES**				
11.1	**Consultants' fees**				
	Works cost estimate:			27,353,243	
	A = Actual agreed fee. Est = Estimated fee.				
11.1.3	*Site investigation fees*				
11.1.3.1	Geotechnical investigation	A	Lump sum		20,000
11.1.3.2	Trial pits	A	Lump sum		17,800
11.1.3.3	Pre-demolition asbestos survey	Est	Lump sum		3,500
11.1.3	**TOTAL – Site investigation fees: to project and design team fees summary**				**41,300**

Example 28.2 *(Continued)*

(d) Specialist support consultants' fees:

PROJECT AND DESIGN TEAM FEES					
Code	**Element/sub-elements/ components**				**Totals**
				£	**£**
11	**PROJECT AND DESIGN TEAM FEES**				
11.1	**Consultants' fees**				
	Works cost estimate:			27,353,243	
	A = Actual agreed fee. Est = Estimated fee.				
11.1.4	*Specialist support consultants' fees*				
11.1.4.1	Planning consultant	A	Lump sum		43,698
11.1.4	**TOTAL – Specialist support consultants' fees: to project and design team fees summary**				**43,698**

(e) Consultants' fees – summary:

PROJECT AND DESIGN TEAM FEES			
Code	**Element/sub-elements/ components**		**Totals**
		£	**£**
11	**PROJECT AND DESIGN TEAM FEES**		
11.1	**Consultants' fees**		
11.1.1	Project team and design team fees:		2,933,414
11.1.2	Other consultants' fees:		127,211
11.1.3	Site investigation fees:		41,300
11.1.4	Specialist support consultants' fees:		43,698
11.1	**TOTAL – Consultants' fees: to project and design team fees summary**		**3,145,623**

Where fees are based on a percentage they have been calculated as a percentage of the works cost estimate. However, it should be noted that, in practice, percentage fees for certain consultants (e.g. building services engineer, structural engineer and landscape architect) are calculated as a percentage of the cost of the work which they have designed and not as a percentage of the total cost of the works.

Concerning fees for architectural services, it is common for architects to quote a fee for architectural services that excludes all or certain disbursements (e.g. additional costs paid out for travelling, printing, document searches, etc.). The cost of such disbursements can amount to a significant cost, so the cost manager must make certain that an estimated sum is included in the architect's fee for disbursements.

28.3 Estimation of main contractor's pre-construction fee (element 11.2)

Modern procurement methods advocate early main contractor involvement in the design development process. The early appointment of a main contractor offers considerable scope for the contractor to contribute expertise and innovation to both design development and method of construction, as well as increase the level of supply chain integration. In such cases, the main contractor is appointed under a Pre-Construction Services Agreement. The intention of a Pre-Construction Services Agreement is also to secure at the earliest opportunity the project team and key personnel submitted as part of the main contractor's proposals. Without this early engagement there is a very real risk that the main contractor's key personnel (e.g. project director, construction manager, procurement manager, design manager, building services engineering manager and senior project surveyor) will move to other projects and this might be detrimental to the success of the building project.

The use of a Pre-Construction Services Agreement can enable the main contractor to:

- contribute to the design process itself;
- advise on buildability, sequencing and construction risk;
- advise on the packaging of the works (and the risks of interfaces between work packages);
- advise on the selection of specialist subcontractors;
- help develop the cost plan and construction programme;
- help develop the method of construction;
- obtain prices for work packages from subcontractors or suppliers on an open book basis;
- prepare a site layout plan for the construction stage showing temporary facilities;
- draft the preliminaries for specialist and works contractors' bid documents; and
- assist with any planning conditions on matters concerning the build phase, such as waste disposal proposals, construction traffic movements, tree preservation protection and the like.

This early involvement of the contractor should improve the buildability and cost-certainty of the design as well as creating a better integrated project team and reducing the likelihood of disputes.

Although the earlier appointment of a main contractor can offer considerable scope for better value, it is important to get the right timing. The earlier the appointment, the more scope there is for the main contractor to contribute expertise and innovation, but the time period to construction should not be too long. There would be a risk that, if a main contractor were appointed too early, they would not be motivated to contribute their best staff. A long period before construction could also make it difficult to maintain enthusiasm and to retain key staff.

Where the employer, in consultation with the project team, has decided to employ a main contractor (or specialist contractors) to provide pre-construction advice, an allowance for the fee for providing such services is to be determined and included in the cost plan. The estimated cost of main contractor's pre-construction fees can be calculated by using a percentage of the works cost estimate or a derived lump sum.

Main contractors often infer that they will reduce their eventual tender price to take account of their pre-construction fees. In reality, this seldom happens. Therefore it is recommended that the cost manager advises the employer to maintain a discrete budgetary allowance for main contractor's pre-construction fees should it be decided to involve the main contractor early in the design development process.

Main contractor's overheads and profit attributable to pre-construction fees is not covered under group element 10 (Main contractor's overheads and profit), as this relates solely to building works. Therefore, care needs to be taken by the cost manager when estimating the cost of pre-construction fees, to ensure that sufficient allowance has been made for main contractor's overheads and profit on the pre-construction fee.

A typical estimate build-up of main contractor's pre-construction fees for inclusion in a cost plan is illustrated in Example 28.3.

Example 28.3 Calculation of cost target for main contractor's pre-construction fee

Assumptions:

- Pre-construction period: 12 weeks.
- Unit rates and percentage allowances: Interpolated from analyses of past pre-construction fee agreements.

(a) Management and staff fees:

PROJECT AND DESIGN TEAM FEES					
Code	**Element/sub-elements/ components**	**Qty**	**Unit**	**Rate**	**Totals**
				£/p	**£**
12	**PROJECT AND DESIGN TEAM FEES**				
11.2.1	*Management and staff fees*				
11.2.1.1	Project manager	12	wks	1,637.00	19,644
11.2.1.2	Construction manager	12	wks	1,425.00	17,100
11.2.1.3	Commercial manager	6	wks	1,637.00	9,822
11.2.1.4	Quantity surveyor (senior)	12	wks	1,462.00	17,544
11.2.1.5	Estimator (senior)	8	wks	1,580.00	12,640
11.2.1.6	Estimator (assistant)	8	wks	998.00	7,984
11.2.1.7	Planner/programmer	9	wks	1,462.00	13,158
11.2.1.8	Design manager	12	wks	1,610.00	19,320
11.2.1.9	Building services engineering manager	12	wks	1,490.00	17,880
11.2.1.10	Secretary/administrative support	12	wks	634.00	7,608
11.2.1	**TOTAL – Management and staff fees: to Main contractor's pre-construction fees summary**				**142,700**

Example 28.3 *(Continued)*

(b) Specialist support services fees:

PROJECT AND DESIGN TEAM FEES					
Code	**Element/sub-elements/ components**	**Qty**	**Unit**	**Rate**	**Totals**
				£/p	**£**
12	**PROJECT AND DESIGN TEAM FEES**				
11.2	**Main contractor's pre-construction fees**				
11.2.2	*Specialist support services fees*				
11.2.2.1	Legal advice		Lump sum		15,000
11.2.2	**TOTAL – Specialist support services fees: to Main contractor's pre-construction fees summary**				**15,000**

(c) Temporary accommodation, services and facilities charges:

PROJECT AND DESIGN TEAM FEES					
Code	**Element/sub-elements/ components**	**Qty**	**Unit**	**Rate**	**Totals**
				£/p	**£**
12	**PROJECT AND DESIGN TEAM FEES**				
11.2	**Main contractor's pre-construction fees**				
11.2.3	*Temporary accommodation, services and facilities charges*				
11.2.3.1	Offices – rental of temporary office space, including rates and service charges	12	wks	1,800.00	21,600
11.2.3.2	Furniture and equipment, including photocopier	12	wks	250.00	3,000
11.2.3.3	IT systems	12	wks	450.00	5,400
11.2.3.4	Office consumables	12	wks	100.00	1,200
11.2.3.5	Cleaning	12	wks	175.00	2,100
11.2.3	**TOTAL – Temporary accommodation, services and facilities charges: to Main contractor's pre-construction fees summary**				**33,300**

(d) Main contractor's overheads and profit on pre-construction fees:

PROJECT AND DESIGN TEAM FEES			
Code	**Element/sub-elements/ components**		**Totals**
		£	**£**
12	**PROJECT AND DESIGN TEAM FEES**		
11.2	**Main contractor's pre-construction fees**		
11.2.4	*Main contractor's overheads and profit on pre-construction fees*		
	Management and staff fees:	142,700	

Example 28.3 *(Continued)*

(d) Main contractor's overheads and profit on pre-construction fees:

PROJECT AND DESIGN TEAM FEES						
Code	**Element/sub-elements/ components**					**Totals**
					£	£
	Specialist support services fees:				12,000	
	Temporary accommodation, services and facilities charges:				33,300	
	Total: Main contractor's pre-construction fees (excluding overheads and profit):				188,000	
11.2.4.1	Main contractor's overheads and profit on pre-construction fees:	4.00	%	of	188,000	7,520
11.2.4	**TOTAL – Main contractor's overheads and profit on pre-construction fees: to Main contractor's pre-construction fees summary**					**7,520**

(e) Main contractor's pre-construction fees – summary:

PROJECT AND DESIGN TEAM FEES			
Code	**Element/sub-elements/ components**		**Totals**
		£	£
12	**PROJECT AND DESIGN TEAM FEES**		
11.2	**Main contractor's pre-construction fees**		
11.2.1	Management and staff fees:		142,700
11.2.2	Specialist support services fees:		15,000
11.2.3	Temporary accommodation, services and facilities charges:		33,300
11.2.4	Main contractor's overheads and profit on pre-construction fees:		7,520
11.2	**TOTAL – Main contractor's pre-construction fees: to project and design team fees summary**		**198,520**

28.4 Estimation of main contractor's design fees (element 11.3)

This element is normally only used where a contractor-led design contract strategy is to be used, as it results in design liability being transferred to the main contractor from the employer (who would otherwise be taking the responsibility and risk of design team performance). It will also result in all, or some, of the design team being novated to the main contractor. Contractor-led design contract strategies include design and build, design and manage, and any other contractor-led design strategy.

As soon as possible after it has been decided to use a contractor-led design contract strategy, the designers who are to be novated to the main contractor, and the timing of the novation, needs to be identified by the cost manager. This will enable the residual fees relating to the novated designers to be transferred from the cost centre for consultants' fees (element 11.1) to the cost centre for main contractor's design fees (element 11.3).

The estimate for main contractor's design fees will usually include fees in connection with residual design services, fees for additional design services and main contractor's overheads and profit. Like main contractor's pre-construction fees, main contractor's overheads and profit on main contractor's design fees is not covered under group element 10 (Main contractor's overheads and profit). Therefore, the cost manager needs to make sure that sufficient allowance has been made for main contractor's overheads and profit on the main contractor's design fees.

- **Residual design services** – fees in connection with design services included in the consultant's original scope of services, agreed with the employer, which are still to be completed.
- **Additional design services** – fees for services which are not covered by the consultants' existing scope of services with the employer (e.g. detailed reinforcement design and provision of bar-bending schedules – the cost of carrying out detailed reinforcement design would not normally be included in scope of services for the structural engineer, unless specifically requested by the employer).
- **Overheads and profit** – on main contractor's design fees.

When pricing a tender, main contractors invariably incorporate a premium in their tender price to cover the cost of consultants' and in-house fees for additional design services. This premium is usually based on a percentage addition and is often referred to as the 'main contractor's design development risk'. In the early stages of cost planning, potential costs in connection with additional design services can be dealt with by way of a risk allowance – refer to Chapter 30 ('Setting and managing risk allowances').

Example 28.4 illustrates a typical build-up of an estimate for main contractor's design fees:

Example 28.4 Calculation of cost target for main contractor's design fees

Assumption:

Only the architect and the structural engineer are to be novated to the main contractor.

(a) Main contractor's design consultants' fees:

PROJECT AND DESIGN TEAM FEES					
Code	**Element/sub-elements/components**	**Qty**	**Unit**	**Rate**	**Totals**
				£/p	**£**
11	**PROJECT AND DESIGN TEAM FEES**				
11.3	**Main contractor's design fees**				
11.3.1	*Main contractor's design consultants' fees*				

Example 28.4 *(Continued)*

(a) Main contractor's design consultants' fees:

PROJECT AND DESIGN TEAM FEES					
Code	**Element/sub-elements/components**	**Qty**	**Unit**	**Rate**	**Totals**
				£/p	**£**
	Notes: 1. The architect and structural engineer are to be novated to the main contractor on execution of the contract 2. 65% of the architect's fee will have been expended prior to contract execution 3. 70% of the structural engineer's fee will have been expended prior to contract execution 4. Allowance made to cover services required of architect and structural engineer by contractor, but not included in consultant appointments				
	Novated design consultants (residual services)				
11.3.1.1	Architect	35%	of	984,717	344,651
11.3.1.2	Structural engineer	30%	of	423,975	127,193
11.3.1.3	Additional services not covered by existing consultant appointments (Allowance)				50,000
	Other design consultant input required by main contractor				
	Assumption:				
	Main contractor's total design fees will equate to 3.90% of the works cost estimate. Therefore:				
	Architect fees			344,651	
	Structural engineer's fees			127,193	
	Additional architectural and structural engineering services			75,000	
	Total: Novated design consultants' fees			546,844	
		3.90%	of	27,353,243	
	Total: Main contractor's design fees			1,066,776	
	Less: Novated design consultants' fees			546,844	
11.3.1.4	Other design fees				519,932
11.3.1	**TOTAL – Main contractor's design consultants' fees: to Main contractor's design fees summary**				**1,066,776**

Example 28.4 *(Continued)*

(b) Main contractor's overheads and profit on design consultants' fees:

PROJECT AND DESIGN TEAM FEES					
Code	**Element/sub-elements/components**				**Totals**
				£	£
11	**PROJECT AND DESIGN TEAM FEES**				
11.3	**Main contractor's design fees**				
11.3.2	*Main contractor's overheads and profit on design consultants' fees*				
11.3.2.1	Main contractor's overheads and profit on design fees:	4.00%	of	1,066,776	42,671
11.3.2	**TOTAL – Main contractor's overheads and profit on design consultants' fees: to Main contractor's design fees summary**				**42,671**

(c) Main contractor's design fees – summary:

PROJECT AND DESIGN TEAM FEES			
Code	**Element/sub-elements/components**		**Totals**
		£	£
11	**PROJECT AND DESIGN TEAM FEES**		
11.3	**Main contractor's design fees**		
11.3.1	Main contractor's design consultants' fees:		1,066,776
11.3.2	Main contractor's overheads and profit on design fees:		42,671
11.3	**TOTAL – Main contractor's design fees: to project and design team fees summary**		**1,109,447**

28.5 Subcontractors' design fees

NRM 1 recommends that subcontractors' design fees are included in the unit rates for building works items. Suggested methods of dealing with subcontractors' design fees are illustrated in Chapter 23 ('Deriving unit rates for building components, sub-elements and elements').

28.6 Establishing the cost target for project and design team fees

The cost target for project and design team fees is calculated by simply adding together the individual cost targets for consultants' fees, main contractor's pre-construction fees and main contractor's design fees. However, the content of the calculation will depend on the contract strategy selected, which in turn influences the design strategy.

The following examples illustrate two common scenarios. Example 28.5 is based on a designer-led contract strategy, where the employer directly employs the design

Example 28.5 Calculation of cost target for project and design team fees where a designer-led design strategy to be used (i.e. where employer directly employs all consultants)

PROJECT AND DESIGN TEAM FEES			
Code	**Element/sub-elements/components**		**Totals**
		£	**£**
11	**PROJECT AND DESIGN TEAM FEES**		
11.1	Consultants' fees:		3,145,623
11.2	Main contractor's pre-construction fees:		0
11.3	Main contractor's design fees:		0
11	**TOTAL – Project and design team fees: to cost plan summary**		**3,145,623**

consultants from inception to completion of the building project (e.g. a lump sum contract based on a bill of quantities). Whereas Example 28.6 is based on a contractor-led design strategy (e.g. a lump sum design and build contract), which will most likely result in certain design consultants being novated to the main contractor. In this example it has been assumed that both the architect and the structural engineer will be novated by the employer to the contractor. Therefore, the cost target for consultants' fees will need to be adjusted in respect of these two consultants in respect of residual services and transferred to the cost target for main contractor's

Example 28.6 Calculation of cost target for project and design team fees where contractor-led design strategy is to be used (i.e. where employer novates certain consultants to the contractor)

PROJECT AND DESIGN TEAM FEES			
Code	**Element/sub-elements/components**		**Totals**
		£	**£**
11	**PROJECT AND DESIGN TEAM FEES**		
	Consultants' fees estimate:	3,145,623	
	Less: Main contractor's design consultants' fees estimate (excluding main contractor's overheads and profit on consultants' design fees)	1,066,776	
	Adjusted: Consultants' fees estimate	2,078,847	
11.1	Consultants' fees:		2,078,847
11.2	Main contractor's design fees (including overheads and profit):		1,109,447
11.3	Main contractor's pre-construction fees (including overheads and profit):		198,520
11	**TOTAL – Project and design team fees: to cost plan summary**		**3,386,814**

Note:
Both the main contractor's pre-construction fees (code 11.2) and the main contractor's design fees (code 11.3) include allowances for main contractor's overheads and profit.

design fees. The same applies to other consultancy services not completed that will become the responsibility of the contractor under a contractor-led design strategy.

It should also be noted that the main contractor's design fees are most likely to attract overheads and profit, which is for managing and taking responsibility for the consultants.

CHAPTER 29

Other development and project costs (group element 12)

Introduction

This chapter:

- defines building-related costs;
- describes other development and project costs in the context of initial costs;
- outlines the concept of whole life cost planning;
- emphasises the need for communication between cost manager and employer about the treatment of other development and project costs within the cost estimate or cost plan; and
- explains the approach to estimating other development and project costs.

Figure 29.1 highlights where this part of the order of cost estimate fits within the

Estimate	Process	Facet	Chapter reference
Estimate 1	Works cost estimate	Total cost of building works (including main contractor's preliminaries and main contractor's overheads and profit)	Chapters 6 and 27
	+		
Estimate 2	Project and design team fees estimate		Chapter 28
	+		
Estimate 3	**Other development and project costs estimate**		
	+		
Estimate 4	Risk allowances estimate		Chapter 30
	+		
Estimate 5	Inflation estimate		Chapter 31
	=		
	COST LIMIT		Chapter 32
	VAT assessment		Chapter 33

Figure 29.1 Other development and project costs estimate in the context of the cost plan framework

order of cost estimate framework and provides a quick cross-reference to the other relevant chapters.

29.1 What are other development and project costs?

NRM 1 deals specifically with the initial costs of a building project (i.e. the initial capital costs). However, cost planning is often concerned with the time stream of costs and revenues that flow throughout the life of a building project, from its inception until the end of the building project's life. This is referred to as 'whole life cost planning'. Building-related costs usually fall into the following categories:

- Initial costs – acquisition and construction costs;
- Annual costs – fuel costs, operation costs, maintenance costs, finance charges and loan interest payments;
- Intermittent costs – repair costs, replacement costs and costs in connection with future changes; and
- Residual costs and values – resale, salvage values or disposal costs.

Non-monetary benefits or costs also need to be considered. For example, a new high-quality office building, both in terms of aesthetics and ergonomics, is likely to provide a physical work environment that attracts high-quality staff, helps retain them and might also improve productivity.

The focus of the rules for order of cost estimating and elemental cost planning is on the initial costs of site acquisition and building construction. Future annual costs, intermittent costs, residual costs and values or non-monetary benefits are not tackled by the rules. To those who wish to better understand the concept of whole life cost planning, it is recommended that a text on whole life costs be purchased.

29.1.1 Initial costs – other development and project costs

NRM 1 recognises that the initial costs of a building project are not simply the design and construction costs. Initial costs, in relation to a building project, are associated with the planning, design, procurement and construction of the building project. That is, all costs prior to a building or facility being brought into use.

Initial costs which are not directly associated with the cost of constructing the building, but form part of the total cost of delivering the building project are referred to as other development and project costs by the rules. Checklists are given in the rules (group element 13) for the following cost categories:

- land acquisition costs – the cost of acquiring the site;
- employer finance costs – the costs in connection with funding of the building project;
- fees – including fees in connection with planning, building control, oversailing licences, party wall awards, rights of light agreements, NHBC Buildmark registration fees, agreements between the employer and neighbours to facilitate the building project and, with licences, permits and agreements;

- charges – adoption charges in connection with highways, adoption charges in connection with services (e.g. sewers, water, electricity and gas), maintenance costs in connection with services;
- planning contributions – direct financial contributions in connection with planning consent, environmental improvement works;
- insurances – insurance for the works, including insurance premium tax (IPT);
- archaeological works – archaeological works financed by the employer;
- decanting and relocating costs – including temporary relocation costs, fit-out of temporary accommodation, and rents and other running costs;
- fittings, furnishings and equipment – items which do not form part of the building contract;
- tenant's costs and contributions;
- marketing costs – including launch events, site-based advertising (e.g. sales hoardings), show units and marketing suites (i.e. separate or within building to be built), operating costs associated with show units and marketing suites, marketing literature; and
- other employer costs in connection with the building project.

The checklists are not intended to be exhaustive lists that cover the needs of every employer or building project. They are given solely as a starting point and as an indication of the most significant type of costs that are likely to occur in relation to most building projects.

An overview of future building-related costs (i.e. annual costs, intermittent costs and residual costs and values) is provided in the ensuing sub-paragraphs.

29.1.2 Annual costs

Costs that occur each year throughout the life of the building are referred to as 'annual costs'. They can also include certain employer costs in operating the building for the purpose of conducting their business or fulfilling the building's function (often called 'operating costs'). Annual costs include:

- routine maintenance of building:
  - components; and
  - mechanical and electrical services.
- inspections;
- day-to-day operation of mechanical and electrical services;
- energy consumption for heating, cooling, power and lighting;
- rates;
- insurances;
- computer-related charges;
- communications;
- cleaning – internal and external;
- staff related to the building
- security staff;
- landscape maintenance; and
- employer's operating costs.

29.1.3 Intermittent costs

Costs that tend to occur at cycles of more than one year, throughout the life of the building project in use, are referred to as intermittent costs. Many of these costs are related to the maintenance of elements of the building fabric or services that will require maintenance or replacement during the life of the building project. These can include:

- external decorations;
- internal decorations;
- relamping luminaires;
- rewiring electrical installations;
- replacement of heat source and other mechanical plant and equipment;
- replacement of roof coverings;
- replacement of windows and external doors;
- renewal of fittings, furnishings and equipment;
- replacement of floor and ceiling finishes; and
- reconfiguration or refitting of the building for new working practices or change of use, including costs incurred through decanting occupants.

29.1.4 Residual costs and values

Residual costs and values are the costs that might be incurred and the revenue that might be received when the building or facility has reached the end of the life of the investment. Any revenue received as a result of the sale of land and buildings at the end of the project's life will be included as credits in the whole life cost plan. Costs and revenues might include:

- legal fees;
- demolition of buildings;
- removal of toxic and hazardous materials;
- treatment of contaminated land;
- site clearance;
- site security;
- sale of land;
- sale of buildings; and
- taxation implications of disposal.

26.2 Communication

It is essential that the cost manager determines how the employer wishes to deal with the financial management of other development and project costs, as well as the extent of the cost manager's involvement. Many employers do not have the in-house capability or capacity to maintain financial control of their building projects. Consequently, it is becoming more and more common for employers to ask cost managers to assist in setting budgets and maintaining up-to-date estimated outturn costs for other development and project items in addition to maintaining cash flows, validating payments, monitoring expenditure, and reporting on the financial status of such items. Where required by the employer, the cost manager and the employer will need to agree the scope of items to be included in the cost plan and establish the cost targets (budgetary allowances) for each.

26.3 Estimation of other development and project costs

Owing to the diverse nature of other development and project items, the method of estimating costs will be dependent on the nature of the works or services to be procured.

CHAPTER 30

Setting and managing risk allowances (group element 13)

Introduction

This chapter:

- describes risk management;
- summarises the key principles of risk management;
- discusses the management of risk in the context of building projects;
- explains the application of risk management to building projects;
- defines the concept of risk allowance;
- outlines the categories of risk defined by NRM 1;
- sets out the essential steps and techniques for risk identification and risk assessment at important stages of the building project, including:
  - content of the base estimate;
  - identification of potential risk;
  - qualitative assessment; and
  - quantitative assessment.
- describes methods, with worked examples, of setting appropriate levels of risk allowance for accurate estimating and control of expenditure arising from the employer's residual risk exposure;
- explains how to calculate individual risk allowances and the total risk allowance;
- describes risk response strategies; and
- explains the role of the cost manager in setting and controlling the risk allowance.

Figure 30.1 highlights where this part of the order of cost estimate fits within the order of cost estimate framework and provides a quick cross-reference to the other relevant chapters.

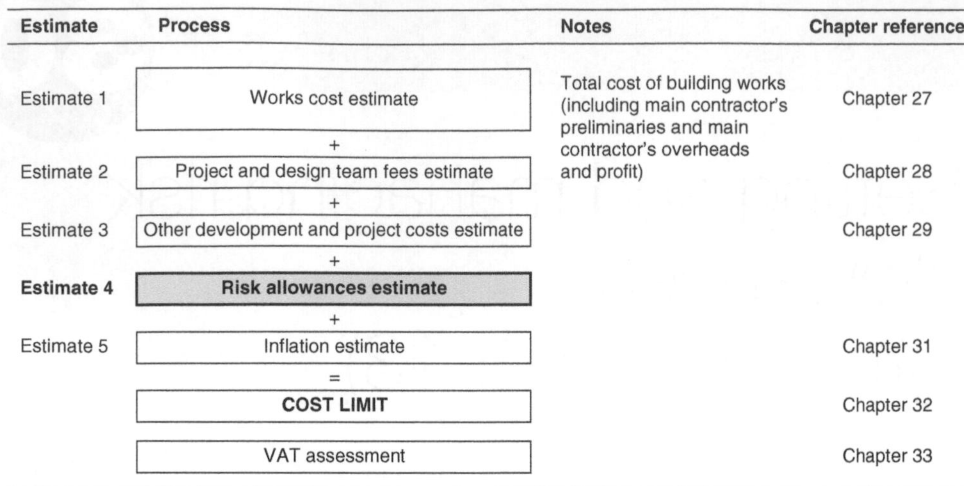

Estimate	Process	Notes	Chapter reference
Estimate 1	Works cost estimate	Total cost of building works (including main contractor's preliminaries and main contractor's overheads and profit)	Chapter 27
	+		
Estimate 2	Project and design team fees estimate		Chapter 28
	+		
Estimate 3	Other development and project costs estimate		Chapter 29
	+		
Estimate 4	**Risk allowances estimate**		
	+		
Estimate 5	Inflation estimate		Chapter 31
	=		
	COST LIMIT		Chapter 32
	VAT assessment		Chapter 33

Figure 30.1 Risk allowance estimate in the context of the cost plan framework

30.1 Risk management

Risk management can greatly assist in securing the success of any building project. Risk management ensures that the employer and the project team are working to common objectives, and that the constraints of the employer (cost, time, functionality, image or accountability) and the prevailing physical and social-political environment are acknowledged. Risk management is a tool for the project team to identify and fully understand the uncertainty within these aspects. It should not be used to merely analyse information or lists of risks, but to provide objective guidance to the team to enable decisions to be made. A risk management system will allow the project team members to appreciate the potential impact of risks on the building project and will monitor management actions deployed to reduce their effect.

30.2 Principles of risk management

The principles of risk management are:

- The management of risk must not remove incentive.
- The party bearing the risk must have the ability to control it.
- Information and perception of risk are fundamental to its assessment and acceptance.
- Allocation of risk is to be gauged by the various parties' abilities to bear the risk.
- All risks change with time and any action or inaction taken upon them.

Risk management is not solely concerned with avoiding risk but with assessing and managing the uncertainty, which inevitably exists when striving to achieve a building project's goal.

30.3 Risk management and NRM 1

NRM 1 departs from the non-specific use of contingencies and promotes the more meaningful application of risk allowances. Notwithstanding this, it is still evident that some cost managers are reluctant to consider the consequence of potential risks on building projects. Common excuses, and responses, include:

- 'The process takes time and costs money' – proper application saves time (allows early action and planning) and saves money by reducing unexpected costs.
- 'Responses cost money' – responses (e.g. on exploratory investigations) are an investment in the future (i.e. spending to save or spending to gain).
- 'Risk management doesn't work properly' – do it properly and it will be effective.
- 'Risk management is just scaremongering' – finding the real risks (uncertainties that matter) should always include the positives (opportunities).
- 'Managing issues is more fun' – key performance indicators (KPIs) should be developed that measure the effectiveness of risk management and reward those who do it properly (e.g. through the contract conditions).
- 'It's too late' – it is never too late; failing to identify risks does not make them go away.
- 'Too busy dealing with issues' – risk management will 'prevent' issues, so starting the process will make for a better future.
- 'It's just common sense' – it is not to everyone. The NRM 1 framework of risk management will help those with less common sense.
- 'Can't prove it works' – the benefits are demonstrable and can be shown by emphasising the management opportunities.

Risk management forms a key part of most business activities. The aim of risk management is not to avoid or eliminate risk, but to manage the risks involved in all activities to maximise opportunities and minimise adverse effects.

Risk management provides both soft and hard benefits. The soft benefits of risk management include:

- demonstrates a responsible approach to employers;
- improves communication;
- leads to common understanding and improved team spirit;
- helps distinguish between good luck/management and bad luck/management;
- helps develop the ability of those involved in the project to assess risks; and
- focuses attention on the most important issues.

The hard benefits of risk management include:

- enables better informed and more believable budgets;
- increases the likelihood of a project adhering to its budget;
- leads to the use of the most suitable procurement and contract strategy;
- allows for a meaningful assessment of risk allowances;
- discourages the acceptance of financially unsound projects;
- enables a more objective comparison of alternatives; and
- identifies and allocates responsibility to the best risk owner.

Risk management is not a passive activity, and there is a cost associated with the 'up-front' risk process, i.e. the cost of assessing and quantifying risk. Moreover, risk responses inevitably cost money, but failing to respond to risk through planned response activities will mean that risks go unmanaged, the risk exposure will not change and the risk management process will be ineffective.

However, when applied properly, risk management will save time and money, and will help provide greater cost certainty. It is the responsibility of the quantity surveyor/cost manager to advise their customers of the benefits of effective risk management – not doing so is remiss and does not demonstrate a responsible approach to cost management.

Risk management is not solely concerned with avoiding risk but with assessing and managing the uncertainty, which inevitably exists when striving to achieve a building project's goal.

The advantages of risk management include:

- reductions in unexpected problems – cost and/or time overruns and/or forced compromises of performance objectives will usually result in crisis management (i.e. fire fighting) and the ineffective application of urgent remedial measures;
- appreciation of excessive caution (e.g. onerous management systems, duplicating insurances, and/or high-cost performance bonds) and the potential to derive opportunities by their relaxation; and
- appreciation of actual confidence levels for achieving objectives with monitoring to illustrate improving or diminishing chances of success.

NRM 1 contends with these challenges. Now it is for cost managers to put into action the best practice guidance afforded by NRM 1 and demonstrate a responsible approach to the cost management of building projects.

30.4 Risk management of building projects

The relevance of risk management of building projects is the high-risk nature of the construction industry.

Designing and constructing a building is rarely straightforward. It involves bringing together a number of different organisations, each with different skills and experience (who may not have worked together before) to create an often unique and complex undertaking, often in an inhospitable working environment and usually within a short timeframe. What's more, these organisations will have their own requirements for making a return on the building project, which can conflict with the objectives of the employer. In addition, each building project itself is probably unique – even if the substance of the construction has been created before, its physical, social, financial and/or political environment will probably be completely new. Fully appreciating this inherent facet of building projects is the first important aspect of effective risk management.

Every building project involves risk, some obvious, some less so. Therefore, it is important to identify the risks, when and how they might occur and what action can be taken to minimise their effect whilst maximising value for money. Risk exposure (the potential effect of risk should it materialise) changes as the building project progresses; continuously managing risks throughout the building project is therefore essential. However, risks cannot be managed in isolation; effective risk management must be an integral part of the project management process.

Research has demonstrated that poorly managed risk adversely affects the employer's ability to achieve the project's objectives, whilst planned systematic risk management will assist the employer in achieving a successful building project. The level of risk can often be reduced and sometimes eliminated if an appropriate response is implemented early enough. But this needs to be regularly reviewed to ensure that appropriate action is taken in light of changing circumstances.

Many of the major decisions that will have the greatest impact on a building project are made early during the appraisal and design development stages. It is at these stages that changes can be made with least disruption and at minimum cost. However, this is also when information on which to base those decisions is most likely to be incomplete or inaccurate. Estimates made before physical components have been defined accurately can often lead to optimistic bias (with its associated false sense of well-being). Therefore, they rarely reflect the true position. To ensure that the right value-for-money decisions are made in the best interests of the employer, all important potential risks must be identified and assessed from the earliest stage of a building project.

As the building project progresses, new risks will emerge and the relative importance of known risks will change (e.g. as the substructure is completed any risks associated with the ground conditions will be resolved). Because of the changing nature of the risk environment, the employer should plan to revise the formal risk analysis at all important stages of the building project to correspond with updates of the cost estimate or cost plan for the building project. The objective of each risk analysis is to provide confidence that all potential risks have been recognised and taken into account, and the accuracy of the revised cost estimate is verified. An assessment of potential risk might require the use of complex analytical tools and a high degree of technical skill. Consequently, the employer should consider commissioning a specialist risk manager.

During the execution of a building project it is most likely that the employer, certain project team members, the main contractor and the main contractor's subcontractors will each manage some part of the total risk. Generally, responsibility for risk should be allocated to the party best placed to exercise the most effective control over it. Where another party can properly manage and control the risk, then the total cost should be less than if the entire risk were retained by the employer. There may well be a number of different risk response measures available. The objective should be to identify those measures that reduce the total cost of risk and give best value for money to the employer.

The remaining risk (after action has been taken to reduce risk through appropriate risk response measures) is referred to as the employer's residual risk exposure. The best estimate of the most likely cost or time effect of this residual risk exposure must be provided at each cost-estimating and cost-planning stage – this is called the risk allowance.

The cost manager is responsible for estimating the cost impact of a risk should it materialise. The materialisation of some risks will also impact on time – time estimates are normally undertaken by a programmer or planner.

As the building project progresses, more of the requirements are defined, risks pass or are transferred to other parties and the consequences are incorporated into the base cost estimate. As that happens, both the residual risk exposure and the risk allowances at the cost-estimating and cost-planning stage will change and should steadily reduce. Thus, risk allowances within an order of cost estimate can be a significant percentage of the total cost estimate; after completion (when all accounts are settled) the requirement for risk allowances pertaining to the construction phase will be zero.

Proper risk analysis and control of risk allowances is therefore a prerequisite of realistic estimates and for minimising the consequential loss arising from the employer's residual risk exposure. Where these consequential costs do not result in the full expenditure of the risk allowances (by managing the risk effectively), any savings must be reflected in revised cost limits and not spent on enhancing the building project – unless sanctioned by the employer.

30.5 Application of risk management to building projects

Risk management should be applied to all building projects. Risk management is now seen as an integral part of good project and programme management of building projects. Risk management is the planned and systematic process of identifying, recording, prioritising, assessing, responding to, monitoring, re-evaluating and controlling risks. Most importantly, risk management is an iterative process, with risks being reviewed at each stage in the development of a building project. The effect of good risk management should be measurable as cost or time savings, both in the capital costs of a building project and throughout the life cycle of a building.

Risk management activities (sometimes referred to as a formal risk analysis) comprise:

- **Risk identification** – to determine what could go wrong so as to identify the risks.
- **Risk assessment** – understanding how the important risks might occur, quantifying their possible effect on the project outturn.
- **Monitoring, updating and controlling risks** – investigating whether the nature and level of risk is acceptable, in terms of value for money and to:
  - ensure that potential risks are recognised, quantified and understood;
  - control expenditure on the employer's residual risks; and
  - improve value for money.
- **Risk response** – responding to contain risks within acceptable limits, including taking countermeasures to mitigate risks and communication of risk responses.
- **Feedback** – on how well risks were managed and lessons learned.

These activities are illustrated in Figure 30.2.

30.6 What is a risk allowance?

A risk allowance is a financial provision for dealing with potential risks should they materialise. The only satisfactory way to ensure that risk allowances provide for the risks to the building project is to determine the size of the allowances from the results of the risk analysis. However, NRM 1 recognises that limited or no information about the site conditions will be available at the time of preparing an order of

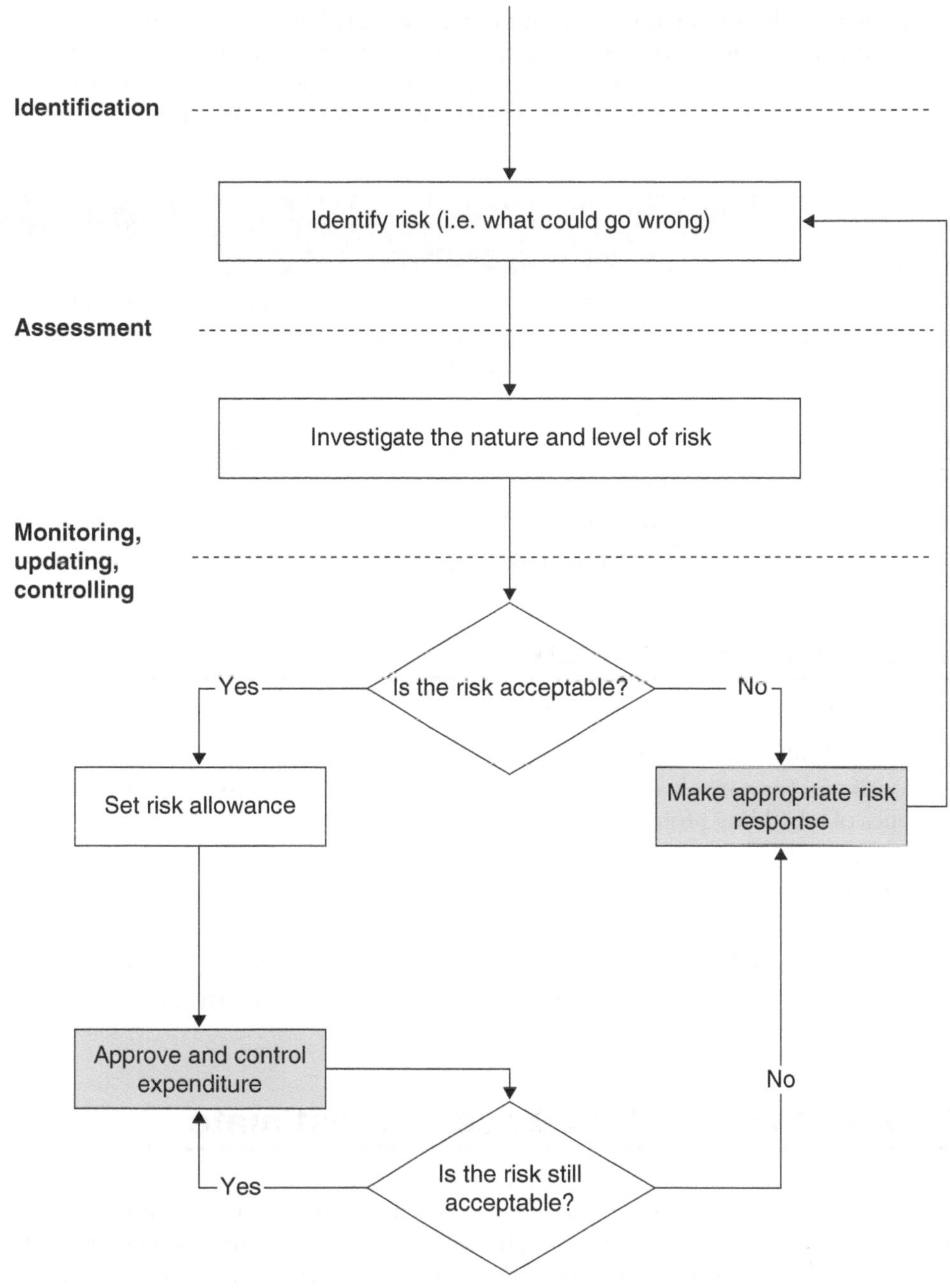

Figure 30.2 Risk management process

cost estimate. For this reason, the rules advise that risk allowances can initially be based on a percentage of the base cost estimate. Notwithstanding this, the rules emphasise that risk allowances should not be a standard percentage, but a properly considered assessment, taking into account the completeness of the design and other uncertainties such as the amount of site investigation done. Moreover, the rules further recommend that appropriate risk assessment techniques be applied as the employer's brief and design are developed.

Successive risk assessments should show decreasing risk due to reducing uncertainty as a consequence of the increasing definition of the project itself and decreasing uncertainty as a result of decisions that are made as the building project progresses. However, it should be noted that risk does not always decrease.

30.7 Formal risk analysis

Risk analysis is a formal review of the employer's residual risk exposure at important stages of a building project. It is concerned with assessing the qualitative and quantitative impact of potential risks. The objectives of a risk analysis are to:

- identify all potential risks;
- identify important risks at the current stage of the building project;
- identify any unusual risk characteristics; and
- determine the most likely project outturn.

The risk analysis process is illustrated in Figure 30.3.

30.7.1 Risk identification

Risk identification usually consists of four parts:

- understanding the content of the base cost estimate;
- identifying (or re-evaluating) the likely sources of potential risk at the important stages of a building project;
- identifying (or re-evaluating) all potential risks at the important stages of a building project; and
- compiling (or updating) the risk register.

The result of the risk identification process should be a schedule of those potential risks that could affect the project outturn or otherwise jeopardise the project objectives.

30.7.2 Content of the base cost estimate

It is important that the scope and criteria used to calculate the base cost estimate are understood before moving on to identify potential risks and to review their likely impact on the building project. The base cost estimate is the total of cost and time (including the cost of risk allocated to the main contractor), but with no risk allowance for the possibility that the employer's residual risk exposure will affect the outturn.

30.7.3 Sources of potential risks

NRM 1 groups sources of potential risk under four headings: design development risks, construction risks, employer's change risk and employer's other risks (the rules also recommend that each of these sources of potential of risk be treated as separate cost centres):

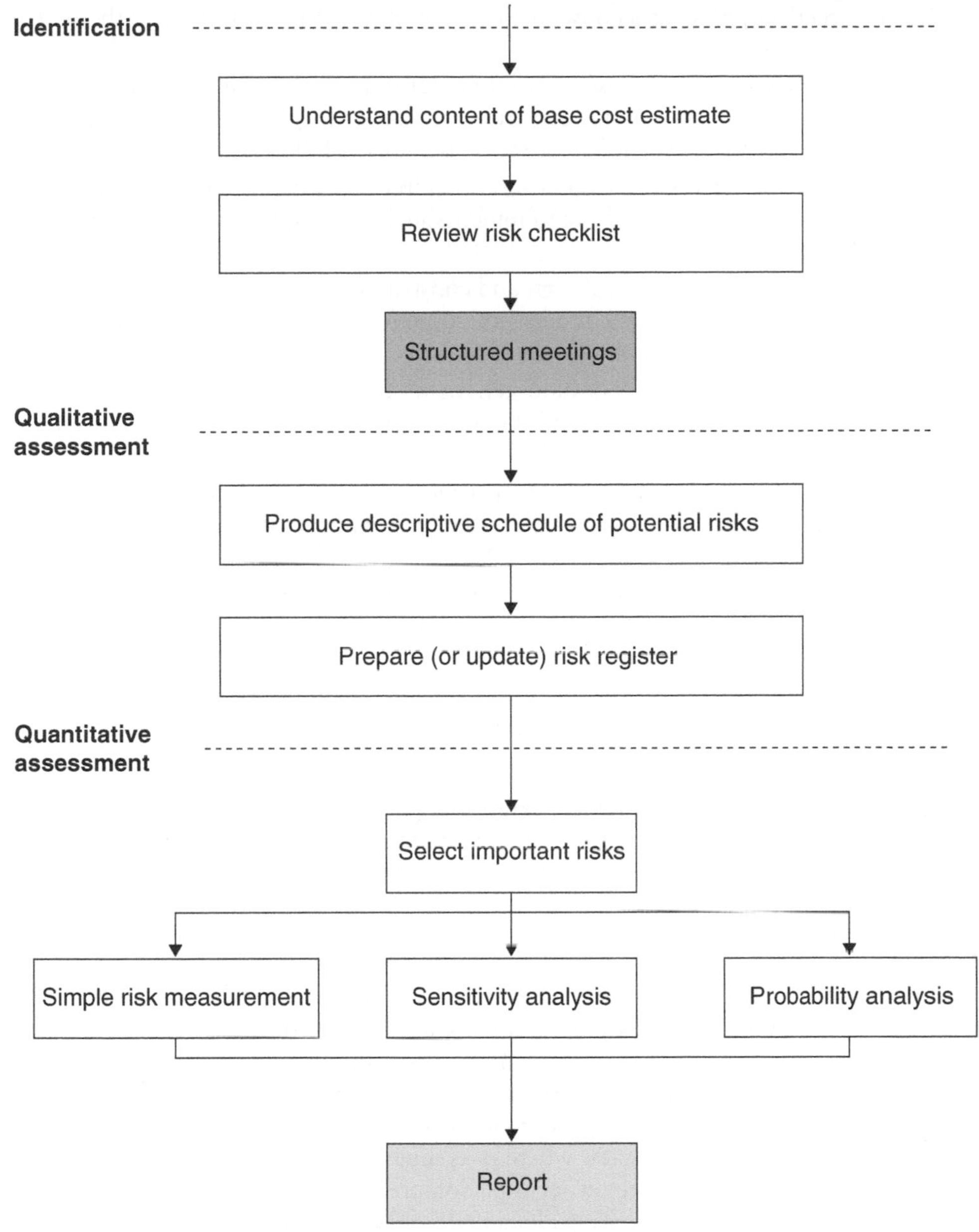

Figure 30.3 Risk analysis process

- **Design development risks** – which might arise during design development as a result of changes to design and in estimating data.
- **Construction risks** – which might be caused by the nature of the site conditions (e.g. access restrictions and limitations, existing buildings, boundaries, existing services, existing occupants and users), that arise during the construction process (e.g. ground conditions, weather, existing services and delays by statutory undertakers) or as a result of construction and commissioning.
- **Employer change risks** – could occur at any time during design development, or

the construction phase, as a consequence of employer-driven changes to the scope of works or brief.

- **Employer other risks** – address sources of potential risk that cannot be categorised under design development risks, construction risks or employer's change risks. They are subdivided by the rules into six categories as follows:
  - project brief – e.g. changes to end-user requirements, inadequate or unclear brief, and employer's specific requirements such as compliance with functional standards;
  - timescales – e.g. unrealistic design and construction programmes imposed by the employer, unrealistic tender periods, effect of phased completion requirements and timescales for decision making;
  - financial – e.g. availability of funds, changing interest rates, amendments to standard contract conditions and the introduction of onerous supplementary contract conditions;
  - management – e.g. competence of the project team, unclear definition of project team responsibilities, ineffective risk management strategy and ineffective change control procedures;
  - third party – e.g. planning refusal, legal agreements, works arising out of party wall agreements and requirements relating to environmental impact assessments; and
  - other – availability of labour, materials and plant, significant changes in market conditions, and changes in legislation.

30.7.4 Identification of potential risks

Successful risk management depends on accurate risk identification. Methods for identifying potential risks range from using standard checklists, previous experience, structured review meetings, interviews, questionnaires and workshops, to brainstorming sessions. On larger or more complex projects, the employer might appoint a risk manager with specific responsibility for co-ordinating risk management. When identifying potential risks, it is important to distinguish between the origin of a risk and its effect. Furthermore, it is important that the potential risks identified are clearly defined rather than defined in general terms so that the precise nature of the risk is understood and appropriate action taken.

It can be advantageous to review where risks might manifest themselves, which can be used to elicit appropriate ownership. Comprehensive lists of causes of potential risks can be found in group element 13 of NRM 1, which relate to each source of potential risk defined by the rules. These lists are not exhaustive but provide a useful aide-memoire when considering potential risks that may have a significant impact on the project. Consequently, the method of identifying potential risks should not be restricted to the use of these lists as this might result in a 'blinkered' approach. When identifying potential risks, it is normally advisable to adopt a combination of methods.

30.7.5 Risk register

Potential risks identified should be entered on a risk register (sometimes called a risk log). A risk register is a means of recording the identified risks, their severity, and the actions and steps to be taken. It can be a simple document, spreadsheet, or a database system. The most effective format for this document is a table, because it will allow a great deal of information to be conveyed in a few pages. There is no standard list of components that should be included in the risk register. However, typical components will include:

- risk identification number (ID) – provides point of reference;
- description of the risk – a phrase that explicitly describes the potential risk;
- potential impact – a phrase that clearly describes the impact of the potential risk should it materialise;
- source of risk – classification of the source of risk;
- risk probability – assessment of how likely it is that the risk will occur (e.g. high, medium or low likelihood of occurrence);
- high, medium or low impact;
- risk response – identifies countermeasures to be undertaken in response to risk;
- risk owner – individual responsible for ensuring the risk is appropriately managed and countermeasures are undertaken; and
- status – indicates whether risk is current or if risk can no longer arise, as well as the impact on the building project.

Other columns, such as quantitative values (e.g. numerical probability of the risk materialising, and likely cost and time impacts should the risk materialise) can be added if appropriate.

An example of a typical risk register is shown in Figure 30.4.

The risk register has an important role to play in the ongoing management of risk and should not only be used to record the initial potential risks identified, but also to actively work to reduce them and review progress regularly.

Maintenance of the risk register should be the responsibility of the risk manager. Regular reviews should ensure that any new risks are captured and that mitigating actions are undertaken to contain risks that remain current. Consideration should also be given to reassessing the impact and probability scores for risks as the building project progresses and as mitigating actions are completed – as they are likely to change over time and risks should be reranked accordingly. Management effort can then be redirected to always focus on the most important risks at any point in time.

30.7.6 Risk assessment

The purpose of risk assessment is to understand and quantify the likelihood of occurrence and the potential impacts on the project outturn.

Various analytical techniques are available, but the key approaches are:

- Qualitative risk assessment – a descriptive review of the potential risks to understand each risk and gain an early indication of the more significant risks.
- Quantitative risk assessment – to measure the possible effect of the significant risks to determine the variability of the project outturn, and to estimate both the maximum likely and most likely project outturn.

Risk Register							
Project: Woollard Hotel, Holborn Viaduct, St Leonards-on-Sea, East Sussex					**Date:** 27 September 2013		
ID	**Potential risk (or event)/ effect(s)**	**Source**	**Risk probability**	**Potential impact**	**Risk response**	**Risk owner**	**Status**
E001	Delay in decanting existing occupants/delay possession of the site and delay to construction works (time and cost impact)	Employer	M	M	Retention	RMB	Employer has prepared draft decanting/reoccupation strategy in consultation with each affected department
E002	Party wall awards/delay the progress of construction works (time and cost impact)	Adjoining properties	L	H	Reduction – further	CDB	Party wall survey and rights of light investigations to be instructed
C001	Japanese knotweed/delay the progress of construction works (cost impact)	Existing site conditions	H	H	Reduction – eradication by specialist prior to commence-ment of construction works	RMB	Specialist survey instructed; to ascertain full extent of infestation to aid decisions
D001	Existing underground services/delay the progress of construction works (time and cost)	Existing site conditions	H	M	Reduction – further site investigation	DS	Specialist survey instructed; to ascertain full extent of over ground and underground services to aid decisions
C002	Exceptional adverse weather conditions/delay the progress of construction works (time and cost impact)	Weather	M	M	Sharing – via contract conditions	RMB/ Contractor	To be dealt with under standard contract conditions
D002	Change to external facade treatment/delay project (time, cost and performance impact)	Planning	L	H	Reduction – redesign (if necessary)	RKT	Await outcome from consultation with planners, which is due w/c 28 October 2013
C003							
C004							
C005							

Figure 30.4 Example of a typical risk register

30.8 Qualitative risk assessment

The objective of a qualitative risk assessment is to understand the nature and impact of a potential risk exposure and to determine the relative importance of each risk in terms of how likely it is that things might go wrong and the possible consequences if they do. Potential risks should be classified in terms of their likely impact on the building project. An acceptable method of formalising this process is to use a qualitative risk assessment schedule, which should accompany the risk register. For each potential risk, the following issues should then be considered:

- the stages of the project when the risk could occur;
- the elements of the project that could be affected;
- the factors that could cause the risk to occur (sometimes referred to as risk triggers);
- any relationship or interdependency to other risks;
- the likelihood of it occurring (probability); and
- how it could affect the building project (potential impact).

The actual qualitative risk assessment can then be undertaken. This comprises three stages:

- risk impact assessment;
- risk probability assessment; and
- risk severity and ranking.

30.8.1 Risk impact assessment:

The impact of the risk, should it occur, needs to be considered in terms of time, cost and performance (or quality) dimensions:

- time impact – the impact on the building project programme;
- cost impact – the impact on the project outturn cost; and
- performance (or quality) impact – the impact on outcomes or outputs (or the impact on quality).

For each potential risk, the likely effect of each impact type is rated and scored. There are a number of ways in which this can be done, one method being:

Very low	1
Low	2
Medium	3
High	4
Very high	5

When carrying out a risk impact assessment, it is important to set some parameters to define what each of these terms means. This is because different project team members will invariably have differing views on the severity of the impact. The parameters might be based on percentage impact or absolute (e.g. 5% to 10% of the base cost estimate, or £5,000 to £10,000). Generally, a score of 1 will be given where the risk will have no impact or an insignificant impact, whilst risks that will have project-critical impact (i.e. the risk is so severe that it threatens to stop the project) will score 5 across one or more dimensions.

The total impact score can be obtained by simply adding up the scores given to the time, cost and performance (or quality) dimensions (refer to Figure 30.5). The value of the impact score will be between 3 and 15.

30.8.2 Risk probability assessment

This considers the likelihood of the risk occurring and is rated in terms of percentage chance. The same five-point scoring system is used.

Very low	1
Low	2
Medium	3
High	4
Very high	5

A score of 1 would imply that the risk is very unlikely to occur; a medium score might imply a 50/50 chance, and a score of 5 would apply to risk that is very likely to materialise.

30.8.3 Risk severity and ranking

The overall severity and ranking of each potential risk is ascertained by assessing the combined effect of impact and probability. The severity score is calculated by multiplying the impact and probability scores. The higher the resulting severity score, the more significant the risk. The severity score is used to rank and prioritise and rank the risks (refer to Figure 30.5).

30.9 Quantitative risk assessment

The detailed quantitative assessment of risk attempts to apply meaningful and objective probabilities to risks and subsequently consider and then quantify the potential impact of risks in terms of time, cost and performance (or quality).

- Increased time – additional time, beyond the base estimate of the completion date for the building project.
- Increased cost – additional cost, above the base cost estimate for the building project.
- Reduced performance (or quality) – the extent to which the project would fail to meet the user requirements for standards and performance.

There are a number of methods and techniques that are available for measuring the probability and consequences of risk (the effect of risk should it materialise). By its nature, however, the effect of risk on the project outturn cannot be calculated

Qualitative Risk Assessment Schedule													
Project: Woollard Hotel, Holborn Viaduct, St Leonards-on-Sea, East Sussex										**Date:** 27 September 2013			
ID	**Potential risk/potential impact**	**Source**	**Stage**			**Time**	**Cost**	**Perfor-mance**	**Impact**	**Proba-bility**	**Severity**	**Rank-ing**
			Pre-Con	Con	Post-Con	(t) 1 to 5	(c) 1 to 5	(q) 1 to 5	(i) t + c + q	(p)	i × p	
E001	Delay in decanting existing occupants/delay possession of the site and delay to construction works (time and cost impact)	Employer		X		3	2	3	8	3	24	4
E002	Party wall awards/delay the progress of construction works (time and cost impact)	Adjoining properties	X			3	3	3	9	5	45	1
C001	Japanese knotweed/delay the progress of construction works (cost impact)	Existing site conditions	X			3	4	1	8	4	32	3
D001	Existing underground services/delay the progress of construction works (time and cost)	Existing site conditions	X			3	4	2	9	4	36	2
C002	Exceptional adverse weather conditions/delay the progress of construction works (time and cost impact)	Weather		X		4	1	1	6	3	18	5
D002	Change to external facade treatment/delay project (time, cost and performance impact)	Planning		X		4	2	2	8	3	24	4
C003												
C004												
C005												

Figure 30.5 Typical qualitative assessment risk schedule, showing calculation of the impact score, together with the risk severity and ranking.

accurately – if it could, then the event is no longer a risk, and its cost or time value should be incorporated in the base cost estimate. It is important to remember that a quantitative assessment is an approximation, and however sophisticated the technique may appear to be, it cannot result in a precise figure.

Each important risk will affect the project outturn with different levels of probability and consequence. It is very unlikely that all the risks will affect the project outturn to the same extent. In order to determine how the risks might occur and the resulting range of possible consequences, it is necessary to carry out an evaluation of the data either by inspection or by statistical calculation to estimate the likely or probable effect of a combination of all the important risks on the overall project outturn.

The aim of quantitative risk assessment is, therefore, to set a risk allowance at a level that realistically reflects the risk (should it materialise) at the time. The size of the risk allowance should be based on the results of the formal cost analysis; if it is too high, the building project might not proceed; too low, and the building project could be compromised at a later date.

30.10 Methods of quantifying the risk allowance

Methods of quantifying risk range from the simple method of assessment to complex statistical methods such as Monte Carlo Simulation. The methods of quantitative risk assessment discussed in this guide are:

- percentage addition (only suitable for order of cost estimates);
- simple method of assessment;
- probabilistic method; and
- sensitivity analysis.

It is not intended to advocate one method over another but simply to give a brief introduction to the simpler methods of quantitative risk assessment used. The more complex statistical methods are outside the scope of this guide (e.g. Monte Carlo Simulation, Latin Hyper Cube Sampling, MERA (multiple estimating using risk analysis), decision trees and stochastic dominance). To those who wish to understand some of the more statistically based methods of risk assessment, it is recommended that a text explaining the statistical principles of these methods be purchased.

30.10.1 Based on a percentage of the base cost estimate

For the purpose of order of cost estimates, NRM 1 advises that risk allowances can be based on a percentage of the base cost estimate (see Figure 30.6 and Example 30.1). However, the rules advocate that this method should only be used in preparing rough and initial order of cost estimates.

Using this approach, the steps to calculating the risk allowance are as follows:

Step 1: Determine the base cost estimate (i.e. risk-free estimate).

Note:

The equation for calculating the base cost estimate is therefore:

d = a + b + c

Where:

a = works cost estimate

b = estimated cost of project and design team fees

c = estimated cost of other development and project costs

d = base cost estimate.

Step 2: Determine the percentage to be applied in respect of design development risks.

Step 3: Determine the percentage to be applied in respect of construction risks.

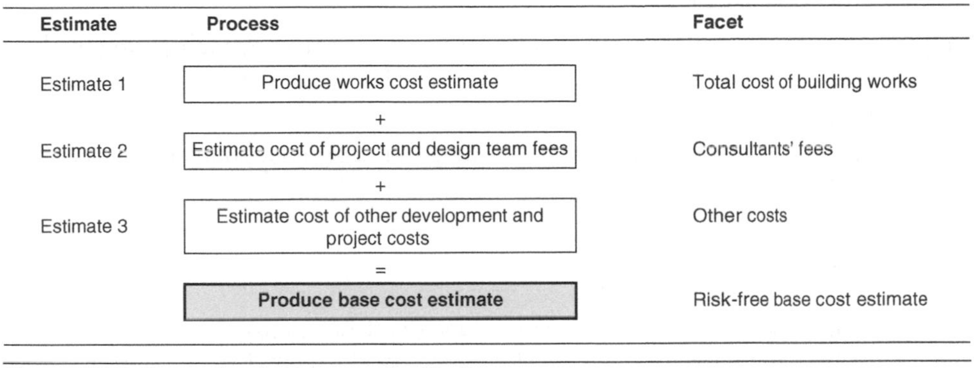

Estimate	Process	Facet
Estimate 1	Produce works cost estimate	Total cost of building works
	+	
Estimate 2	Estimate cost of project and design team fees	Consultants' fees
	+	
Estimate 3	Estimate cost of other development and project costs	Other costs
	=	
	Produce base cost estimate	Risk-free base cost estimate

Figure 30.6 Computing the base cost estimate

Example 30.1 Quantifying the risk allowance based on a percentage of the base cost estimate

RISKS					
Code	**Element/sub-elements/components**				**Totals**
				£	**£**
13	**RISKS**				
	Calculation of base cost estimate:				
	Works cost estimate			22,930,268	
	Project and design fees estimate			2,636,981	
	Other development and project costs estimate			0	
	Total: base cost estimate			25,567,249	
13.1	Design development risks	1.75%	of	25,567,249	447,427
13.2	Construction risks	2.25%	of	25,567,249	575,263
13.3	Employer change risks				No Allowance
13.4	Employer other risks				No Allowance
13	**TOTAL – Risk: to cost plan summary**				**1,022,690**

Note:
Method only suitable for rough and initial order of cost estimates

Step 4: Determine the percentage to be applied in respect of employer's change risks.

Note:
It should be noted that not all employers would want an allowance for employer change risks to be included in an order of cost estimate. For this reason, it is important for the cost manager to establish if the employer wishes an allowance for employer's change risks to be included in the cost estimate.

Step 5: Determine the percentage to be applied in respect of employer's other risks.

Note:
As with employer's change risks, the cost manager will need to establish if the employer wishes an allowance for employer's other risks to be included in the order of cost estimate.

Step 6: Ascertain the allowance for design development risks by multiplying the base cost estimate by the percentage applicable for design development risks.

Note:
The equation for calculating the allowance for design development risks is therefore:

$$R1 = d \times p1$$

Where:

d = base cost estimate
p1 = percentage allowance for design development risks
R1 = risk allowance estimate for design development risks.

Step 7: Ascertain the allowance for construction risks by multiplying the base cost estimate by the percentage applicable for construction risks.

Note:
The equation for calculating the allowance for construction risks is therefore:

$$R2 = d \times p2$$

Where:

d = base cost estimate
p2 = percentage allowance for construction risks
R2 = risk allowance estimate for construction risks.

Step 8: Ascertain the allowance for employer change risks by multiplying the base cost estimate by the percentage applicable for employer change risks.

Note:
The equation for calculating the allowance for employer change risks is therefore:

$$R3 = d \times p3$$

Where:

d = base cost estimate
p3 = percentage allowance for employer change risks
R3 = risk allowance estimate for employer change risks.

Step 9: Ascertain the allowance for employer other risks by multiplying the base cost estimate by the percentage applicable for employer other risks.

Note:
The equation for calculating the allowance for employer other risks is therefore:

$$R4 = d \times p4$$

Where:
d = base cost estimate
p4 = percentage allowance for employer other risks
R4 = risk allowance estimate for employer other risks.

Step 10: Ascertain the total risk allowance by adding the allowances for design development risks, construction risks, employer change risks and employer other risks.

Note:
The equation for calculating the total risk allowance is:

RA = R1 + R2 + R3 + R4

Where:
R1 = risk allowance estimate for design development risks (estimate 4A)
R2 = risk allowance estimate for construction risks (estimate 4B)
R3 = risk allowance estimate for employer change risks (estimate 4C)
R4 = risk allowance estimate for employer other risks (estimate 4D)
RA = total risk allowance (estimate 4).

30.10.2 Simple method of assessment

The simplest form of quantitative analysis is to consider each important risk separately and then to investigate their possible combined effect by inspection. Usually (in some cases without trying to quantify the probability of occurrence) an estimate (often subjective judgement is made of the most likely consequence – expected value – of each risk), these are then added together to give an estimation of the projected outcome, which will become the risk allowance. This simple method of assessment is sometimes called the basic simple method of assessment. The method will usually give a good idea of how important risks could affect the building project and is most commonly used for small and medium-sized straightforward building projects. (See Example 30.2.)

The simplest method of quantitative analysis can be enhanced by quantifying the probability of occurrence for each risk, which is multiplied by the maximum likely

Example 30.2 Quantifying the risk allowance using the basic simple method of assessment

Construction Risk Allowance Estimate		
ID	Consequence (of risk materialising)	Expected value £
C001	Eradication of Japanese knotweed	30,000
C002	Removal and disposal of asbestos	25,000
C003	Ground water	10,000
C004	Removal and disposal of underground obstructions	18,000
C005	Extra cost of disposing of non-hazardous and hazardous material with ground	350,000
C006	Diversion of existing sewer	25,000

Example 30.2 *(Continued)*

Construction Risk Allowance Estimate		
ID	Consequence (of risk materialising)	Expected value
		£
C007	Diversion of existing electrical services	60,000
C008	Statutory undertakers (i.e. performance)	60,000
C009	Adjacent structures (i.e. requiring special precautions)	100,000
C010	Effect of changes/variations on construction programme	100,000
	Total construction risk allowance estimate – to cost plan summary:	**778,000**

cost of the risk, should it materialise (i.e. maximum likely cost). The probability of occurrence of each risk is usually made based on a subjective judgement of the most likely consequence of each risk. This method can only be used if a reasonable number of potential risks exist. This is because it assumes that some risks will materialise and others will not (i.e. a swings and roundabout approach). Clearly, the danger with this method is that, if there is a high impact risk that does materialise, it could result in the overall risk allowance being wiped out. This method is often referred to as the enhanced simple method of assessment (see Example 30.3).

Example 30.3 Quantifying the risk allowance using the enhanced simple method of assessment

Construction Risk Allowance Estimate – Summary				
ID	Consequence (of risk materialising)	Maximum likely cost of risk	Probability (of occurrence)	Expected value
		£	% (Factor)	£
		(c)	(p)	(c × p)
C001	Eradication of Japanese knotweed	30,000	90% (0.90)	27,000
C002	Removal and disposal of asbestos	25,000	70% (0.70)	17,500
C003	Ground water	10,000	75% (0.75)	7,500
C004	Removal and disposal of underground obstructions	18,000	70% (0.70)	12,600
C005	Extra cost of disposing of non-hazardous and hazardous material with ground	350,000	65% (0.65)	227,500
C006	Diversion of existing sewer	2,5000	80% (0.80)	20,000
C007	Diversion of existing electrical services	60,000	90% (0.90)	54,000
C008	Statutory undertakers (i.e. performance)	60,000	50% (0.50)	30,000
C009	Adjacent structures (i.e. requiring special precautions)	100,000	50% (0.50)	50,000
C010	Effect of changes/variations on construction programme	100,000	90% (0.90)	90,000
	Total construction risk allowance estimate – to cost plan summary:			**536,100**

30.10.3 Probabilistic analysis

The method of probabilistic analysis relies upon applying meaningful, although somewhat subjective, probabilities to cost estimates. A probabilistic analysis starts by calculating the cost or time consequence of each important risk for a limited number of different assumptions. This is then expressed as a range of values of consequence (expected values). Usually each important risk can only be quantified realistically by calculating possible values of cost (or time) for three reasonably foreseeable cases. These are:

- the worst case scenario (maximum effect or pessimistic);
- the most likely case scenario (anticipated or realistic effect) and
- the best case scenario (minimum effect or optimistic).

This approach is sometimes referred to as 'three-point risk estimating'. Using probabilistic analysis, the risk allowance for a risk item is calculated as follows:

Step 1: Determine the estimated cost of the worst case, most likely case and best case scenarios.

Step 2: Determine the probability of occurrence for the worst case, most likely case and best case scenarios.

Note:
The sum of these probabilities must equate to 1 (or 100%). For example:

Worst case =	0.15 (15%);
Most likely case =	0.50 (50%); and
Best case =	0.35 (35%);
Therefore:	
Total of probabilities =	0.15 (15%) + 0.50 (50%) + 0.35 (35%) = 1 (100%).

Step 3: Ascertain the expected value (value of consequence) for:

- the worst case scenario by multiplying the estimated cost of the worst case scenario by the probability of occurrence for the worst case scenario.
- the most likely case scenario by multiplying the estimated cost of the most likely case scenario by the probability of occurrence for the most likely case scenario.
- the best case scenario by multiplying the estimated cost of the best case scenario by the probability of occurrence for the best case scenario.

Step 4: Ascertain the risk allowance of the risk item by adding together the expected values of the worst case, most likely and best case scenarios.

Examples of probabilistic analysis using estimated costs and unit rates are given in Examples 30.4 and 30.5, respectively, whereas the method of summarising the risk allowances ascertained is illustrated in Example 30.6.

30.10.4 Sensitivity analysis

A sensitivity analysis (sometimes called a 'what-if' scenario or analysis) is a practical method of investigating the risks on a building project by varying the values of key factors and measuring the outcome. By creating a given set of scenarios, the cost manager can determine how changes in one or more variables will impact the target variable (e.g. the impact of changes in inflationary allowances on the cost target for

Example 30.4 Quantifying the risk allowance using probabilistic analysis – for tasks

ID: C001: Eradication of Japanese knotweed			
Scenario	Estimated cost	Probability (of occurrence)	Expected value
	£	% (Factor)	£
	(c)	(p)	(c × p)
Worst case	30,000	40% (0.30)	12,000
Most likely case	20,000	50% (0.60)	10,000
Best case	10,000	10% (0.10)	1,000
Expected value for C001 – to construction risk allowance estimate summary:			**23,000**

Example 30.5 Quantifying the risk allowance using probabilistic analysis – for unit rates

ID: C005: Extra cost of disposing of non-hazardous and hazardous material with ground					
Scenario	Consequence (of risk materialising)	Quantity	Unit rate	Probability (of occurrence)	Expected value
		m³	£/m³	% (Factor)	
		(q)	(r)	(p)	(q × r × p)
Worst case	Disposal of hazardous material	1,000	350	25 (0.25)	87,500
Most likely case	Disposal of non-hazardous material	1,000	120	40 (0.40)	48,000
Best case	Disposal of inert material	1,000	50	35 (0.35)	17,500
Expected value for C005 – to construction risk allowance estimate summary:					**153,000**

Example 30.6 Construction risk allowance estimate summary

Construction Risk Allowance Estimate – Summary		
ID	Consequence (of risk materialising)	Expected value
C001	Eradication of Japanese knotweed	23,000
C002	Removal and disposal of asbestos	17,500
C003	Ground water	7,500
C004	Removal and disposal of underground obstructions	12,600
C005	Extra cost of disposing of non-hazardous and hazardous material with ground	153,000
C006	Diversion of existing sewer	20,000
C007	Diversion of existing electrical services	54,000
C008	Statutory undertakers (i.e. performance)	30,000
C009	Adjacent structures (i.e. requiring special precautions)	50,000
C010	Effect of changes/variations on construction programme	90,000
	Total construction risk allowance estimate – to cost plan summary:	**457,600**

inflation). This method does not use subjective estimates of probability, but aims to provide estimate-based information upon which decisions might be made (sensitivity factors). As such it does not provide mathematical results which are themselves used in the preparation of estimates, but highlights key factors which could have a significant impact on the overall project outturn should they be varied. In practice, a sensitivity analysis is carried out on the more important risks in order to establish if any have a potentially high impact on the project outturn. They might also be used as an aid when investigating the effect of risks by inspection, or as a step in a formal risk assessment

Example 30.7 illustrates the impact of changing inflation rates on the cost of a building project (i.e. the sensitivity of the cost of the building project to changes in the rate of inflation).

Example 30.7 Sensitivity analysis of the effect of changes to the rate of inflation

Base cost estimate	£30,740,057
Risk estimate	£1,075,902
Cost limit (excluding inflation)	£31,815,959

Inflation rate	Cost limit (excluding inflation)	Inflation estimate	Cost limit (including inflation)	Incremental change
%	£	£	£	£
(a)	(b)	(c) (a) × (b)	(d) (b) + (c)	(e)
−1.50	31,815,959	−(477,240)	31,338,719	−(79,540)
−1.25	31,815,959	−(397,700)	31,418,259	−(79,540)
−1.00	31,815,959	−(318,160)	31,497,800	− (79,540)
−0.75	31,815,959	−(238,620)	31,577,339	− (79,540)
−0.50	31,815,959	−(159,080)	31,656,879	− (79,540)
−0.25	31,815,959	− (79,540)	31,736,419	− (79,540)
0.00	**31,815,959**	0	**31,815,959**	0
0.25	31,815,959	79,540	31,895,499	79,540
0.50	31,815,959	159,080	31,975,039	79,540
0.75	31,815,959	238,620	32,054,579	79,540
1.00	31,815,959	318,160	32,134,119	79,540
1.25	31,815,959	397,700	32,213,659	79,540
1.50	31,815,959	477,240	32,239,199	79,540
1.75	31,815,959	556,780	32,372,739	79,540
2.00	31,815,959	636,320	32,452,279	79,540
2.25	31,815,959	715,860	32,531,819	79,540
2.50	31,815,959	795,400	32,611,359	79,540
2.75	31,815,959	874,940	32,690,899	79,540
3.00	31,815,959	954,480	32,770,439	79,540

Example 30.7 *(Continued)*

Inflation rate	Cost limit (excluding inflation)	Inflation estimate	Cost limit (including inflation)	Incremental change
%	£	£	£	£
(a)	(b)	(c) (a) × (b)	(d) (b) + (c)	(e)
3.25	31,815,959	1,034,020	32,849,979	79,540
3.50	31,815,959	1,113,560	32,295,519	79,540
3.75	31,815,959	1,193,100	33,009,059	79,540
4.00	31,815,959	1,272,640	33,088,599	79,540
4.25	31,815,959	1,352,180	33,168,139	79,540
4.50	31,815,959	1,431,720	33,247,679	79,540
4.75	31,815,959	1,511,260	33,327,219	79,540
5.00	31,815,959	1,590,800	33,406,759	79,540

30.11 Risk response and risk allocation

It is essential that the employer and the project team members understand risk response and risk allocation strategies in order to reduce the employer's residual risk exposure and improve value for money.

30.11.1 Total cost of risk

A successful risk management culture will promote the concept of 'the total cost of risk' when reviewing what actions to take to reduce the employer's residual risk exposure and to achieve value for money.

The total cost of risk to the employer will include:

- the cost of reducing the risk;
- the cost of transferring the risk (e.g. the cost of insurance or premium to be paid);
- the cost of any retained risk (or the cost of the risk that was uninsurable or not insured);
- all 'down time' or 'claims' that arise as a result of the risk from contractual relationships (or other obligations) between the employer and others; and
- all management and administrative time, consultants' fees and other charges in connection with managing and dealing with the risk.

The objective of risk management is to reduce the total cost of risk whilst achieving value for money for the employer. This will involve making decisions in the light of an appraisal of the costs and benefits of different options. It is often difficult to quantify costs and benefits accurately, and they may need to be assessed in a judgemental way.

30.11.2 Risk response strategies

As the design evolves, more of the building project requirements are defined, and a risk response can be decided. Typical risk response strategies can be considered under the following headings: risk avoidance, risk reduction, risk transfer, risk sharing and risk retention as shown in Figure 30.7.

- **Risk avoidance** – where risks have such serious consequences on the project outcome that they are totally unacceptable. Risk avoidance measures might include a review of the employer's brief and a reappraisal of the project, perhaps leading to an alternative development mix, alternative design solution or its cancellation.
- **Risk reduction** – where the level of risk is unacceptable. Typical action to reduce risk can take the form of:
  - redesign – combined with improved value engineering;
  - more detailed design or further site investigation – to improve the information on which cost estimates and programmes are based;
  - different materials or engineering services – to avoid new technology or unproven systems or long delivery items;
  - different methods of construction – to avoid inherently risky construction techniques;
  - changing the project execution plan – to package the work content differently, or to carry out enabling works;
  - changing the contract strategy – to allocate risk between the project participants in a different way.

Risk reduction measures are often considered to be worthwhile, as they lead to a more certain project outturn. They usually result in a direct increase in the base cost estimate, and a reduction in risk allowances. A risk reduction measure should only be adopted where the costs are less than the reduction in the total cost of the risk to the employer.

Figure 30.7 Risk response strategies

- **Risk transference** – where accepting the risk would not give the employer best value for money. The object of transferring risk is to pass the responsibility to another party able to better control the risk. Risk is usually transferred from:
  - the employer to the main contractor;
  - the main contractor to a subcontractor;
  - the employer to an insurer in the form of insurance cover; or
  - the main contractor or a subcontractor to a bank or a surety in the form of warranties, bonds and guarantees.

Whenever risk is transferred there is usually a premium to be paid (i.e. the receiving party's valuation of the cost of the risk). To be worthwhile, risk transfer should give better overall value for money to the employer (the total cost of the risk to the employer is reduced by more than the cost of the risk premium). Risk transfer measures include taking out insurance cover where appropriate.

- **Risk sharing** – occurs when risk is not entirely transferred and the employer retains some element of risk. This approach might be adopted where risk exposure is beyond the control of any one party and a realistic way of sharing can be agreed. It is important that each party understands the value of the portion of risk for which it is responsible (risk communication). The object should be to improve control and to reduce or limit the most likely cost of the risk to the employer.
- **Risk retention** – risks retained by the employer that are not necessarily controllable. This remaining risk is called the residual risk exposure.

30.11.3 Controllable and uncontrollable risks

Risks arise as a result of actions or events that are either within or outside the control of the participants of the building project. Thus, they are termed controllable or uncontrollable risks.

- **Controllable risk** – a risk that can be controlled by the participants of the building project. For example, the price of a curtain walling system increases. The outcome of this occurrence is that the project team can explore alternative (more cost-effective) curtain walling systems, or change the specification. The situation is within the control of the project team.
- **Uncontrollable risk** – a factor that is outside the control of the participants of the building project and cannot be influenced by them in any way. Such risks very often originate from external sources. Examples include unexpected changes in market conditions, government legislation, political conditions, and adverse weather conditions.

30.11.4 Risk allocation

To allocate risk that cannot be controlled or underwritten means that the employer retains that risk, but loses the opportunity to exercise control over it. Therefore, before allocating risk, the following factors should be considered:

- Which party is best able to control the events that might lead to the risk occurring?
- Which party can control the risk if it occurs?

- Is it preferable for the employer to be involved in the control of the risk?
- Which party (or parties) should be responsible for a risk if it cannot be controlled?
- If the risk is transferred to a party:
  - Is the total cost to the employer likely to be reduced?
  - Will the recipient be able to bear the full consequences if the risk occurs?
  - Could it lead to different risks being transferred back to the employer?
  - Would the transfer be legally secure (i.e. will the transfer to accepted under law)?

Risks should be allocated at the appropriate time. Allocation of risks too early can lead to unnecessary attempts to reduce the risk before the risk is fully understood. This can increase costs and waste time. But late allocation can also increase costs by compromising the recipient's flexibility to take avoiding action or to exercise effective control over the events that might lead up to the risk occurring.

The concept of risk ownership can be useful when deciding how to allocate and control risk. The employer and the project team members must have a clear understanding of where the ultimate authority and responsibility for controlling and managing the important risks lie.

30.11.5 Risk and contract strategy

Risk and contract strategy are interrelated. Selecting the right contract strategy is one of the most important decisions for the employer. The chosen strategy and the forms of contract conditions influence the allocation of risk, the project management requirements, the design strategy, the employment of consultants and contractors, and the way in which the employer's organisation, the various designers, other project team members, main contractor and subcontractors work together. Each form of contract uses different ways of allocating risks to the contracting parties and the degree of control retained by the employer over the design and construction processes. With allocation of risk goes the authority and responsibility for controlling it and underwriting the possible consequences. Consequently, the contract strategy should be the most appropriate to meet the project objectives.

30.12 Setting and controlling the risk allowance (estimate 4)

Risk allowances are used to:

- improve the overall accuracy of estimates; and
- control expenditure on the employer's residual risks.

Risk allowances, based on the results of a formal risk analysis, are to be included in each formal cost plan. Risk allowances are to reflect the employer's risk exposure. In setting the amount of the risk allowances, the possible consequences of the employer's residual risk must be taken into account. The only satisfactory way to ensure that risk allowances provide for the risks to the project is to determine the size of the allowances from the results of risk analysis.

Historically, risk allowances (historically referred to as contingencies) have been often calculated on an arbitrary percentage (e.g. 5% for new buildings and, say, 10% for refurbishment projects). NRM 1 advocates the use of a risk management approach to ensure that risk allowances realistically reflect the risks inherent in a building project. The rules further stipulate that risk allowances should not be a standard percentage, but a properly considered assessment of the risk, taking into account the completeness of the design and other uncertainties such as the amount of site investigation done. Used correctly, risk allowances facilitate an improvement in the overall accuracy of cost estimates and ensure that expenditure against risks is controlled.

30.12.1 Estimating stages

As a design becomes more detailed and parts of the building project are completed, the project requirement becomes better defined and risks are gradually absorbed. This means that, as the building project progresses, the employer's residual risk exposure and risk allowance should reduce.

The rules recognise that it may still be necessary to base aspects of risk allowances on a percentage addition. Where any aspects of risk allowances for design development risks, construction risks and employer risks are to be based on a percentage addition, the allowances are to be calculated by multiplying the base cost estimate by the selected percentage additions (refer to sub-paragraph 30.7.1 for method of calculation).

30.12.2 Estimating the risk allowance and the tolerance

The risk allowance (the allowance to cover the employer's residual risk exposure), when added to the base cost estimate, should give the total financial provision (i.e. cost limit) most likely to be required by the employer. This allowance should cover all the project risks that can reasonably be expected to materialise.

30.12.3 Setting risk allowances (estimate 4)

Risk allowances should reflect the employer's risk exposure. In setting the amount of each allowance, the possible consequence of the employer's residual risk must be taken into account. The only satisfactory way to ensure that risk allowances provide for the risks to the building project is to determine the size of the allowances from the results of the risk analysis.

NRM 1 recommends that there should be separate risk allowances for:

- **Design development risks** – an allowance for use during the design process to provide for the risks of changes due to design development or in estimating data.

  Notes:

  (1) Costs in connection with additional design services should also be dealt with by way of a risk allowance under this category (i.e. fees for services that are not covered by the consultants' existing scope of services with the employer).

  (2) Where the contract strategy is likely to entail contractor-led design,

additional design services are often referred to as the 'main contractor's design development risk' – refer to paragraph 28.4 (Estimation of main contractor's design fees – element 11.3) in Chapter 28.

- **Construction risks** – an allowance for use during the construction phase to provide for the risks of changes due to site conditions or as a result of construction and commissioning.
- **Employer change risks** – an allowance to provide for the risks of employer changes to the building project.
- **Employer other risks** – an allowance to provide for the employer's residual risks associated with the project or design brief, timescales, financial matters (e.g. funding availability), management, third party and any other which cannot be categorised under design development risks, construction risks or employer's change risks.

Note:
Employer change risks and employer's other risks are sometimes referred to as the employer's risk allowance.

These risk allowances correspond to the main categories of potential risk described in paragraph 30.4.3 and used in the rules (group element 13). The total risk allowance is the sum of the risk allowances for design development risks, construction risks, employer's change risks and employer's other risks (see Figure 30.8).

Thus, the equation for calculating the total risk allowance is:

$$RA = R1 + R2 + R3 + R4$$

Where:

R1 = risk allowance estimate for design development risks (estimate 4A)
R2 = risk allowance estimate for construction risks (estimate 4B)
R3 = risk allowance estimate for employer change risks (estimate 4C)
R4 = risk allowance estimate for employer other risks (estimate 4D)
RA = total risk allowance (estimate 4).

Estimate	Process	Facet
Estimate 4A	Estimate cost of design development risks	Design development risk
	+	
Estimate 4B	Estimate cost of construction risks	Construction risk
	+	
Estimate 4C	Estimate cost of employer's change risks	Employer's change risks
	+	
Estimate 4D	Estimate cost of employer's other risks	Employer's other risks
	=	
Estimate 4	**Produce risk allowance estimate**	Risk allowance

Figure 30.8 Estimating the risk allowance

Example 30.8 shows a typical summary of the risk allowance estimate, which is incorporated in the cost plan. In this example, no allowances have been made for either employer change risks or employer other risks. It has been assumed that the cost manager has still to discuss these aspects of risk with the employer.

Example 30.8 Risk allowance estimate – summary

RISKS			
Code	**Element/sub-elements/components**		**Totals**
		£	£
13	**RISKS**		
13.1	Design development risks		618,302
13.2	Construction risks		457,600
13.3	Employer change risks		No Allowance
13.4	Employer other risks		No Allowance
13	**TOTAL – Risk: to cost plan summary**		**1,075,902**

30.12.4 Controlling the expenditure of risk allowances

The cost manager should manage the risk allowance, with support and advice from the project team. Essentially, management of the risk allowance consists of a procedure to move costs out of the risk allowance into the base cost estimate for the building work, or other applicable cost centre, as risks materialise or actions are taken to manage the risks. There must be formal procedures for controlling quality, cost, time and changes. Risk allowances should only be expended when the identified risks to which they relate occur. When risks occur that have not previously been identified, they should be treated as changes to the project. Similarly, risks that materialise but have insufficient risk allowance made for them will also need to be treated as changes.

Risks and the risk allowance should be reviewed on a regular basis, particularly when formal estimates are prepared, but also throughout the design, construction and equipping stages. As more firm commitments are entered into and the work is carried out, so the risks in future commitments and work are reduced. The estimate for the risk allowance should reflect this.

Box 30.1 Key definitions

- **Base cost estimate** – an evolving estimate of known factors without any allowances for risk or uncertainties.
- **Expected value** – the most likely outcome.
- **Probability** – the degree of uncertainty of an event happening measured on a scale where zero equals impossibility and one equals certainty. By making statistical analysis of cost (or time) against probability, the most likely value (or expected value) may be determined.
- **Project outturn** – in the context of cost, the total cost of the building project.
- **Residual risk** – the remaining risk after all appropriate risk response measures have been taken.
- **Risk** – used to describe the possibility of more than one outcome occurring (literally the possibility of something going wrong). 'Uncertainty' in everyday

language is used to mean that an event cannot be predicted or determined precisely (i.e. a course of action is subject to doubt). Uncertainty becomes risk when constraints are applied (such as a cost ceiling or a time deadline). Throughout this book 'risk' is used in both senses. Risks will include those that can be reduced or eliminated whilst value for money is improved as an appropriate risk response, and those that cannot.

- **Risk allowance** – an estimate of the cost of dealing with an individual risk should it materialise. The summation of all risk allowances provides the most likely cost estimate of the employer's residual risks. The size of the overall risk allowance should be determined from risk assessment.
- **Risk analysis** – a formal review (carried out at important stages of the building project) for the identification and assessment of the employer's residual risk exposure after risk control action has been taken to identify all the potential risks and then the important, or significant, risks, to determine the most likely risk estimate, to identify any unusual risk characteristics, and to determine the variability of the project outturn.
- **Risk assessment** – the qualitative review and quantification of risk.
- **Risk exposure** – the potential impact or effect of a risk, should it materialise, on the project outturn.
- **Risk register (or risk log)** – used to describe the descriptive schedule of potential risks that are thought likely to occur at different stages of the building project.
- **Risk response** – a management action to avoid, reduce, transfer, share or retain risk in a way that improves value for money.
- **Risk trigger** – an event that is identified as being the cause of a potential risk materialising.
- **Sensitivity analysis** – used to highlight the effect on the project outturn of a change in the assumptions of one of the risks. In practice, a sensitivity analysis is carried out on the more important risks in order to establish if any have a potentially high impact on the project outturn. They might also be used as an aid when investigating the effect of risks by inspection, or as a step in a formal risk assessment.
- **Optimistic bias** – used to describe the risk that the predicted outturn does not fully allow for the likelihood of things going wrong. Optimistic bias can be very common in cost and time estimates (especially for new technology and those prepared during the early stages of a building project).
- **Uncertainty** – see definition for risk.

CHAPTER 31

Estimating the possible effects of inflation (group element 14)

Introduction

This chapter:

- defines the concept of inflation;
- explains how inflation in the context of building projects is dealt with by NRM 1;
- defines the time periods used by NRM 1 to measure the possible effect of inflation; and
- contains worked examples showing how to calculate allowances for inflation.

Figure 31.1 highlights where this part of the order of cost estimate fits within the order of cost estimate framework and provides a quick cross-reference to the other relevant chapters.

Estimate	Process	Notes	Chapter reference
Estimate 1	Works cost estimate	Total cost of building works (including main contractor's preliminaries and main contractor's overheads and profit)	Chapter 27
	+		
Estimate 2	Project and design team fees estimate		Chapter 28
	+		
Estimate 3	Other development and project costs estimate		Chapter 29
	+		
Estimate 4	Risk allowances estimate		Chapter 30
	+		
Estimate 5	**Inflation estimate**		
	=		
	COST LIMIT		Chapter 32
	VAT assessment		Chapter 33

Figure 31.1 Inflation estimate in the context of the cost plan framework

31.1 What is inflation?

The cost manager should view inflation as another risk item – although a specific risk that is outside the control of the employer and the project team. The only way that an employer can mitigate the risks of inflation is by deferring the commencement of the construction of the building project, or by bringing the building project forward in times of deflation to obtain cheaper construction prices. However, there will be other factors that also affect the employer's decision when to build, such as the availability of tenants.

Inflation is defined as a sustained increase in the general level of prices for goods and services. It is measured as an annual percentage increase using a price index. As inflation rises, every pound (£GBP), or euro (€) that a person owns buys a smaller percentage of a good or service.

The value of a £GBP does not stay constant when there is inflation. The value of a pound is observed in terms of purchasing power (i.e. the amount of tangible goods that money can buy). As inflation goes up, there is a decline in the purchasing power of money. For example, if the inflation rate is 2% annually, then theoretically 1,000 bricks at £330.00/1,000 today will cost £336.60/1,000 (£330.00 × 1.02) in one year's time. After inflation, a pound cannot buy the same goods that it could do beforehand.

There are several variations on inflation:

- Deflation – when the general level of prices is falling. This is the opposite of inflation.
- Hyperinflation – often referred to as rapid inflation. In extreme cases, this can lead to the breakdown of a country's monetary system. One of the most notable examples of hyperinflation occurred in Germany in 1923, when prices rose by 2,500% in a single month.
- Stagflation – the combination of high unemployment and economic stagnation with inflation. This happened to industrialised countries during the 1970s, when a bad economy was combined with OPEC (Organization of the Petroleum Exporting Countries) raising oil prices. OPEC is an intergovernmental organisation dedicated to stability in and shared control of the petroleum markets.

31.1.1 Causes of inflation

There are two accepted theories on the causes of inflation:

- Demand-pull inflation – 'too much money chasing too few goods'. In other words, if demand is growing faster than supply, prices will increase. This usually occurs in growing economies such as the UK.
- Cost-push inflation – when companies' costs rise, they need to increase prices to maintain their profit margins. Increased costs include things such as salaries and taxes.

31.1.2 Costs of inflation

Almost everyone thinks inflation is evil, but that is not necessarily so. Inflation affects different people and different companies in different ways. It also depends on whether inflation is anticipated or unanticipated. If the inflation rate corresponds to what the majority of individuals are expecting (anticipated inflation), then they can compensate and the cost is not considered to be high. Problems arise when there is unanticipated inflation:

- Creditors lose and debtors gain if the lender does not anticipate inflation correctly. For those that borrow, this is similar to getting an interest-free loan.
- Uncertainty about what will happen next makes consumers and corporations less likely to spend. This hurts the economic output in the long run.
- Individuals living off fixed income see a decline in their purchasing power and, consequently, their standard of living.

If the inflation rate is greater than that of other countries, domestic products become less competitive.

Inflation is a sign that an economy is growing. In some situations, little inflation (or even deflation) can be just as bad as high inflation. The lack of inflation may be an indication that the economy is weakening. As it can be seen, it is not so easy to label inflation as either good or bad – it depends on the overall economy as well as an individual's or a company's own situation.

31.2 Estimating the effects of inflation on a building project

The works cost estimate, the estimate of project and design team fees, the estimate of other development and project costs, and the risk allowances estimate are prepared using rates and prices current at the time the estimate is produced – referred to as the estimate base date. Therefore, it is necessary to consider the possible future effects of inflation on these estimates over a period of time (i.e. from the estimate base date to construction completion). Inflation is an upward movement in the average level of prices. Its opposite is deflation, a downward movement in the average level of prices. The boundary between inflation and deflation is price stability.

NRM 1 contends with inflation, or deflation, under the following two headings:

- tender inflation; and
- construction inflation.

The rules also stipulate that tender inflation and construction inflation be treated as two separate cost targets.

Figure 31.2 shows the processes involved in calculating the inflation estimate.

Cost targets for both tender inflation and construction inflation are based on the difference in forecast price levels between two defined time periods:

Estimate	Process	Facet
Estimate 5A	Estimate cost of tender inflation	Tender inflation
	+	
Estimate 5B	Estimate cost of construction inflation	Construction inflation
	=	
Estimate 5	**Produce inflation estimate**	Inflation (tender and construction periods)

Figure 31.2 Calculation of inflation estimate

- For tender inflation – the period measured is from estimate base date to the date of tender return.
- For construction inflation – the period measured is from the date of tender return to the mid-point of the construction period.

The difference in forecast price levels is measured as a percentage change, using an appropriate time-based price index. The most commonly used index is the 'all-in TPI'. The index is based on a random sample of schemes and represents general trends of tender prices across all sectors. It covers new building work and reflects all sectors of the construction industry (e.g. residential, hotels, commercial offices, retail, warehouses, hospitals and schools), as well as both private and public sectors. Tender price indices are also produced in a number of other formats, including private sector TPI (covering non-housing schemes in the private sector and public sector schemes where funded from the private sector), public sector TPI (covering non-housing schemes in the public sector), housing TPI and housing refurbishment TPI.

Using indices, the equation for calculating the percentage change between two defined time periods is as follows:

$$\frac{(\text{The index at the later date} - \text{the index at the earlier date})}{\text{The index at the earlier date}} \times 100 = \%\ \text{Change}$$

The percentage change will be negative where deflation is forecast.

For order of cost estimates and early cost plans, the inflation allowances for both tender and construction periods are likely to be based on a sequential linear timeframe (i.e. all construction will start simultaneously, and the construction period directly follows on from the tender period). This will result in single percentage change rates being used to calculate the cost allowances for tender inflation and construction inflation. In practice, however, inflation estimates can be far more complex – involving different timeframes and non-sequential procurement. For example, a building project might comprise more than one building, with the construction of each building commencing and completing at different times (i.e. phased). Likewise, the building project might be procured using a number of separate work packages, with packages being procured at different times during the construction phase – not all at once. In such cases, it would be prudent for the cost manager to produce a separate estimate of inflation for each building or works package.

31.3 Estimating the cost of tender inflation (estimate 5A)

Cost estimates for tender inflation are based on a percentage of the cost limit exclusive of inflation (see Figure 31.3 and Example 31.1).

The steps to calculating the allowance for tender inflation are:

Step 1: Ascertain the cost limit, excluding inflation.

Step 2: Determine the estimate base date.

Step 3: Determine the date of tender return.

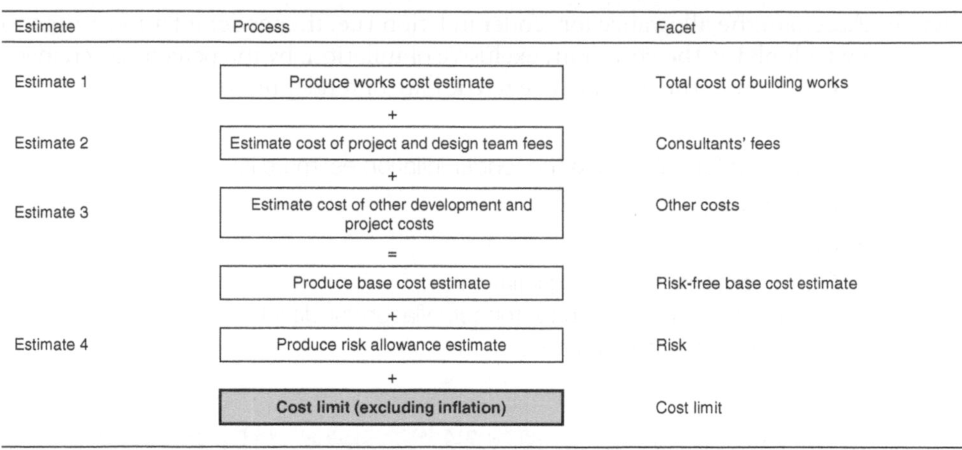

Estimate
Process
Facet
Estimate 1
Produce works cost estimate
Total cost of building works
+
Estimate 2
Estimate cost of project and design team fees
Consultants' fees
+
Estimate 3
Estimate cost of other development and project costs
Other costs
=
Produce base cost estimate
Risk-free base cost estimate
+
Estimate 4
Produce risk allowance estimate
Risk
+
Cost limit (excluding inflation)
Cost limit

Figure 31.3 Calculating the cost limit exclusive of inflation

Example 31.1 Tender inflation estimate

Assumptions:

Cost limit, excluding inflation: £31,815,959

- Estimate base date: September 2013
- Tender return date: March 2013

  Therefore, tender inflation is measured from period September 2013 to March 2014.
- Tender price indices (using All-in TPIs):

  September 2013 = 229 (Forecast)
  March 2014 = 236 (Forecast)

  Hence, the percentage applicable for tender inflation is:

$$\frac{(236 - 229)}{229} \times 100 = 3.06\%$$

Note:
In this case it is forecast that tender prices will increase during this period by 3.06%

Therefore, the tender inflation estimate is:
£31,815,959 × 3.06%
= £973,568 – i.e. inflation in the sum of £973,568

Step 4: Determine the percentage change in costs from estimate base date to the date of tender return (by use of forecast tender-based indices or other suitable indices).

Note:

$$\% \text{ Change} = \frac{\text{The index at the later date} - \text{the index at the earlier date}}{\text{the index at the earlier date}} \times 100$$

Step 5: Ascertain the allowance for tender inflation (i.e. the tender inflation estimate) by multiplying the cost limit, exclusive of inflation, by the percentage change in costs from estimate base date to the date of tender return.

Note:
The equation for calculating the tender inflation estimate is:

$$t = CL1 \times p1$$

Where:
CL1 = cost limit, excluding inflation
p1 = percentage addition for tender inflation (or deflation)
t = tender inflation estimate.

31.4 Estimating the cost of construction inflation (estimate 5B)

Construction inflation allowances are based on a percentage of the sum of the cost limit, excluding inflation, and the tender inflation estimate (see Example 31.2). The steps to calculating the allowance for construction inflation are:

Step 1: Ascertain the cost limit, excluding inflation.

Step 2: Ascertain the tender inflation estimate.

Step 3: Ascertain the cost limit inclusive of tender inflation.

Note:

Cost limit (excluding inflation) + tender inflation estimate

Step 4: Determine the date of tender return.

Step 5: Determine the mid-point of the construction period.

Step 6: Determine the percentage change in costs from the date of tender return to mid-point of the construction period (by use of forecast tender-based indices or other suitable indices).

Note:

$$\%\ \text{Change} = \frac{\text{The index at the later date} - \text{the index at the earlier date}}{\text{the index at the earlier date}} \times 100$$

Step 7: Ascertain the allowance for construction inflation (i.e. the construction inflation estimate) by multiplying the cost limit, inclusive of tender inflation, by the percentage change in costs from the date of tender return to mid-point of the construction period.

Note:
The equation for calculating the construction inflation estimate is:

$$c = (CL1 + t) \times p2$$

Where:
CL1 = cost limit, excluding inflation

t = tender inflation estimate
p2 = percentage addition for construction inflation (or deflation), as a decimal
c = construction inflation estimate.

Example 31.2 Construction inflation estimate

Assumptions:

- Cost limit, excluding inflation: £31,815,959
- Estimated cost of tender inflation: £973,568
- Tender return date: 6 March 2014
- Construction phase commencement: 12 May 2014
- Construction phase completion date: 21 2015
- Construction period: 65 weeks
- Mid-point construction phase occurs at end of week 32 (33 weeks = (approx.) 7½ months)
- Programme from tender return date:

Tender evaluation, reporting, post-tender negotiations, contract award:	7 weeks
Mobilisation period:	4 weeks
Mid-point construction phase occurs after:	33 weeks
Period of construction inflation:	44 weeks

- Mid-point of construction period occurs weeks after the tender return date. Thus, the mid-point of construction period is 8 January 2015.
  Therefore, construction inflation is measured from the period March 2014 to January 2015.
- Tender price indices (using All-in TPIs):
  March 2014 (Forecast) = 236
  January 2015 (Forecast) = 240

  Hence, the percentage applicable for construction inflation is:

$$\frac{(240-236)}{236} \times 100 = 1.69\%$$

Note:
In this case it is forecast that tender prices will increase during this period by 1.69%.

Therefore, the construction inflation estimate is:
(£31,815,959 + 973,568) × 1.69%
= £32,789,527 × 1.69%
= £554,143

31.5 Producing the inflation estimate (estimate 5)

The total allowance for inflation is the sum of the tender inflation estimate and the construction inflation estimate (see Example 31.3). Hence, the equation for calculating the inflation estimate is:

Example 31.3 Inflation estimate

Assumptions:

- Estimated cost of tender inflation: £973,568
- Estimated cost of construction inflation: £554,143

Therefore, the inflation estimate is:
£973,568 + £554,143
= £1,527,711

$$i = t + c$$

Where:

t = tender inflation estimate
c = construction inflation estimate
i = inflation estimate.

In this example, it is forecast that average price levels will be subject to inflation, resulting in a forecast increase in construction prices from the estimate base date.

31.6 Estimating inflation for multiple time periods

Figure 31.4 illustrates a typical method for calculating inflation when a number of different time periods, and inflation rates, are applicable.

The content is as follows:

Column (a): *Code*: self-explanatory.

Column (b): *Works package description*: self-explanatory.

Column (c): *Cost limit (excluding inflation)*: this is the latest approved cost estimate at the defined estimated base date.

Column (d): *Tender inflation (%)*: this is the percentage applicable for tender inflation.

Column (e): *Tender inflation (£)*: this is the total sum allocated to the cost centre for tender inflation.

Column (f): *Cost limit (including tender inflation)*: this is the latest approved cost estimate at the defined estimated base date, which includes a forecast of inflation up to the date of tender return.

Column (g): *Construction inflation (%)*: this is the percentage applicable for construction inflation.

Column (h): *Construction inflation (£)*: this is the total sum allocated to the cost centre for construction inflation.

Column (i): *Cost limit (including tender and construction inflation)*: this is the latest approved cost estimate at the defined estimated base date, which includes the forecast of inflation to project completion. This also represents the current cost target for the cost centre – i.e. the 'control estimate'.

Column (j): *Inflation estimate*: This is the total sum allocated to the cost centre for inflation.

Code	Works Package Description	Cost Limit (excluding Inflation)	Tender Inflation	Tender Inflation Estimate	Cost Limit (including Tender Inflation)	Construction Inflation	Construction Inflation Estimate	Cost Limit (including Tender and Construction Inflation)	Inflation Estimate
(a)	(b)	(c)	(d)	(e) (c) × (d)	(f) (c) × (e)	(g)	(h) (f) × (g)	(i) (f) + (h)	(j) (e) + (h)
		£	%	£	£	%		£	
1	Main contractor's preliminaries	3,025,081	3.06	92,567	3,117,648	1.69	52,688	3,170,336	145,255
2	Groundworks	379,566	3.06	11,615	391,181	0.00	0	391,181	11,615
3	Piling	1,287,900	3.06	39,410	1,327,310	0.00	0	1,327,310	39,410
4	Concrete works (including precast components)	840,200	3.06	25,710	865,910	0.70	6,061	871,971	31,771
5	Structural steelwork	1,149,907	3.06	35,187	1,185,094	4.39	52,026	1,237,120	87,213
6	Carpentry	365,000	3.06	11,169	376,169	2.03	7,636	383,805	18,805
7	Masonry (brickwork and blockwork)	126,326	3.06	3,866	130,192	2.03	2,643	132,835	6,509
8	Roof systems and rainwater goods	223,158	3.06	6,829	229,987	2.31	5,313	235,300	12,142
9	Joinery (including internal doors, toilet cubicles and vanity units)	579,661	3.06	17,738	597,399	3.06	18,280	615,679	36,018
10	Windows and external doors	85,523	3.06	2,617	88,140	2.03	1,789	89,929	4,406
11	Curtain walling	2,629,284	3.06	80,456	2,709,740	2.03	55,008	2,764,748	135,464
12	Modular units (guest rooms)	7,548,517	3.06	230,985	7,779,502	1.69	131,474	7,910,976	362,459
13			3.06			1.69			
14	Dry linings and partitions	176,211	3.06	5,392	181,603	2.89	5,248	186,851	10,640
15	Tiling	1,821,684	3.06	55,744	1,877,428	2.89	54,258	1,931,686	110,002
16	Painting and decorating	205,163	3.06	6,278	211,441	2.89	6,111	217,552	12,389
17	Floor coverings	415,163	3.06	12,704	427,867	2.89	12,365	440,232	25,069

Figure 31.4 Reprofiling of inflation projection based on works packages

Code	Works Package Description		Cost Limit (excluding Inflation)	Tender Inflation	Tender Inflation Estimate	Cost Limit (including Tender Inflation)	Construction Inflation	Construction Inflation Estimate	Cost Limit (including Tender and Construction Inflation)	Inflation Estimate
(a)	(b)		(c)	(d)	(e) (c) × (d)	(f) (c) × (e)	(g)	(h) (f) × (g)	(i) (f) + (h)	(j) (e) + (h)
			£	%	£	£	%		£	
18	Suspended ceilings		166,384	3.06	5,091	171,475	2.89	4,956	176,431	10,047
19	Mechanical and electrical services installations (including sanitary appliances)		3,691,050	3.06	112,946	3,803,996	1.97	74,939	3,878,935	187,885
20	Lifts		342,595	3.06	10,483	353,078	1.80	6,355	359,433	16,838
21	Fittings, furnishings and equipment		591,458	3.06	18,099	609,557	1.69	10,302	619,859	28,401
22	Architectural metalwork		281,000	3.06	8,599	289,599	1.80	5,213	294,812	13,812
23	External drainage		0	0.00	0	0	0.00	0	0	0
24	Works by statutory undertakers – mains connections		302,364	3.06	9,252	311,616	0.00	0	311,616	9,252
25	External works – hard landscape works		68,000	3.06	2,081	70,081	3.09	2,166	72,247	4,247
	Works Package (excluding main contractor's overheads and profit):		26,301,195		804,818	27,106,013		514,831	27,620,844	1,319,649
26	Main contractor's overheads and profit	4.00%	1,052,048	–	32,193	1,084,241	–	20,593	1,104,834	52,786
	Works Package Subtotal:		27,353,243		837,011	28,190,254		535,424	28,725,678	1,372,435
27	Project and design team fees		3,386,814	3.06	103,637	3,490,451	0.00	0	3,490,451	103,637
28	Other development and project costs		0	0.00	0	0	0.00	0	0	0
29	Risk allowances		1,075,902	3.06	32,923	1,108,825	1.69	18,739	1,127,564	51,662
	Totals		**31,815,959**		**973,571**	**32,789,530**		**554,163**	**33,343,693**	**1,527,734**

Figure 31.4 Reprofiling of inflation projection based on works packages

CHAPTER 32

Establishing the cost limit

Introduction

This chapter:

- Provides step-by-step guidance on how to calculate the cost limit.
- Explains the purpose of the 'cost limit'.

Figure 32.1 highlights where this part of the order of cost estimate fits within the order of cost estimate framework and provides a quick cross-reference to the other relevant chapters.

Estimate	Process	Notes	Chapter reference
Estimate 1	Works cost estimate	Total cost of building works (including main contractor's preliminaries and main contractor's overheads and profit)	Chapter 27
	+		
Estimate 2	Project and design team fees estimate		Chapter 28
	+		
Estimate 3	Other development and project costs estimate		Chapter 29
	+		
Estimate 4	Risk allowances estimate		Chapter 30
	+		
Estimate 5	Inflation estimate		Chapter 31
	=		
	COST LIMIT		
	VAT assessment		Chapter 33

Figure 32.1 Cost limit in the context of the cost plan framework

32.1 What is the purpose of the cost limit?

Cost managers often fall into the trap of estimating the cost of the design rather than establishing a budget, in conjunction with the employer, for the designers to design to. This invariably results in the cost manager reporting a change to the estimated cost every time the design changes or is developed. More often than not, the latest costs reported tend to take an upward trend. Not only does this exasperate the employer, it introduces conflict between the employer and the cost manager, as well as with other members of the project team. Cost managers must educate themselves to set the cost parameters to which the designers must design – as most employers do not possess never-ending financial resources. Furthermore, it becomes extremely difficult to judge whether the employer is obtaining value for money when the costs of building projects are designer-led. Hence the need for a cost limit, which includes cost targets for each of the elements and sub-elements of a building, together with cost targets for risk. Cost managers must learn to properly estimate the costs of risk allowances, as well as manage them, so that the building costs remain within the cost limit.

NRM 1 defines the cost limit as 'the maximum expenditure that the employer is prepared to make in relation to the completed building'. The cost limit becomes the employer's authorised budget that, unless agreed by the employer, is not to be exceeded. Figure 32.2 shows how the cost limit is established. The only exception to this rule is if the employer changes the scope of the building project – when a revised cost limit will need to be ascertained by the cost manager and set by the employer.

Box 32.1 Key definitions

- **Cost limit** – the maximum expenditure that the employer is prepared to make in relation to the completed building. Sometimes referred to as authorised budget or approved estimate.

32.2 Determining the cost limit

Figure 32.1 shows how the cost limit is calculated.

The steps to calculating the cost limit are:

Step 1: Determine the cost limit (excluding inflation).

Step 2: Determine the tender inflation estimate.

Step 3: Determine the construction inflation estimate.

Note:
The method of determining the cost limit, excluding inflation, the tender inflation estimate and the construction inflation estimate are described in Chapter 31 ('Estimating the possible effects of inflation')

Estimate	Process	Facet
	Cost limit (excluding inflation)	Cost limit
	+	
Estimate 5	Estimate cost of inflation	Inflation
	=	
	Cost limit (including inflation)	Cost limit

Figure 32.2 Computing the cost limit inclusive of inflation

Step 4: Ascertain the cost limit, including inflation, by adding the tender inflation estimate and the construction inflation estimate to the cost limit (excluding inflation).

The equation for calculating the cost limit, including inflation, is therefore:

$$CL2 = CL1 + Ti + Ci$$

Where:

CL1	=	cost limit (excluding inflation)
CL2	=	cost limit (including inflation)
Ti	=	tender inflation estimate
Ci	=	construction inflation estimate.

Note:

Ti + Ci = total inflation estimate

The cost limit is calculated for the Woollard Hotel scheme as follows:

Cost limit (excluding inflation)	=	£31,816,450
Tender inflation estimate	=	£973,571
Construction inflation estimate	=	£554,163

Therefore, the 'cost limit' is:

Cost limit (excluding inflation) =	£31,816,450
Tender inflation estimate =	£973,571
Construction inflation estimate =	£554,163
	£33,344,184

CHAPTER 33

Taxes and incentives

Introduction

This chapter:

- identifies the different types of taxes applicable to building projects;
- defines value added tax (VAT);
- discusses VAT in the context of building and construction works;
- explains how to assess VAT liability;
- outlines the merits of capital allowances, land remediation tax incentives and grants.

Figure 33.1 highlights where this part of the order of cost estimate fits within the order of cost estimate framework and provides a quick cross-reference to the other relevant chapters.

33.1 Value added tax

Estimate	Process	Notes	Chapter reference
Estimate 1	Works cost estimate	Total cost of building works (including main contractor's preliminaries and main contractor's overheads and profit)	Chapter 27
	+		
Estimate 2	Project and design team fees estimate		Chapter 28
	+		
Estimate 3	Other development and project costs estimate		Chapter 29
	+		
Estimate 4	Risk allowances estimate		Chapter 30
	+		
Estimate 5	Inflation estimate		Chapter 31
	=		
	COST LIMIT		Chapter 32
	VAT assessment		

Figure 33.1 Value added tax assessment in the context of the cost plan framework

33.1.1 What is VAT?

Value added tax is a tax on most business transactions that involve a transfer of goods and services in the United Kingdom, as well as in other countries, charged by the supplier on goods and services provided and in turn paid by the supplier in the UK to HM Revenue & Customs. It is also charged on goods and some services that are imported from countries outside the European Union (EU), and brought into the UK from other EU countries. VAT was introduced to the UK property and construction industries through the Finance Act 1972 (now the Finance Act 2013).

To charge or reclaim VAT, a company must be a VAT-registered business. Once registered for VAT the company charges VAT on sales invoices (output VAT) and pays VAT on company purchases (input VAT). Each quarter the difference between output VAT and input VAT is paid to or reclaimed from HM Revenue & Customs when completing a VAT return (i.e. the form used to identify the amount of VAT payable to or recoverable from HM Revenue & Customs in respect of the accounting period).

VAT is charged when a supplier sells to either another business or to a non-business customer (e.g. members of the public or 'consumers'). When VAT-registered businesses buy goods or services from a supplier they can generally reclaim the VAT they have paid. However, if a business is not VAT-registered then it is unable to reclaim the VAT paid on purchased goods and services.

There are different VAT rates in the UK, depending on the goods or services that are being provided. At the moment there are three rates used in the UK:

- standard – 20% (since 4 January 2011);
- zero – 0%; and
- reduced – 5%.

There are also some goods and services that are:

- exempt, so no VAT is charged on them (because the law stipulates that VAT must not be applied); or
- outside the UK VAT system altogether (e.g. fees that are fixed by law – known as 'statutory fees').

The standard rate of VAT is the default rate – this is the rate that is charged on most goods and services in the UK unless they're specifically identified as being reduced- or zero-rated. Reduced- or zero-rated construction works might include: ordinary domestic dwellings, installing energy-saving materials in certain types of buildings, installing grant-funded heating equipment and security goods, installing mobility aids for elderly people and connecting or reconnecting properties to mains gas supply.

Not surprisingly, due to the ease of payment and its relative comprehensibility, VAT systems have been adopted by different nations across the world.

33.1.2 VAT in the context of building and construction works

Not all goods and services in connection with building and construction works attract VAT and those that do may attract differing rates.

Collection and payment of VAT to HM Revenue & Customs is, therefore, particularly difficult in the context of building and construction works. It might be that works of repair and alteration and certain new buildings could be standard-rated whilst new residential dwellings and communal residential dwellings and some new buildings for charitable use might be zero-rated or entirely exempt from VAT. Thus, a complicated building project might involve a combination of exempt, zero-, reduced- and standard-rated works. The situation is made yet more complicated by the fact that contractors charge for works in instalments (i.e. interim payments), as the building project proceeds, thus raising the difficult tasks of correctly apportioning the VAT and applying the appropriate VAT rate to each aspect of the building works on a month-by-month basis as interim certificates and payments fall due. Added to this, the VAT regulations are in any event constantly being revised.

For these reasons, contract sums are usually treated as VAT-exclusive and special VAT provisions – such as those for use with the JCT (Joint Contracts Tribunal) standard forms of contract – are used to enable VAT to be treated separately from other contractual payments. It is therefore important to check on the position before commencing any building works and before concluding any contract to ensure that the latest such provisions are used.

33.1.3 Assessing liability for VAT

NRM 1 recognises that value added tax in relation to building projects is a notoriously difficult area and, unless specifically required by the employer, recommends that VAT assessments be excluded from cost estimates and cost plans. Even so, it is essential that the cost manager ascertains how the employer wishes VAT to be treated in the context of a cost estimate or cost plan.

Where an assessment of VAT is required by the employer, the rules advise that specialist advice be sought to ensure that the aspects of the building project on which VAT can be charged are identified, and that the correct VAT rates are applied.

The amount of VAT liability likely to be payable by the employer is calculated by simply multiplying the estimated cost of the aspect of a building project to which VAT is applicable by the relevant VAT rate. The equation for estimating the total amount of VAT payable is:

$$vt = \Sigma\ (c1 \times p1) + (c2 \times p2) + (c3 \times p3)$$

Where:

c1 = estimated cost of the aspect of a building project at standard-rated VAT
c2 = estimated cost of the aspect of a building project at reduced-rated VAT
c3 = estimated cost of the aspect of a building project at zero-rated VAT
p1 = standard rate VAT as a percentage
p2 = reduced rate VAT as a percentage
p3 = zero rate VAT as a percentage
vt = total VAT assessment.

Note:
The calculation of zero-rated VAT is simply shown for completeness.

A VAT assessment is shown in Example 33.1.

Example 33.1 VAT assessment

Assumptions:

Estimated cost of the aspect of the building project at:
Standard-rated VAT = £31,343,693
Reduced-rated VAT = £1,500,000
Zero-rated VAT = £500,000

VAT rates:
Standard rate = 20.00%
Reduced rate = 5.00%
Zero rate = 0.00%

The amount of VAT liability can be assessed as follows:

Estimated cost £	VAT rate %	VAT assessment £	
(c)	*(p)*	*(c × p)*	
31,343,693	20.00	6,268,739	(c1 × p1)
1,500,000	5.00	75,000	(c2 × p2)
500,000	0.00	0	(c3 × p3)
Total VAT assessment:		£6,343,739	vt = Σ (c1 × p1) + (c2 × p2) + (c3 × p3)

33.2 Capital and incentive allowances

33.2.1 Capital allowances

Capital allowances are tax allowances that the owner of a commercial asset can offset against taxable income or profits. They provide a deduction for tax purposes in lieu of the depreciation charged in accounts – turning an expense into income.

For companies incurring capital expenditure, capital allowances provide a valuable form of tax relief for certain parts of buildings (initial allowance). This valuable form of tax relief is, in most cases, either under-claimed or not claimed at all due to lack of understanding or application of the legislation and case law governing the availability of the relief. Tax relief is spread over a number of years and provides cash flow benefits by reducing tax payments.

Capital allowances can also provide tax relief for the depreciation in value of the capital assets over time (writing-down allowance) and upon the items' disposal (balancing allowances and charges). The law relating to capital allowances in the UK, and the items for which allowances can be claimed, is contained in the Capital Allowances Act 2001. However, the ascertainment of tax relief for items of capital expenditure is an extremely complicated matter. To take full advantage of the tax relief opportunities requires a specialist who can advise about the latest changes to legislation, case law and other practical issues. For this reason, the cost manager should recommend that the employer obtain specialist advice so that the availability and quantum of capital allowances can be maximised.

NRM 1 recommends that the treatment of capital allowances be excluded from cost plans. This is because capital allowances will be realised in the future, after the building project is completed. Therefore, any capital allowances should be dealt with as a future income stream – shown in juxtaposition to any whole life cost plan. Should the employer require capital allowances to be included in the whole life cost plan, such allowances are maybe best allocated to group element 12 (Other development and project costs) – with the allowances set against the year in which they are due.

33.2.2 Land remediation tax incentives

Contaminated land can have significant negative impacts on the environment, local households and companies. Consequently, they provide a significant barrier to redevelopment of the land. It is for these reasons that the UK government encourages the clean-up of contaminated land by introducing an enhanced tax relief for the costs incurred by companies (including property owners, investors and developers) in cleaning up land they acquire in a contaminated state. However, the relief is not available where the land is contaminated due to the actions of the acquiring company or someone connected with that company, or where the company has failed to prevent contamination. Such relief is not available to individuals or partnerships.

To qualify for tax relief, expenditure must be incurred on the prevention, remediation or mitigation of the effects of pollutants, or in tackling Japanese knotweed infestation or in bringing long-term derelict land back into productive use. General site clearance will not qualify. Preparatory works such as site investigations and incidental professional fees can be included in the claim for tax relief. Where remediation work is subcontracted, then this cost will form the basis of the claim. However, any claim for relief must be made within two years of the end of the accounting period in which the qualifying expenditure was incurred.

Furthermore, there are tax-planning opportunities to structure building projects such that the waste from qualifying remediation works is exempt from landfill tax or the aggregates levy.

Current legislation in the United Kingdom for land remediation relief (Finance Act 2013) gives companies that are liable to corporation tax a deduction of 150% for qualifying expenditure on removing or mitigating the effect of contamination. Financial assistance offered includes:

Derelict land

Under the Finance Act 2013, expenditure qualifies for tax relief if the remediated land had been derelict since 1 April 1998. However, relief is only available where the land was already derelict when it was acquired by the claimant.

Qualifying costs are expenditure on the removal of:

- post-tensioned concrete heavyweight construction;
- foundations of buildings or other structures or machinery bases;
- reinforced concrete pile caps;
- reinforced concrete basements; and
- underground pipes or other apparatus for drainage or sewerage or for the supply of electricity, gas, water or telecommunication services.

Contaminated land

The legislation concentrates on tackling the blight caused by the legacy of previous industrial use. This means that, in general, expenditure on naturally occurring contamination does not qualify.

Japanese knotweed, radon and arsenic

Japanese knotweed, radon and arsenic are specified naturally occurring contaminants which can cause a market failure. Legislation deals with relief for the costs of treating land contaminated by such contaminants where infestation has occurred whilst the land has been in the ownership of the claimant. Notwithstanding this, the UK government now excludes methods of remediation that are considered inappropriate from the relief – particularly methods that involve the removal of Japanese knotweed to landfill sites (e.g. using dig and dump methods) – the aim being to significantly reduce the movement of contaminated material to landfill sites.

Land remediation tax relief is much the same as capital allowances – complicated and realised in the future, after the building project is completed! Accordingly, NRM 1 recommends that specialist advice be sought on the availability, and method to maximise the quantum, of the tax relief. It is important, therefore, that the cost manager is mindful of opportunities for the employer to obtain land remediation tax incentives, and advises that specialist advice be sought where appropriate. The rules also recommend that allowances in connection with land remediation tax incentives be excluded from cost plans. For the purpose of whole life cost planning, it is recommended that such incentives be treated in the same way as capital allowances and allocated to group element 12 (Other development and project costs).

33.3 Grants

Financial assistance is often necessary in order to encourage the development of building projects in unattractive locations, or in circumstances where there appears to be little obvious economic benefit to a developer. This assistance might be obtainable from a number of different sources, but especially from government agencies through discretionary grants – including the European Union (EU). Grants are methods of financial assistance which are not a loan, and do not need to be paid back – as long as they are used for their intended purpose.

Typical reasons for bestowing grants on building projects include:

- regeneration of industrial areas;
- urban development – used for promoting job creation, inward investment and environmental improvement, by developing vacant, derelict or underused land or buildings in priority areas (i.e. urban renewal programmes);
- property improvement;
- land reclamation schemes; and
- slum clearance and derelict land clearance.

The aim of grants is to encourage the development or redevelopment of areas either as a means of improving existing standards or through investing in building development projects that will help create wealth for the developer at the same time as creating jobs, promoting inward investment and contributing to environmental improvement in a specific area.

In addition, grants can often be obtained for specific aspects of construction such as:

- sustainable construction – used to encourage certain types of green construction; and
- preservation – used to support the conservation of existing buildings, in particular listed buildings.

Grants provide important financial aid to funding, in addition to acting as a stimulus to continuing with, certain building projects. Such allowances can be significant. However, one of the biggest difficulties for developers is that government policy on grants is constantly under review and change, in order to meet new demands and help generate improved economic and environmental conditions. The nature of construction grants is constantly changing.

Understanding what grants are available, the sources of grants and how to successfully apply for grants is a convoluted business. Therefore, it is recommended that specialist advice is sought to identify the types of grant available and to obtain advice on the approach to grant applications to increase the likelihood of success.

NRM 1 recommends that allowances in connection with grants be excluded from cost plans. However, where the employer requires grant funding to be included in the cost plan, such allowances might be best allocated to group element 12 (Other development or project costs). Where grants are payable in the future, the allowances are to be set against the year in which they are to be paid.

PART 5

Writing cost estimate and cost plan reports

CHAPTER 34

Reporting of cost estimates and cost plans

Introduction

This chapter:

- explains the purpose of a cost estimate and cost plan report;
- conveys the basic rules for good report writing;
- highlights the characteristics of a good report;
- describes the content of cost estimate and cost plan reports;
- contains tips on how to write cost estimate and cost plan reports;
- provides a progressive worked example of a cost plan report;
- discusses how to convey the cost estimate or cost plan report to the employer.

34.1 What is the purpose of a cost estimate or cost plan report?

Cost estimate reports are an essential part of cost management. Their purpose is to provide an up-to-date record of the likely cost of a building project to the employer. It is an extremely important document, as the employer will use the cost manager's advice to obtain funding for the building project or for incorporating into a development (or investment) appraisal to ascertain if the building project is a viable business proposition. Therefore, the cost manager needs to think carefully about why the report is being written – what is to be achieved?

By and large, the four main purposes of a cost estimate or cost plan report are for:

- information – to give the employer, or other readers, information as clearly and concisely as possible;
- interpretation – to help the employer, or other readers, grasp the relative significance of different pieces of information;
- persuasion – to obtain agreement to a course of action;
- recording – to provide a record of how things are at a point in time.

Cost estimate and cost plan reports are likely to be for a combination of the four different purposes.

34.1.1 Information

Above all, cost estimate and cost plan reports need to inform. If they fail to provide relevant information, then there is little point in them. Information needs to be presented clearly and appropriately. This can be done in a number of ways – continuous prose, tables or various graphical presentations.

34.1.2 Interpretation

Cost estimate and cost plan reports are not just factual statements; they also have to explain the information they contain. To do this they should present arguments and recommendations, normally in continuous prose. This tells the employer, and other readers, how important individual pieces of information are and how they relate to each other.

After reading cost estimate and cost plan reports the employer, or other reader, should have a clear mental grasp of the essentials of the report. To achieve this, the cost manager has to give very careful thought to the structure of the report.

If the main advice contains too much detailed information, it can become difficult if not impossible to follow. A common practice is to place detailed information in one or more annexes and appendices to which the reader is referred. This can work well, provided the report does not become wearisome, moving forwards and backwards in a bulky document. For example, a summary of an order of cost estimate might be included within the body of the report, whilst the underlying detailed calculations and methodology are incorporated as an annexe to the cost estimate report. The same approach can be used for a cost plan report. However, cost plans themselves can tend to be bulky documents and by including them as an annexe might make the report unwieldy – particularly if the building project consists of more than one building and, as a consequence, a number of separate cost plans. In such instances, it might be best for the cost manager to treat a detailed cost plan as a separate document and issue it as a supplementary document to the cost plan report.

If the cost estimate or cost plan report is to be published electronically – for example, as an Acrobat™ pdf file – then this kind of cross-referring is much simpler. The cost manager can place a hyperlink at the relevant point in the narrative of the document: clicking on this will take the reader to the related information or data. A 'return' hyperlink will then take the reader back to the point of narrative that they have just left.

Should the cost manager be in any doubt as to whether or not information should be included in a cost estimate or cost plan report, then it should either be left out or incorporated as an annexe or appendix.

34.1.3 Persuasion

It is only a short step from interpretation to persuasion. It is usually possibly to interpret information in more than one way. Naturally, the cost manager will believe that his interpretation is the correct one, so he will want to 'sell' it to the employer, and other readers.

Where there are two interpretations, both of which have merit, which makes it possible for one or more decisions to be made, the cost manager has a responsibility to explain the merits of both solutions (giving the advantages and disadvantages for each option) – even if the cost manager then goes on to explain why one of them is preferable, or strongly urges one course of action. There is nothing more irritating than the kind of writing that constantly sits on the fence and refuses to take a clear line. Readers are left feeling that if they have to do all the decision making they might as well have written the report themselves. Therefore, it is essential for cost managers to express clearly their interpretations and preferences.

Cost estimate or cost plan reports often have to include recommendations for action. This is another step on from interpretation. In an order of cost estimate report, for example, the employer will be told what needs to be done in order to improve cost certainty. Cost estimate and cost plan reports inform and recommend.

34.1.4 Recording

A final purpose of a cost estimate or cost plan report, and one that can be overlooked, is to provide a picture of how things are at a point in time – a record. People joining the project team or new stakeholders who do not have the background information that the cost manager possesses might read it. The information in the report will be needed at some point in the future.

34.2 Basic rules for writing cost estimate or cost plan reports

There are a number of basic rules that should be followed for all types of cost estimate and cost plan report. These basic rules are shown in Table 34.1

Table 34.1 Basic rules for all types of cost estimate and cost plan report

Brevity	Do not write 200 words if what needs to be said can be said in a dozen words. Thin reports are attractive, easy to handle, cheap to produce, environmentally friendly, easy to change and usually much more readable. Nobody normally complains of a report being too short.
Simplicity	Only use words that everyone will be able to read and understand. A report is a means of communication. It is not the place to prove the cost manager's technical knowledge or command of the English language.
Purpose	Remember why the report is being written and try and put yourself in the reader's shoes. It is all too easy to get lost in your writing. We all review our own writing as writers, not as readers, and consequently we often hide the message amongst too many words.
Begin	Have a meaningful beginning that explains the reason for the report and introduces the reader to the subject. If the reader is aware of the report and the subject, he can always skip the beginning.
End	Always end by summarising how the report has achieved its objectives. You don't need to actually write, 'the end', but the reader should be quite clear that he has reached the end.
Content	Make sure that the main content of the report is readable with points of interest well spaced throughout the report (e.g. every 500 words).

Table 34.1 *(Continued)*

Jargon, abbreviations and acronyms	These, defined as 'the unnecessary use of technical terms', should be vigorously avoided. If technical terms are necessary, make sure that they are defined. However, wherever possible technical terms should be kept outside the main body of the report. Many people believe that words or acronyms developed to cover technical expressions should be used to simplify reports. However, the report writer must consider the readers – will they understand them? It is best to assume not! Therefore, it is best to avoid them.
Title	Give the report a meaningful name.
Author	Always name the writer or writers.
Date	Always date the report.

34.3 Characteristics of a good cost estimate or cost plan report

Because of the nature of information that needs to be covered by a cost report, the cost manager might end up presenting information in an unwieldy manner. Reports must be clear, concise, complete and correct (i.e. the 4Cs rule), and presented in the most appropriate format (i.e. text, table, graph, diagram). Good reports, therefore, follow the 4Cs rule – which is:

- Clear – well written. Information has to be understood at the first reading. The report has to be easy to read with legible writing and a clear message.
- Concise – sufficiently comprehensive. The report needs only to be as long as needed to include the necessary information. Short, simple sentences should be used. Jargon or words that are not required should not be included.
- Complete – relevant, up to date and meets user needs. Ensure all required information is in the report.
- Correct – reliable. Every piece of information must be accurate and verifiable.

Ultimately, therefore, a cost estimate and cost plan report will be most effective if they are: clear and readable, concise but thorough, complete and correct – that is, useful to the employer and other readers!

34.4 Organising the content of a cost estimate or cost plan report

Collecting and organising the content of a cost estimate or cost plan report is vital. However well written a report, it will fail if the basic facts and figures upon which it is founded are incorrect, inadequate or irrelevant. Every comment in the report must be well researched, accurate, meaningful and relevant to the purpose of the report.

Moreover, a report should present information to the reader in a logical order. Wherever possible, items should be sequenced so that one flows neatly on to the next. Generally, the cost manager will be trying to present a lot of information in the cost report. It is important, therefore, that the information builds as a set of logically linked items, rather than a heap that has just been thrown together.

The following guidelines on presentation apply to each type of report:

- Use annexes and appendices to remove detail from the body of the report.
- Ensure the report can be read from introduction to conclusion/recommendation without the need to refer to annexes and appendices.
- Provide a method of referencing to enable the reader to use the report.
- If the advantages are included, make sure that the disadvantages are also included, and vice versa.
- Use the minimum of words to make the point.

34.5 Structure and content of cost estimate and cost plan reports

A typical structure for cost estimate and cost plan reports is as follows:

- cover page;
- executive summary;
- contents page;
- body of the report;
- conclusions and recommendations; and
- annexes.

34.5.1 Cover page

On the cover should be given the title of the report, a short building project title, the project number, the name of the cost management company responsible for the report, the date of the report and the issue number of the report.

Should a cost management company be certified to the ISO 9001 standard or another similar standard which relates to quality management systems, then a document control panel can be incorporated on the back of the cover page.

A typical example of a cover page for a cost plan report is illustrated in Example 34.1.

34.5.2 Executive summary

The purpose of an executive summary is to summarise the key points of a document for its readers, saving them time and preparing them for the upcoming content; it is an overview and, whenever possible, should be no longer than one to two pages. It should provide a statement of the headline costs, and briefly explain any important issues. If there are any recommendations for actions at the end of the report, these should also appear at the end of the executive summary, usually as a bulleted or numbered list. An illustration of an executive summary is provided in Example 34.2.

Example 34.1 Cover page for a cost plan report

Bright Kewess Partnership

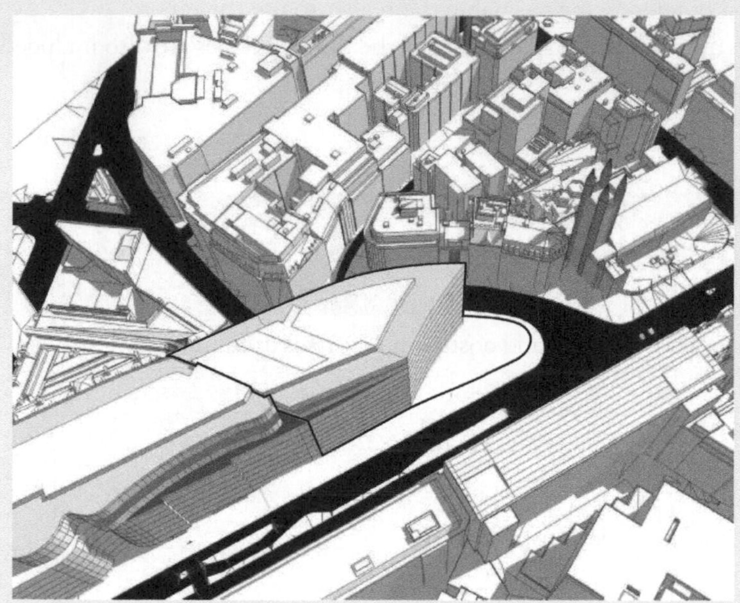

COST PLAN REPORT

WOOLLARD HOTEL,
HOLBORN VIADUCT, ST LEONARDS-ON-SEA, EAST SUSSEX

For GMB Developments Limited

Project No. BKP-0693/13

Date: 27 September 2013

Issue No. 1.01

Bright Kewess Partnership
Mancini House
1 City Place
Maine Road
Hastings,
East Sussex Cl36 9TY

t: +44 (0) 1424 693693 www.brightkewess.com

Example 34.2 Executive summary for a cost plan report

EXECUTIVE SUMMARY

1. The total estimated outturn construction cost for the proposed Hotel building is in the sum of £33,344,000, excluding VAT; based on demolition works commencing in February 2014, and construction commencing on site in June 2014.
2. The cost plan includes a cost premium of £430,000 for bringing forward certain aspects of the piled foundation system works and temporary work to the tunnel – to make use of existing temporary Network Rail track/tunnel possessions.
3. The cost plan includes allowances for the following:
   - Physical construction works;
   - Main contractor's preliminaries;
   - Main contractor's overheads and profit;
   - Subcontractors' allowances for preliminaries, overheads and profit, design fees and risk;
   - Project and design team fees;
   - Design development risks;
   - Construction risks; and the
   - Potential effects of inflation.
4. Notwithstanding the risk allowances for design development and construction included in the cost plan, no costs have yet been ascertained for any specific risk. They are simply initial risk allowances, which will be reassessed in conjunction with the employer and project team members as the design develops and as action is taken to reduce the risk exposure.
5. No allowances have been made in the cost plan for the following:
   - Other development and project costs;
   - Employer change risks;
   - Employer other risks;
   - Value added tax (VAT); and
   - Taxation and incentives (e.g. capital allowances, land remediation relief or grants).
6. The project team will continue to review the design with the aim of reducing construction costs through value engineering the design wherever possible without impacting on the employer's brief.

Recommendations

A. Make sufficient budgetary allowances in investment and development appraisals for the following:
   - Employer change risks; and
   - Employer other risks.

B. Carry out a proper assessment of the potential risks associated with the building project. Consider employing a risk manager to facilitate the risk management process.

C. **Obtain specialist advice to ensure that the correct** VAT rates are applied to the various aspects of the scheme.

D. **Obtain specialist advice to maximise the availability and quantum** of capital allowances, land remediation relief and grants.

E. Instruct project team **to continue to explore value improvement opportunities** through value engineering during design development and look to reduce costs wherever possible without impacting on the employer's design brief and quality standards.

34.5.3 Contents page

The contents page is a list of the headed sections and annexes that comprise the report. It can comprise the headings of the main sections, or both main headings and sub-headings (or paragraph headings). A contents page is illustrated in Example 34.3.

Example 34.3 Contents page for a cost report, showing the main headings only

CONTENTS **Page No.**

34.5.4 Body of the report

Dividing the body of the cost estimate or cost plan report into headed sections and sub-sections enables readers to find their way around more easily. The headed sections listed below comprise the body of the report used in this chapter to illustrate the structure and content of a cost estimate and cost plan report:

- Introduction
- Section 1: Description of the building project
- Section 2: Basis of the cost plan (or order of cost estimate)
- Section 3: Cost plan summary (or order of cost estimate summary)
- Section 4: Conclusions and recommendations.

Introduction

The introduction to the report sets out the origin of the instruction and its precise nature. It is designed to ensure that there is no misunderstanding between the cost manager and the employer, or other readers, about the report's purpose.

Example 34.4 Introduction for a cost plan report

INTRODUCTION

This report presents the cost plan prepared at RIBA Work Stage C (Concept Design) for the proposed Hotel development at Holborn Viaduct, St Leonards-on-Sea, East Sussex. The Bright Kewess Partnership has prepared it in response to an instruction from GMB Developments Limited on behalf of City Investments Limited.

The cost plan only addresses the estimated cost of the capital works. No consideration or allowances have been made in connection with future maintenance, operation or replacement costs (i.e. whole life costs).

Sometimes this section needs elaboration. If the instruction was not clear or if the cost manager can point to areas outside the brief that are relevant, say so in this section. Normally the brief can be expressed in a single paragraph. Example 34.4 provides a sample introduction to a cost plan report which can be easily adopted for an order of cost estimate report.

Section 1: Description of the building project

This section provides a brief description of the building project. It normally consists of two aspects: (1) a description of both the existing site and the existing buildings; (2) a description of the new building works to be completed. A description of the site and the existing buildings occupying the site can normally be obtained from the architect or from the planning deeds. A typical description is provided in Example 34.5.

Example 34.5 Brief project description

SECTION 1: DESCRIPTION OF THE BUILDING PROJECT

1 Description of the site, the existing buildings and the works

1.1 Description of the existing site and buildings

The site, located at 63–69 Holborn Viaduct, St Leonards-on-Sea, East Sussex, covers an area of approximately 0.20 hectares (2,144m^2). It is occupied by a large 'T-shaped' office development known as United House. The building was designed by M. Busby & Partners and built between 1956 and 1960.

The office building takes the form of an 11-storey slab block at the east end of the site and an adjoining six-storey block behind it to the west of the same construction. The eastern part of the building comprises basement, lower ground, ground and 10 upper floors with its main facade oriented 90 degrees to Holborn Viaduct. The western part of the building comprises lower ground, ground and five upper floors and faces directly on to Holborn Viaduct.

The building was reclad in 1980–81 with limestone panels and metal windows with brown frames and spandrel panels.

The building is currently unoccupied and is to be demolished to enable construction of the proposed Hotel and Office schemes. The last tenants of the building (A. Ferguson & Partners) left the premises in 2010 and the soft strip of the interior of the building has been completed.

The existing building is classified for B1 office use; with a gross external floor area (GEA) 7,710m^2, including plant, ancillary and circulation space.

Example 34.5 *(Continued)*

1.2 **Description of the Works**

The proposed works comprise the demolition of an existing office building and the design and construction of a Hotel, which will afford 246 guest rooms together with associated front and back of house facilities over 11 floors; with a GIFA of 8,399m^2.

Where more than one option is being reported, a brief description of each scheme will need to be provided by the cost manager, with each option being clearly delineated.

Section 2: Basis of cost plan

This section of the report describes the principal factors considered, and the assumptions made, by the cost manager in preparing the cost estimate or cost plan. It reassures the employer, or other reader, that the cost manager has considered all the main factors. A heading and typical introductory text to section 2 is provided in Example 34.6.

Example 34.6 Basis of the cost plan

SECTION 2: BASIS OF COST PLAN

The cost plan has been based on the following:

(a) Estimate base date

Here, at the very beginning of the report, the reader is informed that the estimate has been based on rates current at a certain date – i.e. the date on which the estimate or cost plan was produced: referred to as the estimate base date. The estimate base date is also clearly stated. Likewise, the reader is told that the potential effect of inflation has been addressed elsewhere in the report. (See Example 34.7.)

Example 34.7 Estimate base date statement

2.1 **Estimate Base Date**

2.1.1 The base cost estimate and the risk allowance estimate have been prepared using rates and prices current at the time the estimate is produced – referred to as the estimate base date.

2.1.2 The estimate base date is September 2013.

2.1.3 The potential effects of inflation during the tender and construction phases of the project are addressed in paragraph 2.14 of this report.

(b) Procurement strategy

The method of procurement can impact on the final outturn costs of a building project, as it involves the combination of a multitude of different factors. These include the:

- building project work breakdown structure – e.g. separate contracts for enabling works and main construction works;
- phasing requirements – including any sectional completion requirements;

Example 34.8 Procurement strategy statement

2.2 Procurement Strategy

The cost plan is based on:

(1) Enabling works prior to demolition works: comprising temporary works and the insertion of steel tubular pile sleeves (which pass through the tunnel) – to make use of extant Network Rail temporary railway possessions – with the selected piling works contractor being novated to the main contractor on execution of the main contract.

(2) Demolition works being carried out as a separate contract prior to the main construction works.

(3) Piling works being completed after the completion of the demolition works – as the second stage of the enabling works; with the piling contractor being novated to the main contractor – on the appointment of the main contractor.

(4) The contract strategy for the main construction works being based on a single stage design and build contract.

(5) No phasing or sectional completion in respect of the contract for the main construction works.

(6) The tender price/contract sum being obtained through competitive tendering.

(7) The use of a standard form of contract with no significant amendments.

(8) Liquidated damages being set at levels that are commercially acceptable to the demolition contractor, the piling contractor and the main contractor.

- tender strategy – including the methods of tenderer and contractor selection (i.e. through competitive tendering or negotiation);
- contract strategy – e.g. discrete lump sum contract, design and build, build and construct, management contracting, construction management or hybrid strategy;
- contract conditions – e.g. standard conditions, amended standard conditions or bespoke conditions; and non-onerous or onerous conditions. If the actual contract conditions to be used are known, then these should be referred to;
- liquidated damages – these are to be set at levels that are commercially acceptable.

In view of the complexities and choices there are in respect of procuring a building, it is essential that the cost manager sets out the assumptions about the procurement strategy on which the cost estimate is based (see Example 34.8.)

(c) Programme

This part of the report consists of a summary explanation of the pre-construction and construction phase programmes on which the cost plan was based, clearly stating the key dates and time periods (see Example 34.9.) These dates and time periods will also have been the basis of predicting future effects of inflation on costs.

Example 34.9 Programme statement

2.3 Pre-construction and Construction Phase Programme

The key dates on which the cost plan is based were taken from the Pre-construction and Construction Phase Programme produced by the project manager – Bunkum, Twaddle & Miss Management Limited. A copy of the programme is at Annexe I of this Report.

Example 34.9 *(Continued)*

(1) Pre-Construction Phase – present to 27 September 2013
(2) Construction Phase 1: demolition of existing structures – from 11 February 2014 to 31 May 2014 (approximately 4 months' duration);
(3) Construction Phase 1: installation of piled foundation system – from 1 June 2014 to 30 September 2014 (approximately 4 months' duration);
(4) Construction Phase 2: main construction works – commencing on 1 October 2014; duration of Construction Phase: assumed: 65 weeks.

If considered absolutely necessary by the cost manager, a copy of the detailed programme can be included as an annexe to the report. Cross-reference of the annexe should be included in the body of the report – otherwise the detailed programme has no link to the report and is isolated.

(d) Information

The employer is going to use the information contained in the report as a basis of decision making – for example, to instruct the project team to proceed to the next stage of design development, to obtain funding or to procure the project. Therefore, it is essential for the cost manager to clearly and explicitly inform the employer of the information on which the cost estimate or cost plan was based. For an order of cost report or option study, it might be possible for the cost manager to list the information used in the body of the report. However, because of the large amount of information on which cost plans are based, it is recommended that a schedule of all the documents referred to in producing the cost plan is annexed to the report. (See Example 34.10.)

Example 34.10 Information statement

2.5 Information

2.5.1 The cost plan is based on the outline specification information contained within the *Hotel Design & Construction Manual* produced by Woollard Hotel Group Limited.

2.5.2 The schedule of drawings, reports and other documents on which the cost plan is based is at Annexe D of this report.

(e) Schedule of gross internal floor areas

Details of the gross internal floor areas on which the cost estimate or cost plan was based should be made in the report, together with the method used to measure them. In the case of GIFAs, they should have been measured in accordance with the latest edition of the RICS Code of Measurement: A Guide for Property Professionals – so this should be stated. The source of the areas should also be stated (e.g. the architect).

The schedule can either be included in the body of the report, as part of the detailed cost estimate or cost plan (i.e. in Section 2 of the report) or as an annexe to the report.

In the example report provided, the area schedule has been incorporated as an annexe to the report (see Example 34.11).

Example 34.11 Statement about gross internal floor areas

2.6 Schedule of Gross Internal Floor Areas

2..6.1 The Area Schedule on which the cost plan is based was prepared by R. K. Tex Limited; and the areas have been measured in accordance with the latest edition of the RICS Code of Measurement: A Guide for Property Professionals. The areas have been verified.

2.6.2 The Area Schedule is at Annexe A of this report.

(f) Building works

This part of the report comprises a brief statement that highlights the method of measurement used by the cost manager to quantify the building work, as well as the key assumptions taken by the cost manager. A sample statement is provided in Example 34.12.

Example 34.12 Statement about building works

2.7 Building Works

2.7.1 The detailed cost plan describes the building works items included in the detailed cost plan (refer to Annexe B of this report).

2.7.2 Building works have been measured and described in accordance with the RICS *New Rules of Measurement: Order of Estimating and Cost Planning for Capital Building Works* (NRM 1) (2nd edition).

2.7.3 Following demolition of United House, new foundations are to be constructed through the existing railway tunnel – supplementing the existing foundations of the existing building.

2.7.4 Once the foundations are complete the structural steel core elements can be erected, together with the lower floors of the Hotel. Once the cores are complete, the bedroom accommodation will be delivered in fully finished prefabricated units complete with windows and corridor modules.

2.7.5 The Hotel is then made wind- and water-tight in parallel with the completion of the fit-out works.

2.7.6 In order to verify the design solution and construction methodology for the piled foundation system and temporary works to the tunnel, several meetings were held with Tootles Structural Limited (Structural Engineers) and latterly with Chrismiester Foundations Ltd, on which the cost plan is based. The foundation system comprises 8 no. steel-lined concrete piles from street level, through the viaduct roof and floor to a depth of 35m below (approximately 39m total length); with 1,200mm rotary bored cast in-situ piles – with the upper 12m to 15m of pile being permanently steel cased. The use of a 'Bentonite' drilling fluid is advocated which will be used to provide stability to the bore as it is drilled into the Hastings sand layer. Suitably designed support platforms will be put in place in advance to provide a level working area and to protect the viaduct roof from piling activities (i.e. to act as a crash deck). Both the concrete and 'Bentonite' slurry will be delivered remotely and piped down to the workface. This cost plan is reflective of the pile sleeves and piling works being carried out as a separate enabling works contract. All steel pile sleeves will be installed prior to the demolition of the existing building being undertaken – making use of extant Network Rail 'track possessions'. The extra cost for carrying out these works as discrete enabling works contracts has been estimated and included in the cost plan.

Example 34.12 *(Continued)*

In addition, allowances have been included in the cost plan for costs in connection with Network Rail Asset Agreement, movement monitoring, working conditions and work sequencing. These allowances have been based on the costs received from TIM Limited – dated 6 August 2013; however, because the costs received are variable (i.e. subject to the extent and duration of monitoring required), we have included a 'realistic' cost scenario – with the difference in costs between the 'realistic' and 'worst' case scenarios included in the cost plan as a specific construction risk.

2.7.7 The estimated mass of structural steelwork (421 tonnes) was provided by Tootles Structural Limited, with an allowance of 10% added to cover fixings, connections and holding down bolt assemblies.

2.7.8 With regard to cladding of the building, R. K. Tex Limited have produced a design intent, which we have discussed with a number of cladding specialists, and the figures used are based upon advice received on 6 August 2013 and 23 September 2013 from Clad U Right (UK). The cost reflects the complexity of the design that will be refined during the next design stage.

2.7.9 The cleaning strategy for the building is also complex and we have received specialist advice and prices from Building Access (International) Ltd that has been used in the cost plan.

2.7.10 With regard to the modular bedroom units, we have had meetings with Modern-Methods Ltd and discussed the content of this package in some detail. Modern-Methods Ltd has provided an indicative quotation that has been used to inform the cost plan (a copy of the indicative quotation is at Annexe J of this report). One deviation to this is that the cost plan is based on painted plasterboard to the corridor walls, not High Pressure Laminate (HPL) as described in the *Hotel Design & Construction Manual* (Appendix II).

2.7.11 The unit rates take account of strengthening of the steel carcasses to resist wind loadings on the projections and general bracing of units. In addition to Modern-Methods Ltd's pricing, we have made an allowance for the fan coil units (FCUs) to be provided in this works package. This requirement is to be confirmed.

2.7.12 The fit-out of the front and back of house areas has been based on the Indicative Schedule of Finishes set out in *Hotel Design & Construction Manual* (Appendix II), whereas details of the fitted furniture and equipment have been taken from historic information provided by GMB Development Limited (dated 11 February 2012). Allowances for fittings, furnishings and equipment have been based on Woollard Hotel Limited's schedule of requirements dated 11 February 2013, with a 10% addition for items relating to guest rooms to reflect the uplift from 223 rooms to 246 rooms.

2.7.13 Owing to limited design information being available for building engineering services – sanitary appliances and lift installations have been enumerated, whereas other installations have been based on the GIFA with appropriate unit rates applied (based on market research).

(g) Unit rates

As to the unit rates applied to measured components, it is recommended that a brief explanation be given of what has been included and excluded (see Example 34.13). NRM 1 treats main contractor's preliminaries, and overheads and profit as separate

Example 34.13 Statement about unit rates

2.8 Unit Rates

2.8.1 Unit rates for building works items exclude any allowance for main contractor's preliminaries and main contractor's overheads and profit, which have been treated as separate cost centres.

2.8.2 Unless otherwise indicated in the cost plan, unit rates for building works make allowances for preliminaries, overheads and profit, design fees, risk allowances and all other on-costs in connection with works executed by subcontractors and specialist subcontractors.

cost centres. In contrast, subcontractors' preliminaries and overheads and profit (together with their design fees, risk allowances and any other costs in connection with their works) are to be incorporated within the unit rates applied to measured components.

(h) Main contractor's preliminaries

In this part of the report a brief explanation is given about the way in which the level of main contractor's preliminaries has been assessed (see Example 34.14).

Example 34.14 Statement about main contractor's preliminaries

2.9 Main Contractor's Preliminaries

2.9.1 Main contractor's preliminaries have been calculated on the basis of a percentage addition, which has been determined from an assessment of building projects of a similar nature.

2.9.2 An allowance of 13.00% of the building works estimate has been made for main contractor's preliminaries. In addition, an extra allowance has been included for the complexities of dealing with Network Rail and other adjoining property owners.

(i) Main contractor's overheads and profit

Similarly, a brief explanation about the way in which the expected level of main contractor's overheads and profit has been judged is provided (see Example 34.15).

Example 34.15 Statement about main contractor's overheads and profit

2.10 Main Contractor's Overheads and Profit

2.10.1 Main contractor's overheads and profit have been calculated on the basis of a percentage addition, which has been determined from an assessment of building projects of a similar nature, and an assessment of current market trends.

2.10.2 An allowance of 4.00% of the combined total of the building works estimate and main contractor's preliminaries estimate has been made for main contractor's overheads and profit.

(j) Project and design team fees

Here, the extent of project and design fees that have been included in the cost plan is simply listed (see Example 34.16). If no fees or only certain fees have been included, then this needs to be stated.

If no fees have been included in the cost estimate or cost plan, then the cost manager should consider including a statement such as:

> *An indicative list of project and design fees that will need to be considered by [insert name of the employer] in any development or investment appraisal is at Annexe [insert Annexe reference] to this report. The list is not comprehensive, but just provided as an aide-memoire.*
>
> *It is recommended that [insert name of the employer] make sufficient budgetary allowances for project and design team, as well as any other fees, in their development or investment appraisal for the scheme.*

The second paragraph is a recommendation. Therefore, it should also be reiterated by the cost manager in the recommendations section, and in the executive summary.

Example 34.16 Statement about project and design team fees

2.11 Project and Design Team Fees

2.11.1 Allowance has been made in the cost plan for costs in connection with the following:

(1) project team and design team fees;
(2) other consultants' fees;
(3) site investigation fees;
(4) specialist support consultants' fees;
(5) main contractor's pre-construction fees; and
(6) main contractor's design fees.

2.11.2 The allowances for project and design team fees included in the cost plan were advised by GMB Developments Limited.

(k) Other development and project costs

As with project and design team fees, a statement is to be included in the report listing what other development and project costs have been included in the cost estimate or cost plan. Likewise, if no costs, or only costs for certain employer's requirements, have been included then this needs to be stated. (See Example 34.17.)

Again, the last paragraph (2.12.4) in the example is a recommendation and, therefore, it should be reiterated in the recommendations section, and in the executive summary, by the cost manager.

Example 34.17 Statement about other development and project costs

2.12 Other Development and Project Costs

2.12.1 Specific allowances, advised by GMB Developments Limited, for the following have been included in the cost plan:

(1) planning contributions (e.g. in respect of Section 106 Agreements);
(2) provision of public utility services (not included in main contract works);

Example 34.17 *(Continued)*

(3) planning fees;
(4) building regulation fees;
(5) Network Rail fees; and
(6) insurances.

2.12.2 No allowance has been made in the cost plan for any other development and projects costs (e.g. land acquisition costs, finance costs and other employer costs in connection with the building project).

2.12.3 A list of other development and project costs that should be considered by GMB Developments Limited in any development or investment appraisal is at Annexe G to this report. The list is not comprehensive, but just provided as an aide-memoire.

2.12.4 It is recommended that GMB Developments Limited make sufficient budgetary allowances for all other development and project costs considered necessary in their development or investment appraisal for the scheme.

(I) Risk allowances

The quantum of risk allowances included in the cost estimate or cost plan is included in this part of the report (see Example 34.18). For order of cost estimates or for formal cost plan 1, the percentage allowances should be stated. It should also be made clear by the cost manager that, at this stage of design development, no formal risk assessment has been carried out.

Where no allowance has been included in the cost plan for any particular category of risk, for the avoidance of doubt, it is important that the cost manager makes a

Example 34.18 Statement about risk allowances

2.13 Risk Allowances

2.13.1 Risk Allowances have been determined and included in the cost plan as follows:

(1)	Design Development Risks:	0.00% (No allowance)
(2)	Construction Risks:	3.50%
(3)	Employer's Change Risks:	0.00% (Excluded)
(4)	Employer Other Risks:	0.00% (Excluded)

2.13.2 It has been assumed that the Employer will be looking to achieve cost reductions through appropriate value engineering during further stages of design development. In view of this, no risk allowance has been included in the cost plan for increased costs due to such design development.

2.13.3 Notwithstanding the risk allowance for construction-related risks included in the cost plan, no costs have yet been ascertained for any specific construction-related risks. The percentage addition is simply an initial risk allowance determined from our knowledge and experience of previous similar building projects. The risk allowance will be reassessed in conjunction with GMB Developments Limited and the project team members as the design develops, further survey results become available and action is taken to reduce risk exposure.

Example 34.18 *(Continued)*

2.13.4 No allowances have been included in the cost plan for either employer's change risks or employer's other risks. It is therefore recommended that GMB Developments Limited make sufficient budgetary allowances for both categories of risk in any investment or development appraisal. An indicative list of the potential risks that fall under these categories that should be considered by GMB Developments Limited is at Annexe H of this report.

clear distinction between 'no allowance' and 'excluded'. 'Excluded' infers that the employer still needs to consider such risks, whereas 'no allowance' conjectures that the area of risk has been considered and it has been decided that no costs will arise.

However, if the risk allowances included in the cost plan relate to a formal risk assessment – where the cost impact of risks have been estimated by the cost manager – then a copy of the risk register, or risk treatment plan, should be provided as an annexe to the cost plan report. Reference to the annexe should also be made in the text of the report.

(m) Inflation

A statement setting out the allowances included in the cost estimate or cost plan also needs to be included in the report (see Example 34.19).

Example 34.19 Statement about inflation

2.14 Inflation

2.14.1 An allowance has been included in the cost plan for both tender inflation (i.e. inflation from the estimate base date to the date of tender return) and construction inflation (i.e. inflation from the date of tender return to the contract completion date). These are as follows:

(1) Tender inflation: 3.06%

(2) Construction inflation: 1.69%

2.14.2 The effect of inflation has been calculated using the data and time periods referred to in paragraph 2.3 (Programme) of this report.

(n) Value added tax (VAT)

The report should highlight whether or not VAT has been assessed and included in the cost estimate or cost plan. If VAT is included, then the basis of calculating the amount of tax should be stated. (See Example 34.20.)

Example 34.20 Statement about VAT

2.15 Value Added Tax (VAT)

2.15.1 No allowance has been included in the cost plan for VAT.

2.15.2 VAT in relation to buildings is a complex area. Therefore, it is recommended that specialist advice be sought to ensure that the correct rates are applied to the various aspects of the scheme.

(o) Other considerations

It is recommended that a brief commentary about other aspects of costs and incentives be given in the report, together with any recommendations (see Example 34.21).

Example 34.21 Statement about other considerations

2.16 Other Considerations

2.16.1 No allowance has been included in the cost plan for the following:

(1) capital allowances for taxation purposes
(2) land remediation relief
(3) grants.

2.16.2 Taxation allowances, taxation relief and grants can provide valuable financial aid to an Employer on certain types of building project. Therefore, it is recommended that specialist advice be sought to maximise the availability and quantum of capital allowances, land remediation relief and grants.

Section 3: Cost plan summary

This section of the report comprises a high-level summary of the cost plan. As previously explained, the detailed cost plan itself is included in the report as an annexe, due to its bulkiness. A typical cost plan summary is shown in Example 34.22.

Example 34.22 Cost plan summary

SECTION 3: COST PLAN SUMMARY

Project: Woollard Hotel, Holborn Viaduct, St Leonards-on-Sea, East Sussex

Estimate base date: September 2013

Gross internal floor area (GIFA): 8,399m²

Number of Guest rooms (keys): 246

Code	Group element/Element	Element totals	Group element	Cost/m² of GIFA	Cost/Key
		£	£	£	£
0	Facilitating works		0	0.00	0.00
1	Substructure		2,480,300		
1.1	Substructure	2,480,300		295.31	10,082.52
2	Superstructure		5,841,312		
2.1	Frame	1,149,949		136.91	4,674.59
2.2	Upper floors	776,526		92.45	3,156.61
2.3	Roof	413,158		49.19	1,679.50
2.4	Stairs and ramps	281,000		33.46	1,142.28
2.5	External walls	2,629,284		313.05	10,688.15
2.6	Windows and external doors	85,523		10.18	347.65
2.7	Internal walls and partitions	176,211		20.98	716.30
2.8	Internal doors	329,661		39.25	1,340.09

Example 34.22 *(Continued)*

Code	Group element/Element	Element totals £	Group element £	Cost/m² of GIFA £	Cost/Key £
3	Internal finishes		2,447,882		
3.1	Wall finishes	1,666,129		198.37	6,772.88
3.2	Floor finishes	415,163		49.43	1,687.65
3.3	Ceiling finishes	366,590		43.65	1,490.20
4	Fittings, furnishings and equipment		591,458		
4.1	Fittings, furnishings and equipment	591,458		70.42	2,404.30
5	Services		4,033,645		
5.1	Sanitary installations	25,701		3.06	104.48
5.2	Services equipment	235,000		27.98	955.28
5.3	Disposal installations	68,575		8.16	278.76
5.4	Water installations	375,939		44.76	1,528.21
5.5	Heat source	230,049		27.39	935.16
5.6	Space heating and air conditioning	288,506		34.35	1,172.79
5.7	Ventilation	466,984		55.60	1,898.31
5.8	Electrical installations	498,806		59.39	2,027.67
5.9	Fuel installations	447,583		53.29	1,819.44
5.10	Lift and conveyor installations	342,595		40.79	1,392.66
5.11	Fire and lightning protection	532,413		63.39	2,164.28
5.12	Communication, security and control systems	261,797		31.17	1,064.22
5.13	Special installations	150,006		17.86	609.78
5.14	Builder's work in connection with services	109,691		13.06	445.90
6	Prefabricated buildings and building units		7,548,517		
6.1	Prefabricated buildings and building units	7,548,517		898.74	30,685.03
7	Works to existing buildings	0	0	0.00	0.00
8	External works		333,000		
8.1	Site preparation works	0		0.00	0.00
8.2	Roads, paths, pavings and surfacings	123,500		14.70	502.03
8.3	Soft landscaping, planting and irrigation systems	0		0.00	0.00
8.4	Fencing, railings and walls	0		0.00	0.00
8.5	External fixtures	0		0.00	0.00
8.6	External drainage	20,000		2.38	81.30
8.7	External services	189,500		22.56	770.33

Example 34.22 *(Continued)*

Code	Group element/Element	Element totals £	Group element £	Cost/m² of GIFA £	Cost/Key £
8.8	Minor building works and ancillary buildings	0		0.00	0.00
	Building works estimate	23,276,114	23,276,114	2,771.28	94,618.35
9	Main contractor's preliminaries	3,025,081	3,025,081	360.17	12,297.08
	Subtotal	26,301,195	26,301,195	3,131.45	106,915.43
10	Main contractor's overheads and profit	1,052,522	1,052,522	125.32	4,278.54
	Works cost estimate	27,353,717	27,353,717	3,256.77	111,193.97
11	Project/design team fees estimate	3,386,814	3,386,814	403.24	13,767.54
12	Other development/project costs estimate	0	0	0.00	0.00
	Base estimate	30,740,531	30,740,531	3,660.01	124,961.51
13	Risk allowance estimate		1,075,919		
13.1	Design development risks	0		0.00	0.00
13.2	Construction risks	1,075,919		128.10	4,373.65
13.3	Design development risks	0		0.00	0.00
13.4	Construction risks	0		0.00	0.00
	Cost limit (excluding inflation)	31,816,450	31,816,450	3,788.11	129,335.16
14	Inflation estimate		1,527,734		
14.1	Tender inflation	973,571		115.92	3,957.61
14.2	Construction inflation	554,163		65.98	2,252.70
	Cost limit (including inflation)	33,344,184	33,344,184	3,970.01	135,545.47
	Cost limit (including inflation) – to 3 significant figures	**33,344,000**	**33,344,000**	–	–
15	VAT assessment	0	0	0.00	0.00

Section 4: Conclusions and recommendations

This should state the cost manager's conclusions succinctly, so that the employer and other readers are left in no doubt about the meaning (see Example 34.23).

Example 34.23 Conclusions and recommendations

SECTION 4: CONCLUSIONS AND RECOMMENDATIONS

4.1 At 34m² per key, the hotel is not as efficient as required by brand standards, but is a product of the planning restraints of the site and ultimate space planning of the building to maximise efficiency.

4.2 We recommend that GMB Developments Limited:

(1) Make sufficient budgetary allowances in investment and development appraisals for the following:

Example 34.23 *(Continued)*

(a) Employer change risks; and
(b) Employer other risks.

(2) Carry out a proper assessment of the potential risks associated with the building project. Consider employing a risk manager to facilitate the risk management process.
(3) Obtain specialist advice to ensure that the correct VAT rates are applied to the various aspects of the scheme.
(4) Obtain specialist advice to maximise the availability and quantum of capital allowances, land remediation relief and grants.
(5) Instruct project team to continue to explore value improvement opportunities through value engineering during design development and look to reduce costs wherever possible without impacting on the Employer's design brief and quality standards.

If the conclusion includes recommended actions that are written in annexes or appendices, the cost manager should refer to them in the conclusion also.

Note how the Bright Kewess Partnership, which provides a number of construction consultancy-related services use the recommendations to bring other services that they provide to the attention of the employer.

34.5.5 Annexes

Annexes and appendices should be used to remove detail from the body of the report. There might be two, three or more annexes attached to the main report. For example, one might be a detailed schedule of areas; another about the specification on which the order of cost estimate or cost plan was prepared; another illustrating phasing requirements diagrammatically; another listing the pros, cons and cost impact of various options that have been derived through value management and value-engineering exercises.

Typical annexes are illustrated in Table 34.2.

Table 34.2 Typical content of annexes

Ref.	*Title*	*Purpose*
A	Area Schedule	Provides details of the areas on which the order of cost estimate or cost plan have been based. Comprises GIFAs and, when required, NIAs, GEAs, site area and number of functional units. Sometimes the cost manager will integrate the Area Schedule into his or her order of cost estimate or cost plan. In this instance, further copies need not be included in the report – but the Schedule's location must be identified in the report.
B	Detailed Order of Cost Estimate or Cost Plan	Provides the employer and other readers with a detailed breakdown of the cost targets for each element and shows the detailed assumptions that the cost manager has made in generating the order of cost estimate or cost plan. It also sets out the cost limit.

Table 34.2 *(Continued)*

Ref.	*Title*	*Purpose*
C	Items Included and Excluded from Order of Cost Estimate/Cost Plan	Clarifies what the cost manager has and has not included in the order of cost estimate or cost plan. See Figure 34.1.
D	Schedule of Drawings, Specifications, Reports and Other Documents on which Cost Plan Based	Details the information on which the order of cost estimate or cost plan was based.
E	Outline Specification on which Cost Plan Based	Provides a brief description of the specification for each element and key sub-elements.
F	Transfers and Adjustments since Previous Cost Plan	Provides transparency of any changes to previous cost targets and sets out the principal reasons why the changes have occurred. It should be remembered by the cost manager that the 'cost limit' agreed previously with the employer must not be exceeded. If it does, the cost manager must work with the designers and other project team members to look for ways to bring costs back within the authorised cost limit. See Figure 34.2.
G	Schedule of Value Improvement Opportunities/Options	Provides a schedule of items that could reduce costs or increase revenue. The schedule of items is to be discussed with the employer and other project team members. Obviously, some suggested value improvements would impact on other areas of design, so collaboration is required. Value engineering can be used to evaluate technical aspects of value improvement opportunities. See Figure 34.3.
H	Schedule of Project/Design Team Fees	Where some or no fees have been included in the order of cost estimate or cost plan, a schedule setting out the consultancy services that may be required by the employer to complete the building project should be included in the report by the cost manager. The list of 'Project and Design Team Fees' in group element 11 of NRM 1 provides such a list, which the cost manager can incorporate in his or her report.
I	Schedule of Other Development/ Project Costs	As for Annexe H above. Similarly, the list of 'Other Development and Project Costs' in group element 12 of NRM 1 provides such a list, which the cost manager can incorporate in his or her report.
J	Schedule of Potential Causes of Risk	In the unlikely situation that no allowance has been made for risks in the order of cost estimate or cost plan, the cost manager can include a schedule of the potential causes of risk in his or her report. However, it is more common for allowances in connection with Employer Change Risks and Other Employer Risks not to have been included. Consequently, the cost manager should include a list of causes of these risks. The list of 'Risks' in group element 13 of NRM 1 provides such a list, which the cost manager can incorporate in his or her report.

Table 34.2 *(Continued*

Ref.	*Title*	*Purpose*
K	Risk Register	Provides details of the risks identified and assessed, together with the mitigation method selected. Risk allowance ascertained by the cost manager should also be included in the register. Obviously, the risk allowances given in the order of cost estimate or cost plan must correlate with those given in the risk register.
L	Cash Flow Forecast	Not essential unless specifically requested by the employer – who may require a cash flow forecast to present to potential funders.
M	Pre-construction and Construction Phase Programme	Not essential, but can be included for completeness. Provides a detailed programme of activities and the key dates on which the order of cost estimate or cost plan was based. However, detailed programmes are unlikely to be available to the cost manager when preparing an order of cost estimate.
N	Quotations	Provides details of quotations that have been obtained and used in compiling the cost estimate or cost plan (e.g. modular building units, cladding, lifts and facade-cleaning systems).
O	Glossary	Glossary of technical terms, abbreviations and acronyms used in the report. Only required if a number of technical terms and phrases have been used in the report without explanation of their meaning in the body of the report.

ANNEXE C

ITEMS INCLUDED AND EXCLUDED FROM ORDER OF COST ESTIMATE/COST PLAN

Ref	Item	Included	Excluded	Not Applicable	Comments
		✓	✗	✗	
1.					
2.					
3.					
4.					
5.					
6.					
7.					
8.					
9.					
10.					
Etc.					

Figure 34.1 Example schedule of 'Items included and excluded from order of cost estimate/ cost plan'

ANNEXE F

TRANSFERS AND ADJUSTMENTS SINCE FORMAL COST PLAN NO. 1

Notes on completion:

1. Code – self-explanatory.
2. Group element/element – level of detail provided under headings can be collapsed or expanded to suit requirements. Notwithstanding this, information required to explain changes since the previous cost plan must be maintained to demonstrate reasons for change to the employer, other stakeholders and the project manager.
3. Approved cost plan – these are the cost targets from the previously agreed cost plan.
4. Transfers and adjustments – shows the amount of any adjustment. There should always be a balancing adjustment to the risk allowances, so by definition the column total is 'zero'. That is, the 'cost limit' should not change unless the employer has changed the scope of project.
5. Current control cost plan – 'approved cost plan' plus or minus 'transfers and adjustments' represents the current cost targets for each element. The 'current control cost plan' must correlate with the 'detailed cost plan' included in the report.
6. Reasons for transfers and adjustments – record and explain the reasons for transfers and adjustments to the previous cost plan. Changes in scope by the employer are to be recorded in this column.

Code	Group element/Element	Approved Cost Plan	Transfers and Adjustments	Current Control Cost Plan	Reasons for Transfers & Adjustments
		£	£	£	
0	**Facilitating works**				
0.1	Toxic/hazardous/contaminated material treatment				
0.2	Major demolition works				
0.3	Temporary support to adjacent structures				
0.4	Specialist groundworks				
0.5	Temporary diversion works				
0.6	Extraordinary site investigation works				
1	**Substructure**				
1.1	Substructure				
2	**Superstructure**				
2.1	Frame				
2.2	Upper floors				
2.3	Roof				
2.4	Stairs and ramps				
2.5	External walls				
2.6	Windows and external doors				
2.7	Internal walls and partitions				
2.8	Internal doors				

Figure 34.2 Example schedule of 'Transfers and adjustments since previous cost plan'

Code	Group element/Element	Approved Cost Plan	Transfers and Adjustments	Current Control Cost Plan	Reasons for Transfers & Adjustments
		£	£	£	
3	**Internal finishes**				
3.1	Wall finishes				
3.2	Floor finishes				
3.3	Ceiling finishes				
Etc.. . .					
	Cost Limit (including inflation)				

Figure 34.2 *(Continued)*

ANNEXE G

VALUE OF VALUE IMPROVEMENT OPPORTUNITIES/OPTIONS

Note:

State in the comments column if there is a potential impact on revenue

Ref	Description	Impact		Comments	Accepted/ Not Accepted
		Time	Cost		
		Weeks	£		
1.					
2.					
3.					
4.					
5.					
6.					
7.					
8.					
9.					
10.					
Etc.					

Figure 34.3 Example of 'Schedule of value improvement opportunities/options'

34.6 Communicating the cost estimate or cost plan

34.6.1 Distribution

It must be remembered by the cost manager that the cost estimate or cost plan report that he or she has produced is for the employer. Therefore the report must be issued to the employer only. It is up to the employer to decide if the report should be issued to other parties.

Should the cost manager consider that it would be advantageous for the report, or parts of the report (or cost plan), to be disseminated to other members of the project team, then it is recommended that he or she obtains the employer's consent in writing. Such consent should clearly set out, or agree to, the extent of information to be disclosed to others.

The employer will specify the amount of information that the cost manager is allowed to pass to the project team members. Some employers might be content for the whole cost plan to be given to other project team members, but others might only wish the cost manager to inform consultants of the cost targets of the elements for which they have design responsibility.

Good practice is for the cost manager to issue the report under cover of a letter (or e-mail if sent electronically). The letter should offer the employer the opportunity to meet the cost manager to discuss the cost estimate or cost plan report. Obviously, the employer will need some time to digest the information provided before meeting the cost manager.

34.6.2 Follow-up

In most instances, simply sending the cost estimate report to the employer will be insufficient.

The cost manager saying, 'I told you, it's all in my report', is not good management, in addition to being unacceptable to the employer. Moreover, there is a significant risk of misunderstanding by the employer or other stakeholder reading the report.

> 49% of the cost manager's effort is used to prepare the cost estimate, whilst the other 51% of effort is used to explain and champion why the cost estimate is at the correct level.
>
> Stuart Earl, Chairman of the RICS Measurement Initiative Steering Group and Director of Gleeds Cost Management Limited

As well as being extremely professional, it is essential for the cost manager to meet with the employer, and any other stakeholders in the building project, to present and discuss the cost estimate or cost plan report, and not just issue it to the employer as job done! The aim of the meeting is for the cost manager to explain the cost estimate or cost plan in detail, so that the employer fully understands why certain allowances and assumptions have been made by the cost manager. It also provides the employer the opportunity to ask the cost manager questions about the report. Invariably compromise between the employer and the cost manager is required, which will require adjustment of the cost estimate and an amended report.

Designing pricing documents using NRM 1

CHAPTER 35

Pricing documents

Introduction

This chapter:

- explains how the NRM 1 cost plan framework can be used to formulate a pricing document;
- provides examples of different aspects of a contract sum analysis.

35.1 Using the NRM 1 cost plan framework as a pricing document

The cost plan framework in NRM 1 can be used as a pricing document, for the purpose of obtaining prices from contractors for completing a building project. In particular, it provides a sound basis for construction of a contract sum analysis for use on a design and build contract. However, it can also be used as a pricing framework for other lump sum contract arrangements where bill of quantities have not been used.

Contract sum analyses are being used more and more since the decline in the use of bill of quantities. A contract sum analysis is simply an analysis of the contract sum in accordance with the requirements of the employer. The term is specifically referred to in the JCT suite of contract conditions, which are in common use in the UK, where the contractor is to take liability for all or part of the design. It should be noted that a contract sum analysis is not a contract document, except under the JCT contract conditions – although in all circumstances where one is used, it is the priced document for valuation purposes. Notwithstanding this, with the incorporation of a suitable amendment or supplementary condition, a contract sum analysis can be subsumed into any other conditions of contract.

A key benefit of using the cost plan framework is that contractors' tenders will be in the form of a cost plan that aligns with the NRM 1 elemental breakdown structure. Thus, a cost analysis of the tenders received is readily available to the cost manager on conclusion of the tender evaluation process. It is also becoming normal for the cost manager to include his or her detailed cost plan in support of the contract sum analysis. Obviously, the cost manager's unit rates and pricing data is deleted, so that only the description of the components and quantities are visible to the tendering contractors. Issuing the detailed cost plan aids the tendering contractors in pricing the building project. However, the cost manager must make it clear in the instructions to tenderers that he or she accepts no liability as to the accuracy of the cost plan, including the quantities, and that the contractor shall be responsible for satisfying himself as to its accuracy before submitting his tender.

By producing cost plans in accordance with NRM 1, the cost manager will give confidence to the contractor that the cost plan has been properly constructed.

35.2 Contract sum analysis

Using the cost plan framework, the content and structure of a typical contract sum analysis will be as follows:

Part 1: Form of tender

Part 2: Contract sum analysis:

- Section 1: Tender price summary (see Figure 35.1)
- Section 2: Main contractor's preliminaries – summary (see Figure 35.2)
- Section 3: Main contractor's preliminaries (see Figure 35.3)
- Section 4: Contract sum analysis (see Figure 35.4)
- Section 5: Provisional sums (see Figure 35.5)
- Section 6: Schedule of main contractor's pre-construction phase services (see Figure 35.6)
- Section 7: Schedule of main contractor's design fees and survey charges (see Figure 35.7)
- Section 8: Schedule of contractor's design and construction risks (risk analysis) (see Figure 35.8)
- Section 9: Schedule of option prices (see Figure 35.9)
- Section 10: Schedule of value improvement proposals (see Figure 35.10).

City Investments Limited UK
Woollard Hotel, Holborn Viaduct, St Leonards-on-Sea, East Sussex

PRICING DOCUMENT (INCORPORATING FORM OF TENDER AND CONTRACT SUM ANALYSIS)

NAME OF TENDERER:

TENDER PRICE SUMMARY: TOTAL PRICE FOR DESIGN AND CONSTRUCTION WORKS

Code	Item	£/p
1.	**MAIN CONTRACTOR'S PRELIMINARIES** (Brought forward from Main Contractor's Preliminaries – Summary):	0.00
2.	**BUILDING WORKS, including Building Engineering Services** (Brought forward from Contract Sum Analysis):	0.00
3.	**PROVISIONAL SUMS (Provisional)** (Brought forward from Schedule of Provisional Sums):	2,085,000.00
	Sub-total	2,085,000.00

Figure 35.1 Tender price summary

Code	Item		£/p
4.	**CONTRACTOR'S OVERHEADS AND PROFIT** (Insert required: % adjustment)	**0.00%**	0.00
	Sub-total		2,085,000.00
5.	**MAIN CONTRACTOR'S PRE-CONSTRUCTION SERVICES** (Brought forward from Schedule of Main Contractor's Pre-Construction Services)		0.00
	Sub-total		2,085,000.00
6.	**MAIN CONTRACTOR'S DESIGN FEES AND SURVEY CHARGES** (Brought forward from Schedule of Main Contractor's Design Fees and Survey Charges)		0.00
	Sub-total		2,085,000.00
7.	**MAIN CONTRACTOR'S DESIGN AND CONSTRUCTION RISKS** (Brought forward from Schedule of Contractor's Design and Construction Risks)		0.00
	Sub-total		2,085,000.00
8.	**MAIN CONTRACTOR'S FIXED PRICE ADDITION** (Insert required: % adjustment)	**0.00%**	0.00
	Sub-total		2,085,000.00
9.	**DIRECTOR'S ADJUSTMENT (Insert required: adjustment)**		0.00
	TOTAL TENDER PRICE (AND PRICED OPTIONS), exclusive of VAT (Carried to Form of Tender (Page FoT/1, Paragraph 1):		**£2,085,000.00**

Figure 35.1 *(Continued)*

City Investments Limited UK
Woollard Hotel, Holborn Viaduct, St Leonards-on-Sea, East Sussex

PRICING DOCUMENT (INCORPORATING FORM OF TENDER AND CONTRACT SUM ANALYSIS)

PRICE ANALYSIS:

NAME OF TENDERER

MAIN CONTRACTOR'S PRELIMINARIES – SUMMARY

Notes for tenderer:

1. Tenderer to set out his total price for Contractor's Preliminaries under the heads listed below. Each heading is defined in Group Element 9 of the RICS New Rules of Measurement: Order of Cost Estimating and Cost Planning for Capital Building Works.
2. Tenderer to return with his tender a full and detailed breakdown to show how his total price for Contractor Preliminaries and each item forming his total price has been calculated.

Figure 35.2 Main contractor's preliminaries – summary

3. Tenderer to annex his detailed breakdown of Preliminary items and charges included in his Tender Price to the Pricing Document.
4. Costs relating to items that are not specifically identified in the Tenderer's full and detailed breakdown shall be deemed to have no cost implications or have been included elsewhere within the Tenderer's rates and prices.

Code	Description	Fixed Charge		Time-Related Charge
		£		£
9.1	**Employer's Requirements**			
9.1.1	Site accommodation	0.00		0.00
9.1.2	Site records	0.00		0.00
9.1.3	Completion and post-completion requirements	0.00		0.00
9.2	**Contractor's Cost Items**	0.00		0.00
9.2.1	Management and staff	0.00		0.00
9.2.2	Site establishment	0.00		0.00
9.2.3	Temporary services	0.00		0.00
9.2.4	Security	0.00		0.00
9.2.5	Safety and environmental protection	0.00		0.00
9.2.6	Control and protection	0.00		0.00
9.2.7	Mechanical plant	0.00		0.00
9.2.8	Temporary works	0.00		0.00
9.2.9	Site records	0.00		0.00
9.2.10	Completion and post-completion requirements	0.00		0.00
9.2.11	Cleaning	0.00		0.00
9.2.12	Fees and charges	0.00		0.00
9.2.13	Site services	0.00		0.00
9.2.14	Insurance, bonds, guarantees and warranties	0.00		0.00
9.2.15	Contract conditions – obligations, liabilities and services	0.00		0.00
SUB-TOTAL: FIXED CHARGES	**£**	**0.00**		
SUB-TOTAL: TIME-RELATED CHARGES				**0.00**
Brought forward – Fixed Charge:			**£**	**0.00**
TOTAL MAIN CONTRACTOR'S PRELIMINARIES carried to TENDER PRICE SUMMARY:			**£**	**0.00**

Figure 35.2 *(Continued)*

City Investments Limited UK
Woollard Hotel, Holborn Viaduct, St Leonards-on-Sea, East Sussex

PRICING DOCUMENT (INCORPORATING FORM OF TENDER AND CONTRACT SUM ANALYSIS)

PRICE ANALYSIS:

NAME OF TENDERER

MAIN CONTRACTOR'S PRELIMINARIES – SUMMARY

Notes for tenderer:

1. Tenderer to set out his total price for Contractor's Preliminaries under the heads listed below. Each heading is defined in Group Element 9 of the RICS New Rules of Measurement: Order of Cost Estimating and Cost Planning for Capital Building Works.
2. Tenderer to return with his tender a full and detailed breakdown to show how his total price for Contractor Preliminaries and each item forming his total price has been calculated.
3. Tenderer to annex his detailed breakdown of Preliminary items and charges included in his Tender Price to the Pricing Document.
4. Costs relating to items that are not specifically identified in the Tenderer's full and detailed breakdown shall be deemed to have no cost implications or have been included elsewhere within the Tenderer's rates and prices.

Code	Description	Fixed Charge		Time-Related Charge
		£		£
9.1	**Employer's Requirements**			
9.1.1	**Site accommodation**			
9.1.1.1	Site accommodation	0.00		0.00
9.1.1.2	Furniture and equipment	0.00		0.00
9.1.1.3	Telecommunications and IT systems	0.00		0.00
	Sub-total: Carried to Main Contractor's Preliminaries Summary:	0.00		0.00
9.1.2	**Site records**			
9.1.2.1	Site records	0.00		0.00
	Sub-total: Carried to Main Contractor's Preliminaries Summary:	0.00		0.00
9.1.3	**Completion and post-completion requirements**			
9.1.3.1	Handover requirements	0.00		0.00
9.1.3.2	Operation and maintenance services	0.00		0.00
	Sub-total: Carried to Main Contractor's Preliminaries Summary:	0.00		0.00

Figure 35.3 Main contractor's preliminaries

Code	Description	Fixed Charge		Time-Related Charge
		£		£
9.2	**Contractor's Cost Items**			
9.2.1	**Management and staff**			
9.2.1.1	Project-specific management and staff	0.00		0.00
9.2.1.2	Visiting management and staff	0.00		0.00
9.2.1.3	Extraordinary support costs	0.00		0.00
9.2.1.4	Staff travel	0.00		0.00
	Sub-total: Carried to Main Contractor's Preliminaries Summary:	0.00		0.00
9.2.2	**Site establishment**			
9.2.2.1	Site accommodation	0.00		0.00
9.2.2.2	Temporary works in connection with site establishment	0.00		0.00
9.2.2.3	Furniture and equipment	0.00		0.00
9.2.2.4	IT systems	0.00		0.00
9.2.2.5	Consumables and services	0.00		0.00
9.2.2.6	Brought-in services	0.00		0.00
9.2.2.7	Sundries	0.00		0.00
	Sub-total: Carried to Main Contractor's Preliminaries Summary:	0.00		0.00
9.2.3	**Temporary services**			
9.2.3.1	Temporary water supply	0.00		0.00
9.2.3.2	Temporary gas supply	0.00		0.00
9.2.3.3	Temporary electricity supply	0.00		0.00
9.2.3.4	Temporary telecommunications systems	0.00		0.00
9.2.3.5	Temporary drainage	0.00		0.00
	Sub-total: Carried to Main Contractor's Preliminaries Summary:	0.00		0.00
9.2.4	**Security**			
9.2.4.1	Security staff	0.00		0.00
9.2.4.2	Security equipment	0.00		0.00
9.2.4.3	Hoardings, fences and gates	0.00		0.00
	Sub-total: Carried to Main Contractor's Preliminaries Summary:	0.00		0.00
9.2.5	**Safety and environmental protection**	0.00		0.00
9.2.5.1	Safety programme	0.00		0.00

Figure 35.3 *(Continued)*

Code	Description	Fixed Charge		Time-Related Charge
		£		£
9.2.5.2	Barriers and safety scaffolding	0.00		0.00
9.2.5.3	Environmental protection measures	0.00		0.00
	Sub-total: Carried to Main Contractor's Preliminaries Summary:	0.00		0.00
9.2.6	**Control and protection**	0.00		0.00
9.2.6.1	Surveys, inspections and monitoring	0.00		0.00
9.2.6.2	Setting out	0.00		0.00
9.2.6.3	Protection of works	0.00		0.00
9.2.6.4	Samples	0.00		0.00
9.2.6.5	Environmental control of building	0.00		0.00
	Sub-total: Carried to Main Contractor's Preliminaries Summary:	0.00		0.00
9.2.7	**Mechanical plant**			
9.2.7.1	Generally: Common-user plant	0.00		0.00
9.2.7.2	Tower cranes	0.00		0.00
9.2.7.3	Mobile cranes	0.00		0.00
9.2.7.4	Hoists	0.00		0.00
9.2.7.5	Access plant	0.00		0.00
9.2.7.6	Concrete plant	0.00		0.00
9.2.7.7	Other plant	0.00		0.00
	Sub-total: Carried to Main Contractor's Preliminaries Summary:	0.00		0.00
9.2.8	**Temporary works**			
9.2.8.1	Access scaffolding	0.00		0.00
9.2.8.2	Temporary works	0.00		0.00
	Sub-total: Carried to Main Contractor's Preliminaries Summary:	0.00		0.00
9.2.9	**Site records**	0.00		0.00
9.2.9.1	Site records	0.00		0.00
	Sub-total: Carried to Main Contractor's Preliminaries Summary:	0.00		0.00
9.2.10	**Completion and post-completion requirements**			
9.2.10.1	Testing and commissioning plan	0.00		0.00
9.2.10.2	Handover	0.00		0.00
9.2.10.3	Post-completion services	0.00		0.00

Figure 35.3 *(Continued)*

Code	Description	Fixed Charge		Time-Related Charge
		£		£
	Sub-total: Carried to Main Contractor's Preliminaries Summary:	0.00		0.00
9.2.11	**Cleaning**			
9.2.11.1	Site tidy	0.00		0.00
9.2.11.2	Maintenance of roads, paths and pavings	0.00		0.00
9.2.11.3	Building clean	0.00		0.00
	Sub-total: Carried to Main Contractor's Preliminaries Summary:	0.00		0.00
9.2.12	**Fees and charges**			
9.2.12.1	Fees	0.00		0.00
9.2.12.2	Charges	0.00		0.00
	Sub-total: Carried to Main Contractor's Preliminaries Summary:	0.00		0.00
9.2.13	**Site services**			
9.2.13.1	Temporary works	0.00		0.00
9.2.13.2	Multi-service gang	0.00		0.00
	Sub-total: Carried to Main Contractor's Preliminaries Summary:	0.00		0.00
9.2.14	**Insurance, bonds, guarantees and warranties**			
9.2.14.1	Works insurance	0.00		0.00
9.2.14.2	Public liability insurance	0.00		0.00
9.2.14.3	Employer's (contractor's) liability insurance	0.00		0.00
9.2.14.4	Other insurance	0.00		0.00
9.2.14.5	Bonds	0.00		0.00
9.2.14.6	Guarantees	0.00		0.00
9.2.14.7	Warranties	0.00		0.00
	Sub-total: Carried to Main Contractor's Preliminaries Summary:	0.00		0.00
9.2.15	**Contract Conditions – obligations, liabilities and services**			
9.2.15.1	Allow for all obligations, liabilities and services described in the Contract Conditions, and amendments thereto	0.00		0.00
	Sub-total: Carried to Main Contractor's Preliminaries Summary:	0.00		0.00

Figure 35.3 *(Continued)*

City Investments Limited UK
Woollard Hotel, Holborn Viaduct, St Leonards-on-Sea, East Sussex

PRICING DOCUMENT (INCORPORATING FORM OF TENDER AND CONTRACT SUM ANALYSIS)

NAME OF TENDERER

CONTRACT SUM ANALYSIS

Code	Element	Total Cost of Element	Total Cost of Group Element	Cost/m² of GIFA
		£	£	£
0.1	Toxic/hazardous material treatment	0.00		0.00
0.2	Major demolition works	0.00		0.00
0.3	Temporary support to adjacent structures	0.00		0.00
0.4	Specialist groundworks	0.00		0.00
0.5	Temporary diversion works	0.00		0.00
0.6	Extraordinary site investigation works	0.00		0.00
0	**Facilitating works**	**0.00**	**0.00**	**0.00**
1.1	Substructure	0.00		0.00
1	**Substructure**	**0.00**	**0.00**	**0.00**
2.1	Frame	0.00		0.00
2.2	Upper floors	0.00		0.00
2.3	Roof	0.00		0.00
2.4	Stairs and ramps	0.00		0.00
2.5	External walls	0.00		0.00
2.6	Windows and external doors	0.00		0.00
2.7	Internal walls and partitions	0.00		0.00
2.8	Internal doors	0.00		0.00
2	**Superstructure**	**0.00**	**0.00**	**0.00**
3.1	Wall finishes	0.00		0.00
3.2	Floor finishes	0.00		0.00
3.3	Ceiling finishes	0.00		0.00
3	**Internal finishes**	**0.00**	**0.00**	**0.00**
4.1	General fittings, furnishings and equipment	0.00		0.00
4	**Fittings, furniture and equipment**	**0.00**	**0.00**	**0.00**
5.1	Sanitary installations	0.00		0.00
5.2	Services equipment	0.00		0.00
5.3	Disposal installations	0.00		0.00
5.4	Water installations	0.00		0.00
5.5	Heat source	0.00		0.00

Figure 35.4 Contract sum analysis

Code	Element	Total Cost of Element	Total Cost of Group Element	Cost/m² of GIFA
		£	£	£
5.6	Space heating and air conditioning	0.00		0.00
5.7	Ventilation	0.00		0.00
5.8	Electrical installations	0.00		0.00
5.9	Fuel installations	0.00		0.00
5.10	Lift and conveyor installations	0.00		0.00
5.11	Fire and lightning protection	0.00		0.00
5.12	Communication, security and control systems	0.00		0.00
5.13	Special installations	0.00		0.00
5.14	Builder's work in connection with services	0.00		0.00
5	**Services**	**0.00**	**0.00**	**0.00**
6.1	Prefabricated buildings and building units	0.00		0.00
6	**Prefabricated buildings and building units**	**0.00**	**0.00**	**0.00**
7.1	Minor demolition works and alteration works	0.00		0.00
7.2	Repairs to existing services	0.00		0.00
7.3	Damp-proof courses/fungus and beetle eradication	0.00		0.00
7.4	Facade retention	0.00		0.00
7.5	Cleaning existing surfaces	0.00		0.00
7.6	Renovation works	0.00		0.00
7	**Work to existing building**	**0.00**	**0.00**	**0.00**
8.1	Site preparation works	0.00		0.00
8.2	Roads, paths, pavings and surfacings	0.00		0.00
8.3	Soft landscaping, planting and irrigation systems	0.00		0.00
8.4	Fencing, railings and walls	0.00		0.00
8.5	Site/street furniture and equipment	0.00		0.00
8.6	External drainage	0.00		0.00
8.7	External services	0.00		0.00
8.8	Minor building works and ancillary buildings	0.00		0.00
8	**External works**	**0.00**	**0.00**	**0.00**
TOTAL: BUILDING WORKS, including Building Engineering Services carried to TENDER SUMMARY		0.00		0.00
	Provisional sums	0.00	0.00	0.00
	Sub-total	0.00	0.00	0.00
9	**Main contractor's preliminaries** (Brought forward from Tender Summary)	**0.00**	**0.00**	**0.00**
	Sub-total	0.00	0.00	0.00

Figure 35.4 *(Continued)*

Code	Element	Total Cost of Element	Total Cost of Group Element	Cost/m² of GIFA
		£	£	£
10	**Main contractor's overheads and profit** (Brought forward from Tender Summary)	**0.00**	**0.00**	**0.00**
	Sub-total	0.00	0.00	0.00
11	**Main contractor's design fees and survey charges** (Brought forward from Tender Summary)	**0.00**	**0.00**	**0.00**
	Sub-total	0.00	0.00	0.00
12	**Main contractor's design and construction risks** (Apportioned for purpose of analysis only)	**0.00**	**0.00**	**0.00**
	Sub-total	0.00	0.00	0.00
13	**Main contractor's fixed price addition** (Brought forward from Tender Summary)	**0.00**	**0.00**	**0.00**
	Sub-total	0.00	0.00	0.00
14	**Director's adjustment** (Brought forward from Tender Summary)	**0.00**	**0.00**	**0.00**
TOTAL: DESIGN AND CONSTRUCTION COSTS carried to TENDER PRICE SUMMARY		**0.00**	**0.00**	**0.00**

Figure 35.4 *(Continued)*

City Investments Limited UK
Woollard Hotel, Holborn Viaduct, St Leonards-on-Sea, East Sussex

PRICING DOCUMENT (INCORPORATING FORM OF TENDER AND CONTRACT SUM ANALYSIS)

NAME OF TENDERER

PROVISIONAL SUMS

Code	Description of Provisional Sum	£
P-01	Note: Provisional sum descriptions and sums are inserted by the cost manager	0.00
P-02		0.00
P-03		0.00
P-04		0.00
P-05		0.00

Figure 35.5 Provisional sums

Code	Description of Provisional Sum	£
P-06		0.00
P-07		0.00
P-08		0.00
P-09		0.00
P-10		0.00
Etc.		0.00
		0.00
		0.00
TOTAL PROVISIONAL SUMS carried to TENDER PRICE SUMMARY:		**0.00**

Figure 35.5 *(Continued)*

City Investments Limited UK
Woollard Hotel, Holborn Viaduct, St Leonards-on-Sea, East Sussex

PRICING DOCUMENT (INCORPORATING FORM OF TENDER AND CONTRACT SUM ANALYSIS)

PRICE ANALYSIS:

NAME OF TENDERER

SCHEDULE OF MAIN CONTRACTOR'S PRE-CONSTRUCTION SERVICES

Notes to tenderer:

1. Tenderer to insert below details of ALL resources and other items/services that they require to deliver ALL Pre-Construction Services listed in [Insert Reference] of the Schedule of Amendments to the Contract Conditions; together with financial provision that they have made in their Tender Price/Contract Sum.
2. Tenderer to append their detailed breakdown of all prices for Pre-Construction services to the Pricing Document:

Code	Resource/Item/Service	Price
		£
PCF-01	Insert resource or other item/service required to deliver pre-construction services	0.00
PCF-02	Insert resource or other item/service required to deliver pre-construction services	0.00
PCF-03	Insert resource or other item/service required to deliver pre-construction services	0.00
PCF-04	Insert resource or other item/service required to deliver pre-construction services	0.00
PCF-05	Insert resource or other item/service required to deliver pre-construction services	0.00
PCF-06	Insert resource or other item/service required to deliver pre-construction services	0.00
PCF-07	Insert resource or other item/service required to deliver pre-construction services	0.00
PCF-08	Insert resource or other item/service required to deliver pre-construction services	0.00
PCF-09	Insert resource or other item/service required to deliver pre-construction services	0.00
PCF-10	Insert resource or other item/service required to deliver pre-construction services	0.00
PCF-11	Insert resource or other item/service required to deliver pre-construction services	0.00

Figure 35.6 Schedule of main contractor's pre-construction services

Code	Resource/Item/Service	Price
		£
PCF-12	Insert resource or other item/service required to deliver pre-construction services	0.00
PCF-13	Insert resource or other item/service required to deliver pre-construction services	0.00
PCF-14	Insert resource or other item/service required to deliver pre-construction services	0.00
PCF-15	Insert resource or other item/service required to deliver pre-construction services	0.00
PCF-16	Insert resource or other item/service required to deliver pre-construction services	0.00
PCF-17	Insert resource or other item/service required to deliver pre-construction services	0.00
PCF-18	Insert resource or other item/service required to deliver pre-construction services	0.00
PCF-19	Insert resource or other item/service required to deliver pre-construction services	0.00
PCF-20	Insert resource or other item/service required to deliver pre-construction services	0.00
	Sub-total	0.00
	ADD	
	Main Contractor's Overheads and Profit on Main Contractor's Pre-Construction Services	0.00
	TOTAL MAIN CONTRACTOR'S PRE-CONSTRUCTION SERVICES carried to TENDER PRICE SUMMARY:	0.00

Figure 35.6 *(Continued)*

City Investments Limited UK
Woollard Hotel, Holborn Viaduct, St Leonards-on-Sea, East Sussex

PRICING DOCUMENT (INCORPORATING FORM OF TENDER AND CONTRACT SUM ANALYSIS)

PRICE ANALYSIS:

NAME OF TENDERER

SCHEDULE OF MAIN CONTRACTOR'S DESIGN FEES AND SURVEY CHARGES

Notes to tenderer:

1. Design fees not specifically identified in the Tenderer's full and detailed breakdown shall be deemed to have no cost implications or have been included elsewhere within the Tenderer's rates and prices.
2. Tenderer to append his detailed breakdown of Design Fees and Survey Charges to his Pricing Document.

Code	Construction/Design Risk		Price
			£
11.1	**Novated Consultants' Fees:**		
11.1.1	Architect – R. K. Tex Limited		344,651.00
11.1.2	Structural Engineer – Tootles Structural Limited		127,193.00

Figure 35.7 Schedule of main contractor's design fees and survey charges

Code	Construction/Design Risk		Price £
11.2	**Fees and Charges in connection with Other Design Services and Surveys** (Please insert applicable Design Services and Surveys)		
11.2.1	Insert applicable Design Service/Survey		0.00
11.2.2	Insert applicable Design Service/Survey		0.00
11.2.3	Insert applicable Design Service/Survey		0.00
11.2.4	Insert applicable Design Service/Survey		0.00
11.2.5	Insert applicable Design Service/Survey		0.00
11.2.6	Insert applicable Design Service/Survey		0.00
11.2.7	Insert applicable Design Service/Survey		0.00
11.2.8	Insert applicable Design Service/Survey		0.00
11.2.9	Insert applicable Design Service/Survey		0.00
11.2.10	Insert applicable Design Service/Survey		0.00
11.2.11	Insert applicable Design Service/Survey		0.00
11.2.12	Insert applicable Design Service/Survey		0.00
11.2.13	Insert applicable Design Service/Survey		0.00
11.2.14	Insert applicable Design Service/Survey		0.00
11.2.15	Insert applicable Design Service/Survey		0.00
	Sub-total		0.00
	ADD		
	Main Contractor's Overheads and Profit on Main Contractor's Design Fees and Survey Charges	0.00%	0.00
	TOTAL MAIN CONTRACTOR'S DESIGN FEES AND SURVEY CHARGES carried to TENDER PRICE SUMMARY:		

Figure 35.7 *(Continued)*

City Investments Limited UK
Woollard Hotel, Holborn Viaduct, St Leonards-on-Sea, East Sussex
PRICING DOCUMENT (INCORPORATING FORM OF TENDER AND CONTRACT SUM ANALYSIS)

PRICE ANALYSIS:

NAME OF TENDERER

MAIN CONTRACTOR'S RISK ANALYSIS

Notes to tenderer:

1. Contractor to insert below details of ALL cost-significant construction risks which they have made financial provision for in their Tender Price/Contract Sum, together with the lump sum fixed price that they have included in their Price.

Figure 35.8 Main contractor's risk analysis

2. Costs in connection with risks materialising during the Works which are not identified below shall be deemed to have no cost implications or have been included elsewhere within the Contractor's rates and prices.			
Code	**Construction/Design Risk**		**Price**
13.1	**Design Risks:**		
13.1.1	Define Design Risk		0.00
13.1.2	Define Design Risk		0.00
13.2.3	Define Design Risk		0.00
13.2.4	Define Design Risk		0.00
13.2.5	Define Design Risk		0.00
13.2.6	Define Design Risk		0.00
13.2.7	Define Design Risk		0.00
13.2.8	Define Design Risk		0.00
13.2.9	Define Design Risk		0.00
13.2.10	Define Design Risk		0.00
13.2	**Construction Risks:**		
13.2.1	Define Construction Risk		0.00
13.2.2	Define Construction Risk		0.00
13.2.3	Define Construction Risk		0.00
13.2.4	Define Construction Risk		0.00
13.2.5	Define Construction Risk		0.00
13.2.6	Define Construction Risk		0.00
13.2.7	Define Construction Risk		0.00
13.2.8	Define Construction Risk		0.00
13.2.9	Define Construction Risk		0.00
13.2.10	Define Construction Risk		0.00
13.2.11	Define Construction Risk		0.00
13.2.12	Define Construction Risk		0.00
13.2.13	Define Construction Risk		0.00
13.2.14	Define Construction Risk		0.00
13.2.15	Define Construction Risk		0.00
	Sub-total		0.00
	<u>**ADD**</u>		
	Main Contractor's Overheads and Profit on Design Risks and Construction Risks	0.00%	0.00
	TOTAL DESIGN RISKS AND CONSTRUCTION RISKS carried to CONTRACT SUM ANALYSIS:		0.00

Figure 35.8 *(Continued)*

City Investments Limited UK
Woollard Hotel, Holborn Viaduct, St Leonards-on-Sea, East Sussex

PRICING DOCUMENT (INCORPORATING FORM OF TENDER AND CONTRACT SUM ANALYSIS)

PRICE ANALYSIS:

NAME OF TENDERER

SCHEDULE OF OPTION PRICES

Notes to tenderer:

1. Contractor to advise the cost implications, together with any time implications, of the option prices listed below for consideration by the Employer.
2. These prices are NOT TO BE INCLUDED, and shall be deemed not included, in the Contractor's Tender Price inserted in the Form of Tender.
3. Prices to be inclusive of all costs in connection with Main Contractor's Preliminaries, Main Contractor's Design Fees and Survey Charges, Main Contractor's Overheads and Profit, and Main Contractor's Fixed Price Addition (as required).

Note:

Insert a 'minus' sign if Cost Reduction or Omission (i.e. to confirm that figure is a cost reduction)

Code	**Description of Option Price Required**	**Price Adjustment**
		£
A.	**Design Options:**	
A-01	Use Contractor-selected Structural Engineer in lieu of 'Novated' Structural Engineer. Define Design Risk	0.00
A-02		0.00
A-03		0.00
B.	**Works Options**	
B-01	EXTRA COST of achieving BREEAM rating of 'Outstanding'	0.00
	Notes:	
	1. An explanation of how the Contractor can achieve a BREEAM rating of 'Outstanding' cost effectively is to be incorporated in the 'Contractor's Proposals'	
	2. Breakdown of costs to be appended to Contract Sum Analysis	
B-02	EXTRA COST OR COST REDUCTION Alternative Atrium Roof design – Based on drawing no. REB_1440_TIMCHRIS30 Rev B.	0.00
B-03		0.00
B-04		0.00
B-05		0.00
Etc.		

Figure 35.9 Schedule of option prices

City Investments Limited UK
Woollard Hotel, Holborn Viaduct, St Leonards-on-Sea, East Sussex

PRICING DOCUMENT (INCORPORATING FORM OF TENDER AND CONTRACT SUM ANALYSIS)

PRICE ANALYSIS:

NAME OF TENDERER

SCHEDULE OF VALUE IMPROVEMENT PROPOSALS

Notes to tenderer:

1. In addition to the submission of a 'Compliant' Tender in accordance with the Employer's Requirements and accompanying Invitation Documents, we submit the following 'value improvement' proposals in relation to the Works for consideration by the Employer.
2. Tenderer to insert value improvement proposals (if any) on the 'Schedule of Value Improvement Proposals'.
3. The dependency and independency of items on one another must be clearly stated. For example, Item VI-01 can be considered in isolation; or Items VI.05 and VI.07 are required to be considered together.

Code	Description of Proposed Value Improvement	Independent Saving	Dependent Saving	What is Value Improvement dependent on?	Accepted/ Rejected [To be completed by Employer]
		£	£		
VI-01	Define value improvement proposal	0.00	0.00		
VI-02	Define value improvement proposal	0.00	0.00		
VI-03	Define value improvement proposal	0.00	0.00		
VI-04	Define value improvement proposal	0.00	0.00		
VI-05	Define value improvement proposal	0.00	0.00		
VI-06	Define value improvement proposal	0.00	0.00		
VI-07	Define value improvement proposal	0.00	0.00		
VI-08	Define value improvement proposal	0.00	0.00		
VI-09	Define value improvement proposal	0.00	0.00		

Figure 35.10 Schedule of value improvement proposals

Analysing bids and collecting data using NRM 1

CHAPTER 36

Analysing bids and collecting cost data

Introduction

This chapter:

- discusses the change from designer-led design to contractor-led design contract strategies;
- discusses the problems with collecting historical cost data;
- explains how the NRM 1 cost plan framework and elemental structure can be used as a basis for collecting cost data that can be analysed and used for the cost estimating of future building projects; and
- considers the way in which benchmarking is used by cost managers.

36.1 Changing contract strategies

In the past, the main method of obtaining tender prices from contractors was through the use of bill of quantities (BQs), which supported a designer-led contract strategy (i.e. where design liability remained with the employer). BQs provide a co-ordinated list of items that comprise the building works, together with their identifying descriptions and quantities, which enable contractors to prepare tenders efficiently and accurately. Today, more and more building projects are being tendered using contractor-led design strategies using contractual arrangements such as design and build, design and construct, and design and manage (i.e. where the contractor becomes liable for design). Originally intended for industrial building schemes and other building projects of a simple design, contractor-led design strategies are now used for any type of building project – including new build and refurbishment, irrespective of the level of construction costs.

When properly prepared by cost managers, the use of BQs is beneficial for a number of reasons:

- They save the cost and time of several contractors measuring the same design in order to calculate their bids.
- They provide a consistent basis for obtaining competitive bids.
- They provide an extensive and clear statement of the work to be executed.

- They provide a very strong basis for budgetary control and accurate cost reporting of the contract (i.e. post-contract cost control), including:
  - the preparation of regular cash flow forecasts;
  - a basis for valuing variations; and
  - a basis for ascertaining progress payments (i.e. interim payments) for the contractor.
- When BQ items are codified, it allows reconciliation with and any necessary transfers and adjustments to be made to the cost plan.
- When priced, BQs provide cost data in a form that aids the calculation of tax benefits (i.e. capital allowances and value added tax (VAT), as well as helping support entitlement to these benefits.
- When priced, BQs provide data to support grant applications.
- They provide one of the best sources of real-time cost data that can be readily analysed and used to inform future cost estimates of building projects (i.e. they provide historical cost data).

BQs, to enable contractors to prepare tenders efficiently and accurately, need to be based on the full and detailed design and specification of the building to be built. Consequently, BQs should be prepared using the completed technical design information produced by the design team (at RIBA Work Stage 4: Technical Design). Herein lies the problem, which became even more exacerbated in the 1990s, as the need to commence construction early became increasingly important to employers. Invariably, the cost manager would receive incomplete design information on which to produce the BQ – meaning that a significant portion of the building works could not be measured in accordance with the measurement rules for building works (i.e. the Standard Method of Measurement (SMM), now NRM 2: *Detailed rules of measurement for building works*). This resulted in the overuse of provisional sums and contractor-designed works. It was not unusual, and still is not, to see more than 40% of the cost of building works being the subject of provisional sums. This undermines one of the principal benefits of a lump sum contract based on BQs, which is to give significant cost certainty of the outturn cost of a building project to the employer – well, at least as much cost certainty an employer could expect on a building project. Obviously, a large part of the building project costs being based on provisional sums meant that the outturn costs were at risk of escalating, which was, unfortunately, very common.

Decline in the use of BQs has also resulted in a lack of easily collectable cost data. Whereas BQs provided readily available cost data that could be quickly analysed to produce cost analyses (cost/m^2 of GIFA, functional unit rates and element unit rates) that could also be used as benchmark data, cost managers struggle to gather the same data from other forms of pricing documents (e.g. from contract sum analyses and other forms of price analysis). Cost managers have been able to collect high-level cost data (e.g. cost/m^2 of GIFA, functional unit rates, main contractor's preliminaries and overheads and profit) from main contractors' bids and, whenever the opportunity arose, by analysing subcontract tenders (unit rates of components, and subcontractors' preliminaries and overheads and profit). However, compared with the quality of cost data available from BQs, cost data has become difficult to collect and this impacts heavily on the cost manager's ability to accurately forecast the costs of future building projects.

Who was instrumental in undermining the use of BQs as a principal pricing document? Employers were influential in the decline of the use of BQs; however, quantity surveyors and cost managers also significantly contributed to the decline by not reacting quickly enough to the emerging changes in employers' needs. One of the primary causes of the decline in the use of BQs in the private sector has been the increased cost to employers of short-term borrowing of monies. The consequence of this has been that employers often now need construction to commence as quickly as possible after they have obtained permission to build from the planning authorities. This gives little opportunity for the designers to develop the building design from concept to technical. This resulted in incomplete design information being available to cost managers on which to prepare the BQ. Thus, incomplete BQs were prepared and used to obtain tender bids from contractors. The BQs produced by cost managers were not fit for their purpose. Hence, there was a move away from designer-led design strategies to contractor-led design strategies. This, together with the move to better integrate design and construction, led to the prominence of design and build contractual arrangements by employers. Employers also saw such contractual arrangements as being a one-stop shop, with most risks being passed to the main contractor. However, it is contended that design and build contracts are actually high-risk contract strategies – particularly if the employer's requirements have not been accurately and clearly documented in the 'Employer's Requirements' document.

As mentioned in Chapter 35 ('Pricing documents'), contract sum analyses and other forms of price analysis are being used more and more since the decline in the use of BQs. Despite this, cost managers have been rather poor at preparing suitable pricing analyses that will make available suitable, or adequate, cost data that can be readily analysed and used for estimating the cost of future building projects. More often than not, cost managers will only request the contractor to complete a very high-level cost breakdown – normally just to elemental level, or to a bespoke breakdown. Not only do these provide little useful historical cost data, it also makes the post-contract cost management and control extremely difficult for the cost manager; in particular, it can put the cost manager at risk of overvaluing the building works when correctly completed in accordance with the contract when carrying out a valuation, or when certifying a contractor's interim payment application – and the employer overpaying the contractor. The cost manager preparing a contract sum analysis or price analysis based on the NRM 1 cost plan framework can overcome such problems.

36.2 Framework for collecting cost data

The NRM 1 elemental work breakdown structure, together with its cost plan framework, provides a best practice solution for analysing, collecting and storing actual cost data that can be easily retrieved and reprocessed for use in future order of cost estimates and elemental cost estimates – to enable the viability of various options to be determined and to set the cost limit for the selected option. However, this will only work properly if the NRM 1 elemental work breakdown structure and cost plan framework are used as the basis of the cost manager's pricing document (see Chapter 35: 'Pricing documents'). When used, contractors' bids will be returned in a form that can be readily analysed for the purpose of providing historical cost data.

36.3 Analysing bids and collecting cost data

When using historical cost data, it is essential that cost managers understand what has been included in the rates. NRM 1 overcomes this problem by providing a common and transparent method for analysing cost information obtained from contractors' bids. This is the framework provided for preparing an elemental estimate, which is the same as the framework given in the BCIS Standard Form of Cost Analysis (NRM edition).

The method of measuring element unit quantities given in Table 2.1: Rules of measurement for elemental method of estimating in Part 2 (Measurement rules for order of cost estimating) of NRM 1 defines the basis for preparing cost analyses of contractors' bids. However, to be able to appropriately analyse costs, it is necessary for the cost manager to obtain contractors' bids in an elemental form – which was explained in Chapter 35 ('Pricing documents'). By obtaining contractors' bids in elemental form, the following can be quickly established:

- total cost of the building project;
- total cost of each group element;
- total cost of each element;
- cost/m^2 of GIFA – for entire building, group elements and elements – by dividing the total cost of the building project by the GIFA;
- the element unit rate for each element – by dividing the total cost of each element by the relevant element unit quantity;
- the percentage rate of main contractor's preliminaries;
- the percentage rate of main contractor's overheads and profit; as well as
- other use cost data.

In addition, contractors should be requested by the cost manager, in the invitation to tender, to provide adequate build-up to demonstrate how they have arrived at each sum included in their contract sum analysis or price analysis. Contractors are usually instructed to submit such information in the form of a BQ – being responsible for their own quantities and unit rates.

Obtaining contractors' bids in elemental form is not such an important issue when BQs are used, as the cost manager, with a little bit of effort, should be able to produce a cost analysis from the data obtained in the same form.

The process is straightforward and extremely simple; however, there is a need for cost managers to change their approach to formulating pricing documents to benefit from the best practice guidance given by NRM 1. What is interesting, from the author's experience, is that it is contractors who are comprehending the benefits of the cost management cycle ('the Benge Cycle') provided by NRM 1, ahead of cost managers working in consultancies. In particular, contractors are realising the benefits of using NRM 1 as part of their Building Information Modelling initiatives.

36.4 Benchmarking

Benchmarking is a very useful comparison tool, yet most cost managers fail to use benchmarking properly – often misusing the term.

What is benchmarking? In its widest context, it is an improvement process used to discover and incorporate measurable best practices into an organisation: the process of identifying, learning and adapting outstanding practices and processes globally to improve organisational performance.

To all intents and purposes, cost managers use benchmarking simply as a cost-comparator tool, with the term benchmark analysis being commonly used when a cost plan for a new building scheme is compared against the costs of a number of other similar previous schemes.

Box 36.1 Key definitions

- **Benchmarking** – an improvement process used to discover and incorporate measurable best practices into an organisation; the process of identifying, learning and adapting outstanding practices and processes globally to improve organisational performance. In cost management, benchmarking is used to see how the estimated costs compare with the cost of similar completed building projects.
- **Benchmark analysis** – comparing metrics of organisational performance against those of other organisations. In cost management, benchmark analysis is used to see how the estimated costs compare with the cost of similar completed building projects.
- **Cost analysis** – a full appraisal of the costs of a building project that has been tendered or previously constructed that aims mainly at providing reliable information that will assist in accurately estimating the cost of future building projects. It also provides a basis to cost planning.

Bibliography

Ashworth, A., *Cost Studies of Buildings* (5th edition), Prentice Hall, Essex, 2010.
Building Costs Information Service, *Standard Form of Cost Analysis* (4th edition), BCIS, London, 2012.
Burke, R., *Project Management: Planning and Control Techniques* (4th edition), John Wiley & Sons, 2003.
Central Unit on Procurement (CUP), *Guidance Note No. 25: Cost Management for Works Projects*, HM Treasury, London, 1991.
Central Unit on Procurement (CUP), *Guidance Note No. 41: Managing Risk and Contingency for Works Projects*, HM Treasury, 1996.
Central Unit on Procurement (CUP), *Guidance Note No. 54: Value Management*, HM Treasury, 1996.
Chudley, R. and Greeno, R., *Building Construction Handbook* (8th edition), Butterworth-Heinemann, Oxford, 2010.
Egan, J., *Rethinking Construction: Report of the Construction Task Force*, DETR, London, 1998.
Latham, M., *Constructing the Team: Final Report of the Government/Industry Review of Procurement and Contractual Arrangements in the UK Construction Industry*, HMSO, London, 1994.
Published by the Office of Government Commerce, ITIL, 2007:

- *OGC Gateway™ Process Review 0: Strategic Assessment.*
- *OGC Gateway™ Process Review 1: Business Justification.*
- *OGC Gateway™ Process Review 2: Delivery Strategy.*
- *OGC Gateway™ Process Review 3: Investment Decision.*
- *OGC Gateway™ Process Review 4: Readiness for Service.*
- *OGC Gateway™ Process Review 5: Operational Review and Benefits Realisation.*

Property Services Agency, *Cost Planning and Computers*, HMSO, London, 1981.
Purba, S. and Zucchero, J., *Project Rescue: Avoiding a Project Management Disaster*, McGraw-Hill/Osbourne, USA, 2004.
Royal Institute of British Architects, *RIBA Outline Plan of Work 2013*, RIBA, London, 2013.
Royal Institute of British Architects, *RIBA Plan of Work 2013 Overview*, RIBA, London, 2013.
Royal Institution of Chartered Surveyors, *Code of Measuring Practice: A Guide for Property Professionals* (6th edition), RICS, London, 2007.
Royal Institution of Chartered Surveyors, *Cost Management in Engineering Construction Projects*, RICS Holdings Limited, London, 2007.
Royal Institution of Chartered Surveyors, *New Rules of Measurement: Order of Cost Estimating and Cost Planning for Capital Building Works* – NRM 1 (2nd edition), RICS, Coventry, 2012.
Royal Institution of Chartered Surveyors, *New Rules of Measurement: Detailed Measurement for Building Works* – NRM 2, RICS, Coventry, 2012.
Royal Institution of Chartered Surveyors, *New Rules of Measurement: Order of Cost Estimating and Cost Planning for Building Maintenance Works* – NRM 3, RICS, Coventry, unpublished.
Seeley, I. H., *Building Economics* (4th edition), McMillan, London, 1996.
Smith, J. and Jagger, D., *Building Cost Planning for the Design Team* (2nd edition), Butterworth-Heinemann, Oxford, 2007.

INDEX

I

L